T0133897

Handbook of Automated Scoring

Theory into Practice

Statistics in the Social and Behavioral Sciences Series

Recently Published Titles

Multilevel Modelling Using Mplus
Holmes Finch and Jocelyn Bolin

Applied Survey Data Analysis, Second Edition
Steven G. Heering, Brady T. West, and Patricia A. Berglund

Adaptive Survey Design
Barry Schouten, Andy Peytchev, and James Wagner

Handbook of Item Response Theory, Volume One: Models
Wim J. van der Linden

Handbook of Item Response Theory, Volume Two: Statistical Tools
Wim J. van der Linden

Handbook of Item Response Theory, Volume Three: Applications
Wim J. van der Linden

Bayesian Demographic Estimation and Forecasting
John Bryant and Junni L. Zhang

Multivariate Analysis in the Behavioral Sciences, Second Edition
Kimmo Vehkalahti and Brian S. Everitt

Analysis of Integrated Data
Li-Chun Zhang and Raymond L. Chambers, Editors

Multilevel Modeling Using R, Second Edition
W. Holmes Finch, Joselyn E. Bolin, and Ken Kelley

Modelling Spatial and Spatial-Temporal Data: A Bayesian Approach
Robert Haining and Guangquan Li

For more information about this series, please visit: https://www.crcpress.com/go/ssbs

Handbook of Automated Scoring
Theory into Practice

Edited by
Duanli Yan, André A. Rupp, and Peter W. Foltz

CRC Press
Taylor & Francis Group
Boca Raton London New York

CRC Press is an imprint of the
Taylor & Francis Group, an **informa** business

A CHAPMAN & HALL BOOK

CRC Press
Taylor & Francis Group
6000 Broken Sound Parkway NW, Suite 300
Boca Raton, FL 33487-2742

International Standard Book Number-13: 978-1-1385-7827-2 (Hardback)

Visit the Taylor & Francis Web site at
http://www.taylorandfrancis.com

and the CRC Press Web site at
http://www.crcpress.com

To my daughter Victoria, a young and intelligent professional.
You gave me a new meaning of life. Thanks for your love and support.

Duanli Yan

To my amazing wife Brooke and my awesome son Jean-Marie – no machine
could ever come close to replicating their uniqueness and impact on my life.

André A. Rupp

To A and M

Peter W. Foltz

Contents

Part III: Practical Illustrations

Foreword

"The tutorial system is a method of university teaching where the main focus is on regular, very small group teaching sessions. It was established by and is still practised by the University of Oxford and the University of Cambridge in the United Kingdom. In addition to attending lectures and other classes, students are taught by faculty fellows in groups of one to three on a weekly basis. [...] One benefit of the tutorial system is that students receive direct feedback on their weekly essays or work in a small discussion setting."*

One of the earliest formal concepts that shaped Western education is the dialectic method of Socrates that influenced the tutorial system.† As we can see in the quote above, the model of a complex educational experience where the students receive direct feedback on their essays or work collaboratively still represents an aspiring model for educating the next generation. Despite this long history, the central question remains: how can we teach the young at scale to think for themselves? We must solve this challenge in order to help teachers provide this type of personalized feedback to much larger groups of students than tutorial systems were designed to accommodate.

Alongside technological and computational advances, progress in *artificial intelligence* (AI) and *natural language processing* (NLP) has made it possible to use a machine to analyze an essay according to specific features identified by the subject-matter experts, score the essay according to a mathematical model, and eventually provide feedback with respect to these features. In the 1990s, one of the editors of this handbook, Peter W. Foltz, launched one of the first *automated scoring* (AS) engines and many of the contributors to this handbook have developed their own software packages for scoring essays.

In this *Handbook of Automated Scoring: Theory into Practice* edited by Duanli Yan, André A. Rupp, and Peter W. Foltz, the editors provide a snapshot of the use of AS in education: which successes and challenges these systems have encountered, how they have been improved, and how they maintain their relevance for educational assessment as reliable, fair, and valid tools.

The handbook is a timely one. Valid assessments and learning systems are essential for the evaluation of a nation's educational standing and for the implementation of educational reforms. Policy makers and education experts have always been interested in fair and accurate ways of designing assessments and comparing students' achievement in authentic ways. Because scoring essays or performance tasks consume time that otherwise

---

* Tutorial System. (n.d.). In Wikipedia. Retrieved August 28, 2019, from https://en.wikipedia.org/wiki/Tutorial_system
† https://www.greenes.org.uk/greenes-education/our-history/the-history-of-the-tutorial/

could be devoted to instruction, it is important to consider tools that are efficient. AS engines provide a good solution. Doing so requires a careful balancing of the contributions of technology, NLP, psychometrics, artificial intelligence, and the learning sciences. The present handbook is evidence that the theories, methodologies, and underlying technology that surround AS have reached maturity, and that there is a growing acceptance of these technologies among experts and the public.

This book examines AS from the point of view of innovative AS theory, the latest developments in computational methodologies, and several real-world large-scale applications for AS for complex tasks. It also provides a scientifically grounded description of the key research and development efforts that it takes to move automated scoring systems into operational practice. The book is organized into three parts covering (1) theoretical foundations, (2) operational methodologies, and (3) practical illustrations for the use and deployment of AS systems, with a commentary at the end of each section.

It is relevant to mention that despite the heavy use of edtech vocabulary, any AS system (and any AI-based system, for that matter) requires a significant amount of human intervention, from scoring the "training set" of responses to designing the models to evaluating the results. All AS-systems are rather hybrid systems than pure machines, which is true even for the most advanced AI-based scoring systems. This is actually a wonderful state of affairs, one that reminds us that the machines are here to help and support us, not replace or render us obsolete.

Most recently, some of us have been working on blending learning and assessment in ways that would allow students to demonstrate their mastery of skills and knowledge as part of a (digital) learning and assessment system. Can AS become a part of these learning and assessment systems? Can they be extended to match the diagnostic assessment to recommendations of instructional resources using crosswalks of taxonomies and mastery levels as von Davier, Deonovic, Yudelson, Polyak, and Woo (2019) proposed? Or, more generally, do we expect educational experience to change in the age of AI and, therefore, expect AS engines to thrive and expand their roles in supporting new educational needs: mobility, access, real-time feedback, voice-based coaching? Accordingly, a few chapters in this handbook are dedicated to illustrating how these technologies can be used to support tutoring as a digital experience.

Another line of current research is focused on developing appropriate methodologies for assessing hard-to-measure 21st-century skills like collaborative problem-solving and creative thinking. The *Programme for International Student Assessment* (PISA),* for example, became a leader in experimenting with the measurement of innovative domains across multiple countries and cultures at scale. Can we develop AS engines that analyze not only text, but also video, audio, and personal interactions? To examine these

---

* see http://www.oecd.org/pisa/aboutpisa/

questions, several chapters in this handbook focus on complex competencies and the analysis of multimodal data.

Can we ensure that the resulting tools are fair to all? Can we rely on them for comparing students' achievement? Are they harming us or helping us? To return to Socrates' theories and then to his own fate, we are reminded that society can be unforgiving and, therefore, we need to carefully test our algorithms and models to ensure that the fairness of our tools is upheld.

This handbook takes an important step in addressing these challenges by examining foundational issues that reflect different types of theoretical and operational situations and by delving into challenges that are particularly relevant for different applications of AS.

As often happens after I read a good book, I found myself pondering many of the questions and research ideas posed in this edited handbook. I recommend readers to build and develop their own questions and research agendas based on the broad range of topics and insights provided in this handbook.

Alina A. von Davier
Princeton, NJ, USA

Editors

Duanli Yan is Director of Data Analysis and Computational Research in the Psychometrics, Statistics, and Data Sciences area at the Educational Testing Service, and Adjunct Professor at Fordham University and Rutgers University. She is a co-author of *Bayesian Networks in Educational Assessment* and *Computerized Adaptive and Multistage Testing with R*, editor for *Practical Issues and Solutions for Computerized Multistage Testing*, and co-editor for *Computerized Multistage Testing: Theory and Applications*. Her awards include the 2016 AERA Division D Significant Contribution to Educational Measurement and Research Methodology Award.

André A. Rupp is Research Director in the Psychometrics, Statistics, and Data Sciences area at the Educational Testing Service. He is co-author and co-editor of two award-winning interdisciplinary books entitled *Diagnostic Measurement: Theory, Methods, and Applications* and *The Handbook of Cognition and Assessment: Frameworks, Methodologies, and Applications*. His synthesis- and framework-oriented research has appeared in a wide variety of prestigious peer-reviewed journals.

Peter W. Foltz is Vice President in Pearson's AI and Products Solutions Organization and Research Professor at the University of Colorado's Institute of Cognitive Science. His work covers machine learning and natural language processing for educational and clinical assessments, discourse processing, reading comprehension and writing skills, 21st Century skills learning, and large-scale data analytics. He has authored more than 150 journal articles, book chapters, and conference papers as well as multiple patents.

List of Contributors

Russell G. Almond
Florida State University
Tallahassee, FL

Malcolm I. Bauer
ETS
Princeton, NJ

John T. Behrens
Pearson
Boulder, CO

Brad Bolender
ACT
Iowa City, IA

Michelle Boyer
NCIEA
Dover, NH

Brent Bridgeman
ETS
Princeton, NJ

Jill Burstein
ETS
Princeton, NJ

Aoife Cahill
ETS
Princeton, NJ

Zhiqiang Cai
University of Memphis
Memphis, TN

Brian E. Clauser
NBME
Philadephia, PA

Carolyn Connelly
ETS
Princeton, NJ

Paul Deane
ETS
Princeton, NJ

Kristen DiCerbo
Pearson
Boulder, CO

Sidney K. D'Mello
University of Colorado
Boulder, CO

Scott Dooley
Pearson
Boulder, CO

Keelan Evanini
ETS
Princeton, NJ

Peter W. Foltz
Pearson
Boulder, CO

Janice Gobert
Apprendis
Mountain View, CA

Arthur C. Graesser
University of Memphis
Memphis, TN

Kyle Habermehl
Pearson
Boulder, CO

Susan Hines
ETS
Princeton, NJ

Nick Hoefer
Independent Consultant
Iowa City, IA

Xiangen Hu
University of Memphis
Memphis, TN

Yuchi Huang
ACT
Iowa City, IA

Saad M. Khan
ACT Next
Iowa City, IA

Emily Lai
Pearson
Boulder, CO

Sue Lottridge
AIR
Washington, DC

Anastassia Loukina
ETS
Princeton, NJ

Melissa J. Margolis
NBME
Philadephia, PA

Daniel McCaffrey
ETS
Princeton, NJ

Rick Meisner
ACT
Iowa City, IA

Robert J. Mislevy
ETS
Princeton, NJ

Aditya Nagarajan
Pearson
Boulder, CO

Michael Sao Pedro
Apprendis
Mountain View, CA

Kathryn L. Ricker-Pedley
ETS
Princeton, NJ

Brian Riordan
ETS
Princeton, NJ

André A. Rupp
ETS
Princeton, NJ

Vasile Rus
University of Memphis
Memphis, TN

Christina Schneider
NWEA
Portland, OR

Dan Shaw
ACT
Iowa City, IA

Mark D. Shermis
University of Houston Clear Lake
Houston, TX

Matthew Ventura
Pearson
Boulder, CO

Alina A. von Davier
ACTNext
Iowa City, IA

David M. Williamson
ETS
Princeton, NJ

Edward W. Wolfe
ETS
Princeton, NJ

Scott W. Wood
Pacific Metrics
Monterey, CA

Duanli Yan
ETS
Princeton, NJ

Diego Zapata-Rivera
ETS
Princeton, NJ

Klaus Zechner
ETS
Princeton, NJ

Mo Zhang
ETS
Princeton, NJ

1

The Past, Present, and Future of Automated Scoring

Peter W. Foltz, Duanli Yan, and André A. Rupp

CONTENTS

Automated scoring (AS) can be broadly conceived as using computers to convert students' performance on educational tasks into characterizations of the quality of performance. AS can be applied to *simple tasks* such as a student's response to a multiple-choice question or a word filled into a blank space. However, for the purposes of this handbook, we distinguish these simple tasks from *complex tasks*, which are cognitively complex in that they encompass multiple *cognitive* and *psychosocial processes* that occur in the mind of the student while performing the task. These processes may include reasoning, problem solving, arguing, seeking consensus with teammates, or integrating prior knowledge with information from multiple sources.

Complexity also comes from the output produced by the student. Rather than making a single choice or performing a single button click, the student's actions result in *multifaceted output* such as a written or spoken response, an interaction with a tutor, a video of a fluent conversation, or a pattern of problem-solving steps. Such output may contain many *pieces of evidence* about a variety of interacting cognitive and psychosocial processes. The raw output, nevertheless, may not be easily interpretable although it may be decipherable by skilled instructors armed with rubrics and domain knowledge and a lot of time. The output may also have dependencies on the task and the prior states of the learning environment that the student has encountered, as well as on the affective or motivational state of the student. Thus, the goal of AS is one of *evidence identification*; that is, AS processes must be able to extract and transform relevant evidence from student output in order to make useful inferences about the students' abilities.

The Growth of Automated Scoring

Applying automated technologies to assess students is not new. Indeed, early research and development efforts in the late 1960s and early 1970s showed the promise of assessing student performance for personalized assessment and student pathways customized to a student's knowledge deficiencies. For example, Whitlock (1964) wrote about the use of AS to tabulate, analyze, and report scores using a "test scoring machine." Similarly, Johnson (1971) opined that "test scoring is a time-consuming task" and that "AS can reduce the time given to the task and provide the teacher with statistics" (p. 70). These early uses focused primarily on simple tasks such as scoring multiple-choice responses with the goal of saving teacher effort rather than expanding the complexity of item types.

However, other researchers focused on more complex tasks. For example, Carbonnel's SCHOLAR (1970) used *natural language processing* (NLP) to interact with learners by asking and answering questions as well as assessing the correctness of learner responses. Ellis Page (1966b) wrote an article titled "The imminence of grading essays by computer" in which he viewed computers as a means to relieve the burden of teachers and to allow them to assign more writing tasks to students. Page and his colleagues (Ajay, Tillet, & Page, 1973) developed the first computerized essay scoring system – albeit running on a mainframe with punchcards – which reduced an essay to counts of a number of statistical *features* that, when combined through regression, could predict instructor scores with correlations of about 0.5 to 0.7.

While these early efforts illustrated the promise of applying technology to education, computers had neither sufficient power nor availability to be easily deployed in the classroom, and the dream of AS in large-scale use was not fully realized. Across the following decades, however, the field of AS has experienced strong growth, as evidenced by more recent volumes that have compiled work from a number of researchers in the field. For example, the volume *Automated Scoring of Complex Tasks in Computer-Based Testing* (Williamson, Mislevy, & Bejar, 2006) broadly covered best practices and approaches for developing valid assessments through automated scoring. In addition, Shermis and Burstein's 2003 and 2013 edited volumes focused on a range of techniques and applications applied for scoring and feedback of student essays.

The growth in the field is further evidenced by the amount of research in this area as indexed by *Google Scholar*, which indicates that there are now over

8100 research articles matching "automated scoring" and "education," with about 40% of those articles published within the last five years. Similarly, if we look at the use of the phrase "automated scoring" in the *Google English Book Corpus* between 1970 and 2008 (see Figure 1.1), we see an inflection point around 1998 and rapid growth starting in the 2000s.

The aim of research in this field is to have an impact on education, which can best be accomplished by translating research into practice. Consequently, in the past two decades, AS engines are being deployed into wide use in a variety of educational and assessment contexts. Example systems cover a range of complex tasks including writing, speaking, and intelligent tutors in both formative and summative contexts. These applications cover a range of use cases and users from grade school to university level to professional certification. Automated writing assessments embedded within high-stakes testing contexts include *The Intelligent Essay Assessor*®, *e-rater*®, *Intellimetric*®, and *Autoscore*®. Similarly, automated writing assessment has become part of formative writing programs via applications such as *Writing Mentor*®, *WritetoLearn*®, *WriteLab*, and *PEGWriting*®.

Moreover, AS of speech for training and assessing language proficiency is used within assessments such as the *Pearson Test of English Academic, TOEFL iBT,* and the *TOEFL Practice Online* tests, a wide range of specialized corporate language tests, as well as within practice and benchmark formative systems such as *Pearson Benchmark, Duolingo,* and *ELSA*. Mature *automated tutoring systems* track spoken and written language to assess knowledge, affect, and attention during dialogues with *intelligent agents* (e.g., *AutoTutor*); some even evaluate open-ended responses and provide personalized feedback in areas such as math and physical life sciences (e.g., *ASSISTments, Inq-ITS*).

What has led to the exponential growth in research on and implementations of AS? We see a convergence of factors in data analytics, hardware, software, and measurement theory, as well as needs from educational institutions to measure students' performance in more realistic contexts. Throughout the chapters in this handbook, these same factors will appear, interwoven with

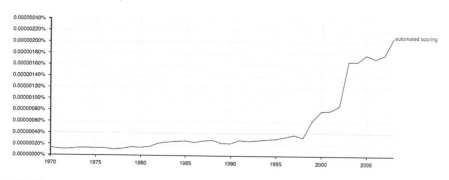

FIGURE 1.1
Mentions of "automated scoring" in the *Google English Book Corpus* between 1970 and 2008.

each other and applied across a range of research approaches, domains, and applications. Below we elaborate on these key factors.

Factors in the Growth of Automated Scoring

Advances in Data Capture

AS had not been practical to implement before data were collected digitally. Before the early 2000s, digital devices were expensive, not networked, and not well integrated into schools. With the advent of relatively *low-cost computing* (e.g., mobile phones, tablets, and laptops), it became possible for devices to be easily deployed in educational settings. Indeed, a 2017 study reported that over 50% of classrooms in the U.S. had a 1:1 student-to-device ratio (*EdTech Magazine*, 2017).

However, having devices in a classroom is not sufficient because they must also be able to collect the right kinds of data. Fortunately, modern devices can be instrumented to capture a wealth of complex student interactions including writing (e.g., essays, short answers, equations), speech (responses in a dialogue, short speeches, pronouncing a sentence), and rich *log files* of actions taken by students as they use learning environments on the device (e.g., interacting in a digital chemistry lab or showing steps to solve a complex problem). With the growth of *learning management systems*, these data can be collected and managed. This "digital ocean" of educational data (e.g., Dicerbo & Behrens, 2012) is rich and complex but must be analyzed appropriately to make inferences about performance.

Advances in NLP and Speech Processing Technologies

For many complex assessment tasks, the expression of language provides a window into a person's cognitive and psychosocial processes. Writing and speaking can reveal a person's level of knowledge, their fluency with language, their ability to organize and convey knowledge coherently, and their ability to maintain topics while they hold conversations. Speech can reveal additional language skills unique to this medium such as their accents, pronunciation, prosody, and emphasis. As we noted earlier, in order to make inferences about these abilities, the resulting language data must be recorded, converted, and reduced to sets of statistical features relevant to the educational experience being measured.

NLP and speech processing technologies have greatly advanced in the past few decades. We see consumers becoming more familiar with speaking into devices through commercial applications such as *Siri* and *Alexa* that can convert their speech to text and comprehend the language sufficiently to complete basic chores and answer questions. Similar approaches provide the bases for assessing language-based tasks. Through training on a large sample of text or speech – with sizes often in the thousands to millions – modern *statistical models* and *deep learning-based approaches* (e.g., Goodfellow, Bengio, & Courville, 2016) have transformed the field. Modern methods do not merely count *surface features* of language such as 'the number of nouns' or 'the number of words that humans categorize as emotional words' but, instead, can convert streams of language into accurate measures of *construct-relevant characteristics* such as semantic meaning, coherence, syntactical structure, and variations in prosody. These features provide the basis to characterize a rich set of language abilities in students.

Advances in Machine Learning (ML) Algorithms

As noted above, reducing data from a complex performance to sets of features can only be effective if one can then combine and weight those features to convert them into construct-relevant performance measures; this process is typically called "building an AS model." The field of ML focuses on algorithms that can process large amounts of data to analyze the patterns of across and within the features and infer relationships to criteria of interest. The world has experienced an explosion of online data, which include emails, chats, inputs to search engines, clicks on advertisements, and credit card transactions. Variants of ML-based methods including generalized linear models, support vector machines, decision trees, and neural networks continue to be developed and refined to exploit this data. Along with providing great commercial benefits for many fields that exploit online data, the same algorithms can equally be applied to educational data.

Advances in Computer Hardware and Distributed Computing

Using advanced algorithms to analyze large amounts of data requires extensive processing power. Ever-increasing power in computer processors as well as *parallelization* allow ML modeling to be performed simultaneously across multiple processors. *Cloud-based computing* services further enable harnessing many computers at any time anywhere in the world. Modern computational architectures allow development and deployment for efficient and highly scalable processing distributed on servers around the world. For assessment of complex performances, these advances allow rapid building of complex ML models and real-time processing of student input. By being able to do measurement on the fly, an AS system can provide students with feedback immediately after they submit their responses and instructors can efficiently monitor students' progress.

Scientifically Grounded Methods for Human Performance Measurement

Scoring any kind of performance – whether by human raters or by computer – rests on the assumption that the performance in a task can be decomposed into a set of targeted skills that are brought to bear in some combination during task completion. As a result, computer-based scoring cannot advance without a firm understanding of the skills being assessed, their alignment to the items / tasks / activities, and how human raters evaluate these skills. Indeed, many AS techniques continue to use *supervised learning* in which they are trained on scores generated by human raters. Additionally, without a good understanding of the skills and measures, it would be impossible to validate the performance of an automated system. Thus, AS is dependent on a firm foundation of effective human performance measurement.

Performance measurement is a large field of its own, with its own constituent societies (e.g., *National Council on Measurement in Education, Society for Industrial and Occupational Psychology*) and journals (e.g., *Journal of Educational Measurement; Assessing Writing; Educational and Psychological Measurement; Applied Psychology*). A number of recent developments in educational performance measurement align with the computational advances in AS.

One advancement is the development and refinement of methods and frameworks for the design of assessments based on evidentiary arguments. For example, *evidence-centered design* (ECD) (Almond, Steinberg, & Mislevy, 2002; Mislevy, Almond, & Lukas, 2003) provides a framework for designing assessments that provide explicit construct-driven processes for aligning the knowledge, skills, and other abilities of a student to the intended constructs, the assessment items / tasks / activities, and the evidence derived from the student's performance on those items. This work is critical for the design of methods for AS because, by using construct-relevant features to derive a prediction model, the model provides an explicit linkage between the features being scored and the constructs. In this manner, AS models can support validity arguments for how combined scoring features represent the constructs of interest (see, e.g., Williamson, Xi, & Breyer, 2012).

A second advancement in the field of educational measurement is the development of frameworks of skills. Driven by the need to have more complex tasks in assessments, a variety of organizations have employed domain and educational experts to build frameworks that identify skills and proficiencies of higher-level cognitive processes. One example of a framework is the *Common Core State Standards,* which was designed to help grade school students graduate with the skills needed for the college and the workforce. These standards emphasize skills such as being able to read complex non-fiction texts, write while citing evidence, and solve problems with mathematics.

Other frameworks have been developed by organizations such as OECD, ATC21s, and the *Partnership for 21st-Century Skills* to address complex skills such as collaboration, communication, critical thinking, and creativity (see, e.g., OECD, 2015; Griffin, McGaw & Care, 2012). These frameworks provide insights into the skills in complex tasks and support better design and evaluation of AS. For example, OECD's 2015 *Framework for Collaborative Problem Solving* (OECD, 2015) refined sets of collaborative skills that could be assessed, developed methods in which students interacted with intelligent agents, and created new approaches to analyzing log data of their interactions to characterize the students' collaborative skills (e.g., Graesser et al., 2018).

A third advancement is the ongoing research that has produced refined understandings on how humans assess performance. Research has evaluated factors such as features that raters use in their evaluation of student work, the methods that are used in order to train raters (e.g., to have raters focus on features tied to the assessments constructs to ensure high interrater reliability), and ways to reduce bias and drift in scoring complex performance. Understanding human performance allows new insights into how best to apply ML methods to model humans as well as provide valid, reliable data to serve as gold standards to evaluate AS performance.

A Growing Need for Assessment of Complex Tasks

The final factor that is driving the growth for AS comes from the changing needs of the students, instructors, workforce organizations, and educational institutions. In particular, there has been broad recognition of the need for training and assessing *higher-level thinking skills* in order to prepare students for higher education and the workforce. *Computerization* in jobs has changed the kinds of skills of the workforce from performing manual and routine cognitive tasks to greater problem-solving and complex communication tasks (e.g., Levy & Murnane, 2005). Organizations such as those described in the previous section have recognized this need and accordingly have adjusted frameworks and focus in assessments and learning activities to help drive the development.

AS provides organizations the advantage of cost-effective ways of implementing mechanisms that can support assessment by scoring performances on complex tasks quickly and reliably in order to make inferences about construct-relevant complex skill sets. In fact, a number of high-stakes assessments have now adopted AS, which enables the introduction of more advanced performance tasks without the need for as many human raters as in the past. However, a second advantage is that AS can foster better learning of complex skills through the automated evaluation of performances in learning environments. If we go back to Ellis Page's 1968 monograph, his primary goal was to provide automation so that students would spend more time writing to foster better thinking. Indeed, we see this growth in the adoption of applications that provide rapid, actionable feedback to students while keeping instructors and decision-makers better informed about the students' progress.

The Multidisciplinarity of Automated Scoring

AS requires a *multidisciplinary approach*. Research and applications are built by combining information from diverse domains including educational measurement, computational linguistics, machine learning, cognitive science, user-interface design, and gaming and require a firm understanding of the content domains being assessed. Each discipline contributes critical parts to the whole. The chapters in this handbook reflect this multidisciplinary approach and include contributions from authors with diverse backgrounds across the different fields of study.

The multidisciplinary approach is further reflected in the domains and skills to which AS is applied. Much of the research and applied use has focused on formative and summative assessment in grades K–12 and higher education where it has covered aspects of domains such as science, math, and the humanities and skill sets such as problem-solving, reasoning, collaboration, and communication. There has also been increasing work in supporting second and foreign language learners as well as skills practice and professional certification in professional domains such as medicine and business. In other words, the theories, principles, and methods developed in AS can be applied broadly to new domains for varied use cases at the time of this writing.

Goals and Organization of the Handbook

The growth in the field of AS for complex tasks necessitates building common understanding across the multiple disciplines involved. The future of AS lies in building a field where there is continued innovation and, at the same time, creation and use of established methods and best practices that can assure effective and trusted performance. When we created this handbook, we wanted to further the common multidisciplinary understanding of this work by examining AS from the points of view of innovative AS theory, the latest development of the computational methodologies, and illustrative real-world large-scale applications. The handbook provides scientifically-grounded descriptions of the key research and development efforts that are required to move AS systems from theory to research and from research into operational practice.

In alignment with this vision, we have organized the chapters into three parts:

(1) Theoretical Foundations

(2) Operational Methodologies

(3) Practical Illustrations

A special chapter is provided at the end of each section, which provides a commentary of the constituent chapters in order to provide a broad overview of the implications of the approaches described in each section.

The handbook is designed to serve practitioners who develop educational technology, researchers in machine learning, educational measurement, and educational design, as well as decision-makers who may use or interpret results from AS. For a practitioner, this handbook can help increase understanding of how best to apply the latest methodologies in valid and effective ways. For a researcher, this handbook can help understand how to extend theories of measurement and evaluation in order to create new approaches to AS. For a decision-maker involved in educational design and evaluation, this handbook can help them understand the practical implications of the process to design, implement, and deploy AS and the kinds of information that are needed in order to make informed decisions about implementing AS. Since AS is a multidisciplinary field, we do not expect that the reader will be familiar with all the terms from the variety of fields that inform automate scoring in education; as such, we have provided a glossary at the end of the handbook to define key terms and we discuss a few additional avenues for staying abreast of current developments in the concluding synthesis chapter of the handbook.

Part I: Theoretical Foundations

2

Cognitive Foundations of Automated Scoring

Malcolm I. Bauer and Diego Zapata-Rivera

CONTENTS

What is the relation between human scoring and automated scoring? What are the properties of the knowledge structures (categories) that raters use in scoring constructed-response items? Do *automated scoring* (AS) approaches make use of these same properties, and if so how? These are some of the questions that guide our review in this chapter.

Typically, human raters involved in constructed-response scoring are given *scoring rubrics* that outline the characteristics of different score levels. They are also provided with rubrics and *exemplars* or *benchmark responses* to review, which are prototypical members of each score level category. Furthermore, they are given sample responses on which to practice scoring and are given feedback as they do so. Raters are then tested in a *calibration session* to ensure they have reached an acceptable level of mastery in their ability to categorize responses (Walsh, Arslan, & Finn, 2019).

Bejar's (2012) review of human scoring focused on rater cognition and provided the following definition: "scoring of constructed responses in its most abstract form is simply the process of assigning an ordered categorization to responses elicited by a test question" (p. 3). We agree with this high-level view and propose that cognitive processes associated with *categorization* are central to understanding human scoring. Specifically, constructed-response scoring is related to two of the four types of human category learning tasks described in Ashby and Maddox (2005), specifically, *rule-based categorization* and *information-integration categorization*, which, in part, depend upon explicit reasoning, working memory, and executive functioning.

In the following, we use the literature on categorization and category learning to propose several important properties of this type of human scoring. Specifically, we develop principles of scoring that are consistent with both human and AS to propose a more general framework for scoring, much as studying how both engineered planes and birds fly lead to a deeper understanding of the principles of flight and specifically lift (Armer, 1963).

Accordingly, we have divided the chapter into two major sections. In the first section, we focus on human scoring and make the case to frame scoring as categorization, which includes a brief review of some of the literature on categorization and the associated area of category learning. We end this section with an integrative framework that maps aspects of categorization to constructed-response scoring. In the second section we focus on AS, review different types of AS algorithms, and explore how validity questions can be asked from the perspective of categorization and category learning in human scoring.

Human Scoring

Cognitive science attempts to understand how humans perceive, think, and act. Within this broad scientific field, several areas such as learning, categorization, and decision-making can help play a role in explaining how human raters score test responses. While much previous research on human scoring has taken a judgment and decision-making perspective (Suto, 2012), we will focus on scoring as categorization and category learning as mentioned in the introduction. Unlike prior approaches, this allows us to explore the knowledge around the responses to be scored and the scoring categories in more detail. Since AS must address the same response patterns as human raters, viewing scoring as categorization provides a strong foundation for comparing and contrasting human scoring and AS.

This process of learning to score is very similar to what is called *supervised category learning* (Pothos, Edwards, & Pearlman, 2011) in the cognitive literature in that they practice scoring by assigning levels to a sample set of responses until they reach a required degree of mastery. AS has the same goal, which is to categorize responses into correct levels; however, while the approaches used in AS are related to those used in human scoring, they are not the same.

To provide a basis for making comparisons between scoring and categorization, we briefly review the core ideas and principles of human categorization.

Rosch (1999) proposed two overarching ideas that drive human categorization, namely that (a) "the task of category systems is to provide maximum information with the least cognitive effort" (p. 190) and that (b) "the perceived world comes as structured information rather than as arbitrary or unpredictable attributes" (p. 190). In other words, humans actively develop a rich internal conceptual structure that, to the extent possible, mirrors the patterns, features, and covariance that exist in the world that underlies our understanding of words, concepts, and relations among them.

Categorization Properties

There is a substantial literature in cognitive psychology that characterizes principles of human natural categories. Hanson and Bauer (1989, p. 344) summarized some principles of human categorization research, which include the following three:

Principle 1. Natural categories are typically defined through *family resemblance* among members rather than strict logical rules. In other words, categories tend not to be defined by features among members that are all necessary and sufficient for membership but, instead, by the correlational structure of the features across category members defines membership (i.e., categories possess *polymorphy*) with some degree of trade-off among features and allowing category members with a range of family resemblance.

Principle 2. Not all category members are equal; instead, categories contain a distribution of members. Some are more representative or prototypical of the category (i.e., are at the center of the category), while others are less representative (i.e., are closer to the category boundary). For example, in the category "birds," a robin is very representative of the category for many people in the U.S. and a penguin is less representative. Consequently, people are faster in deciding the truth of the statement "a robin is a bird" compared to the statement "a penguin is a bird," for instance.

Together, these two principles are consistent with the idea that being a member or not being a member of a category is a question of degree and not of an exact match to a logical rule, which is in keeping with theories in psycholinguistics as well (e.g., Lakeoff, 1987).

Principle 3. Categories are not built in a vacuum but are developed as contrasts between one another. That is, people tend to develop categories that maximize the *within-category cohesion* among category members (i.e., how similar the members of a category are to one another) and the *between-category discriminability* across all the categories (how dissimilar the categories are to one another) while keeping the number of categories to a minimum (see Smith & Medin, 1981).

This idea has been the basis for *machine-learning algorithms* such as the Witt category learning algorithm (e.g., Hanson & Bauer, 1989) and for *quantitative modeling* in psychology such as the closed-form definition of *category utility* (e.g., Corter & Gluck, 1992). Creating a category structure that meets all three criteria involves finding a balance. One could, for example, create one large category (i.e., minimize the number of categories), but the category would likely have low within-category cohesion. At the other extreme, one could put each object (or, for scoring, each response) in its own category. This would be good for maximizing within-category cohesion, but poor for minimizing the number of categories and for maximizing between-category discriminability if there were many responses that were too similar to one another. Usually the best set of categories involves balancing these criteria so that category members that are similar are in a single category while keeping categories as far apart as possible. In a sense, categories are a best fit with respect to a set of features over a set of objects (or set of responses in the case of scoring).

Human-created scoring categories and constructed-response category members have these three properties as well. Scoring rubrics are typically developed as contrasts to distinguish among different score levels so between-category discriminability is especially important for scoring (as in Principle 3). However, within-category cohesion plays a role in scoring as well. Some responses are more central to a scoring level (i.e., are more representative or prototypical of a scoring category), whereas others are near the boundaries of two levels (i.e., are at the fringe of a scoring category).

For example, in scoring short essays on a scale of '1' to '5,' some essays are more central to the central '3' category and are a better fit to the rubric while other essays scored as a '3' are near category boundaries (i.e., near a '2' or a '4') and are not as obviously a fit with the '3' category, are harder to score, and/or take longer to score (as in Principle 2). While rubrics express criteria for score levels, there tend to be windows of performance allowed for each criterion and often some trade-off among them. An essay can be a category '4' because it meets a minimum for one criterion but is especially good on another. The range of performance creates the property of family resemblance among members rather than an exact match to specific features (as in Principle 1).

Illustrative Example

To make these ideas more concrete, we turn to a brief example from operational practice. In a recent discussion about how she uses rubrics, a rater noted that "responses are rarely an exact match, I'm always looking more for the better fit of a response to the two rubrics that seem closest to the response." This notion of relative fit underscores that score-level categories exist in relation to others (as in Principle 3). If one score level and associated descriptors in the rubric were removed, a response could be categorized using the next best fit.

The example also suggests that distances between the centers of some adjacent scoring categories may be psychologically further apart than those of other adjacent categories. For example, the same rater reported that "sometimes the difference between a '1' and a '2' seems much larger than the difference between a '2' and a '3.'" The example also illustrates that relative fit is related to within-category cohesion. That is, some responses are better fitting (i.e., more prototypical) to a specific scoring category than others even though both may be assigned by the rater to the same category (as in Principle 2). Haswell (1998) reports that holistic scoring for writing placement results in categories having prototypical members much like natural categories.

Overall then, what we have seen so far is that we can view scoring as a human categorization process guided by a set of rubrics as well as associated scored exemplar responses and sets of training responses. This defines a psychological space consisting of categories that may be different psychological distances apart with individual categories that are of different sizes and have central and less central members.

Category Learning

Ashby and Maddox (2005) reviewed four types of categorization tasks that have been studied extensively. As noted earlier, rule-based tasks involve learning categories for which a simple rule, often focusing on one dimension of the stimulus, is adequate and can be learned via an explicit reasoning process (p. 152). The authors provided examples of rule-based categorization tasks such as "respond A if the object is small." In this case, the single dimension is size, and an object is in category A if it is small. The authors then expand the notion of rule-based tasks by saying that, as long as the category can be easily described verbally, it may include two or more dimensions such as a conjunction rule like "respond A if the object is small and round" (p. 153). For a physics diagrammatic problem rubric, for instance, this may take on the form "score a '2' (full credit) if the force diagram represents all needed forces with correct magnitudes and directions." It is considered a rule-based task since the rubric is easily described verbally for two dimensions (i.e., magnitude and direction) that are considered separately, and exemplar responses are not necessarily required to fully understand the basis of scoring these responses.

In contrast, for information-integration tasks, people must combine information across multiple dimensions for categorization in ways that are not easily articulated. Again, Ashby and Maddox (2005) provide an example of a difficult perceptual categorization task for experts, namely the identification by radiologists of tumors in X-rays in which multiple perceptual features and dimensions must be taken into account. While doctors can provide a post-facto rationale describing the tumor, it does not provide the full basis for their identification of the tumor.

Holistic essay scoring is also an example of an information-integration task, with prototypical descriptors such as "score the essay at level 4 if the essay consists of a clear claim with a well-developed argument including supporting evidence and reasoning and rebuttals of major counter arguments." In order to assign this score, there are many higher-level dimensions that must be integrated along with numerous lower-level linguistic features of writing. While the rubric is an important guide, practice in rubric application through many exemplar responses is needed to fully learn the scoring category.

These categorization tasks are connected to a leading neuro-cognitive model of human category learning called COVIS (COmpetition between Verbal and Implicit Systems). The COVIS model postulates that there are at least two systems involved in category learning. The first system is a *declarative system* that employs logical reasoning and has access to working memory; rule-based tasks draw upon the declarative system. The second system is a *procedural system* that learns slowly and incrementally, uses implicit statistical processes for categorization, and is much more responsive to feedback and its timing; information-integration tasks draw much more heavily on the procedural system. Delays in feedback by as little as 2.5 seconds affected some types of category learning significantly (Ashby & Valentin, 2017). The COVIS model leads to the articulation of a fourth principle:

Principle 4. Human category learning draws upon two separate systems, one focused on logical reasoning (the declarative system) and the other on an implicit statistical process (the procedural system).

Table 2.1 summarizes these psychological principles across the categorization and category learning review above and includes their interpretation and implications for constructed-response scoring.

Automated Scoring

The process of AS involves the use of *computational algorithms* to process selected data features present in the students' responses to assign a holistic or trait score, provide diagnostic feedback, and/or perform internal routing tasks (e.g., Deane & Zhang, 2015).

General Approaches

AS algorithms have been implemented for a variety of constructed-response items. Bejar (2011) described sample features that have been used to automatically score graphical responses, textual responses, and spoken responses. For example, for graphical responses these include presence and count of relevant graphical objects, relative orientation, proximity to other objects, and qualitative characteristics of objects (e.g., Braun, Bejar, &Williamson, 2006; Forbus, Chang, McLure, & Usher, 2017; Katz & James, 1998). For textual responses, these include presence and count of grammar errors, structural

TABLE 2.1

Psychological Principles and Implications for Constructed-response Scoring

Psychological Principle	Interpretation	Implication for Training
Principle 1	A set of responses within a score level can help define important features and family resemblances that provide understanding of the score level since rubrics are not complete by themselves	Training materials should include rubrics and example responses that highlight critical features and family resemblances
Principle 2	Some responses are considered a better fit to a score category than others	Training should include opportunities to make judgments about how good a fit a response is to a single category since responses that are more central to a scoring category will be easier and faster to score
Principle 3	Adjacent scoring levels can be different psychological distances with some pairs closer than others	Rubrics and prototypical exemplars should maximize between-category distance and many "borderline" examples should be included since these will be the hardest to score
Principle 4	Scoring tasks may draw upon one or both the declarative and procedural categorizations systems depending upon the degree of complexity and explicitness of the task and rubric	If the scoring draws strongly upon the procedural system, there will be more need for timely and targeted feedback for complex / integrated rating tasks

attributes at the sentence and paragraph levels, and qualitative elements of a response including lexical sophistication and content aspects (Burstein, Chodorow, & Leacock, 2004; Cahill & Evanini, Chapter 5, this handbook; Sukkarieh & Blackmore, 2009). For spoken responses, these include rate of speech, pauses, pronunciation, and affect (e.g., Zechner & Loukina, Chapter 20, this handbook; Zechner, Higgins, Xi, & Williamson, 2009). Other types of responses that have been explored include mathematical responses (Bennett, Steffen, Singley, Morley, & Jacquemin, 1997; Fife, 2011), collaborative responses (Trentin, 2009), conversations with a virtual agent (Graesser, Forsyth, & Foltz, 2017; Graesser et al., Chapter 21, this handbook; Liu, Steinberg, Qureshi, Bejar, & Yan, 2016; Zapata-Rivera et al., 2014), and musical responses (Pearlman, Berger, & Tyler, 1993).

AS approaches have drawn upon many of the principles and ideas of human categorization, category learning, and scoring as a foundation and have drawn inspiration from human categorization systems as described in the COVIS model. AS researchers have developed algorithms using the two categorization approaches in human cognition described above in Principle 4.

That is, like the declarative system, early in the development of AS, explicit logical reasoning represented as rules such as those in production systems were used for scoring constructed responses (e.g., Braun, Bejar, & Williamson, 2006; Braun, Bennett, Frye, & Soloway, 1990; Leacock & Chodorow, 2003). For these systems, the basis for scoring a response is explicit in the rules used for categorization. In other AS systems, the features to be evaluated and how they should be weighted were pre-determined with features drawn from existing human scoring guides, again providing an explicit basis for scoring responses (Burstein, Chodorow, & Leacock, 2004). Both approaches have strong similarities to the human categorization process that uses logical-reasoning processes and draws on working memory described above (i.e., COVIS's declarative system).

More recently, machine-learning approaches that learn incrementally over many steps have been applied to AS. For these systems, the basis for specific categorizations is harder to articulate because it is represented as a more complex statistical structure (Burrows, Gurevych, & Stein, 2015; Taghipour & Ng, 2016). Consequently, this approach has similarities with the second human categorization system described in Principle 4 (i.e., COVIS' procedural system). For example, machine-learning approaches such as variants of *deep neural networks* show remarkable performance on classification tasks using a variety of compositional structures, but it is hard to describe the basis for categorizations in verbally represented rules (see Khan & Huang, Chapter 15, this handbook).

In addition to Principle 4, AS systems draw upon the other cognitive principles of categorization as well. Almost universally, AS systems make predictions about the degree of match between a response and each putative score category (as in Principle 2) and then choose the category that is the best match (as in Principle 3). Estimates of a score for a response may be near the center of one score level – which would then presumably be a prototypical category member – or halfway between two score levels – which would then presumably mean that it was near a score-level boundary. In other words, AS approaches result in a distribution of members within a score level with more and less typical members (as in Principle 2). AS researchers may be able to improve scoring models by exploring these distributions, comparing them to distributions of similar rater judgments, and creating measures of error that take these distributions into account.

AS systems are also consistent with cognitive categorization processes (as in Principle 3). These systems estimate scores from responses initially either as a bounded real-valued number (e.g., a probability) or as an unbounded real-valued number (e.g., an estimate of 2.63 for a discrete scale with score levels '1' to '5', which gets rounded up to a score of '3'). The initial estimates imply the properties of within-category cohesion (i.e., how close the score estimates within a category are to each other) and between-category discrimination (i.e., how far apart the score estimates are for responses in different categories). A set of score categories with substantial distances between category boundaries (i.e., sharp category contrasts) makes the categorization process easier for both human raters and computer-based algorithms.

Validity Evidence

Automated scores are usually compared to those produced by human raters in order to evaluate their alignment (i.e., there should typically be a close match of the categories produced given the definition of the rubric). However, there is no guarantee of a direct mapping between the features used by the computational algorithm to produce the scores and the classification processes used by expert humans to score the responses. Thus, there is a strong need to validate the interpretations around automated scores to ensure that they can be used defensibly for the targeted use contexts without being compromised.

Various researchers have addressed these issues in the literature. For example, Bejar (2017) elaborated on the need for accounting for possible threats to assessment validity when machine scores do not reflect proficiency levels for the intended constructs (e.g., construct-irrelevant score strategies); he suggests quality-control processes involving human experts to identify and address possible vulnerabilities of automated scores (see Yan and Bridgeman, Chapter 16, this handbook).

Like human raters, AS engines usually benefit from a clear definition of categories (e.g., the rubric or, more commonly, a gold standard set of scored responses) and "enough" data (e.g., exemplars) to train the models that are used to distinguish among these categories. Since AS engines tend to use the statistical structure of relations among scored responses to identify categories (as in Principle 1), without a statistically sufficient number of accurately scored responses representing all the key family resemblances within and across categories, the algorithms employed by AS engines will create impoverished models that will be prone to a large amount of prediction error. If the scored responses systematically underrepresent kinds of responses or the human-scored data contains systematic scoring errors, biases will be included in the AS engine as well.

Put differently, as machine-learning approaches are increasingly used within AS engines, it is important to explore mechanisms for interpreting the results of these systems. Assessment systems should be able to support assessment claims with sound argument structures including the evidence necessary to back such claims (Kane, 1992; Mislevy & Riconscente, 2011). Interpretation of new patterns found by a machine-learning approach that makes use of features that are not directly related to the construct becomes a challenge to the use of these approaches for scoring purposes in higher stakes assessment contexts.

This problem is common to many areas in which *artificial intelligence* approaches are being applied. Research under the name of *explainable artificial intelligence* (Doran, Schulz, & Besold, 2017) is offering methodologies for helping interpret the so-called black box models (Pardos, Fan, & Jiang, 2019). For example, new patterns of evidence in responses identified by machine-learning algorithms based on response corpora and process data can potentially be used to refine existing rubrics and new response patterns can result in recommendations for a new score level or changes to the rubric description to cover the new pattern (see Almond, Chapter 23, this handbook). Hybrid approaches in which humans and computers inform each other can produce advanced AS systems that can accurately score responses that contain evidence of complex skills. Since this approach combines the use of statistical relationships among features within scored responses (like COVIS's procedural system) with easily interpretable logical descriptions of categories (like COVIS's declarative system), this effort draws upon Principle 4.

Table 2.2 summarizes relations between cognitive categorization principles, AS properties, and implications for scoring processes, validity, and explainability.

Comparing and Contrasting Human Scoring and AS

Zhang (2013) compared the strengths and weaknesses of human scoring and AS of essays. The strengths of human scoring include the ability of human raters to attend to more complex aspects of reasoning to make holistic judgments on the quality of essays (e.g., identifying deeper layers of meaning,

TABLE 2.2

Relation between Cognitive Categorization Principles and AS

Cognitive Categorization Principle	Relation to AS Development	Implication for AS Quality
Principle 1	AS approaches are usually evaluated against human-scored responses that inform the development and results of the AS approach	Bias present in human ratings may influence the results of AS processes
Principle 2	AS typically results in score level categories with prototypical and less-typical members since estimates are based upon degree or likelihood of match to category	AS researchers may be able to make use of the distributions of estimates within a score level to improve AS engines
Principle 3	AS systems typically attempt to identify statistical relations that are consistent with the notions of within-category cohesion and between-category discrimination	AS models may help create new subcategories that may not have been identified at the time of development
Principle 4	AS systems have typically used approaches consistent with either human categorization system or a hybrid of both	AS engines may identify new, relevant data patterns that need to be identified, explained, and operationalized in the rubric

artistic/ironic/rhetorical styles, audience awareness, creativity, critical think-ing). The weaknesses of human scoring include the tendency of human rat-ers to be prone to error, bias, and other undesirable response characteristics (e.g., severity/leniency, inaccuracy, differential dimensionality, halo effects, temporal drift [e.g., Wolfe, 2004]).

The strength of AS include objective, consistent, reproducible, and tractable scoring, especially when evaluating structural features (e.g., development, grammar, mechanics, organization, syntactics, word usage), content-rel-evance, plagiarism, and some aspects of style and coherence for example. However, AS is unlikely to capture evidence about the more complex aspects of writings that human raters can identify well and may propagate biases/errors from human raters that are present in the data used to train the AS engines. Although the processes for training human raters and machine algorithms are different, both benefit from a variety of example responses that helps them acquire a better sense of the rubric (e.g., understanding the different forms in which each of the categories could be observed in the data).

One could argue, in fact, that both human- and machine-produced catego-ries are susceptible to bias. For example, without good training and moni-toring mechanisms, human ratings may be biased due to the attitudes and experiences that raters bring to the rating task (Amorim, Cançado, &Veloso, 2018; Ricker-Pedley et al., Chapter 10, this handbook; Wolfe, Chapter 4, this handbook). Similarly, machine-produced scores may reflect possible bias included in the training dataset (Holstein, Wortman Vaughan, Daumé, Dudik, & Wallach, 2019). For example, based upon the results of 35 semi-structured interviews and a survey with 267 machine-learning practitio-ners, Holstein et al. (2019) identified fairness issues with datasets and the algorithms used, possible bias issues with humans in the loop (e.g., when crowdsourcing participants), the need for proactive auditing processes (i.e., encouraging practitioners to monitor and promptly follow-up on possible unfairness issues), and holistic auditing methods that include fairness as one of the key monitoring indicators.

Summary and Future Work

In this chapter, we considered human constructed-response scoring as an instance of cognitive categorization processes. We reviewed several cogni-tive principles and properties of categorization and category learning, most notably that natural human categories have a graded structure with proto-typical and less-typical members, exist as contrasts, and that human cate-gorization involves two very different neurocognitive systems that support categorization in different ways. We then applied these principles to human raters scoring constructed-responses items and explored their implications.

Finally, we used the same principles as a foundation for describing and examining aspects of AS, which included the nature of AS algorithms along with its strengths and weaknesses as well as key facets of validity.

What are the possible implications for validity and fairness? The increasing use of AS engines that employ black-box modeling approaches may raise validity questions such as, "How do we justify results and use results produced by these systems that may not be easily interpretable or directly inspectable by humans?" "How do these results or new categories inform the definition of the construct?" "Are there any benefits from exposing domain experts to these results?" It seems that hybrid AS systems that keep a clear role for domain experts may be part of the future work in this area.

To explore this, we briefly summarize Rupp (2018), who provided a workflow for AS. We expand upon this by identifying the domain expertise needed and how the categorization principles impact each area. The author described methodological design decisions in AS within a workflow of five two-part interacting phases and organizes these into three main groupings of design, evaluation, and deployment. Each phase identifies and explores validity issues and methodological approaches and typically involves different experts and stakeholders; see also Bennett and Zhang (2016).

Specifically, the *design phase* includes the processes of task design, human-rating development, and development of initial-scoring models. This includes identifying the nature of the evidence of the target competencies that tasks can collect, the creation of rubrics, and the design of approaches for human scoring to create an initial scored corpus of responses. The *evaluation phase* covers global and local exploration of the scoring model's specific statistical properties for aggregation of features and inspection of empirical results, along with evaluations of validity with respect to generalizability and cognitive fidelity based on process data. Finally, the *deployment phase* includes system readiness/rollout checks and continued monitoring during ongoing use with needed expertise including psychometrics and information technology. The categorization principles have implications in each of these major phases, which we discuss next and summarize in Table 2.3 (see Lottridge & Hoefer, Chapter 11, this handbook; Schneider & Boyer, Chapter 12, this handbook; Yan & Bridgeman, Chapter 16, this handbook).

Design Phase

For the design phase, Principle 1 suggests that rubrics can be used to describe, but not fully define, polymorphic categories and should be guides for human scoring as well as inform feature identification in early AS work with response corpora. Subject-matter experts, cognitive scientists, specialists in natural language processing, and computer scientists should all be involved in reviewing rubrics with respect to these different uses and each of their areas of expertise.

TABLE 2.3

Cognitive Categorization Principles and their Implications across the AS Workflow

Cognitive Categorization Principle	Implications for Design	Implication for Evaluation	Implications for Deployment
Principle 1	Develop rubrics that can effectively guide human rating and inform feature identification	Analyze the properties and internal structure of responses and support validity claims by extracting explicit descriptions of the categories	Monitor changes in the internal structure of scoring level categories
Principle 2	Collect data that include prototypical and less-typical responses for each scoring category when trialing out tasks and rubrics	Evaluate the fit of prototypical and less-typical responses to the model	Monitor the distribution of prototypical and less-typical responses within a category
Principle 3	Identify outliers in adjacent scoring level categories to refine rubrics and collect data that over-sample responses at category boundaries	In evaluating models, explicitly selecting response sets with different typicality distributions within and between categories may provide stronger tests of sensitivity	Monitor changes in between-category structure and adapt scoring model as larger and different populations engage in test-taking
Principle 4	Provide well-timed feedback to improve scoring for raters	Provide justifications for scores and develop more complex and accurate scoring systems	Increase understanding of score categories through the use of prototypical and less-typical responses

Principle 2 suggests that while creating tasks and rubrics, designers should consider the kinds of responses that are the most prototypical for the scoring category, as well as the multiple ways in which responses could be less typical. Addressing this will involve a collaboration between task designers and subject-matter experts. Principle 2 also suggests that, while collecting response corpora, it might be beneficial to collect human data on prototypical vs less-typical responses for use in the creation of AS models.

Similarly, Principle 3 suggests that typicality judgments may be useful in defining score category boundaries. Identification of outliers that are still in a score category and other responses that are outliers in adjacent categories can prompt discussion and refinement of rubrics to highlight contrasts between score levels. Data sets that oversample responses at boundaries may be helpful for AS model creation.

Lastly, Principle 4 suggests that since written rubrics only address the declarative system in human categorization systems, rater training should directly provide well-timed feedback to improve scoring within the procedural categorization system. For the design of AS systems, feature identification can draw upon both categorization systems in people and in computational systems if collaboration between the members of the interdisciplinary development team is carefully coordinated.

Evaluation Phase

For the evaluation phase, Principle 1 suggests that it may be beneficial to evaluate AS scoring models with respect to the properties and internal structure of the categories of responses. Evaluation might be based upon explicitly or implicitly defined distance metrics in the models where model estimates can express the graded structure of the category inherent in polymorphy rather than transformed rounded right/wrong judgments. For validity claims, explainability could be based upon extracting explicit descriptions of the categories, which would again involve a collaboration between interdisciplinary specialists.

From Principle 2, one idea for improving the evaluation phase is to consider the fit of prototypical vs. less-typical responses to the model, as judged by human raters. This would include addressing questions such as whether the model fit of the scores aligned with the typicality judgments and whether such kind of alignment improves score accuracy.

Principle 3 suggests that explicitly selecting response sets with different typicality distributions within and between categories may provide stronger tests of sensitivity. Principle 4 suggests that evaluation would benefit from analyses that make use of both declarative and procedural categorization systems – declarative systems can provide more clearly described justification for scores (i.e., enhance explainability) but procedural systems (i.e., implicit, statistical systems) can provide more complex and accurate scoring using a richer between- and within-category structure that will need to be part of the validity argument.

Deployment Phase

For the deployment phase, Principle 1 suggests that it could be useful to monitor any changes in the internal structure of the scoring categories through continued collection, scoring, and analyses of responses. Similarly, Principle 2 suggests that ongoing monitoring of the distribution of prototypical and less-typical responses within a category might provide warning of shifts that require adaptation (e.g., perhaps due to changes in test taking).

Principle 3 suggests that it may be useful to monitor the between-category discrimination (i.e., category contrasts) as larger and different populations engage in test-taking. Changes in between-category structure may be a strong indicator that adaptation of the model is needed. Lastly, Principle 4 suggests that, for roll-out, in communicating with stakeholders, descriptions of score categories typically connect only with people's declarative systems. However, there is potential to increase understanding of score categories though the use of prototypical and less-typical responses that may require engagement of the procedural system.

In summary, in this chapter we provided a review and synthesis of some of the research on human categorization as a foundation for AS and used that foundation to propose areas of AS research that could be further explored and enhanced by viewing them through the lens of human cognition. We hypothesized that the richness inherent in natural human categories and their principles could be used to expand the evaluation metrics for AS models and could potentially provide leading indicators of scoring problems before they become apparent through current measures. As more is learned about human categorization through neuro-cognitive science research and as advancements in artificial intelligence continue, we anticipate continued synergy leading to additional advances in the field of AS.

Author Note

We wish to thank Brian Riordan and Matt Mulholland for sharing their perspectives as researchers in automated scoring and natural language processing as we wrote this chapter.

3

Assessment Design with Automated Scoring in Mind

Kristen DiCerbo, Emily Lai, and Matthew Ventura

CONTENTS

Scoring can be defined as the process of translating learners' *work products* into interpretable evidence. *Automated scoring* (AS), then, is the use of computers to do this task. In fixed-response assessments, this information translation is relatively easy to accomplish for both humans and computers. For example, responses on a multiple-choice assessment can easily be turned into 'incorrect' / 'correct' ('0' / '1') indicators based on an objective key. The resulting inference about learner knowledge and skills, however, may not be as strong as we would like because the task learners are doing (i.e., answering multiple-choice questions) is relatively far from the real-world tasks we want learners to be able to do. That is, the real-world skills we want learners to use are often not appropriately represented by the simpler assessment tasks.

The alternative is to have learners engage in performance-based assessment where they engage in more authentic activity. In the past, this has required human scoring, but advances in computing now allow for scoring of more complex work products. Work on AS of open-ended responses began with essay scoring and has now advanced to scoring other types of performance assessment like activity in games and simulations. What should be the considerations in assessment design when using complex AS mechanisms? How much does AS drive other components of design? In this chapter, we lay out five principles of assessment design for AS illustrated with examples of AS of writing and game-based assessment.

Principled Assessment Design for Automated Scoring

Principled assessment design refers to a broad family of frameworks or approaches to designing assessments with the goal of linking constructs of interest to tasks and to evidence; common frameworks include *evidence-centered design* (ECD) (e.g., Risconscente, Mislevy, & Corrigan, 2016), *assessment engineering* (Luecht, 2012), and *construct modeling* (e.g., Wilson, 2005). These approaches have been described in depth elsewhere, including what they share in common (e.g., Ferrara, Lai, Reilly, & Nichols, 2017) and it is not the purpose of this chapter to explore these different approaches. However, at their core, they all seek to:

- Define the *knowledge, skills, and attributes* (KSAs) to be measured in learners
- Identify the types of evidence needed to collect to make inferences about those things
- Create assessment activities that will elicit those kinds of evidence

ECD is probably the most widely used principled design approach and refers to these elements as the *student model(s)*, *evidence model(s)*, and *task model(s)*, respectively. For convenience, we adopt these terms throughout our chapter. Together, the student, evidence, and task models form the *conceptual assessment framework*, and serve to operationalize the assessment argument (Mislevy & Haertel, 2006). In conjunction with the *four-process architecture*, which is concerned with the computational underpinnings and information flows in operational practice, they represent the key assessment architecture (see Almond, Chapter 23, this handbook).

In this chapter, we enumerate the challenges of assessment design when using AS to interpret evidence and provide suggestions for addressing them. First, we briefly introduce threaded examples that we use to illustrate points throughout the chapter. Next, we introduce five principles for designing in AS contexts. In that section, we also identify challenges and offer approaches for addressing them. We conclude with final thoughts about the implications of AS for assessment design.

Threaded Examples

We pull from a number of examples to illustrate points throughout the chapter. In order to demonstrate the breadth of AS issues, we examine both AS for essays and AS in game-based assessment. Specifically, we focus on an essay writing task and two game-based assessments: one that seeks to assess argumentation skills and one that focuses on the geometric measurement of area.

Writing Assessment

Evidence identification in human scoring is often accomplished with the help of a scoring rubric that describes features of the work product that are associated with a particular score on a particular variable. Examples of variables related to writing skills can include topic coverage, grammar, clarity, and explanation of ideas. AS of writing involves machine-learning techniques that identify linguistic features in a set of writing samples that predict the human ratings (e.g., Shermis & Bustein, 2013). To enable AS, we typically need to identify construct-relevant language features that correspond to evidence in the writing produced by the learner. The AS system then learns how to weigh those features to best predict the human rating for each variable we wish to score (e.g., Cahill & Evanini, Chapter 5, this handbook; Foltz, Streeter, Lochbaum & Landauer, 2013; Williamson, Xi, & Breyer, 2012).

While AS of writing is often used to assess a student's abilities in written expression, it is also used to assess learners' grasp of specific concepts related to a particular domain or discipline. As an example, we use a writing prompt from an *Introduction to Sociology* course throughout this chapter. This prompt was included in a mid-term exam for a college level course, which consisted of a combination of various item types and was automatically scored. This prompt was designed to elicit evidence that 'learners analyze culture from a sociological perspective':

> People are so accustomed to their own culture they often don't even notice it. Reflect on the town where you live now and pretend that you are an alien sociologist visiting Earth for the first time. You have just landed and your job is to send a report to the mothership describing the

culture of the people in the city around you. This will be easy because
everything you observe – behaviors, foods, activities – all seems so foreign and strange to you. Be sure to include all of the components of culture, with at least one example of each, in your report.

The rubric for the assessment of learners' ability to analyze culture from
a sociological perspective (as opposed to other writing traits) provides
descriptions of writing quality for five different potential scores that human
raters can use to score learner essays. In order to allow for AS, 500 samples
of learner responses were collected and scored by two human raters with the
rubric. *Machine learning* was then used to create models that could reliably
approximate the human scores.

Game- and Simulation-Based Assessment

We use two games-based assessments for illustration in this chapter, which
we refer to as 'games' for simplicity. *Mars Generation One* is an iPad-based
game developed by *GlassLab* to assess argumentation skills of middle school
learners. It is based on a learning progression for argumentation skills with
the subskills of 'identifying evidence,' 'organizing evidence with claims,'
and 'using evidence in arguments'; the core loop of game play tasks were
designed to elicit evidence of these skills (Bauer et al., 2017). The game is set
on the first colony on Mars where there are a number of disputes as to how
the colony should be run, which are settled by robot battles and robots are
created by linking claims and evidence. Players seek out evidence, identify claims with relevant evidence in order to create robots, and then go
into battle where they identify weaknesses in other's arguments; see Figure
3.1. In contrast to essay scoring development, design of game scoring is a
more iterative process consisting of multiple cycles of *play testing* with 8
to 12 students followed by revision and more testing. This is used to validate how game play elicits intended knowledge and skills. This is followed

FIGURE 3.1
Robots and their inner claim and evidence cores from *Argubot Academy*.

by *alpha testing* with 50 to 100 students and another round of revision, and finally *beta testing* with several hundred students that allows for calibration, in the cases of the games here, of *conditional probability tables* associated with *Bayesian inference networks* (e.g., Almond, Mislevy, Steinberg, Williamson, & Yan, 2015).

The *Alice in Arealand* game (DiCerbo, Crowell, & John, 2015) is focused on assessing elementary age learners' mastery of the concept of 'geometric measurement of area' (e.g., what numbers mean rather than just how they are computed through the application of a formula). Players in the game help their new friend *Alice* and her pal *Flat Cat* navigate through a 2D world resolving challenges by demonstrating mastery of various stages related to manipulation of area units. The game has characters that function as tools to glue units together (character name *Gluumi*), break them apart (character name *Esploda*), and make copies of single or composite units (character name *Multi*). Learners can access hints on each level by clicking a magnifying glass. If their solution to a level is not correct, they are looped back to play the level again. As in *Mars Generation One*, the levels of the game are aligned to a *learning progression*, which also underlies associated performance tasks and non-digital classroom activities (DiCerbo, Xu, Levy, Lai, & Holland, 2017).

Principles for Designing Assessments for Automated Scoring

Much of the focus on AS is typically on the scoring process itself, rather than on the larger view of designing the entire *assessment argument*. As such, there is less guidance on how to approach task or game design for AS. Consequently, in all examples in the previous section, we experienced challenges with assessment design because the tasks were being designed to make use of AS technology; Table 3.1 summarizes these assessment design challenges.

In the following, we articulate the principles stated in the fourth column of Table 3.1 in more detail based on our experience of addressing the challenges described above. These principles are meant to be stimulants for thinking more broadly about how AS impacts all elements of assessment design. We expect these principles to apply to both *formative* and *summative assessment* use cases, although some may be more salient for a particular application and for particular types of activity (e.g., game-based science assessment versus essay-based writing assessment).

Principle 1: Define Student Models for Breadth and Depth of Coverage

Given the computational complexities around AS, there is a challenge in determining how much of a *domain model* can and should be brought into the student model that is targeted in a particular assessment. The domain model

TABLE 3.1

Challenges and Solutions to Design of Assessments when Using Automated Scoring

ID	Challenge	Essay Challenge	Game Challenge	Principle	Solutions/Strategies
1	Determining breadth and granularity of student model	• Human scorers cannot reliably differentiate between similar target constructs leading to highly correlated scores • AS algorithms cannot learn to differentiate constructs that humans cannot differentiate	• A game covering the entire breadth of the domain model would take years to build, cost millions of dollars, and take too much class time to play • There is not sufficient log file evidence related to sub-subskills to estimate proficiency	Define student models at appropriate breadth and depth	• Evaluate the assessment purpose and intended uses to determine how granular the student models should be • Combine constructs with similar evidence
2	Developing tasks that produce good evidence	Some essay prompts produce such a wide variety of responses that AS algorithms cannot tell what is acceptable	• It is easy to get caught up in what is fun at the expense of what will produce good evidence or alternatively to remove the fun from a game in an effort to produce evidence • There is so much data in the log file that it is hard to know what helps to differentiate between students	Create tasks with evidence in mind	• Use features of good writing prompts • Use features of good rubrics • Balance engagement and evidence in games • Develop tasks with *a priori* hypotheses about evidence

(Continued)

TABLE 3.1 (CONTINUED)

Challenges and Solutions to Design of Assessments when Using Automated Scoring

ID	Challenge	Essay Challenge	Game Challenge	Principle	Solutions/Strategies
3	Designing for reusability	Gathering and scoring student examples for each new prompt is time-consuming and costly	Game levels are time-consuming and costly to build	Design for reusability	Use task models to promote reusability
4	Selecting the best approach to evidence identification	Human scorers are considered gold standard for machine scoring	Developers tend to choose either human-identified or machine-identified approaches	Combine human and machine approaches to evidence identification	• Use a priori hypotheses about evidence to guide machine-learning approaches • Use humans to validate machine-learning models
5	Designing for automated feedback	• Need to make feedback granular enough to be actionable, indicating how to improve • Overemphasizing superficial response attributes that are less construct relevant	Need to make feedback granular enough to be actionable, indicating how to improve	Design for actionable feedback	• Provide feedback on student model variable level plus elaborated, task-specific feedback • Consider including common mistakes or misunderstandings within evidence models for this purpose

is the representation of all of the KSAs important to a given domain. The student model is the subset of those KSAs to be assessed in a given assessment. We must determine both the breadth of content (i.e., the number of KSAs the assessment is meant to evaluate) and the depth (i.e., the number of subskills and sub-subskills that are statistically modeled).

Often, a project will begin with a target student model based on a desire to ensure broad content coverage but development and testing reveal that full coverage is not possible in a single assessment. When facing the challenge of determining the content coverage of an assessment, it is critical to consider what can be reliably scored and understand what can be supported in the time available for assessment activity.

Another consideration in identifying *granularity* or depth of coverage is communication to educators, learners, parents, or other stakeholders. If the goal is to provide very coarse-grained information about learners at the end of a period of instruction, then there is no need to build a granular student model. However, if the purpose of the assessment is formative and the intention is to support learning, then student model variables may need to be more granular in order to help learners improve. Again, however, these more granular variables need to be able to be reliably scored by humans.

Writing Assessment

Without reliable human scores, the machine-learning algorithms will not be able to create reliable prediction models to approximate those scores. Unfortunately, it is all too common for those familiar with the hype around artificial intelligence but unfamiliar with the processes of AS to assume that the models are able to identify constructs that humans themselves cannot discern. It is therefore worth making explicit that automated essay scoring, as done today, can only approximate the scoring of the constructs that humans themselves can score.

In the case of evaluating learners' writing ability through essays, which involves component skills related to both writing mechanics and content, every student model variable needs a scoring rubric (i.e., evidence model) that describes the features that are used to translate elements of the learner essays into scores. This is what humans will use to score the writing, and these scores typically form the "ground truth" used to train the AS capabilities unless unsupervised classification algorithms are used. However, the features for different student model variables can be difficult to distinguish by humans due to their conceptual or empirical similarity (Aryadoust, 2010; Bacha, 2001; Bauer & Zapata-Rivera, Chapter 2, this handbook; Lai, Wolfe, & Vickers, 2015; Lee, Gentile, & Kantor, 2008; Wolfe, Chapter 4, this handbook).

When student model variables cannot be differentiated via rubrics, the student model likely needs to be simplified, which of course would be the case even if relying on a fully human scoring approach. This problem can be solved in some cases by combining student model variables with very similar evidence. However, it is also important to consider the *instructional context* in making these decisions. For example, in the culture essay example in the previous section, the overall objective about analyzing culture could potentially be broken down into separate scores for analyzing behavior, beliefs, values, and rules. However, human ratings of the analysis of values and the analysis of beliefs would likely be highly related. Consequently, they might be combined into a broader, more coarse-grained, objective about the analysis of culture generally that humans could then reliably score.

Ultimately, in our example, it was decided that, for a mid-term exam, the broader objective of analyzing culture from a sociological perspective was appropriate. Alternately, if the culture essay was a homework assignment to help the learners gauge what portions of the chapter they should review, more granular scoring would be required and, in this case, multiple tasks might need to be created. This kind of student model re-design prior to implementing any automation can help save training costs, reduce scoring time, and minimize confusion for human raters even though such aggregation analyses could also, technically, be done post-hoc using purely statistical means.

Game-Based Assessment

In the case of evaluating learners' skills through game-based assessments, it is often the case that the domain model contains far more student-model variables than can be assessed in a single game. Even with the promise of collecting many individual pieces of evidence from learners' interactions in a digital environment through process data and associated logfiles, there are limits to the amount of meaningful evidence that can be extracted from a task. For example, the potential domain model for the *Alice in Arealand* game was developed from a review of the literature on how learners acquire the concept of 'geometric measurement of area,' which resulted in 17 potential student model variables.

However, as game design developed, it quickly became apparent that game play also required learning the *mechanics* of the game and that three levels were needed for each student model variable: one to introduce/teach a concept, one to practice it, and one to extend it. Given the time available for learner play and budget constraints, in the end, the game focused on only four student model variables: 'understanding area as made up of individual unit squares,' 'placing unit squares end to end,' 'understanding the squares need to be of the same size,' and 'combining squares into rows and columns.'

One consequence of a relatively more granular student model when using activity evidence from *log files* is the potential lack of evidence to support student model variables. If there is just a single observable to support a student model variable, this could lead one to make incorrect inferences about learners around this variable. Having sparse evidence leads to unreliable inferences. One challenge when attempting to modify existing games is the amount of evidence available to support inferences about granular student model variables. This was the case in creating *SimCityEDU* from the *SimCity* commercial game (DiCerbo, Crowell, & John, 2015). Although there were potentially hundreds of evidence pieces, in the end only a handful ended up in the final model estimating proficiency on the relatively broad construct of 'systems thinking.' Clearly, the breadth and depth of the student model are related, and there are trade-offs in enhancing one over the other.

Principle 2: Create Tasks with Evidence in Mind

Given the cost and complexity involved in AS, it is critical to design activities (e.g., writing prompts or activities / tasks in a game) that actually elicit the evidence needed to score consistently and accurately.

Writing Assessment

Fortunately, there are established guidelines for creating writing prompts to enable easier AS (Foltz, in press). One major goal of these guidelines is to elicit learner responses that fall within a predictable set of guideposts. A wide range of responses is difficult for a computer algorithm to score, particularly if responses at the extremes are acceptable. Creating clear expectations narrows the range of acceptable responses and makes the job of AS easier. Features of good prompts for AS (B. Murray, personal communication, July 11, 2018):

- Instructions are clear and concise
- Everything assessed in the rubric is clearly asked for in the prompt
- Clear direction is provided as to what to write (e.g., 'identify and describe,' 'provide examples and explain,' 'compare and contrast')

Note that most of these are features of good writing prompts in general. However, there are also relatively common features of writing prompts that are more difficult for AS. These include the following:

- Answers that rely on personal experience or subjective interpretation – these will produce a very broad range of answers for which AS routines will find it difficult to discern regularities
- Multipart questions where each answer depends and builds on a previous answer – AS processes will not be able to maintain the scoring rules for later parts dependent on earlier answers
- Short answer questions where the answer is a phrase or just a sentence or two – it is difficult for AS algorithms to detect patterns in short samples
- Highly abstract discussions that require deep knowledge of a subject - most AS engines do not model this kind of deep reasoning

Certain types of prompts may work better for evaluating certain kinds of student model variables. On the one hand, (relatively) open-ended prompts (i.e., those with less prescribed structure) are best suited for assessing general writing skills. Consider, for example, this persuasive writing prompt from Foltz, Streeter, Lochbaum, and Landauer (2013): "Should students upon graduating from high school be required to give a year of compulsory community service?"

On the other hand, prompts intended to elicit evidence about critical thinking skills or mastery of certain concepts may need to be more highly structured in order to elicit the types of evidence we are seeking to observe. Consider, for example, the sociology prompt mentioned at the outset of the chapter, which includes specific language that instructs students what particular aspects to include in their responses (boldface added for emphasis):

> You have just landed and your job is to **send a report** to the mothership **describing the culture** of the people in the city around you. This will be easy because everything you observe – **behaviors, foods, activities** – all seems so foreign and strange to you. **Be sure to include all of the components of culture, with at least one example of each, in your report.**

One way of keeping the evidence model in mind when designing the writing tasks for AS is to include in the task model a rubric identifying specific linguistic and other response features that align to the student model variables. Just as there are criteria for good writing prompts, there are also criteria for good (and bad) rubrics (B. Murray, personal communication, July 11, 2018).

Features of good rubrics include:

- Clear specification of levels of proficiency for the student model variable
- Clear articulation of the evidence needed to support each level of proficiency
- Clarification that the rubric is just a guide and that the human scorer needs to acknowledge there is more than one way to express an idea

Features of poor rubrics include:

- Descriptions of learner answers in subjective, metalevel terms (e.g., "insightful", "vivid")
- Descriptions vary unevenly across scoring levels (e.g., "stellar," "good," "disappointing," "no effort" as there is a large difference between "good" and "disappointing")
- Descriptions are pejorative at the lower levels (e.g., describing the work as "poor" or "inadequate" rather than describing what is missing as in "missing topic sentence")

For example, the sociology prompt rubric refers specifically to "shared systems of beliefs and knowledge (system of meaning and symbols), a set of values, beliefs, and practices, and shared forms of communication" as important features of a response and shows how responses are expected to vary across proficiency levels both in their completeness in addressing all these components and in the quality of the examples provided. While the criteria laid out in good rubrics will generally not be explicitly used to train AS models, they will help humans, who are creating the ground truth those models are based on, be reliable in their scoring.

Game-Based Assessment

Moving from writing assessment to game- or simulation-based assessment highlights different issues in task design. First, in game-based assessment, there is a delicate balance between engagement in the game and evidence needed. In *SimCity*, part of the fun of the game comes from its open-ended nature. Players can build and do an almost infinite number of things to create their city. However, this means that, as an assessment, it is nearly impossible to define evidence that all players, or even a large percentage of players, would have because every player does different things. To address this, the design team for *SimCityEdu* constrained what players could do. In initial designs, play testing made clear these constraints had gone too far and learners reported the game was not fun. Designing with evidence in mind required finding the correct balance of what could be captured and scored with maintaining enough openness to be fun. Finding this balance required iterative design and testing, further described below.

Second, it is important to design tasks with *a priori* hypotheses about evidence. In the case of *Alice in Arealand*, every click of the mouse produced a time-stamped record of learners' actions in the game, producing a massive logfile from which to retrieve evidence of learner knowledge and skill. Without the use of tools such as task models, it would be very difficult to make sense of this data. However, the application of a principled approach to game design in this context meant that game levels were designed to produce evidence about how learners used the functionality to cover spaces – for example, whether they used *Gluumi* before *Multi* or the reverse. This combination of behaviors was linked to whether learners thought about area in terms of individual unit squares or more complex composites of units such as rows and columns. Such explicit *a priori* articulation of specific scoring criteria meant that evaluation of learners' game play was not an unstructured *post-hoc* fishing expedition. During design of each game level, researchers maintained detailed design documents listing the desired evidence pieces for each type of activity (e.g., use of *Gluumi* and *Multi*), which meant that the scenario had to present players with opportunities to construct composites and a toolbox from which to choose. It is important to note that the use of AS does not imply that the initial identification of key evidence pieces must be automated. We will discuss this later in the section on balancing human hypotheses and machine learning.

Principle 3: Design for Reusability

Building AS is costly and time-consuming in both writing and game-based applications. Assessment designers can help recoup some of those costs by creating task models that permit reusability of costly assets. Task models articulate the kinds of activities likely to elicit evidence needed to make inferences about the student model variables.

Writing Assessment

From a single task model, many similar instantiations of the task can be created by manipulating variable task features, although task models do not enable 100% reusability, particularly in regard to AS of writing. In other words, although one may be able to use a task model to generate any number of similar prompts of a particular type, one still has to collect learner responses and human ratings in order to train the scoring engine. However, because task models embody the alignment of student model variables to evidence, use of task models and associated families of rubrics makes it more likely that the responses one collects will ultimately be scorable, particularly if the task models are based on the kinds of instances that have been automatically scored successfully in the past. Thus, there is still a degree of efficiency that can be gained.

For writing assessment with essays, task models might constitute different kinds of prompts aligned to different kinds of constructs or student model variables while drawing on different kinds of response features as evidence. For example, a task model for a 'persuasive writing' task might prescribe the provision of multiple information sources representing conflicting points of view on a given topic along with a prompt that asks the learner to make a claim about the topic and support it with evidence. A task model for a 'summary writing' task might have as *characteristic (fixed) features* the provision of a stimulus like a literary passage along with a writing prompt that directs the learner to summarize the story. *Variable features* for either type of writing task might include the specific writing topic and the level of complexity of the associated stimulus passages or sources.

Game-Based Assessment

In game-based assessments, task models might be organized around core loop activities, which are recurring interactions the player engages in to advance through the game. The interactions are key for observing the targeted evidence and are typically associated with specific functionality. For example, in *Alice in Arealand*, the core loop activity involves using manipulable unit squares to completely cover shapes without overlapping them, testing their solution, and revising as necessary. The characteristic features of the task are the unit squares, the space to be filled, and the ability to test their solution. Variable features of this model included the kinds of shapes to be filled, the number of squares available, the size of the squares available, the tools available (e.g., *Gluumi* and *Multi*), and the theme of the level (e.g., escaping a kraken or building hot air balloons with butterflies). Multiple levels could be built by varying the complexity of the shapes to be filled, the sizes of shapes to fill them, and the tools present, which was associated with the difficulty of the level.

Given the cost associated with enabling AS of complex performances such as writing and game play, it is critical that these task models and their associated evidence rules can be reused to generate additional tasks that can also be scored automatically. In writing assessment, task models represent families of prompt and rubric types, where designers can manipulate variable features to create additional prompts that will likely be successfully scorable automatically. In game play, this may mean the creation of multiple levels using the same task model and many of the same scoring rules coded into the AS engine.

Principle 4: Combine Human and Machine Methods of Evidence Identification

Many assessment projects use either extensive human identification of possible evidence or machine-learning algorithms, seeking unexpected and complex combinations of evidence. In practice, it is generally advantageous to leverage what both have to offer.

Writing Assessment

In AS of writing, human scores generally serve as the "gold standard" used to train the AS models. Consequently, the set of features used in AS is often driven by humans. Foltz, Streeter, Lochbaum, and Landauer (2013) show examples of how various features are used in the assessment of different constructs. For example, 'lexical sophistication' is composed of 'word maturity', 'word variety', and 'confusable words' (among others) while 'style / organization / development' is made up of features like overall 'essay coherence' and sentence-by-sentence 'coherence'. In other words, analysts do not train machine-learning algorithms with all possible language features but, instead, choose features that are relevant to the constructs being assessed. Thus, while the machine does the generalization of the features, humans are still involved in identifying the features for developing the training approach, as well as validating the performance.

Game-Based Assessment

Similarly, the development of both *Mars Generation One* and *Alice in Arealand* relied heavily on human hypothesized evidence features. As described above, the tasks in *Alice in Arealand* were created with specific evidence in mind. During design and analysis iterations, these evidence pieces were examined to identify their distributions and relations to each other, resulting in some evidence being eliminated from consideration. However, little work was done to identify whether there were other nonhypothesized features that might provide evidence for inferences. Similarly, Gobert et al. (2013a, b), for example, in building a science inquiry simulation system, used humans to narrow down the potential set of features (i.e., pieces of evidence) that might indicate learner proficiency. This reduced set was then submitted to the machine-learning algorithms. In the end, this hybrid approach produced a better model fit than either solely human- or machine-created models (see Mislevy et al., Chapter 22, this handbook).

Principle 5: Design for Actionable Feedback

One of the primary benefits of AS is that it allows for real-time performance feedback, which makes the application of AS for formative assessment purposes extremely attractive. Feedback communicates information to support decision-making:

- For teachers, it can support decisions around how to group learners, how to help struggling learners, or what topics to reteach to the whole class
- For learners, it can support decisions around how to study, what learning objectives to revisit, or what aspects of their response they most need to improve

In order to support these kinds of decisions, however, feedback needs to be task-specific, response-contingent, and elaborative.

Writing Assessment

For AS of essays, it is common to provide an overall holistic score as well as separate analytic or trait scores. These represent student model variables like overall writing proficiency plus more granular traits of writing such as organization and fluency. In addition, several writing programs that utilize AS provide even more specific suggestions for improvement such as identifying specific misspelled words, noting grammatical errors, and highlighting redundant or irrelevant text (Foltz, Streeter, Lochbaum, & Landauer, 2013). In *WriteToLearn*™, for example, students can earn a subscore for 'task and focus' (a student model variable assessing the completeness of a response to all parts of the prompt and its alignment to that prompt) as part of their overall argumentative writing prompt score. A student who receives a '2' out of '4' for this attribute because her response did not include all required elements might be encouraged to improve the completeness of her response by making sure to include both a clear claim and counter-claim in her written argument. Such feedback is provided in real time, enabling students to revise their essays on their fly.

Game-Based Assessment

Providing interpretable feedback is also important within a game-based assessment context. In *Alice in Arealand*, players needed to completely fill in the spaces when constructing bridges out of ice squares. When they failed to do so, any cracks in the bridges would blink red to show students where the gaps were. Similarly, students not only needed to build a barrier to block *Alice* from the kraken but also needed to use an efficient strategy to do so. Students completing this level of the game received additional feedback as shown in Figure 3.2, if they used a less-efficient strategy of building their barrier.

The challenge in both contexts is to provide feedback that is detailed enough to scaffold the learner in revising and improving the response without focusing undue attention on superficial attributes of performance. One approach to doing this is to create feedback templates based on specific evidence models, which show how to connect feedback statements to specific evidence model variables. Assessment designers can create standardized feedback statements to correspond to the most common types of construct-relevant errors. In the writing example, a common error is neglecting to answer all parts of the question. In the game-based assessment examples, common errors include leaving spaces in between squares and not taking advantage of composite structures when building shapes.

FIGURE 3.2
Feedback in *Alice in Arealand*.

Conclusion

In this chapter, we highlighted several principles that should be considered when dealing with the complexities around AS within a principled assessment design approach such as ECD. Key challenges around AS can often be remedied through adjustment of the student model, evidence model, or task model. For example, if the evidence for one student model variable is too similar to the evidence for another student model variable, it might be useful to combine the two student model variables. Alternatively, the task can be redesigned to elicit additional evidence to make the distinction between the student model variables more salient. In addition to ECD, we briefly discuss a few additional issues for making principled assessment design with an eye toward AS being successful.

Iterative Design Processes

New forms of digital assessment have allowed us to gather evidence about previous hard-to-measure constructs. Constructs such as 'persistence' were historically assessed through self-report or live observation but can now be assessed through digital interaction (DiCerbo, 2015). The same is true of constructs such as 'confusion' and 'frustration' (Baker, D'Mello, Rodrigo, & Graesser, 2010), 'systems thinking' (Mislevy et al., 2014), and 'scientific inquiry' (Gobert, Kim, Sao Pedro, Kennedy, & Betts, 2015). For many of these constructs, few well-formed hypotheses exist about the elements of a work product that might serve as evidence of the construct.

In addition, in principled design of more traditional assessments, the elements of a *work product* (e.g., a list of options chosen in a multiple-choice test) that serve as evidence for the target construct are relatively well defined. However, in new forms of digital assessment it is typically not as clear which elements of the work product will constitute meaningful evidence. In general ECD terms, one could think that there must always exist explicit *a priori* hypotheses about all relationship between the tasks, evidence, and student models. However, this does not have to be the case; the specification of the scoring model can be and, in many cases, must be an iterative process.

In the context of automated essay scoring, some iteration can also be helpful. Prior to obtaining a large number of learner samples to score, a feasibility analysis is recommended in which humans score a small number of learner essays. The results may require further iteration of the tasks and rubrics in order to get reliable human scores. As machine-learning algorithms are trained, multiple iterations will again be needed to reach acceptable fit. Similarly, game design for assessment begins with tentative hypotheses about what player actions will be important for making inferences. However, before diving directly into confirming these hypotheses, design teams can do more to understand the data obtained from the activity and uncover unexpected patterns in the data that may generate new hypotheses. As design progresses, stronger hypotheses are developed and confirmed regarding which elements of the learner's response constitute evidence. Ideally, what is learned can be fed back to improve the design of the activity itself.

In designing complex digital assessment activities, a challenge arises in incorporating the ECD logic with the quick, *agile design sprints* sometimes needed to fit into software development cycles and production timelines. Specifically, *agile software design* is an approach in which design emerges from a series of iterations where quick prototypes are created, feedback is obtained, and the prototype is modified. It stands in contrast to models where the entire design is specified before anything is created. The desire to combine *agile game design* with assessment logic led to the development of an *evidence-centered game design* (ECgD) process (Mislevy et al., 2014); DiCerbo (2017) presents an example of such an iterative process in practice in the development of *Alice in Arealand*.

Scalability/Replicability

There are ongoing issues with the scalability of many AS solutions. This likely seems counterintuitive since one of the promises of AS is to support the scaling of scoring done by individual instructors. In practice, however, the development of construct-centered AS models is difficult to scale. In open-ended digital environments such as game-based assessments, task design to elicit appropriate evidence, and then evidence identification takes an amount of time that is incompatible with the time needed for large-scale production. For example, the games used as examples in this chapter each took more than eight months to build and, in the end, provide only a few hours of game play and cover just a small slice of curriculum. The completion of many iterations, as described above, was necessary to design tasks that elicit evidence of the constructs being assessed. However, in a situation where an entire curriculum is being created, the timelines for creating even a small number of games to address hard-to-assess areas do not fit with reasonable schedules for curriculum development and release.

In addition to time, there are no widely available *authoring tools* that allow for the specification of evidence models from complex activities. Whenever game- or simulation-based assessments are created, there is a need for tools that allow for specification of actions that count as evidence and scoring rules that define how those actions should be translated. Without these tools, extensive time and effort are needed to code in the extraction of evidence pieces and the application of scoring rules in each instance and then in each iteration of development. This prevents extensions like educator and learner authoring of activities.

There are similar scalability issues with automated essay scoring. If the goal is to score more than writing conventions then each writing prompt requires algorithms to be trained on a relatively large set of diverse scored examples. In many cases, learners and educators are most comfortable using AS for formative feedback. However, formative applications would likely involve more prompts and more variable prompts. At this point, classroom teachers cannot create their own prompts on short notice and have their learners receive substantive diagnostic feedback through AS. More work is needed to reduce the number of exemplars needed to train algorithms.

Expertise

A third ongoing challenge is the level of expertise required to implement a principled approach to assessment design. Due to the time commitment and level of effort involved in taking a principled approach, this is not a process that most subject-matter experts employed to design tasks are trained in or required to use for large-scale production. Nor can the design specifications developed for the student, evidence, or task models be "thrown over the wall" and used in a straightforward way by item writers or game design-ers who are not familiar with ECD. Rather, principled approaches to design require close collaboration during all stages of design between ECD practi-tioners, subject-matter experts, psychometricians, and AS experts. Although it may not be practical to provide extensive professional development and training in the entirety of ECD, it may be possible to make the ideas and vocabulary associated with the key ECD models accessible enough so that it can be understood even by novices.

4

Human Scoring with Automated Scoring in Mind

Edward W. Wolfe

CONTENTS

In this chapter, we identify a broad range of important issues in *human scoring* that those involved in the development, implementation, and use of *automated scoring* (AS) should consider when collecting and evaluating the suitability of human scores for calibrating AS engines. We refer to this group as the 'AS team' in the text that follows. The team constitutes a group of individuals who have relevant expertise regarding not only the generation and use of automated scores but also the collection and evaluation of human scores and test score use in general as well. For operational perspectives on collecting human scoring data with appropriate quality-assurance mechanisms in place see Chapter 10 by Ricker-Pedley et al. (this handbook).

Because automated scores are frequently generated by predicting human scores, members of the AS team should understand the many potential ways that *error* may enter into the production of human scores and what efforts can be taken to minimize the magnitude of those errors. Hence, we have structured this chapter by introducing the terminology associated with human scoring of constructed responses, outlining the process through which human scores are collected, identifying a range of conditions and processes for improving the measurement qualities of human scores, and providing an overview of some of the research that has focused on human scoring.

Scoring Terminology and Process

In addition to using the term *AS team*, throughout this chapter, we refer to *human raters* (AKA scorers, markers, graders, evaluators, readers) who assign *scores* (AKA ratings, marks) to *responses* (AKA scripts, essays, entry) that are created by *examinees* (AKA learners, respondents) in response to *prompts* (AKA items, tasks) based on a set of *scoring criteria* (AKA rubric, marking scheme). During this scoring process, human raters typically identify features of the response that are relevant to determining an appropriate score and map those features onto the *score categories* (AKA levels) that are delineated by the scoring criteria. The scoring criteria may be *holistic* in orientation (i.e., requires a rater to jointly consider multiple qualities of the response and implicitly weight the features of the response that are relevant to each aspect in order to arrive at a single score) or *analytic* (i.e., requires a rater to assign separate scores to each of multiple aspects of the response).

Although we couch the discussion that follows within the context of a large-scale scoring project, the considerations we review are important regardless of the size of the scoring effort; Table 4.1 summarizes these considerations.

Rater Recruitment

An AS team should first consider who will serve as the raters because rater recruitment is the first step in most scoring projects. Typically, prior to recruitment, the AS team will conduct a *task analysis* of the scoring process to determine the types of qualifications and knowledge required to score accurately. Those qualifications may include content and/or pedagogical knowledge as well as relevant educational and/or professional experiences. Once the requirements of the scoring tasks are well understood, the AS team identifies and recruits raters who satisfy those requirements. If the team fails to identify the types of knowledge and experiences required to complete the scoring tasks, raters who are not able to differentiate important nuances in the responses may be recruited, potentially introducing errors into the scores due to gaps in rater knowledge.

TABLE 4.1

Considerations in the Scoring Process

Consideration	Description	Importance
Rater Recruitment	Determine what knowledge and skills raters need	Choosing qualified raters reduces risk of errors due to lack of rater knowledge/skill
Rater Training	Develop materials and processes to prepare raters for the scoring process	Training raters helps raters reduce risk of idiosyncratic interpretations of the scoring criteria
Monitoring and Documentation of Score Quality	Certify that training was effective	Preventing unqualified raters from scoring reduces scoring errors
	Verify that raters maintain accuracy during scoring	Monitoring raters allows scoring leaders to intervene should raters begin to develop idiosyncratic practices and standards
	Document quality of assigned scores	Ensures stakeholders that they can have confidence in the score interpretations and subsequent decisions they make
Rater Feedback	Provide information to raters about their scoring performance	Providing corrective feedback helps raters score more accuracy and avoids increases in scoring costs caused by needing to correct erroneous scores

Rater Training

An AS team should next consider the manner in which raters will be trained to assign the scores. To this end, the AS team needs to develop training materials and specify a training process, typically in conjunction with a pre-existing team of *scoring leaders* and *expert raters*. The content of the training materials and the process will vary depending on the form of the responses (e.g., extended versus short answer responses, written versus spoken responses, unstructured versus pre-structured responses), but the team should supply some form of training using a standardized set of materials to raters to orient and prepare them for the scoring process. Training materials typically contain descriptions and documentations of: the purpose of the scoring effort, the types of decisions that will be drawn based on the scores, the characteristics of the examinee population; the prompts to which examinees responded; the scoring criteria in the form of written descriptions of the score levels; and an overview of the logistics of the scoring process and score documentation.

In most situations, the scoring criteria are supplemented with example responses that raters review during the training process, illustrating important nuances of how response features are mapped onto score levels and providing raters with practice and feedback regarding how well they have learned to apply the scoring criteria. Expert raters typically assign *consensus scores* to these training and practice examples, and annotations may be written that explain how the experts arrived at the consensus score. The enterprise of training raters is important because raters may introduce errors into the assigned scores if they apply the scoring criteria in idiosyncratic ways due to inadequate training.

Rater Monitoring

The AS team must also consider how they will monitor and document the quality of the assigned scores and how scoring leaders who oversee the human scoring project will provide feedback to raters. The AS team needs to determine whether the training efforts have been successful, and teams typically employ some form of *testing* (AKA qualification, certification, standardization) at the end of the training process. During this testing, raters blindly score responses that have been assigned expert consensus scores, allowing scoring leaders to determine the accuracy of each rater. Scoring leaders compare a rater's scores to the consensus scores, and raters must typically achieve a predetermined agreement threshold in order to qualify for the scoring project. Scoring leaders may retrain and retest or simply release raters who fail to achieve this criterion from the scoring project.

Once raters have begun scoring, scoring leaders typically employ some form of monitoring and retraining to prevent raters from introducing errors into the scores that they assign due to fatigue or developing idiosyncrasies in their scoring practices (see Wolfe, 2014, for a fuller description of this process). For example, scoring leaders may employ periodic recalibration training and/or recertification testing in order to confirm that raters are maintaining the quality of the scores that they exhibited at the end of training.

Similarly, scoring leaders may implement processes in which they review the scores assigned by raters. *Back-reading*, when scoring leaders review the scores assigned by less-experienced raters, is one such process, which potentially allows scoring leaders to overwrite errant original scores and communicate these errors to the rater. Administration of *validity responses* is another review process, often used to create measures of *validity agreement*, which the AS team may later report in technical documentation as evidence of the quality of the scores. In this process, responses that have been assigned consensus scores are seeded into a rater's queue, raters score those responses blindly, rater scores are compared to consensus scores, and observed errors are communicated back to the raters.

Scoring leaders may also design the scoring task so that multiple raters score a particular response, a practice that can accomplish two things. First, multiple scores allow scoring leaders to compare the scores of two raters so that measures of *interrater agreement* can be computed and reported in technical documentation of the quality of the scores. Second, the resulting multiple scores can be combined (e.g., averaged) to create composite scores that are more reliable and precise than single scores are by themselves.

If the AS team fails to implement rater monitoring and feedback processes, raters are more likely to diverge from the intended scoring criteria over time, thus increasing the risk of errors through inaccuracies or score patterns that are too lenient or severe relative to the targeted standards. Similarly, if the AS team fails to design the scoring process to include opportunities to document rater agreement and accuracy, the team may not be aware of whether errors are creeping into the scores, preventing the team from doing anything about the errors. In addition, stakeholders cannot be assured that the intended interpretations of scores and subsequent decisions are warranted because there is no evidence available regarding the quality of those scores.

Conceptual Framework of Human Scoring Processes

The inputs and outputs of human scoring processes are helpful reference points for thinking about these processes or systems. In this section, we present a conceptual framework that posits three types of inputs, which may potentially impact three important outputs of the consideration described in the previous section. This framework is helpful because it identifies and categorizes the types of variables that AS teams, who seek to understand and improve the enterprise of human scoring, should take into account.

Outputs

To begin with, AS teams should be primarily concerned with three outputs – *rating quality*, *rating speed*, and *rater attitude*. Specifically, they should attempt to balance the competing needs for maintaining a high level of rating quality with the need to complete the scoring project by a given deadline so that scores can be reported to stakeholders while maintaining positive rater attitudes so that attrition is minimized. To complicate the matter, the team attempting to balance these competing needs will likely also want to minimize scoring costs, particularly by minimizing the amount of time that raters spend on tasks other than producing usable scores (e.g., training and monitoring).

Output 1 – Rating quality

Rating quality refers to the accuracy, reliability, and precision of the scores assigned by raters. Rating quality is an important outcome because inaccurate, unreliable, and imprecise scores result in erroneous decision-making on the part of stakeholders. In the context of AS, human scores with poor quality will result in poor engine calibrations and, thus, automated scores with poor quality. The AS team's primary goal should be to maintain the highest level of rating quality possible given the constraints of the scoring system. In discussions of rating quality, it is important to take into account the four concepts shown in Table 4.2.

Frame of Reference

First, AS teams need to consider the *frame of reference* within which a rater's performance is judged. That is, they need to identify the target against which a rater's scores are compared, which is typically another rater, group of raters, or a consensus score. When we compare the scores of a rater to the scores of another rater, we refer to the agreement as interrater agreement, acknowledging the fact that two sets of scores, both containing potential errors, are being compared. When we compare the scores of a rater to a consensus score that is assumed to be a gold standard, we refer to the agreement as validity agreement because the gold standard score is assumed to be valid.

TABLE 4.2

Rating Quality Concepts

Concept	Question Answered	Details
Frame of Reference	Against what target are raters' scores compared?	**Interrater:** compared to other raters **Validity:** compared to a gold standard
Depictions of Rating Quality	How well do raters' scores represent truth?	**Agreement:** how well another set of scores is reproduced **Reliability:** relative agreement with another set of scores **Precision:** reproducibility of measures over measurement contexts
Rater Effects	What error patterns exist in raters' scores?	**Severity/leniency:** decreases/increases in the average assigned score **Centrality/extremity:** decreases/increases in the dispersion of assigned scores **Accuracy/inaccuracy:** increases/decreases in the agreement between assigned scores and target scores **Halo:** spuriously high correlations between multiple scores assigned to a single examinee by a single rater
Rater Effect Duration	Within what time frame do rater effects persist?	**Static:** magnitudes of rater effects are assumed to be constant across time **Dynamic:** magnitudes of rater effects are assumed to vary from one time point to another

Depicting Rating Quality

Second, AS teams need to differentiate *agreement, reliability,* and *precision.* Agreement refers to the correspondence between two sets of scores in absolute terms and, as explained previously, the target for agreement can be another rater (interrater agreement) or a consensus score (validity agreement). Measures of agreement answer the question "to what degree do these two sets of scores match," and they are typically measured in *percentages of exact agreement, adjacent agreement* (i.e., up to one point apart), or *outside of adjacent* (i.e., more than one point apart). In some cases, statistics capturing the level of agreement take into account the potential for chance agreement due to the marginal rates at which score level categories are observed; statistics from the *coefficient kappa family* are most commonly used in this case. Yan and Bridgeman (Chapter 16, this handbook) discuss the details of AS validation.

Reliability most frequently refers to the correspondence between two sets of scores in relative terms. Measures of reliability answer the question "to what degree do these two sets of scores covary," and they are typically indexed with a correlation, most commonly a *Pearson product-moment correlation.* That correlation can be between the scores of two raters (*interrater reliability*) or between a rater and consensus scores (*validity reliability*). Other reliability coefficients may be used which depict absolute agreement (e.g., a *dependability coefficient* or an *intraclass correlation*).

Precision refers to the degree to which a score is likely to vary across instances of measurement, and it is typically indexed with a *standard error* and referred to as a *standard error of measurement.* That standard error is analogous to a *standard deviation,* and it provides a measure of the amount of variability one would expect to see in a hypothetical population of scores assigned over a large number of observation opportunities for an individual examinee (e.g., over several occasions, in response to several prompts, and with responses being scored by several raters). In a *classical test theory* framework, the standard error refers to the standard deviation of potential observed scores around the true score, and in a *latent trait measurement framework* (e.g., *item response theory*), the standard error refers to the standard deviation of potential ability estimates around the true ability.

Rater Effects

AS teams should also consider various *rater effects,* a term that refers to the fact that rater errors may result in recognizable patterns in the scores assigned by individual raters (Wolfe & McVay, 2012). Scoring leaders assume that these errors are caused by states that originate in the raters. Scoring leaders also portray the various rater effects as being positioned on a continuum, even though the labels that are applied to raters who exhibit them imply a dichotomization for the sake of simplifying communication. For each rater effect and the statistics that are used to quantify it, scoring leaders typically establish thresholds that help to identify situations when corrective actions need to be taken to remedy the rating behavior.

The first common rater effect, *severity / leniency*, refers to the degree to which raters assign scores that are lower / higher than would-be scores that accurately depict the responses. This is the most-studied – and arguably most important – rater effect because it causes the scores assigned to examinees by a particular rater to depict that examinee overly negatively or overly positively. Although there are numerous ways to depict a rater's position on the severity / leniency continuum, one of the simplest is to examine the average score assigned by a rater and compare that to the average target score, which is typically either a consensus score assigned by experts or an average of scores assigned by other raters. If the rater's average is less than the target average, then the rater is labeled 'severe.' If the rater's average is greater than the target average, then the rater is labeled 'lenient.' The greater the difference, the more severe or lenient the rater.

A second common rater effect, *centrality / extremity*, refers to the degree to which raters assign scores that are too tightly / too widely clustered around the middle score levels on the score scale. The centrality / extremity rater effect is important because it causes the scores assigned to examinees by a particular rater to depict low- and high-scoring examinee too infrequently/frequently. A simple way to depict rater centrality / extremity is to compute the standard deviation of the scores assigned by a rater and compare that to the standard deviation of the target scores. If the rater's standard deviation is less than the target standard deviation, then the rater is labeled 'central.' If the rater's standard deviation is greater than the target standard deviation, then the rater is labeled 'extreme.' The greater the difference, the more central or extreme the rater.

A third common rater effect, *accuracy / inaccuracy*, refers to the degree to which raters assign scores that are relatively consistent/inconsistent with the target scores. The accuracy / inaccuracy rater effect is important because inaccuracy introduces random error variation. Obviously, perfect rater accuracy is an ideal state. A simple way to depict rater accuracy / inaccuracy is to compute the correlation between the scores assigned by a rater and target scores. If the rater's correlation is near '0', then the rater is labeled 'inaccurate.' If the rater's correlation is near '1', then the rater is labeled 'accurate.'

A fourth common rater effect, the *halo effect* (Lai, Wolfe, & Vickers, 2015), may occur when a single rater is asked to score multiple responses by a single examinee at one time – a human scoring practice that is not recommended for this reason. Specifically, the term 'halo' refers to the degree that raters assign scores on multiple rating dimensions that covary with one another more than do target scores. The psychological phenomenon that produces the halo effect occurs because sometimes the judgments of raters regarding one aspect of the response influence that rater's judgments of other aspects of the response.

For example, suppose a rater does not fully understand how to differentiate voice and style in an essay. If the rater is better able to recognize variation in voice, that rater may conflate interpretations of the quality of the voice in a particular essay with interpretations of the quality of the stylistic aspects of the essay. Alternatively, a rater may inappropriately judge that all aspects of the response are influenced by one or more irrelevant features of the response. For example, a rater may be distracted by the legibility of the handwriting in a particular essay and that rater may allow judgments of that illegibility to cloud judgments of all scored aspects of the essay even though the scoring criteria does not solicit judgments about the quality of the handwriting.

The halo effect is important because it implies that the assigned scores fail to differentiate true differences between the scored aspects of examinee responses. Obviously, a lack of an artificial halo is an ideal state meaning that any observed covariance between response aspect scores (e.g., voice and style) is due to true covariance between them. On the other hand, when a halo effect exists, the observed between-aspect covariance is artificially inflated. A simple way to depict the halo effect is to compute the correlations between the scores assigned by a rater to each pair of aspects in the scoring criteria and compare those correlations to the between-aspect correlations in target scores. Ideally, those target scores were assigned by different expert raters, removing the possibility that the target scores themselves exhibit a halo effect. If the rater's correlations are greater than the target score correlations, then the rater is labeled as exhibiting the halo effect. The closer the between-aspect correlations are to '1,' the stronger the halo effect.

Rater Effect Duration

The time frame within which rater effects are assumed to be usefully interpreted is the fourth and final concept from Table 4.2 relating to rating quality. While most depictions of rater effects portray them as being *static* in nature (i.e., as a characteristic of a rater that is constant across the measurement occasion), in reality it is more likely that they are *dynamic* (i.e., they are changing over time), particularly when the scoring project takes place over a long duration (e.g., weeks or even months or years) (see Myford & Wolfe, 2009, for a more detailed discussion of this issue).

For example, if the initial training process is insufficient and raters learn to better apply the scoring criteria after receiving additional practice and feedback, any observed rater effect that takes place early in the scoring project will likely be reduced in magnitude as the project continues. On the other hand, if raters become fatigued or lose interest in the scoring assignment, any observed rater effect will likely increase in magnitude over time. In fact, it is also possible that raters who were initially well calibrated develop rater effects over time, particularly if insufficient retraining is provided during the scoring project.

In short, all of the rater effects described previously are often treated as if they are immutable characteristics of the rater, while it is more likely that such characteristics vary over time. Thus, while AS teams should be aware of and attempt to document the existence and magnitude of rater effects at a point in time, they should also attempt to determine the trends of those effects within raters over time.

Output 2 – Rating Speed

While rating quality is an extremely important output in the human scoring process, rating speed is also a consideration in most scoring projects because stakeholders impose delivery deadlines so that scores can be reported within a timeframe that makes decision-making possible. As a result, scoring leaders often seek to maximize rating quality while working within the time constraints required by stakeholders and simultaneously holding down scoring costs. Scoring leaders want raters to make their scoring decisions as quickly as possible, particularly when raters are being paid, because the faster raters complete the scoring task, the less the scoring project costs and the quicker scores can be reported. Beyond optimizing the training and scoring processes and work flow so that raters are able to make the quickest accurate decisions that they can, there are a couple of general principles that govern rater speed.

Complexity of Task and Work Flow

First, in general, the more complex the decision-making task and the more convoluted the work flow, the slower the rating speed. For example, consider the difference between having raters assign a single holistic score to each response versus having raters assign multiple analytic scores to each response. When raters assign a single holistic score, at least one rater must review each response and then document a single score. On the other hand, when raters assign multiple analytic scores to each response, at least one rater must review each response and then document each score.

In some cases, the rater may need to review the response more than once because the decision-making process is too complex to make all of the scoring decisions at once. As a result, it is a slower process because multiple analytic scores require a more complex decision-making process and documentation of multiple scores. Alternatively, in order to maximize rating quality (i.e., to minimize the probability of halo effects), scoring leaders may resort to having different raters score each trait covered by the analytic scoring criteria. In that scenario, each response is reviewed by as many different raters as there are analytic traits in the scoring criteria, and each rater must document the single score assigned. This results in a much slower rating speed than if each response is assigned a single holistic score by a single rater.

Nonproductive Time

Second, the more *nonproductive scoring time* that raters are engaged in (i.e., when they complete tasks that do not result in usable scores such as training and monitoring), the slower the rating speed. Consider the fact that the total scoring time required is the sum of scoring time and nonscoring time.

Hence, in order to produce scores as quickly as possible, the proportion of total scoring time that is allocated to nonscoring tasks must be minimized. However, much of the time allocated to nonscoring time is directed toward rater training, monitoring, and feedback, which are all activities that are designed to increase rating quality.

Hence, leaders of scoring projects seek to create the most efficient and effective processes possible for implementing nonscoring tasks. That is, they seek to train raters as thoroughly as possible in as short of a time as possible. Similarly, they seek to recalibrate raters during the scoring project with sufficient frequency to avoid deterioration of rater performance over time. Scoring leaders also seek to create response distribution and score documentation processes that allow raters to review responses and record their scores with minimal waste of time. They also seek to utilize processes for monitoring raters and providing them with feedback that allows for maintaining rating quality while minimizing the amount of time required of both scoring leaders and raters. In fact, because this balance of achieving acceptable rating quality while maximizing rating speed is so important, much of the research concerning human scoring focuses on tasks such as these.

Output 3 – Rater Attitude

Rater attitude is a third important output of the scoring process. By rater attitude, we mean the general disposition of raters toward the training and scoring tasks. While rating quality and rater speed are important, if balancing them degrades rater attitude to the point that they leave the scoring project, then rating speed is likely to suffer. Hence, scoring leaders frequently seek to balance quality and speed without having a negative impact on rater attitude. In addition, *financial incentives* (e.g., payment) and *workflow incentives* (e.g., the ability to work at home or on a convenient schedule) may influence rater attitude.

Finn (2015), in a review of the impact of examinee motivation on test scores in low-stakes contexts, identified three potentially relevant themes from the testing research literature that are likely also relevant when considering rater attitudes. First, she reports that *performance incentives* increase motivation and, thus, performance. In the case of raters, if scoring leaders offer raters suitable compensation and/or other perquisite benefits that are conditional on high rating quality, raters are more likely to be motivated to do their best.

Second, Finn reports that *motivational framing* (i.e., providing a rationale for engaging in a task in earnest) increases motivation. Hence, if scoring leaders explain the importance of providing high-quality ratings and the implications of decisions to raters, it is more likely that raters will take the scoring task seriously and attempt to provide the best ratings that they can. Finally, Finn reports that *motivational filtering* (i.e., identifying and removing cases in which there is evidence that the task was not taken seriously) may improve the quality of the resulting measures. In the case of scoring, this would entail identifying raters who move through the rating task too quickly or who demonstrate very low measures of rating quality, and reviewing and/or replacing the scores assigned by those raters.

Inputs

Leaders of human scoring projects and researchers who focus on that enterprise have sought to understand how the outputs discussed previously are influenced by three types of inputs (i.e., variable clusters) – *rater characteristics, response content,* and *rating context* – as implied by Figure 4.1 (Wolfe & Baird, 2015; Wolfe, Ng, & Baird, 2018). In this section, we identify the variety of variables within each of these categories that may influence the previously discussed outcomes and cite examples of research that focuses on those variables. Note that our intent in doing so is not to provide an exhaustive literature review regarding these variables because such an undertaking would be a monumental task. Rather, our intent is to provide an introduction to the research literature about human scoring via examples of the types of research that may be worth consulting, if applicable to the specific context of their scoring project.

Rater Characteristics

The conceptual framework indicates that the characteristics of the raters who are engaged in the scoring process may influence rating quality, rater speed, and rater attitude. Because discussions of rater characteristics often involve

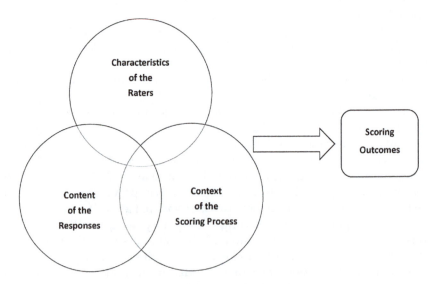

FIGURE 4.1
A conceptual framework for human constructed-response scoring.

comparing raters who exhibit different levels of scoring performance, it is helpful to understand that those levels of performance have been operationalized in several different ways in research that focuses on human scoring of constructed responses. For example, some researchers differentiate proficient raters and less- or nonproficient raters by quantitatively distinguishing between raters who produce ratings with high interrater or validity agreement statistics from raters who produce lower values on such statistics.

Other researchers have differentiated raters who exhibit various rater effects from raters who do not base on quantitative indices of those rater effects. Yet other researchers have chosen to differentiate raters who have considerable scoring experience from raters who are novices. These differences are sometimes difficult to notice at first glance because researchers use a variety of terms to refer to the comparison groups (e.g., 'expert' / 'novice', 'experienced' / 'inexperienced', 'effect' / 'noneffect', and 'proficient' / 'nonproficient'). These variations on labels and lack of consensus regarding which groups are important to compare sometimes make drawing generalizations regarding how rater characteristics influence scoring outputs challenging.

Several rater characteristics have been repeatedly considered by scoring leaders and researchers of human scoring. Clearly, education and content knowledge are important for scoring responses that require communication of factual and conceptual information, and minimum levels of these two rater characteristics are often included in rater selection criteria for scoring projects. Similarly, professional experiences such as teaching and/or scoring activities, may be important considerations in selecting raters.

For example, Meadows and Billington (2013) compared the accuracy of scores assigned by raters who had varying levels of scoring experience, subject knowledge, and teaching experience. They found that raters who had higher levels of subject knowledge and teaching experience were the most accurate in their scoring. In a similar study, Royal-Dawson and Baird (2009) compared scorers who had similar levels of education but different levels of professional experience in teaching and scoring in terms of scoring accuracy. They found little evidence to support the notion that raters with teaching experience score more accurately than do raters without that experience. However, Royal-Dawson and Baird (2009) found more variability between the least experienced teachers on items that might require higher levels of pedagogical knowledge relating to the subject matter.

Other characteristics may be less useful in rater selection and may also influence scoring outputs. For example, some researchers have studied the cognitive style and content focus that raters adopt when they score essays, particularly in terms of how some rater groups differ from others. These researchers engage raters in *think-aloud tasks* and compare the processes raters use and the content upon which they focus during the scoring activity. For example, Wolfe, Kao, and Ranney (1998) reported on such a study in which think-aloud protocols for essay raters with more and less accuracy were compared with respect to the processes utilized during decision-making and the content upon which they focused during those processes. They found that the more proficient raters were more likely to use a *linear approach* to processing an essay (i.e., "read it and assign a score") while less-proficient raters were more likely to use an *iterative approach* (i.e., "read, evaluate, read, evaluate"). In addition, they found that more proficient raters spoke of the content of the essay in more general terms than did less-proficient raters.

Other examples of rater characteristics that may impact scoring outputs include both stable *traits* (e.g., temperament) and fleeting *states* (e.g., affect or mood). Two studies (Carrell, 1995; Rezaee & Kermani, 2011) revealed a potential weak relationship between rater personality and assigned scores. Carrell (1995) found that raters who were classified as 'Sensing and Feeling' on the *Myers-Briggs Type Indicator* scale assigned higher scores to essays than did raters who were classified as 'Intuitive and Thinking.' Similarly, temporary physical states such as fatigue or illness could potentially influence scoring outputs. Specifically, a study by Klein and Pat El (2003) examined patterns in the scores assigned by three raters over a three-hour scoring session. Even during that short session, the researchers found that scores rose to a statistically significant degree over time, presumably as raters became more fatigued.

Response Characteristics

Researchers also study the characteristics of responses that are being scored for evidence that they influence the scoring process outputs. Much of the research of this relationship has employed *feature analysis* of the responses as predictors of rating quality. Feature analysis is the linguistic analysis of the response content through the automated computation and subsequent statistical analysis of variables derived through natural language processing (e.g., Cahill & Evanini, Chapter 5, this handbook). Such studies focus on how the *construct* being measured (e.g., key competencies in writing, speech, math, science), the *content* contained in the response (e.g., scorable and unscorable features), and the *style* with which it is presented (e.g., a writer's voice) influence rater accuracy or assigned score level. Similarly, it may be that the factual information contained in the response influences ratings of quality.

There are numerous examples of research in which feature analysis is used to depict the quality of human ratings. In one such study by Crossley, Roscoe, and McNamara (2014) linguistic feature profiles of essays were used to depict the characteristics of high-quality persuasive essays. They employed several computational tools to derive linguistic, affective, and rhetorical features for a corpus of essays that had been assigned high scores. These features were subjected to a cluster analysis to identify profiles of essays within the corpus of high-quality essays. Their results revealed four profiles, differentiated by higher incidences of various groupings of lexical features of the essays.

However, researchers have expressed significant interest in other aspects of the response. For example, some believe that rating quality may vary by the stated purpose of the examination and intended score use, although there is little research that addresses this relationship. Others have focused on the appearance of the response such as how difference in handwriting or the use of keyboard influences raters' decision. Due to the gradual transition from handwriting to keyboarding as the mode for administering essay examinations, much of the relevant research is older. In one such study, Powers, Fowles, Farnum, and Ramsey (1994) compared the scores assigned to handwritten and keyboarded essays produced by the same student and then transposed to the alternate medium. Those results revealed higher scores for handwritten essays, regardless of the medium in which the essays were composed.

Author clues (e.g., gender, ethnicity) constitute another response characteristic that is assumed to potentially influence rating quality. In one relevant study, Baird (1998) found that, although raters were relatively able to predict the gender of examinees in some cases based on handwriting style, there was no evidence of gender bias based on such interpretations on the part of the rater. In another study, Rubin and Williams-James (1997) fabricated student identities for a set of essays, particularly regarding the student's ethnicity and native language. They found evidence of bias in favor of supposed Asian writers over supposed native English speakers in terms of ratings of overall quality.

Finally, evidence exists suggesting that the *juxtaposition of responses* (i.e., their sequencing) may influence the quality of the assigned ratings. Attali (2011), for example, examined sequential effects in essay ratings by computing the correlation between the scores assigned to an essay and each subsequent essay. He found a general trend toward positive correlations (i.e., scores assigned to subsequent essays tended to be dependent on the scores assigned to their predecessors), suggesting an *assimilation effect*.

Rating Context

Researchers of human scoring also believe in a third class of variables, those having to do with the rating context, influence rating quality, rater speed, and rater attitude. These variables focus on how the process, materials, and activities in which raters are engaged and the manner in which these features are structured influence the scoring process outputs.

In the introduction, we explained that rater selection is conducted at the outset of an operational rating process in order to identify raters who are most likely to produce high-quality ratings in a timely manner. Generally, research regarding rater background has not produced much consensus regarding the influence of rater characteristics on rating quality and speed. Meadows and Billington (2007) conducted a review of this literature, and they concluded that, although inexperienced raters tend to mark more severely than experienced raters, there is little evidence that a rater's background is associated with the quality of the scores produced by a rater.

In a previous section, we also explained that rater-training activities and materials are structured and presented by scoring leaders in a manner that attempts to maximize effectiveness and efficiency. In two studies of rater training, Raczynski, Cohen, Engelhard, and Lu (2015) and Wolfe, Matthews, and Vickers (2010) demonstrated that self-paced rater training is as effective in terms of rating quality but more efficient than a face-to-face training model. In both experimental studies, raters who were trained using a collaborative approach and raters who were trained using a self-paced approach demonstrated comparable levels of accuracy at the completion of training. However, self-paced raters completed their training regimens in somewhere between 67% (Raczynski et al.) and 33% (Wolfe et al.) of the amount of time required for face-to-face training. Additionally, rater certification tests and procedures are believed to influence rating quality, rating speed, and rater attitude, although very little research has been done to examine these relationships.

Once raters begin operational scoring, other variables may become important considerations. For example, the structure, content, and focus of the scoring criteria are thought to impact the scoring outputs. In one relevant study, Zhang, Xiao, and Luo (2015) compared scores from a holistic and an analytic scoring rubric that were applied to the same set of essays. While they found that rater reliability was quite high and similar for the two rubrics, they also found that examinees with lower writing proficiency tended to receive higher scores under the analytic scoring criteria while higher proficiency examinees receive higher scores under the holistic rubric.

Furthermore, the manner in which responses are distributed to raters may influence the outputs. Numerous studies have examined the impact of presentation medium on raters, particularly comparisons of online versus paper-based media. For example, Johnson, Hopkin, Shiell, and Bell (2012) conducted a study to determine whether rating quality is influenced by on-screen marking. They determined that rating accuracy is equivalent for on-screen versus paper presentation, although raters were slightly more likely to be lenient when marking on screen. The environment in which raters are engaged in the scoring activities may also influence their performance, although little research has been directed toward this topic. Similarly, the way that shifts are structured, their duration, and the incentives that are provided to raters were believed to influence the scoring outputs.

Finally, the manner in which raters are monitored and provided with feedback is also likely to influence the scoring outputs. In one study, Attali (2016) provided inexperienced raters with immediate feedback in the form of an indicator of whether they had accurately scored each essay in a set of about 20 practice essays. Attali found that this immediate feedback allowed the inexperienced raters to score with levels of accuracy that rivaled that of expert raters.

Interactions

It is important to note that research that focuses on the human scoring enterprise does not necessarily focus on a single input variable. In fact, it is through examination of the *interactions* of the three classes of variables that the richest depictions of these relationships can be realized. For example, Coniam (2009) compared paper-based marking and onscreen marking (i.e., rating context) by having raters of a second-language writing test who had either had no prior rating experience or had prior paper-based rating experience only (i.e., rater characteristics). Each rater in the study scored responses on two occasions – once paper-based and once onscreen. Analyses focused on whether raters applied different *standards* (i.e., rating quality) and had different *favorability* (i.e., rater attitude) under those scoring conditions. The results indicated that ratings were of comparable quality but that raters who had only scored paper-based responses previously were less positive about onscreen scoring than were raters who had no prior experience.

In another example that examines the intersections of multiple aspects of the framework, Graham and Leone (1987) studied the influence of *disability labels* (i.e., response content) on raters' evaluations essays, some of which were trained more thoroughly than others prior to the scoring task (i.e., rating context). Results indicated that the disability labels did not significantly affect average writing scores (i.e., rating quality) while the rater's training experience did interact with essay quality; specifically, raters with more training assigned a wider range of scores than did those with less training.

Weigle (1999) presented another example of studies that focus on the interaction of multiple inputs in the framework. Specifically, she investigated how experienced and inexperienced raters (i.e., rater characteristics) scored second language essays written in response to two different prompts (i.e., response content). Her results showed that inexperienced raters were more severe (i.e., rating quality) than were experienced raters on one prompt but not on the other prompt and that differences between the two groups of raters were eliminated following rater training.

Psychometric Modeling

Psychometrics is a field that focuses on applications of statistical methods to the measurement of psychological phenomena, and AS teams should be aware of the extensive body of research that focuses on the development and application of measurement models to ratings (e.g., true score test theory, generalizability theory, item response theory, Rasch measurement, signal detection theory). Yan and Bridgeman (Chapter 16, this handbook) discuss the statistical models for AS systems. Within each of these approaches to scaling rating data, researchers have sought to evaluate the accuracy with which rater effects influence decisions that are made about raters and examinees, the sensitivity of indices that are used to detect rater effects, and the relative worth of these varying approaches to measuring raters and rater effects.

Classical test theory focuses on observed scores such as test scores created by computing a composite score of an examinee's performance across multiple observations (Allen & Yen, 2002). It decomposes the observed score into a *true score* (i.e., the expected score over a large number of potential observations) and *error component* (i.e., the deviation of the large number of potential observed scores from the true score).

In a rating context, the error introduced by raters is of primary concern, although any of a number of other potential sources of error exist (e.g., variation in examinee performance across items, across measurement occasions). The important point about this theory is that it deals with observed (composite) scores and the response-level scores that make up the composite. Hence, statistical methods applied in classical test theory focus on the distributions of and relationships among those scores. In the context of the analysis of human and/or automated scores, common statistical indices include the frequency distribution of observed scores for a particular response, summary statistics of that distribution (e.g., mean and standard deviation), the percentage of agreement and/or adjacency (e.g., validity agreement and interrater agreement) and measures of association (e.g., correlation) between observed scores from multiple raters or between observed scores and expert consensus scores (Wolfe, 2014).

Historically, classical test theory has focused on measurement contexts in which there were only two facets of measurement – examinees and test items. To address this shortcoming, *generalizability theory* (e.g., Brennan, 2001; Cronbach, Gleser, Nanda, & Rajaratnam, 1972) decomposes measurement error into multiple systematic sources (e.g., items, raters) and their interactions. This decomposition allows for depiction of the proportion of error variance that is attributable to each source as well as the prediction of how score reliability would be increased if additional observations were made for each examinee.

Item response theory models, on the other hand, focus on the prediction of response-level scores (e.g., what score will a particular examinee get in response to a particular test item when scored by a particular rater) (de Ayala, 2009). That is, these models mathematically portray the probability – more accurately the log of the (cumulative) odds – that a particular response-level score will be obtained under a given set of circumstances. Wolfe and McVay (2012) and Wolfe (2014) provide a simple summary of how rater parameters in one family of latent trait measurement models, the so-called *Rasch model* family, can help detect a variety of rater effects.

Specifically, parameter values for this model are estimated based on observed data and can be used to depict the status of individual raters regarding the degree to which that rater exhibits rater effects. For example, rater severity / leniency is modeled directly as the *location parameter* for a rater while the spread of the *rater-specific thresholds* is a measure of rater centrality / extremity.

Conclusion

In this chapter, we have presented a broad range of considerations for an AS team to review when collecting and evaluating the suitability of human scores for calibrating an AS engine. Clearly, numerous design decisions need to be made, each of which results in tradeoffs regarding a potential increase in costs for the sake of obtaining higher rating quality: selecting highly qualified raters, implementing a rigorous training regimen, implementing a thorough rater monitoring system, and conducting thorough analysis of the quality of scores.

With each of these decisions, the scoring team may make tacit assumptions about the efficacy of a particular practice, and the team should be vigilant to examine the existing research regarding human scoring to determine whether that assumption is supported. In many cases, those assumptions are not supported by research. Rather, they are based on policies and procedures that were put in place years ago out of necessity because the research had not yet been done. Unfortunately, even today, it is difficult to determine the current state of knowledge because few systematic reviews of the human scoring research exist even though such work is currently underway in a few application domains.

5

Natural Language Processing for Writing and Speaking

Aoife Cahill and Keelan Evanini

CONTENTS

Natural language processing (NLP) is an interdisciplinary field of application-focused research that crosses computer science, linguistics, statistics, and digital signal processing. In this chapter, for the sake of convenience, we use a broad definition of the term that covers automated processing of all forms of human language, including the processing of spoken language as well as written language; however, readers should be aware that the term can sometimes refer to the narrower domain of processing written language specifically. We also refer to respondents who provide written or spoken answers mostly as 'test takers' with the exception of formative assessment contexts when we refer to them as 'learners.'

Typically, *automated scoring* (AS) is applied to assessment task types that elicit written and spoken constructed responses. There should be a relevant link between the construct definitions of those assessments and the NLP techniques used in the AS application for the assessments (see Burstein et al., Chapter 18, this handbook; Deane & Zhang, Chapter 19, this handbook). In this introductory section, we will present some of the most common writing and speaking task types for which NLP-based AS has been used and describe the associated constructs that NLP technology aims to evaluate.

NLP for Constructed Tasks

Illustrative Written Constructed-Response Tasks

The idea of AS of written test taker responses was first introduced in the 1960s by Ellis Page (Page, 1968). He proposed the *Project Essay Grade* (PEG) system for the AS of essays written by test takers in high school. At the time, computational resources were not readily available for this solution to be widely adopted. However, with the advent of ubiquitous computing in the 1990s, the field of *automated essay scoring* expanded greatly (see, e.g., Attali & Burstein, 2006; Elliot, 2003; Landauer et al., 2003).

AS of written test taker responses includes the automated assessment of both writing quality as well as content. It has been applied to a wide range of writing tasks across a spectrum of test taker populations. For example, in the K–12 domain, automated assessment of written responses is a common component of *standardized state-level assessments* (e.g., those provided by the PARCC or Smarter Balanced consortia). The kinds of writing tasks that are being evaluated with automated systems include those that assess whether test takers can persuade, explain, or convey experience. AS is applied to both extended written responses (e.g., essays) as well as shorter constructed responses (e.g., responses to reading comprehension items) in assessments that measure writing proficiency or content knowledge.

In addition to *holistic scoring* of extended written responses, AS has also recently been used for *analytic scoring / trait scoring* of these kinds of written responses. Trait scoring involves assigning multiple scores for different dimensions of writing quality to the same response. Traits are specific to a writing genre and grade level. For example, the traits included in the *Smarter Balanced* rubrics for grade 3 narrative items* are 'conventions,' 'elaboration/development,' and 'organization/purpose.'

---

* https://www.ode.state.or.us/wma/teachlearn/subjects/science/assessment/smarter-balanced_scoring_rubrics.pdf

AS is also used as part of the assessment of writing proficiency at the higher education level. For example, the TOEFL® and GRE® tests from Educational Testing Service (ETS) both use AS for the writing sections of those tests. The score from the *e-rater*® AS engine (Attali & Burstein, 2006) is combined with a human score to produce the final score for each essay. If the human and automated scores differ by a pre-set threshold, the automated score may be discarded, or a second human score may be obtained for *adjudication*.

In addition, if the human rater indicates that the essay is nonscorable (e.g., because it is off-topic or blank), the essay is not sent to the AS system. The kinds of writing that are assessed by the GRE are about analyzing an issue ("issue" task) and analyzing an argument ("argument" task).[*] The kinds of writing that are assessed by the TOEFL test are about supporting an opinion ("independent" task) and writing in response to reading and listening tasks ("integrated" task).[†]

AS is also used in formative classroom settings as a way to give immediate feedback to learners. For example, ETS's *Criterion*® system (Burstein et al., 2003) can give both a holistic score as well as detailed feedback for a number of writing aspects. More recently, the *Writing Mentor* application from ETS[‡] (Burstein et al., 2018) provides writing revision assistance via a *Google Docs* add-on. These kinds of systems are used in many domains including K–12, higher education, and adult literacy contexts. They are often tailored to the kinds of writing tasks that typically appear on standardized writing tests, including persuasive, explanatory, and narrative items.

Illustrative Spoken Constructed-Response Tasks

Although state-of-the-art capabilities for AS of spoken responses are less mature than those for scoring written responses, they still have been applied in a wide range of assessments for a variety of speaking task types.

The most straightforward of these is the *read-aloud task* in which the test taker is presented with a text, which can range in complexity from isolated word to connected paragraphs and is then asked to read the text out loud in a fixed time interval. This task has been used widely for assessing oral reading fluency among elementary school test takers who are learning basic literacy skills. When a teacher evaluates a test taker's response to one of these read-aloud tasks, the teacher maintains a tally of which words were read correctly and incorrectly and produces a summary score for the entire passage that can be used to monitor a test taker's progress in acquiring literacy skills over time.

---

[*] https://www.ets.org/gre
[†] https://www.ets.org/toefl
[‡] https://mentormywriting.org

Automatic speech recognition (ASR) can be used to automate this process, and the standard metric that counts the number of words per minute read correctly by the test taker has been shown to correlate very highly with corresponding scores from teachers (Downey et al., 2011). In addition, some researchers have extracted automated metrics related to fluency, intonation, and phrasing as a means to assess a test taker's reading expressiveness and reading comprehension (Schwanenflugel et al., 2004; Duong et al., 2011). Finally, read-aloud tasks have also been used widely in the assessment of nonnative speaker proficiency; for this population, in addition to assessing fluency, intonation, and other features that are relevant for assessing literacy skills, automated speech processing features related to pronunciation proficiency are important (Evanini et al., 2015).

Moving beyond the read-aloud task, automated speech scoring has also been applied to other constrained types of speaking tasks in which the content of the test taker's response is relatively predictable based on the nature of the task. These types of speaking tasks are generally appropriate for assessing the speaking proficiency of nonnative speakers of a language with low proficiency levels. One example of this type of speaking task is the *repeat-aloud task*. In this task, the test taker hears a recording of someone speaking a short utterance and is asked to repeat the utterance aloud immediately after hearing it. The utterance typically consists of a short sentence of approximately 5–12 words, so that the content can be retained in the test taker's working memory. As with the read-aloud task, the repeat-aloud task primarily targets the construct dimensions of pronunciation and fluency. However, grammatical competence and vocabulary are also targeted to a certain extent since these aspects of English proficiency are required to successfully comprehend the recorded utterance and reproduce it correctly.

Another example of a constrained speaking task is the *sentence build task* in which the test taker is presented with a jumbled sequence of words and phrases and is asked to produce a spoken sentence with the words and phrases rearranged to form a grammatical and semantically coherent sentence. In addition to pronunciation and fluency, the sentence-build task also assesses grammatical competence through the test taker's ability to produce the correct word order. The repeat-aloud and sentence-build tasks have both been used effectively with automated speech scoring technology (see, for example, Bernstein et al., 2010).

Similarly, automated speech scoring has been used with constrained speaking tasks that evaluate a test taker's knowledge about how to use particular words and phrases in specific communicative situations. For example, the *Test of English-for-Teaching*™ from ETS included speaking tasks in which the test takers were presented with a few key words and were asked to produce a spoken sentence to achieve a classroom management task such as instructing their students to open their books to a specified page. The human scoring rubrics for these tasks included 'grammatical correctness' and 'semantic appropriateness' in addition to 'intelligibility of speech' and a wide range of automated speech scoring features have been developed to evaluate these aspects of the speaking construct (Zechner et al., 2015; Zechner & Loukina, Chapter 20, this handbook).

Automated speech scoring has also been used in conjunction with relatively unconstrained speaking tasks that elicit extended, spontaneous speech for assessing the speaking proficiency of English learners in an academic context. For example, automated speech scoring has been applied in the TOEFL Practice Online practice test for the TOEFL iBT assessment for scoring the (retired) independent and integrated speaking tasks in the TOEFL iBT Speaking section, which are similar in nature to the analogous TOEFL iBT Writing tasks mentioned earlier. These tasks elicit spoken responses ranging from 45 to 60 seconds in duration and are scored by human raters using rubrics that cover the following three areas of the construct of nonnative English speaking proficiency: 'delivery' (pronunciation, fluency, intonation, and pacing), 'language use' (grammar and vocabulary), and 'topic development' (content appropriateness and discourse coherence). Chen et al. (2018) provide a detailed description of the types of automated speech scoring features that have been applied to address these diverse aspects of the targeted construct.

Fundamentals of Linguistics and Natural Language Processing

In this section, we provide an overview of fundamental concepts in linguistics and NLP that are important for understanding how NLP technology can be used for extracting features to measure diverse aspects of the targeted construct in spoken and written assessment. In addition, we provide an overview of how the field of NLP has evolved over the years and introduce the main general approaches that have been explored for analyzing spoken and written human language.

Linguistic Analysis

In this first part we provide overviews of the main subfields of linguistics that provide a theoretical underpinning for automated processing of spoken and written language. Due to space limitations, we are only able to briefly introduce the main foci of each subfield; further details are available in most introductory linguistics textbooks such as Finegan (2014) and Fasold and Connor-Linton (2014).

- **Phonetics** is concerned with the empirical analysis of speech production and perception. Its two main foci are *articulatory phonetics*, which investigates how the articulators in the human vocal tract (e.g., larynx, tongue, lips) are used to produce speech, and *acoustic phonetics*, which investigates ways of representing and measuring the sounds produced in speech (e.g., by using digital signal processing techniques).

- **Phonology** is concerned with how the sounds of a language are represented as abstract discrete units in a speaker's mind. The phonemes of a language can be determined by finding pairs of words (so-called *minimal pairs*) that differ by only a single sound such as *pit* /pɪt/ and *kit* /kɪt/; this pair indicates that /p/ and /k/ are separate phonemes in English. Phonemes are abstract categories that may be realized as speech sounds with different phonetic characteristics in different contexts in spoken language.

- **Morphology** is concerned with analyzing the internal structure of words wherein a *morpheme* is defined to be the smallest meaning-bearing unit in a word. Words can thus be analyzed in terms of the morphemes that combine to form them such as stems, prefixes, and suffixes.

- **Syntax** is concerned with analyzing the internal structure of clauses and sentences. A core aspect of syntax is understanding how certain sequences of words group together to form phrases, what the allowable order of words in those phrases is, what the function of the phrases is, and what the relationships between the phrases are.

- **Semantics** is concerned with analyzing the meaning of words, sentences, and larger texts. A core aspect of semantics is understanding how the meaning of a text is built up from the meaning of the smaller units in the text (e.g., words or sentences). Other aspects of semantics include understanding and identifying different senses of words, understanding what entities pronouns and other referential words refer to, and discourse analysis.

- **Pragmatics** is concerned with how context interacts with semantics. This includes understanding any pre-existing knowledge relevant for the verbal interaction and the intent of the speaker/writer. Pragmatics aims to understand and explain how we can still communicate efficiently and effectively despite the large amount of ambiguity in natural language.

In the next part, we provide an overview of how the main techniques from the field of NLP for conducting automated analyses of these aspects of linguistic structure have evolved over the years.

Natural Language Processing

Researchers in the constantly evolving field of NLP do not use a unified, shared set of approaches. As computing infrastructure, available data, and new insights from the fields of linguistics and statistics – including *machine* *learning* – have evolved, so too have the prevailing techniques in NLP.* There have been three main paradigms to date: (1) *rule-based techniques*, (2) *statistical techniques*, and (3) *deep learning techniques*.

Rule-Based Techniques

From the 1950s to the 1980s, before large amounts of data were digitally available and before the advent of personal computers, the most common techniques in NLP were based on developing sets of rules by hand. For example, one of the main applications at the time was *automated (machine) translation*. A common approach was to write one set of rules to convert a text written in one language into a more abstract representation and another set of rules to convert the abstract representation back into a text in a different natural language. This approach of writing rules to analyze natural language was the predominant technique in NLP until the mid-1990s. Its popularity has declined steadily since then due to issues with *scalability, maintainability, implementation effort*, and *robustness* when using rule-based systems, though some applications do still rely on rule-based solutions or use them in combination with statistical techniques in hybrid systems.

Statistical Techniques

As data became more readily available in digital format and as computing power increased, the focus in NLP switched from rule-based solutions to those based on statistical models and machine learning. This approach was attractive because it allowed systems to learn directly from data, to be able to indicate a measure of confidence in an analysis, as well as to provide multiple possible analyses for a given utterance. Given the inherent ambiguity in natural language, this offered much more flexibility than the rule-based systems. In addition, empirical analyses demonstrated that statistical systems often achieved higher levels of performance and were more robust when applied to new data sets than rule-based systems.

Statistical systems, though, rely heavily on *feature engineering*, a process during which an expert examines the data to identify and develop potentially relevant linguistic features for a particular task. A feature might be as specific as 'the number of spelling errors in the response' or more general like 'the coherence of the response on a scale from 1–5.' In order to build automated feature detection capabilities, manual, expert annotation of reasonably large datasets is often required in order to provide the data from which the system learns. This large-scale annotation can be time-consuming and expensive.

---

* See Hall et al. (2008) for an interesting overview of the changes in trends in NLP topics (as they appear in conference proceedings) from 1965 to 2008.

Deep Learning Techniques

More recently, since the early- to mid-2010s, deep learning or deep neural-network-based techniques have become increasingly popular in NLP. These techniques are similar to the statistical techniques described above in that they learn directly from the data. The fundamental mathematics behind these techniques was largely developed much earlier in the field of *artificial intelligence,* but computing power has only recently developed to the point where the implementation of these techniques is tractable since the techniques typically require large amounts of data and computing power to train models.

The main advantage of the deep learning approach is that the models learn patterns in the data without any explicit input from the developer (i.e., feature engineering is not required). This then shifts the focus of the expert knowledge required to build useful models from linguistics to neural-network construction and parameter tuning. As this is still an emerging field, it is not yet clear for which applications and domains these techniques are appropriate under what kinds of circumstances, especially for assessment purposes. This is especially true where it is necessary to be able to explain why a particular score was assigned to a response as techniques for interpreting (or explaining) the predictions of deep-learning models are still immature.

Important NLP Techniques

In this part, we provide definitions and examples of several basic NLP techniques that are commonly used in AS applications. It is important to be familiar with these techniques in order to understand the detailed descriptions of NLP features used for AS that we present in the subsequent section. We illustrate the NLP techniques using a sample response to the following speaking prompt*:

> Some people enjoy taking risks and trying new things. Others are not adventurous; they are cautious and prefer to avoid danger. Which behavior do you think is better? Explain why.

The transcription of a sample response is presented below. This sample response is 45 seconds in duration and contains 104 words:

> I definitely think it's better to take risks and try new things. Um, that way you would be able to get new life experiences and learn more about different things in the world rather than if you were just uh doing the same things over and over and not getting any new experiences. Um,

---

* This prompt is a sample TOEFL iBT speaking test question and was taken from the following source: https://www.ets.org/toefl/ibt/prepare/practice_sets/speaking

even though it could be risky to try new things and you might be put in an uncomfortable situation, um, challenging yourself and learning how to grow outside of your comfort zone is definitely an important skill in life and it helps you prepare for new things in the future.

Several standard NLP techniques that would be used to process this response in order to extract information that is required for AS are described below. These techniques could all apply to spoken responses, whereas the first three techniques (ASR, forced alignment, and pitch extraction) would not apply to written responses, since they require an audio input. Figure 5.1 presents two graphical representations – a waveform and a spectrogram – of the few seconds of the digitized audio file containing the sample spoken response; we will discuss this figure further in the first three bullet points below in which we describe NLP techniques that use audio input.

- **Automatic Speech Recognition (ASR)** is the process of automatically converting a digital speech signal into a textual representation (i.e., transcribing the words that were spoken based on an analysis of the characteristics of the speech signal and statistical models of acoustic properties of phonemes and word cooccurrence patterns). Figure 5.1 contains a possible ASR result for the sample response in the 'Words' section of the figure. In this hypothetical ASR result, there is one substitution error ("two" instead of "to"). As is typical in ASR output, the transcription contains no distinctions between uppercase and lowercase letters.

- **Forced alignment** takes as input a spoken response and a transcription of the response, which is either a manual transcription produced by a human or an automatic transcription produced by an ASR engine and produces sequential time stamps for all phonemes

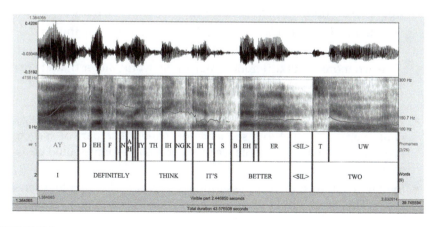

FIGURE 5.1
Sample response showing ASR output, forced alignment, and pitch extraction.

and words included in the transcription. The vertical blue bars in Figure 5.1 show time stamps for the words and phonemes for the first few words in the sample response that would be the result of applying *forced alignment* to the digitized audio and the ASR transcription of the response. The 'Words' section contains the word-level time stamps and the 'Phonemes' section contains the phoneme-level time stamps. It should be noted that time stamps are also available for portions of the spoken response that are detected by the forced alignment procedure to contain no spoken input (i.e., silences, as indicated by the <SIL> portion of the response).

- **Pitch extraction** uses signal processing techniques to extract a sequence of pitch measurements at regular intervals (e.g., every 10 milliseconds) throughout the audio recording in order to determine how the speaker's intonation changes throughout a spoken response. The pitch measurements for the first few words in the sample response are shown by the light blue line on the spectrogram in Figure 5.1. This pitch track indicates that the speaker produced a strong rising pitch on the first word in the response ("I"), a gradual falling pitch on the next three words ("definitely think it's"), and relatively flat, low pitch on the next two words ("better to").

- **Tokenization** is the process of separating a large text into smaller units including paragraphs, sentences, and tokens. A *token* is usually equivalent to a word, though sometimes a word is split into multiple tokens in order to separate components that function differently. For example, the contraction "can't" is typically split into two tokens, e.g., "can" and "'t", so that the negation can be correctly interpreted. Similarly, punctuation is usually separated from the word that it is attached to since its function is completely separate from the word. The tokenization of the first two sentences from the sample response would be as follows (each sentence appears on its own line):

  - I definitely think it 's better to take risks and try new things.
  - Um, that way you would be able to get new life experiences and learn more about different things in the world rather than if you were just uh doing the same things over and over and not getting any new experiences.

- **Part-of-speech tagging** involves assigning a *part-of-speech* (e.g., noun, verb, adjective) to a sequence of tokens. The part of speech tags assigned to the first sentence in the sample response could be as shown in Figure 5.2.*

- **Syntactic parsing** involves assigning a *linguistic structure* to a sentence. This structure is usually a hierarchical grouping of phrases

---

* Annotation and visualization provided by http://nlp.stanford.edu:8080/corenlp/process

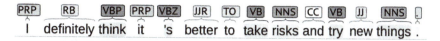

FIGURE 5.2
Part-of-speech tagging for sample sentence.

FIGURE 5.3
Syntactic parsing of sample sentence.

(e.g., noun phrases, verb phrases), sometimes with syntactic functions (e.g., subject, object) assigned. The syntactic structure of the first sentence in the sample response could be as shown in Figure 5.3.[7]

- **Discourse parsing** involves assigning a *discourse structure* (i.e., a representation of the organization of the text) to a sequence of sentences or paragraphs. This includes identifying discourse elements (e.g., spans of text, usually clauses or sentences) and the relationship between them (e.g., contrast, expansion, comparison). Discourse parsing can be shallow, where the requirement that all discourse elements be connected in the form of a tree or graph is removed. Whether the discourse structure is connected or not usually depends on the needs of a particular application. For more detailed discussions about automated speech scoring please see Chapter 20 by Zechner and Loukina (this handbook).

NLP Pipeline for AS Applications

In this section, we provide an overview of the different computational steps that are necessary to process and score both written and spoken responses by conducting linguistic analysis using the NLP approaches that we reviewed in the previous section, following the statistical NLP paradigm. These steps are illustrated in Figure 5.4, which provides an overview of the general pipeline for developing an AS application.

Pre-processing

This concerns data manipulation activities that have to take place before substantive analyses can be conducted. For written responses, initial preprocessing may be required to extract plain text from a response that might contain some formatting markup (e.g., HTML). Then, tokenization is conducted to produce units of text that can be more easily analyzed, and the

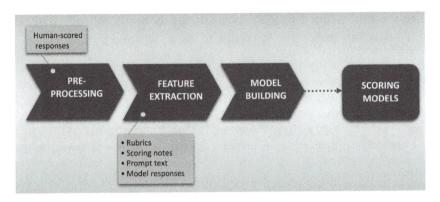

FIGURE 5.4
A pipeline for the development of an AS application.

tokenized text is processed through a variety of NLP modules (e.g., part-of-speech tagging, syntactic parsing) to generate information that will be used in the feature extraction step.

For spoken responses, the pre-processing step typically consists of the ASR component that takes a raw audio signal and extracts a text representation of the test taker's response, the forced alignment component that generates time stamps and additional information based on the text representation and the acoustic models, and additional speech processing components that extract further useful information from the audio signal (e.g., pitch extraction). In addition, the text transcription of a spoken response can be subject to most of the pre-processing modules that are applied to written responses with the exception of text-specific steps such as plain-text extraction and punctuation tokenization, which are not relevant for ASR output.

Feature Extraction

This involves the manual analysis of information used by human raters to make decisions about how to score responses (including scoring rubrics, scoring notes, prompt texts, and exemplar model responses) and then the careful engineering of linguistic features that approximate aspects of the human scoring process. The linguistic features are typically extracted from the output of various pre-processing modules by summarizing construct-relevant information over the entire response; examples include the number of grammatical errors included in a written response (as a measure of grammatical accuracy) or the frequency of long pauses in a spoken response (as a measure of fluency).

Model Building

This involves combining the features extracted in the previous step in a statistical model to predict the original human score for a given response. The modeling techniques used can vary by application, but typically range from simple multiple linear regression to more complex *supervised machine learning* methods such as random forests or ensembles of multiple models. In addition, *unsupervised machine learning* methods (e.g., *k*-nearest neighbors) can be used to cluster similar responses without assigning a label or score; this can be helpful in more exploratory studies such as ones that seek to determine which features differentiate different proficiency group or to validate the human scoring rubrics and scoring scale.

We provide more detail about pre-processing and linguistic feature extraction for both written and spoken responses in the following two subsections.

NLP Analysis for AS of Written Responses

A wide range of linguistically motivated features can be extracted from written responses for the purposes of model building and scoring using current NLP techniques. Some of these features can also be used to provide feedback to learners. In this subsection, we differentiate between (1) *task-independent features*, (2) *task-dependent features*, and (3) *content features*.

Tasks-Independent Features

The most common kinds of features that are used in AS of writing quality include features targeting writing conventions, discourse structure, and vocabulary. Most automated capabilities for evaluating writing quality will include features that measure the structure, style, and content of a response using various NLP and machine learning techniques. For example, ETS's e-rater system computes (1) *writing conventions*, (2) *discourse structure*, and (3) *vocabulary*:

Writing Conventions

This include a wide variety of features that measure errors of grammar, usage, and mechanics at the sentence level (e.g., fragments) as well as at the individual word level (e.g., prepositions). In some contexts, stylistic errors are also included (e.g., overrepetition of the same terms throughout an essay); see Burstein et al. (2003) for a more detailed overview of the kinds of errors that are detected. Most of the errors that are detected can be used both as scoring features as well as indicators to provide feedback to learners.

Discourse Structure

This includes two main features that are used to capture elements of discourse structure in test taker responses: organization and development. Both are based on the automated detection of discourse units in an essay and rely on the traditional five-paragraph essay construction; see Burstein et al. (2003) for more details.

Vocabulary

This includes two main features that aim to address vocabulary sophistication and usage. The first vocabulary feature is based on Breland's standardized frequency index (Breland et al., 1994) while the second is based on the average word length across all words in a response.

The effectiveness of these features in the AS models will usually depend on the nature of the assessment tasks as well as the typical test taker population. For example, an assessment that is typically taken by native English speakers to measure a complex competency such as analyzing an argument may place less emphasis on writing conventions and more emphasis on discourse structure. On the other hand, an assessment designed to measure proficiency of English is likely to place considerable emphasis on writing conventions.

Task-Dependent Features

In addition to the basic set of features typically used in AS of writing, *task-specific features* for evaluating writing quality can also be used for evaluating responses to certain types of writing tasks. Some relevant features used in ETS's e-rater system include (1) *argumentation,* (2) *narrative quality,* and (3) *use of source materials*:

Argumentation

This feature measures the quality of the argumentative structure in an essay and is a very active area of research (e.g., Beigman Klebanov et al., 2017; Nguyen & Litman, 2016; Stab & Gurevych, 2014). The techniques used to identify the structure of an argument often rely on discourse features as well as lexical cues.

Narrative Quality

This feature measures characteristics of writing that are indicative of the test taker's proficiency in a narrative writing task. For example, Somasundaran et al. (2018) show that some important features for capturing aspects of narrative writing include discourse transitions, event cohesion and coherence, and features that can capture the sentiment (positive/negative/neutral) of the writing.

Use of Source Materials

This feature includes evidence about how test takers extract and combine information from multiple stimuli, which may include passages, poems, or even audiovisual stimuli. Typically, test takers will be required to reference the source material in a more complex manner other than simply copying chunks of text verbatim. For example, Beigman Klebanov et al. (2014) evaluate a variety of content importance models that help predict which parts of the source material should be selected by the test taker in order to succeed at a task as part of a test of English proficiency.

Content-Specific Features

AS of content is another growing area of research in NLP as more and more datasets become publicly available. It is sometimes referred to as *short-answer scoring*, since many content-focused items, particularly in the K–12 field, tend to elicit relatively short responses. However, there is no conceptual reason that content scoring cannot also be applied to longer responses (e.g., Madnani et al., 2016).

The main difference between the AS of content and the AS of writing quality is that automated content scoring is typically trying to measure whether a test taker has understood a particular concept; how the test taker expresses this understanding is less important. This means that an automated content scoring model needs to be able to ignore characteristics of the text such as misspelled words and grammatical errors as long as the test taker's intent is clear from the text.

For AS of content, the two most popular approaches are (1) *reference-based scoring* and (2) *response-based scoring*:

Reference-based Scoring

This approach compares the test taker's response to reference texts such as sample responses for each score level or specifications of required content from the assessment's scoring guidelines. Various text similarity methods (Agirre et al., 2013) can be used for the comparison.

Response-based Scoring

This approach uses detailed features extracted from the test taker's response itself (e.g., word or character n-grams) along with supervised machine learning to learn a scoring function from a large dataset of human-scored responses. The features in response-based models tend to center around the meaning of the words in the text itself and the overall organization of the text is largely ignored.

There has also been research to investigate the effect of combining these two approaches (Sakaguchi et al., 2015). That work showed that the size of the available training data influences how much can be gained by adding reference-based features to a response-based system.

NLP Analysis for AS of Spoken Responses

Similar to the NLP features that are used for scoring written responses, a wide range of features can also be extracted using NLP and speech-processing techniques to assess construct-relevant aspects of speaking proficiency. Typically, however, automated speech-scoring systems have been designed specifically for the assessment of nonnative speaking proficiency, whereas automated text-scoring systems have been applied to a wider variety of types of assessments that may include both native and nonnative speakers.

In the following, we briefly summarize some of the approaches that have been taken to use NLP techniques to automatically extract automated speech scoring features. For features that are extracted based on the text transcription produced by the ASR system, the NLP approaches are often similar to ones that are used for processing written responses.

However, in the case of ASR output for spoken responses, the features may need to be designed slightly differently in order to be robust to the presence of ASR errors that may not accurately capture all of the words that were spoken by the test taker. Additionally, ASR transcriptions do not typically contain punctuation automated clause, or sentence boundary detection needs to be performed first in order to extract features that are based on clauses or sentences.

Specifically, the types of features included in automated speech-scoring systems can be categorized into the following three main areas: (1) *delivery*, (2) *language use*, and (3) *topic development.**

Delivery Features

Delivery features evaluate the mechanical aspects of how a spoken response is delivered and are related most closely to the areas of phonetics and phonology; three important subconstructs that are frequently measured include (1) *pronunciation*, (2) *fluency*, and (3) *prosody*.

Pronunciation

The most common approaches for automatically assessing pronunciation in nonnative speech make use of acoustic likelihood scores generated by the ASR system to calculate a 'goodness of pronunciation' score (Witt & Young, 2000). These scores can be used both for detection of individual mispronounced phonemes or words and for holistic pronunciation scoring for an entire utterance; Franco et al. (2014) provide an overview of some of the variations of this general approach.

Additional approaches to detecting pronunciation errors include classifiers that make more explicit use of acoustic-phonetic information and divergences between native and nonnative pronunciation patterns (Strik et al., 2007; Yoon et al., 2010). More recent research has investigated approaches to detecting pronunciation errors using acoustic models based on deep neural networks (Hu et al., 2015). Finally, some systems include automated metrics for assessing pronunciation characteristics beyond the accuracy of segmental phonology such as segment duration (Neumeyer et al., 2000) and vowel dispersion (Chen et al., 2010).

---

* This categorization is based on the three main construct areas that human raters are instructed to pay attention to when scoring TOEFL iBT Speaking tests; see https://www.ets.org/s/toefl/pdf/toefl_speaking_rubrics.pdf

Fluency

A range of fluency features can be extracted using the time stamps for words and phonemes in an utterance obtained through the forced alignment procedure. Examples of these include rate of speech (e.g., words/phonemes per second), frequency of silent pauses, average pause length, number of long pauses (where "long" is defined as a perceptually salient threshold such as 500 milliseconds), frequency of disfluencies (including filled pauses such as *uh* and *um*), and average number of words per run (where a "run" is defined as a stretch of continuous speech separated by silent pauses), among others.

Fluency features are typically robust to ASR errors and have relatively high correlations with human scores and are, therefore, included in most automated speech scoring systems. Two example publications that provide good descriptions of how several different fluency features are calculated are Cucchiarini and Strik (1999) and Zechner et al. (2009).

In addition, some fluency features can take syntactic information into account; for example, studies have shown that nonnative speakers sometimes produce silent pauses within syntactic clauses, whereas native speakers typically do not break up syntactic clauses in this way and primarily produce pauses between clauses (Mizera, 2006); a similar pattern has also been observed for filled pauses (Lauttamus et al., 2010). These findings have motivated the use of automated clause boundary detection methods for extracting fluency features related to the occurrence of within-clause and between-clause pauses (Chen & Yoon, 2012).

Prosody

Different approaches for assessing prosody can be employed depending on whether the content of the spoken response produced by the speaker is known in advance (e.g., in the context of a read-aloud task mentioned earlier) or if the spoken response is unscripted. In the case of predictable, scripted speech, the intonation contour in the learner's speech can be compared to a target model based on patterns observed in responses provided by native speakers (Wang et al., 2015).

Alternatively, manual annotations can be obtained in advance for expected prosodic events in the response (e.g., lexical stress and boundary tones) and the location of automatically detected prosodic events in the nonnative spoken response can be compared to these annotations to calculate speaking proficiency features (Zechner et al., 2011). When the content of the speech is not known in advance, a common approach is to extract distributional statistics for a range of prosodic characteristics (e.g., pitch, intensity, stress, and tone markers) across the entire utterance and to use these statistics (e.g., mean, standard deviation) to predict the target score (Hönig et al., 2012).

Another approach to automated assessment of prosody has been to extract measurements related to the rhythmic characteristics of the spoken response such as how much variability in duration is present for different types of consecutive speech sounds (e.g., vowels, consonants, and syllables); these rhythm features are based on the assumption that different languages have different characteristic rhythmic properties and have been shown to correlate with proficiency ratings of nonnative speech (Lai et al., 2013).

Language Use Features

Language use features evaluate how test takers select words and combine them into larger units to form a response and relate most closely to the areas of morphology and syntax; two important subconstructs include (1) *vocabulary* and (2) *grammar*:

Vocabulary

As discussed above, automated measures for assessing vocabulary proficiency in nonnative spontaneous speech have to contend with potential ASR errors in the automated transcript. Despite this challenge, several studies have demonstrated that features relating to different aspects of vocabulary usage are effective for automated speech scoring. These features are typically related to one of the three dimensions of lexical complexity as defined by Read (2000): *lexical density* (i.e., the proportion of content words in the spoken response), *lexical diversity* (i.e., the variety of words used in the response), and *lexical sophistication* (i.e., the speaker's use of lower frequency vocabulary words).

Relatively simple features relating to lexical diversity such as the number of distinct lexical types present in the response and the type–token ratio have been shown to be robust to ASR errors and have relatively high correlations with human scores (Zechner et al., 2009; Lu, 2010). In addition, Yoon et al. (2012) demonstrated the effectiveness of lexical sophistication features that measure the presence of low-frequency and high-frequency words (based on three different reference corpora) in a spoken response; a high negative correlation was exhibited by a feature that measures the proportion of high-frequency words suggesting that human raters tend to provide lower scores when the response consists primarily of simple, high-frequency words.

Grammar

The two main categories of grammar features used in automated speech scoring are *grammatical accuracy* and *grammatical complexity*. Grammatical accuracy features have primarily been used with speaking task types that elicit restricted speech as in the sentence-build task mentioned earlier, in which a test taker is presented with words and phrases in a random order and is asked to speak a grammatical sentence using these components.

Accurately detecting grammatical errors in responses to tasks that elicit spontaneous speech is challenging due to the presence of disfluencies, false starts, and other difficult characteristics in addition to potential ASR errors. Consequently, most studies of spontaneous speech have limited the assessment of grammatical accuracy to simple features that can be extracted from the ASR system such as the language model score, an indication of how closely the word sequences in the test taker's response match the responses in a large data set of transcribed spoken responses used to train the ASR system (Zechner et al., 2009).

Grammatical complexity in spoken responses can be analyzed in a similar manner as for written responses, namely by generating a syntactic parse of the text produced by the ASR system and computing the frequency of different types of grammatical structures, such as noun phrases, subordinate clauses, etc. (Zechner et al., 2017). Bhat and Yoon (2015) describe an alternative approach to evaluating syntactic complexity that is designed to be more robust to ASR errors. That approach uses automated part-of-speech labels rather than a full syntactic parse of the ASR output and compares the part-of-speech sequences contained in the spoken response to models of part-of-speech patterns for test takers at different proficiency levels.

Topic Development Features

Topic development features evaluate how a test taker presents ideas to form a meaningful response and relate most closely to the areas of semantics and pragmatics; two important subconstructs include (1) *content* and (2) *discourse*:

Content

As described above for written responses, approaches to AS of content for spoken responses can be divided into reference- and response-based approaches. For tasks that elicit restricted speech, such as the read-aloud task mentioned earlier, the words in the ASR transcription can be compared to the expected words based on the prompt that the test taker was supposed to read in order to evaluate the content accuracy by using features such as the number of correct words (Downey et al., 2011).

For spontaneous speech, a similar reference-based approach for source-based speaking tasks is to compare the content of the test taker's response to the content in listening or reading materials that were presented to the test taker (Evanini et al., 2013). Some studies have also compared the content of the spoken response to a model response provided in advance by experts (Xiong et al., 2013).

The standard approach for response-based content scoring for spoken responses is to compare the words in the response to distributions of words contained in sets of responses that received different proficiency scores using similarity metrics such as *pointwise mutual information* (Xie et al., 2012) or *ROUGE* (Loukina et al., 2014). Some studies have attempted to abstract away from the specific words contained in a response by using representations such as semantic categories for each word (Chen & Zechner, 2012) or response-level *semantic vectors* (Tao et al., 2016) to make comparisons between the spoken response and the sets of scored responses.

Discourse

To date, only a few studies have investigated the AS of discourse in spoken responses, likely due to the difficulty in extracting accurate features based on ASR transcriptions, especially since most spoken responses tend to be relatively short (i.e., typically one minute or less).

One study examined the use of surface-oriented features such as the frequency and sequential patterning of discourse connectives in the response, which are designed to be more robust to the presence of ASR errors; the authors demonstrated that they were indeed able to improve the performance of an automated speech scoring system (Wang, Zechner, Evanini, & Mulholland, 2017).

Other studies have adapted approaches that were originally designed for the evaluation of discourse coherence in essays and applied them to spoken responses such as using entity grids to model the relationships between entities in adjacent sentences (Wang et al., 2013) and using features based on how the discourse elements in the response are represented in tree based on *rhetorical structure theory* (Wang, Bruno, Molloy, Evanini, & Zechner, 2017).

Table 5.1 presents a summary of several specific AS features that can be extracted using the AS analysis techniques that we presented in this section and indicates the construct area that they are designed to address, the NLP techniques that can be used to compute them, and the linguistic subfields

TABLE 5.1

Sample AS Features

AS Feature	Construct Area	NLP Techniques	Linguistic Subfields
Average acoustic likelihood score	Speech delivery, pronunciation	ASR, forced alignment	Phonetics, Phonology
Number of clause-internal pauses	Speech delivery, fluency	ASR, forced alignment, syntactic parsing	Phonetics, Phonology, Syntax
Average duration between stressed syllables	Speech delivery, rhythm	ASR, forced alignment, pitch extraction	Phonetics, Phonology
Number of preposition errors	Writing conventions	Tokenization, part-of-speech tagging, syntactic parsing	Morphology, Syntax, Semantics
Sentence variety	Coherence	Tokenization, part-of-speech tagging, syntactic parsing	Morphology, Syntax
Discourse Elements	Organization	Tokenization, part-of-speech tagging, syntactic parsing, discourse parsing	Morphology, Syntax, Semantics, Pragmatics

Note: The first three features in the table are specific to spoken responses, whereas the remaining three features can be used for both spoken and written responses. AS = automated scoring, ASR = automated speech recognition, NLP = natural language processing.

that provide the theoretical foundations for the analysis techniques. It thus illustrates the necessary path to get from a linguistic analysis of a test taker's response to specific AS features using NLP techniques.

Challenges and Recommendations

There are many stakeholders from different backgrounds involved in different stages of the development and application of AS systems (e.g., assessment developers, NLP researchers, testing program administrators, teachers, test takers) and, unsurprisingly, the different stakeholder groups typically have competing priorities. Therefore, it is vital for all stakeholders to understand and incorporate the perspectives of the other parties and work towards a mutually satisfactory solution in order to build AS systems that are accurate, fair, unbiased, and useful (Madnani & Cahill, 2018). Evaluating AS systems along all of these dimensions – not just one or two of them – is an important aspect of ensuring that AS systems do not introduce bias into assessments. One way to encourage comprehensive evaluations is to share advances in evaluation strategies from the field of psychometrics via open-source software packages (e.g., Madnani et al., 2017; Loukina et al., 2019).

The field of NLP has undergone incredibly rapid development in recent years and considerable effort is required to keep abreast of technological advances and to understand whether they can be appropriately applied or adapted for a given AS application. The number of NLP conferences, journals, and workshops have continued to steadily increase along with the numbers of paper submissions. There are now domain-specific workshops for NLP for educational applications (e.g., the *ACL Workshop on Natural Language Processing Techniques for Educational Applications*, the *ACL Workshop on Innovative Use of NLP for Building Educational Applications*, and the *ISCA Workshop on Speech and Language Technology in Education*) as well as special-interest groups to promote the use of NLP in educational applications (such as *SIG-EDU** and *SIG-SLaTE†*). This clearly indicates the growing interest from the field of NLP in applying their techniques to educational applications.

Approaches now exist for extracting detailed information about a test taker's ability from written and spoken responses; this information ranges from simplistic, surface-oriented features such as counting the frequency of a particular grammatical construction to more complex features that require a deeper "understanding" of the response such as automatically detecting metaphorical language used in argumentative writing (Beigman Klebanov & Flor, 2013a). The field has certainly advanced light years beyond

---

* https://sig-edu.org
† https://www.isca-speech.org/iscaweb/index.php/sigs?layout=edit&id=121

the initial NLP approaches that were used for automated essay scoring when the idea was first proposed approximately 50 years ago (Page, 1968).

However, despite these advances, there are still substantial gaps between those aspects of the targeted constructs that can be assessed reliably by human raters and state-of-the-art AS systems and those that cannot, at least for some task types. These gaps make it necessary for assessment organizations and score users to carefully consider the relative stakes of an assessment and the relative maturity of NLP technology for the relevant task types before deciding whether AS systems can be deployed in a valid manner for a given assessment.

For example, a large number of studies have demonstrated that automated speech scoring systems can achieve results that are comparable with human rater agreement levels for scoring responses to read-aloud tasks and a large number of features can be extracted with high degrees of accuracy to cover the main aspects of the targeted construct for this task type (i.e., reading accuracy, pronunciation, fluency, and intonation). Therefore, the use of AS technology for this task type can likely be justified for a wide range of contexts.

On the other hand, the situation is quite different for another common type of speaking assessment task in which the test taker is asked to provide a personal opinion about a topic and support the opinion with relevant information drawn from personal experience or source material. In order to provide a valid score for a response to this task, an automated speech-scoring system would need to be able to accurately recognize the words spoken by the test taker using an ASR system, automatically detect the opinion and supporting reasons provided by the test taker, and determine whether these reasons are relevant to the opinion and provide appropriate support for it (Evanini et al., 2017). While substantial progress is being made to develop NLP and speech-processing technology to solve these problems, achieving such a level of semantic understanding of a test taker's response is still an active area of open research and currently cannot be done completely reliably using current AS technology.

In order to ensure that any AS technology is used in a valid manner given the targeted constructs of the assessment, the intended uses of the scores, and the current state-of-the-art NLP capabilities that are available, it may be appropriate to consider alternative deployment scenarios in addition to sole AS. For example, human and automated scores can be used together in a variety of hybrid configurations including *confirmatory scoring* (in which the automated scores are used as a *quality-control check* for the human scores) and *contributory scoring* (in which the automated scores and human scores are combined using a particular weighting scheme). In many cases, scores produced by a hybrid system that combines human and automated scores are more reliable than scores produced by a single human rater (e.g., Breyer et al., 2017), since the respective strengths of the human and automated raters complement each other to provide more information than either one can by itself.

Another challenge in deploying AS systems is ensuring that they are robust to gaming strategies and other kinds of responses that should not be automatically scored (e.g., keyboard banging, or automatically generated nonsense such as those that come from the Babel* system); see, for example, Yoon et al. (2018) for an overview of NLP techniques for this purpose. The proper treatment of such responses is important for ensuring the validity of an assessment that might use AS.

Conclusions and Future Directions

In this chapter, we have provided a high-level introduction to the field of NLP for AS and have briefly summarized several prevalent techniques that are used to extract information about a test taker's knowledge, skills, and abilities from a written or spoken constructed response. In addition, we have indicated how the NLP features that are used to represent this information relate to the construct of interest for a variety of constructed-response task types.

The field of NLP will surely continue to develop increasingly more sophisticated capabilities and the need for AS technology will likely continue to increase as more assessments are conducted digitally and as assessments become more frequent to facilitate individualized feedback and learning. As we noted earlier, the NLP community in general has recently seen an explosion in popularity of approaches based on deep learning algorithms, and these have also demonstrated promising results for specific areas of automated assessment such as automated essay scoring (Taghipour & Ng, 2016; Dong et al., 2017) and automated speech scoring (Chen et al., 2018).

These approaches are attractive since they typically bypass the feature engineering step which can be difficult, time-consuming, and costly. Instead, statistical models are trained in an "end-to-end" manner using only the constructed response as input, without any subsequent NLP processing such as part-of-speech tagging. However, these deep learning-based systems are often not very "transparent" – in that a test taker or score user cannot easily determine which specific aspects of the test taker's response contributed to the final score – and additional research needs to be conducted to determine how they can be used most effectively for feedback and learning.

Finally, we should mention that NLP technology – as well as machine learning and artificial intelligence technology more generally – will likely be used in the future to score and provide feedback on a much broader range of constructed-response task types than the writing and speaking tasks discussed in this chapter. For example, some researchers are investigating the

---

* https://babel-generator.herokuapp.com

use of spoken dialog technology to deliver interactive and authentic conversation-based tasks to language learners; automated speech scoring technology can be applied to these conversations to produce automated speaking proficiency scores (Ramanarayanan et al., 2017). Another example of next-generation AS technology is the use of NLP and multimodal processing techniques to automatically assess public speaking skills based on videos and 3D motion capture representations of public speaking performances (Chen et al., 2016). As NLP technology is applied to score an increasingly broad range of constructed-response performances in a variety of learning and assessment contexts, researchers should continue to evaluate the reliability, fairness, and validity of the scores for the proposed use cases.

6

Multimodal Analytics for Automated Assessment

Sidney K. D'Mello

CONTENTS

As I write this chapter, seated on an airplane en route to Boston, I find myself contemplating the rich multimodal environment around me. There is the sound of the engines, the uncomfortable feeling of my cramped legs, the aroma of fresh coffee, and the mundane visual environment. I am also exuding multimodal signals via smiles, frowns, sighs, deep breaths, and various gestures. My overt behaviors are restrained, though, because the socio-contextual environment has constrained me to behave within societal norms of what is "normal airplane behavior." There are other imperceptible signals emitted such as the crummy airplane Wi-Fi and radio signals that I hope are not giving me cancer. I am also undoubtedly exuding signals from my brain and body through neurotransmitters, which are largely imperceptible to others. Indeed, the human experience is inherently a tapestry of rich multimodal signals that we exude and perceive.

What does this have to do with automated scoring systems? Not much from a traditional perspective of assessment, which focuses predominantly on responses to carefully curated items or relatively restricted assessment tasks. Perhaps a bit more from a modern perspective on assessment in which patterns of clicks, selections, and extended responses matter and are part and parcel of the assessment. But multimodality matters considerably when the goal is to assess "nontraditional" assessment constructs in "nontraditional" testing contexts. Extending the example above, my annoyance, frowns, sighs, elevated arousal, and overt behaviors are all integral components of my in-flight experience, just as test-taker's panic and anxiety, along with efforts to regulate them, are integral to the assessment of their test-taking competency.

In this chapter, we are concerned with the use of multimodal signals – observable from the perspective of the environment, the individual, and groups of interacting individuals – to assess a variety of cognitive, affective, motivational, behavioral, and meta-cognitive *constructs* (e.g., knowledge, boredom, interest, feelings of knowing). These include cognitive and non-cognitive *traits* (e.g., personality, intelligence), socio-cognitive-affective *states* (e.g., joint attention, expression mirroring), and social *behaviors* (e.g., active participation, negotiation).

Multimodal assessment, in turn, is concerned with identifying and synthe-sizing evidence about these constructs by integrating multimodal signals in order to make principled inferences about individuals or groups of indi-viduals interacting as teams. Furthermore, automated multimodal assess-ment systems aim to fully automate the assessment process in that no human intervention is required once the assessment system has been deployed.

We are, of course, nowhere near a fully automated multimodal assess-ment system that can be fielded even in limited domains, let alone at scale. However, considerable progress has been made over the last two decades in basic research on multimodal interactions and on prototype multimodal systems such that real-world multimodal assessment is within the research and development horizon. In this chapter, we provide an overview of the basic principles of multimodal analytics, present some of the pertinent com-putational methods, illustrate them via a few selected case studies, and dis-cuss a few guidelines for designing multimodal assessment systems. This introduction is necessarily relatively basic and we direct the reader who wants to learn more to the most recent multi-volume *Handbook of Multimodal-Multisensor Interfaces* (Oviatt et al., 2017).

Principles of Multimodal Analytics

Our understanding of multimodal perception, expression, and communication is grounded in classical and contemporary theories in psychology, neuroscience, and communication; see Oviatt (2017) for a detailed discussion. It is important that multimodal assessment systems have a grounding in fundamental prin-ciples derived from these theories, which we briefly summarize here.

The psychological foundation of multimodality traces its roots to foun-dational principles advocated by *Gestalt psychologists* over a century ago who were interested in developing simple principles to explain how people perceive meaning from a complex percept (i.e., stimulus). For example, the key Gestalt principle of *totality,* which posits that the whole is qualitatively different than the sum of its parts, supports the idea that we integrate percepts from different modalities (e.g., linguistic and paralinguistic) to form a coherent percept that goes beyond the unimodal experience. Accordingly, neuroscience suggests that *multisensory integration,* or how information from different senses is integrated by the nervous system, is fundamental to cognition to the extent that it is even observable at the level of individual neurons (Meredith & Stein, 1983).

From a cognitive perspective, *working memory theory, cognitive load theory,* and *multiple resource theory* all posit that working memory has limited capacity with respect to individual modalities. Multimodality plays a critical role by alleviating working memory overload through distributing information across modalities – for example, presenting information aurally when the visual modality is occupied. Multimodality can also strengthen encodings in long-term memory. In fact, many of the principles of multimedia learning such as the *modality principle* (i.e., when processing complex graphics, people learn better from audio narration than from text since the former alleviates visual working memory demands) and the *multimedia principle* (i.e., people learn better from words and graphics than either alone since combined visual and verbal information leads to deeper encoding) are based on these ideas (Mayer, 2005).

In general, multimodal inputs and outputs are expected to influence basic cognitive processing by increasing its capacity relative to one modality. If this is indeed the case, then there should be measurable effects for multimodal signals compared to their unimodal counterparts. To understand this theoretically, consider the taxonomy shown in Table 6.1. It is modeled after multimodal communication patterns observed in nature (Partan & Marler, 1999). This taxonomy considers the various types of effects that might be obtained by integrating two signals A and B, which contain information

TABLE 6.1

Taxonomy of Prototypical Multimodal Patterns

(redundant signals): $A \rightarrow X; B \rightarrow X$
(redundancy) $A + B \rightarrow X$
(redundancy) $A + B \rightarrow X$
(enhancement) $A + B \rightarrow XX$
(nonredundant signals); $A \rightarrow X; B \rightarrow Y$
(independence) $A + B \rightarrow X + Y$
(dominance) $A + B \rightarrow X$ or Y
(modulation: enhancement) $A + B \rightarrow XX$ or YY
(modulation: diminishing) $A + B \rightarrow x$ or y
(emergence or superadditivity) $A + B \rightarrow Z$

units X and Y, respectively. The specific effect depends on the multimodal signals, the information they carry, the integration method, and the interaction context.

Let us begin with effects associated with redundant signals. *Redundancy* is an important feature of multimodal systems since it allows the system to function when one or more of its components fail. Two signals A and B are redundant when they communicate the same information X (A → X; B → X). Their combination might result in X itself or an *enhanced* version of it (XX). Redundancy is not a very interesting case from a multimodal perspective.

In contrast, signals A and B are said to be *nonredundant* when they communicate different information (A → X; B → Y). If the signals are *independent* then the multimodal combination can result in both X and Y. Alternatively, one signal could dominate the other (X or Y) or could *modulate* the other by *enhancing* (XX or YY) or *diminishing* it (x or y). The most interesting case occurs when a multimodal combination results in an entirely new signal (Z), which cannot be explained by the responses of the unimodal signals (X, Y). This property – referred to as *emergence* or *superadditivity* – is closely aligned with the Gestalt principle of totality discussed above in that it maintains that the whole is greater than the sum of the parts.

As noted above, the specific multimodal effect that might be achieved is signal- and context-dependent, factors typically out of the control of the assessment designer. One factor that is within the designer's control pertains to how they integrate (or combine) signals from multiple modalities. We discuss common computational methods for multimodal modeling next.

Computational Methods for Multimodal Modeling

The crux of multimodal modeling involves combining individual modalities, which is called *modality integration* or *modality fusion*. The space of computational approaches for this task is immense so we highlight only the main families of methods in the following.

Whereas the various approaches will become clearer from the case studies discussed in the next section, it helps to consider a working example. Accordingly, consider the task of assessing a person's level of stress from their facial expressions and speech patterns while delivering a talk as part of an assessment of their communication skills. Let us assume that the person's face is recorded using one video camera and his or her audio output is recorded with two microphones (a lapel mic and an audience mic). The modalities are thus 'video' and 'audio' and the task is to infer the speaker's stress by integrating data captured from each modality.

Multimodal fusion typically occurs within a machine-learning framework, particularly *supervised learning*, where the goal is to train a *computational model* to learn how to measure a construct based on provided examples. In the case of our hypothetical example, we would provide the algorithm with several videos (with audio) of a variety of speakers delivering their talks along with *annotations* of their stress levels from self-reports or trained judges. The task of the algorithm is to learn how to accurately estimate the stress level in a generalizable fashion so that it can be applied to videos from a new set of speakers. There are many steps in developing such a computational model and here we focus on how to combine information from multiple modalities as a core part of the modeling effort; see D'Mello, Kappas, and Gratch (2018) for an accessible description.

Basic Methods

The most basic method for fusing modalities is *data-level fusion* or *stream-level fusion*. Here, raw signals themselves are fused prior to any higher-level analysis. Turning to our hypothetical example, it would involve combining the two audio streams via some signal processing technique, the details of which are not relevant for the purpose of this chapter. Subsequent analysis then proceeds on the combined stream and the original signals are discarded.

The next basic method is *feature-level fusion* or *early fusion*, where *features* from different modalities are concatenated prior to analysis. A feature is a higher-order descriptor derived from the raw data. For example, pitch, pauses, and loudness might be pertinent features computed from the audio stream, whereas head nods, eye blinks, and smiles might be features computed from the video. By *feature concatenation* we simply mean merging the features computed from each modality prior to any statistical inference or modeling. In this vein, we would develop a single computational model to make inferences about a person's stress level from combined audio–visual features.

An alternative is *decision-level fusion* or *late fusion*, where intermediate computational models are first independently developed for each modality. The outputs of these individual models are subsequently combined to yield the final model. For example, an acoustic model could make inferences about stress from the audio features, whereas a video-based model could do the same but from video features. The outputs of the two models could then be combined by averaging or some other statistical mechanism.

The primary advantage of the feature-level and decision-level fusion methods lie in their simplicity. Feature-level fusion has an added advantage in that it can capture interactions among features from different modalities. However, it is less robust to cases when data are missing from one modality or when the modalities operate on different time scales. Decision-level fusion has an advantage in these cases. However, both basic fusion methods are limited in that they do not incorporate hierarchical or temporal information across modalities. We need more sophisticated fusion methods that address this limitation, which we discuss next.

Modality Fusion with Graphical Models

Probabilistic graphical models (Koller & Friedman, 2009) can be used to jointly model a range of relationships (e.g., hierarchical, temporal) within a probabilistic framework. *Dynamic Bayesian Networks* (DBNs) are a common graphical model used for modality fusion. Links between variables in DBNs can be used to represent conditional dependencies between features from different modalities as well as associations across time. Figure 6.1 shows a DBN that fuses two modalities along with contextual (top-down) features and models temporal effects; the empty circular node represents the construct being assessed, which is 'stress' in our example. *Top-down features* (e.g., vocabulary knowledge) are modeled as influencing the construct and do not change from one-time step to the next. *Bottom-up features* (e.g., components of facial expressions or vocal pitch) are linked across time. *Bayesian inference* is then used to compute the probability of the output variable (i.e., the statistical estimate for the variable representing the construct) given the top-down (i.e., predictive) and bottom-up (i.e., diagnostic) features. Khan and Huang (Chapter 15, this handbook) also discuss some details on deep learning.

DBNs are quite flexible, allowing for multiple structures of relationships between variables across time (i.e., *temporal relationships*). One such structure is a *hidden Markov model* (HMM), which models the latent construct as an unobserved (hidden) variable that influences observable variables (e.g., a

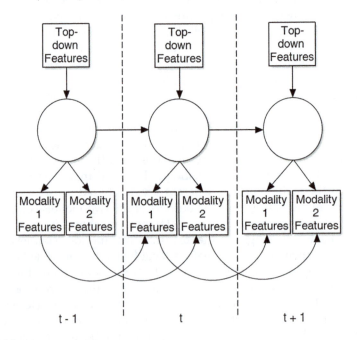

FIGURE 6.1
Dynamic Bayes net model fusing two modalities along with top-down features. Circles = latent construct variable.

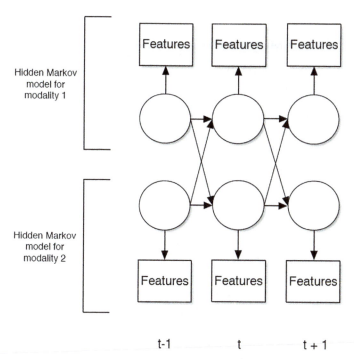

FIGURE 6.2
Coupled hidden Markov model for two modalities. Circles = latent construct variable.

variable reflecting interindividual differences in stress influencing variables reflecting interindividual differences in speech rate and blinks). *Coupled hidden Markov models* (CHMMs) combine two or more HMMs (i.e., one per a modality), so that the hidden states representing the latent construct of the individual HMMs interact across time; this is shown in Figure 6.2. These cross-modal links in CHMMs are chosen to model temporal relationships between modalities that might operate at different time scales (e.g., facial expressions and speech production). These flexibilities open up avenues for representing a wide array of constructs from multiple modalities.

Modality Fusion with Neural Networks and Deep Learning

Neural networks have emerged as another popular approach for modality fusion. In particular, the recent advent of *deep neural networks*, which are neural networks with multiple *layers*, has led to the development of a multitude of architectures for modality fusion, details of which are beyond the scope of this chapter. We refer the reader to Baltrušaitis, Ahuja, and Morency (2019) for a review of methods for modality fusion via deep neural networks. Khan and Huang (Chapter 15, this handbook) also discuss several modeling approaches in the area of *deep learning*.

Case Studies of Multimodal Assessment Systems

We now turn to case studies of multimodal assessment systems which we selected for their illustrative breadth in terms of sensors used, fusion methods applied, and constructs assessed; an overview of the study characteristics is shown in Table 6.2. We mainly selected systems that have been used in real-world classrooms rather than in research lab settings. Because the field of multimodal assessment is still very young, each example serves primarily as a proof-of-concept for a type of multimodal assessment system that has yet to be deployed at scale.

Example 1 – Student Affect

The first example focuses on multimodal affect detection in a computer-enabled classroom (Bosch, D'Mello, Ocumpaugh, Baker, & Shute, 2016). The researchers collected training data from 137 eighth- and ninth-grade U.S. students who interacted with a conceptual physics educational game called *Physics Playground*. Students played the game in two separate 55-minute sessions per day. Trained observers performed real-time annotations of boredom, engaged concentration, confusion, frustration, and delight during gameplay using a validated observational protocol. The annotations were based on observable behavior, including explicit actions toward the interface, interactions with peers and teachers, body movements, gestures, and facial expressions. Videos of students' faces and upper bodies and *log files* from the game were recorded and synchronized with the affect annotations (see Almond, Chapter 23, this handbook).

The videos were then processed using a *computer-vision program* (FACET) that estimates the likelihood of 19 facial action units (e.g., lowered brow, blink) along with head pose and position. Body movements were also estimated from the videos using *motion filtering algorithms*. Supervised learning methods were used to discriminate each affective state from the other states (e.g., boredom vs. [confusion, frustration, engaged concentration, and delight]) and were validated by randomly assigning students into training and testing sets, which is known as *student-level cross-validation*. The classification models yielded an average accuracy of 0.69 measured as the *area under the receiver operating characteristic curve* (AUC), whereas a chance model would yield an AUC of 0.50. Follow-up validation analyses confirmed that the models generalized across multiple days (i.e., training on subset of students from Day 1 and testing on different students in Day 2), class periods, genders (i.e., training on males and testing on females as well as vice versa), and ethnicity as perceived by human coders.

A limitation of video-based measures is that they are only applicable when the face can be detected in the video. This is not always the case outside of a

TABLE 6.2

Overview of Case Studies on Multimodal Assessment Systems

Example	Sample Studies	Constructs	Task	Sensors (Modalities)	Statistical Models	Fusion Methods
1	Bosch, Chen, Baker, Shute, & D'Mello, 2015; Bosch, D'Mello, Ocumpaugh, Baker, & Shute, 2016	Affect (boredom, engaged concentration, confusion, frustration, and delight)	Learning from an educational game	Video (facial expressions and body movement); log files (interaction patterns and contextual cues)	Various standard classifiers (affective states) such as Bayes Nets and support vector machines	Decision-level fusion (stacking with logistic regression)
2	Hutt et al., 2019; Hutt et al., 2017	Attention (mind wandering)	Learning from an intelligent tutoring system	Eye tracking (eye gaze); video (facial expressions)	Bayes nets (eye gaze) and support vector machines (video)	Decision-level fusion (weighted average)
3	Cook, Olney, Kelly, & D'Mello, 2018; Donnelly et al., 2016; Kelly, Olney, Donnelly, Nystrand, & D'Mello, 2018	Instructional activities (question and answer, lecture, small group work, procedures, and directions, supervised seatwork); authentic question detection	Engaging in classroom discourse	Microphone (linguistics, paralinguistics, turn-taking dynamics)	M5P regression trees	Varied
4	Mills et al., 2017	Cognitive load	Learning from an intelligent tutoring system	Dry electroencephalogram electrodes (EEG)	Partial least squares regression	N/A
5	Grafsgaard, Duran, Randall, Tao, & D'Mello, 2018	Nonverbal synchrony (facial expressions, bodily movement, gesturing)	Conversing with a partner in a lab setting	Video (facial expressions; body movements; head movement); microphone (acoustic–prosodic features); bracelet accelerometer (gestural movement)	Recurrent neural network	Long short-term memory recurrent neural networks

research lab where there are occlusions, poor lighting, and other complicating factors. In fact, the face could only be detected about 65% of the time in this study. To address this, the team developed an additional computational model based on interaction / contextual features stored in the game log files (e.g., difficulty of the current game level, the student's actions, the feedback received, response times).

The interaction-based models were less accurate (mean AUC of 0.57) than the video-based models (mean AUC of 0.67 after retraining) but could be applied in almost all of the cases. Separate *logistic regression models* were trained to *adjudicate* among the face- and interaction-based models, essentially weighting their *relative influence* on the final outcome. The resultant multimodal model was almost as accurate as the video-based model (mean AUC of 0.64 for multimodal vs. mean AUC of 0.67 for face only) but was applicable almost all of the time (98% for multimodal vs. 65% for face only). These results are notable given the noisy nature of the real-world environment with students incessantly fidgeting, talking with one another, asking questions, and even occasionally using their cell-phones. The results also illustrate how a multimodal approach addressed a substantial missing data problem despite it not improving measurement accuracy.

Example 2 – Mind Wandering

Eye tracking is perhaps the most effective method of obtaining a window into the locus of a person's visual attention. Until recently, the cost of research-grade eye trackers had limited the applicability of eye tracking in real-world environments at scale. However, the relatively recent introduction of off-the-shelf eye trackers, which are retailing for only about $100 to $150, has ushered forth an exciting era by enabling the application of decades of lab-based research on eye gaze, attention, and learning to real-world class-rooms, thereby affording new opportunities for assessing attention-related constructs.

Accordingly, Hutt et al. (2017) used eye-tracking data to build automated measures of mind wandering (i.e., involuntary attentional lapses or zone outs) during interactions with a conversational *intelligent tutoring system* called *GuruTutor*. The researchers collected data from 135 eighth- and ninth-graders enrolled in a Biology 1 class. Class sizes ranged from 14 to 30 students based on regular class enrollment. The classroom layout remained unchanged from the setup used for standard instruction, with the addition of laptops to run *GuruTutor* and an *EyeTribe* gaze tracker per desk. Each student completed two approximately 30-minute sessions with *GuruTutor*. Mind wandering was measured using auditory thought probes occurring every 90 to 120 seconds during the *GuruTutor* sessions; these data were used to train the *classification models*.

The modeling proceeded by computing eye gaze features in 30-second windows preceding each auditory probe. Global gaze features focused on general gaze patterns (e.g., fixation duration) and were independent of the content on the screen whereas locality features encode where gaze is fixated. A set of context features that encoded information from the session such as time into the session, student performance, response times, and so on were also considered. The features were used to train *static Bayesian networks* to detect mind wandering in a manner that generalized to new students.

The results indicated that global and locality models achieved similar overall accuracy with a slight tradeoff with respect to *precision* (global = 55%; locality = 51%) and *recall* (global = 65%; locality = 70%). Fusing the two sets of features using feature-level fusion resulted in lower performance than using each individual feature set as did adding contextual features. In addition, students completed a post-test assessment after the tutoring session. The predicted mind wandering rates for the global (Spearman's *rho* = –0.112, p = 0.269) and locality (Spearman's *rho* = –0.177, p = 0.076) models were correlated with post-test scores similar to self-reported mind wandering rates (Spearman's *rho* = –0.189, p = 0.058), thereby providing initial evidence of predictive validity of the detector.

In a follow-up study Hutt et al. (2019) investigated whether there were advantages to combining facial features with eye gaze data. They extracted facial features in real time during students' interactions with *GuruTutor* (i.e., simultaneously while gaze was recorded) using an *open-source* facial feature tracking software called *OpenFace*. Real-time processing was done due to privacy considerations, which precluded recording videos of students for offline analysis.

OpenFace provided intensity estimates of 14 *action units* at the time of the study. The researchers also captured *co-occurrence relationships* between action units, which are important for distinguishing among high-level facial expressions (e.g., a genuine smile). The gaze-based mind wandering detectors outperformed the face-based detectors developed via *support vector machines*. There was no advantage in fusing the two models using *weighted-average decision fusion* when only examining cases for which there was valid data from both sensing streams. However, fusion resulted in a small advantage of about 6.4% over the gaze models when data were missing from one of the sensing streams, suggesting some signal in the pattern of missingness.

In short, the results indicate that (a) it is feasible to track eye-gaze using cost-effective gaze-trackers despite the complexities of real-world classrooms, (b) collected data were of sufficient fidelity for real-time mind wandering detection, and (c) there was a small improvement when eye gaze was combined with facial feature tracking when one of the sensing streams was missing. The main general message, however, is that eye tracking no longer needs to be confined to the research lab but can be fielded in more authentic environments.

Example 3 – Teacher Discourse Quality

Most applications of automated assessment in classrooms are focused on students. What is known about what teachers do in the classrooms is typically derived from the occasional classroom observation, which is usually evaluative rather than formative. Current *in-person observational methods* are also logistically complex, require observer training, are an expensive allocation of administrators' time, and simply do not scale well. Further, teachers rarely get regular detailed feedback about their own classroom practice. Given the importance of feedback to learning, it is unsurprising that teaching effectiveness plateaus after five years in the profession despite considerable room for growth (TNTP, 2015).

It is therefore worthwhile asking whether the scoring of teacher instruction can be automated. As an initial proof of concept, Kelly, Olney, Donnelly, Nystrand, and D'Mello (2018) attempted to automate the collection and analysis of classroom discourse. They first experimented with different recording devices to balance a set of technical requirements and constraints. For example, cameras could not be used due to privacy concerns and it was infeasible to have a microphone for each student due to scalability concerns. As a solution, the team selected a wireless vocal headset system (Samson AirLine 77) to record teacher audio and a pressure zone microphone (Crown PZM-30D) to record general classroom activity.

They used this setup to record 7,663 minutes of audio, comprising 132 class sessions of 30- to 90-minute length from 27 classrooms taught by 14 teachers in 7 schools over 5 semesters. At the same time, a trained observer used live coding software to identify *instructional activities* (e.g., question-and-answer, lecture, small group work), teacher / student questions, and their associated properties (e.g., authenticity); these live annotations were refined and verified by a second coder at a later time.

The analytic approach began by using *voice activity detection* to obtain 45,044 teacher utterances with a median length of 2.26 seconds. The spoken utterances were automatically transcribed using *Microsoft Bing Speech* which yielded a *speech recognition accuracy* of 53.3% when word order was considered and 61.5% when it was ignored. Next the team trained *naïve Bayes classifiers* to automatically identify the *instructional activities* (e.g., question-and-answer, lecture, small group work) that collectively comprised a majority (76%) of the data based on *discourse timing* (e.g., patterns of speech and rest, length of utterance), *linguistic features* (e.g., parts of speech, identification of question words), and *paralinguistic features* (e.g., volume, pitch contours) extracted from teacher audio (Donnelly et al., 2016). They achieved F_1 *scores* – a measure of accuracy based on precision and recall – ranging from 0.64 to 0.78, which reflect a substantial improvement over chance; including features from the classroom microphone resulted in a further 8% improvement.

The main research task pertained to estimating 'question authenticity'. Authentic questions in this context are open-ended questions without a pre-scripted response (e.g., "Why do you think that?"), whose use is linked to engagement and achievement growth (Nystrand, 1997). Because these questions comprised only 3% of all teacher utterances, the team focused on estimating the 'proportion of authentic questions per class session' rather than on identifying individual authentic questions. They experimented with both closed- and open-vocabulary approaches for this task.

For the *closed-vocabulary approach*, they extended a hand-crafted set of word features and part-of-speech features useful for classifying question types, which included word-level, utterance-level, and discourse-level features obtained through syntactic and discourse parsing. Examples included *named-entity type* (e.g., LOCATION), *question stems* (e.g., "what"), *word position and order* (e.g., whether the named entity PERSON feature occurred at the first word or the last word), and *referential chains* (connections across utterances via pronouns and their referents). Using *M5P regression trees*, they obtained a 0.602 correlation for predicting authenticity compared to human rater classifications; this model was trained in a manner that generalizes to new teachers to prevent overfitting.

For the *open-vocabulary approach*, they derived frequently occurring words and phrases (e.g., unigrams, bigrams, and n-grams) from the transcripts themselves and again trained M5P regression trees with these features. Although this did not result in performance improvements over the previous closed-vocabulary approach ($r = 0.613$ instead of $r = 0.602$), a combined model that averaged predictions from the two approaches yielded an improved correlation of $r = 0.686$. They replicated these results on a large archival database of text transcripts of approximately 25,000 questions from 451 observations of 112 classrooms (Cook, Olney, Kelly, & D'Mello, 2018).

These results confirm the feasibility of fully automated approaches to analyze classroom discourse despite the noisy nature of real-world classroom discourse, which includes conversational speech, multiparty chatter, background noise, and so on. The next step will be to integrate the models into technologies that provide automated assessment to researchers, teachers, teacher educators, and professional development personnel. In doing so, we can finally make classroom discourse, which has largely been ephemeral, more persistent and actionable.

Example 4 – Cognitive Load

Current learning technologies have no direct objective way to assess students' mental effort. Can brain signals tell whether students are in deep thought, struggling to overcome an impasse, or cognitively overloaded? To help answer this question, we used *electroencephalogram* (EEG) techniques to measure the voltage of coordinated *neural firing* that passes through the scalp. Different patterns of neural firing activity can be indicative of distinct cognitive states such as attentional focus, cognitive load, and engagement. EEG is ostensibly the least invasive and most affordable method of accessing brain activity apart from functional near-infrared spectroscopy; yet, it is rarely used in education, ostensibly due to several complexities involved.

As an exploratory project, Mills et al. (2017) developed an EEG-based cognitive load assessment system. *Cognitive load theory* (Paas & Ayres, 2014) suggests that working memory capacity is limited and cognitive load pertains to the consumption of working memory resources. There are different types of cognitive load (e.g., intrinsic, extraneous, germane load) with distinct implications for learning. In general, however, if cognitive load exceeds working memory resources, learning will generally be stifled; thus, management of cognitive load is essential.

The researchers used a headset that fits on the head like a hat (QUASAR USA EEG). It uses *ultra-high impedance dry-electrode technology*, which has been demonstrated to record high-quality EEG without the need for skin preparation of any kind. It was designed to be a light and relatively low-cost unit, specifically developed for ecological data collection. The system is *self-contained*, which means that it includes battery-operated mechanisms for data acquisition, data storage, as well as cable and wireless data output.

The research team trained a machine-learning model called *Qstates*, which use EEG *spectral features* and *partial least squares regression*, to discriminate between high and low cognitive load using data collected in a separate training phase. Some of the training tasks were *domain-specific* (e.g., one vs. three-digit column addition for low- vs. high-cognitive load conditions, respectively) whereas others were *domain-independent* (e.g., easy vs. difficult versions of *Tetris*). EEG patterns are highly variable between individuals and within individuals across time, so the models were not expected to generalize to new students. Instead, personalized cognitive load models were constructed for each student using their respective training data.

The researchers tested the sensitivity of the EEG-based cognitive load detector to experimental manipulations of difficulty embedded in *GuruTutor* from Example 2 above. Difficulty was manipulated in a *common-ground-building* (CGB) instruction phase during which material was initially introduced and explained and a scaffolding phase during which students answered questions about target concepts and received immediate feedback.

They created easy and difficult versions of CGB instruction, ostensibly corresponding to low and high levels of cognitive load, and manipulated the narrativity, syntactic ease, and referential cohesion of the tutor's spoken content; these dimensions are associated with text complexity. As an example, consider the difficult and easy versions of the following sentence: "Once the brain detects increased heat, it instigates the pumping of blood nearer to the skin" [difficult] vs. "The brain will realize that it is too hot. Then it will begin to pump blood up close to the skin" [easy]. The difficulty manipulation for scaffolding was based on the idea that recall is more effortful than recognition. Accordingly, students in the difficult condition received open-ended questions that required them to recall the answer from memory (e.g., "What term is used to describe the organisms place on the food chain?") whereas those in the easy condition received a true–false question (e.g., "True or false: The second trophic level is composed of primary consumers.").

The cognitive load detector was tested on 12 ninth-grade students from a U.S. high school across a 90-minute session. All students were able to wear the headset and reported no major issues during the learning session. Despite the potential problems that could have occurred (e.g., accidental electrode detachment, headset shifts, hardware issues), the researchers were able to collect an average of 83% usable data.

The researchers developed four partial least squares regression models, each developed using various combinations of the training tasks. Whereas all four models could discriminate among the easy vs. difficult conditions in scaffolding conditions, only one of the models was successful in discriminating among CGB instruction conditions. This latter result is unsurprising because the CGB manipulations were much subtler with self-reported difficulty effects of 0.260 standard deviations for CGB instruction compared to 0.716 standard deviations for scaffolding. Importantly, a domain-independent model was sensitive to manipulations of difficulty for scaffolding, but a domain-specific model was needed for the CGB instruction phase. Thus, domain-specific training data may be needed to detect the subtle differences in cognitive load, especially when humans are less metacognitively aware of these differences. Finally, EEG-based cognitive load estimates were negatively correlated with performance during scaffolding.

In sum, these results showed that EEG can be a viable source of data to assess students' cognitive states across a 90-minute session. Despite lacking a multimodal component, the main lesson of this study is that high-fidelity brain signals can be recorded and used for cognitive state assessment outside of a research lab. That said, there is considerably more research and development to be done before this approach can be deployed more widely.

Example 5 – Nonverbal Synchrony during Social Interactions

Twenty-first-century learning is increasingly collaborative and multiple conceptual frameworks of learning maintain the importance of inter-personal skills to meet the demands of the modern workforce. The ability to coordinate and coregulate thoughts, feelings, and behavior to others is a foundational interpersonal skill. Because thoughts and feelings cannot be communicated directly, they are expressed through conscious or unconscious verbal and nonverbal behaviors. Thus, patterns of behavioral *synchrony* are overlaid on a complex substrate of cognitive and affective states, knowledge and beliefs, and communicative goals.

Traditional analytical approaches typically focus on a particular unimodal channel of synchrony computed on a *per-dyad* basis (i.e., for two individuals at a time). For instance, a researcher may examine how smiles become coupled and decoupled throughout an interaction for each dyad and then average them across dyads in the sample. However, focusing on one channel or one dyad at a time and then averaging results across multiple dyads loses the larger empirical picture. First, behavioral communication is inherently complex with multiple channels interacting (e.g., smiles are synchronized with head movements, which are linked to speech patterns and turn-taking dynamics). Second, independently analyzing individual dyads picks up local patterns of synchrony whereas global patterns might be of greater theoretical interest. Thus, analyses of nonverbal synchrony stand to benefit from multiple data channels (multimodality) jointly analyzed across numerous interlocutors (generalizability).

One way to address the dual concerns of multimodality and generalizability is to turn again to data-driven model-based approaches. *Generative models* can be used to learn patterns of nonverbal behavior that drive cross-modal synchrony, predicting behavior of an interlocutor at any given point of the interaction. By incorporating data across numerous interactions, generalized patterns of nonverbal synchrony that extend beyond any single dyad (or triad or any other groups size) may be inferred. Inspecting how and when behaviors are produced should provide insight into proximal phenomena such as moment-to-moment cognitive and affective processes or interpersonal dynamics.

Grafsgaard, Duran, Randall, Tao, and D'Mello (2018) took an initial step in this direction by examining nonverbal synchrony in a study of 29 heterosexual romantic couples in which audio, video, and bracelet accelerometer were recorded during three six-minute conversations. Although this study did not focus on assessments in educational contexts, the basic methodology is broadly applicable. The researchers first extracted 20 facial expressions (e.g., blinks, smiles), five body movement features (e.g., features derived from gestures, head, and whole-body movement), and seven acoustic-prosodic features (e.g., pitch, loudness, jitter) from each partner. These data were used to train neural network models, specifically *long short-term memory recurrent neural networks*, to predict the nonverbal behaviors of one partner from those of the other (i.e., Actual Male Input → Predicted Female Output; Actual Female Input → Predicted Male Output). The models used multimodal speech (both partners), facial expression (one partner), and bodily and gestural motion features (one partner) as input in order to predict facial expression and motion of the other partner. By including both partners' speech as input, the analytic design accounts for variations in nonverbal behavior due to speech production and turn-taking dynamics (e.g., mouth movements occur during speech while facial expressions typically accentuate utterances). The models were trained using couple-independent nested cross-validation, where model parameters and hyperparameters were learned on an independent subset of couples and then applied to the target test couples.

Results indicated that the models learned behaviors encompassing facial responses, speech-related facial movements, and head movement. However, they did not capture fleeting or periodic behaviors such as nodding, head turning, and hand gestures. Importantly, the models learned to predict behavioral responses above simpler *feed-forward neural networks* and various chance baselines. The models also learned patterns of synchrony (i.e., correlating model-predictions of male and female behaviors) that were unique to what was discovered using a purely analytic approach (i.e., correlating observed male and female behaviors), suggesting that the proposed generative approach can yield new insights into the study of nonverbal synchrony.

In terms of assessment, one specific application would be to gauge sociocommunicative skills like *expression mirroring* and *backchanneling* in an interview setting. This could be achieved by contextualizing the model to each assessment by feeding the interviewer's multimodal time series of expressions, intonations, and so on as input and comping the model's predicted time series for an interviewee to the interviewee's actual responses as a measure of expected fit. Large deviations in fit would suggest closer examination by human judges.

Guidelines to Consider

In this chapter, we discussed multimodal assessment systems that go beyond click streams and beyond assessing more "traditional" knowledge, skills, and abilities. Because there is often a misalignment of expectations and some confusion on what multimodality entails, let us end with a few guidelines around multimodal analytics. We have derived these from the theoretical foundation of multimodal communication, the various methods to integrate multimodal signals, the case studies reviewed in this chapter, and our own extended experiences with designing several multimodal analytic systems:

1. **Multimodal Integrity** – It is advisable to maintain the richness of the multimodal experience if it is a core component of a phenomenon being assessed rather than to reduce it into a series of "clicks" or item responses simply because it is "easy" to collect clicks or item responses. For example, an assessment of collaboration devoid of social interaction is arguably measuring a small part of the construct and its validity might be threatened if the construct is inadequately explicated (Shadish, Cook, & Campbell, 2002).

2. **Multimodal Minimalism** – It is advisable to make principled data-collection design decisions that emphasize only the modalities that are "core" components of the phenomenon of interest, rather than to include all multimodal signals merely because it is "easy" to do so. It is easy to get mixed up in the mire of multimodal data collection thereby risking the integrity of the primary modality of interest.

3. **Added Value** – It is advisable to make realistic assumptions about the potentially limited added value of multimodal signals rather than to

assume a priori that the inclusion of multimodal signals will lead to improved or far superior performance above individual modalities. For example, for measuring affect, multimodal approaches result in about a 6.6% improvement in accuracy than the best unimodal signal (D'Mello & Kory, 2015).

4. **Signal Redundancy** – It is advisable to utilize signal redundancy smartly because it can lead to robustness in the assessment model even though it may not lead to improved measurement accuracy per se because one signal can often compensate when another is unavailable. For example, while measuring affect, facial expressions cannot be tracked when the face cannot be detected in the video (e.g., because it is occluded) and are unreliable when the person is speaking. Vocal information can be used to compensate for the information conveyed by the facial expression.

5. **Data Integrity** – It is advisable to utilize missing data – unless it is missing due to technical issues such as sensor failure – since missing data from one modality might indicate an event of interest that may be picked up from the additional modalities. For example, facial occlusion when a person places their hands over their mouth might indicate deep thought, which can be detected by tracking gestures or hand position. This is obviously preferred rather than just discarding the case due to the occlusion.

6. **Theoretical Sophistication** – It is advisable to utilize more sophisticated theoretical models that can account for multimodal complexity because components of individual signals usually interact dynamically, nonlinearly, and across multiple timescales. In the affective sciences, for example, a traditional view of so-called "basic emotions" with well-defined expressive correlates and coherence among different modalities has been replaced with a more complex, nuanced, and contextual perspective where relationships are ill defined and there is loose coupling among modalities (see D'Mello, Kappas, & Gratch, 2018 for discussion).

7. **Computational Sophistication** – It is advisable to invest in advanced computational approaches to integrate multiple modalities in order to address nonlinearity, interactivity, hierarchies, and multiple timescales, all inherent features of multimodal expression.

In addition to the these seven principles that are specific to multimodal analytics, we will add two further principles that are important to the overall sensor-based and machine-learning approaches:

8. **Interpretability / Explainability** – This characteristic of the underlying computational models is typically a key requirement for the use of information for assessment purposes, especially as the stakes for individual stakeholders increase. Statistical models in this area

have always been relatively complex but are nowadays getting even more complex, especially with the general increased interest in deep learning methods. There is no clear answer to the problem of ensuring relative transparency around evidence identification and accumulation through these models, other than to await the fruits of efforts in the field of *explainable artificial intelligence*, which aims to improve the transparency of the models.

9. **Bias and Fairness** – Addressing these aspects of the underlying computational modeling approach is an old but increasingly recognized problem in the field of machine learning. Specifically, biased models that lead to unfair conclusions of particular subgroups of stakeholders arise, for example, when training data are skewed toward a particular subset of the population and/or when the use of discriminatory information in the modeling approach (e.g., race, gender) can lead to model outputs that favor one group over another. Fortunately, increased awareness of this issue has led to efforts on evaluating and mitigating sources of bias and unfairness in machine learning (e.g., Kusner, Loftus, Russell, & Silva, 2017).

In conclusion, even though we are still at a relatively early stage in research around multimodal assessment with many conceptual and technical problems still waiting to be solved, we are hopeful and optimistic that the field of multimodal analytics overall can contribute to the design of reliable, valid, and fair assessments of complex constructs or so-called 21st-century skills, which are challenging to capture using more traditional approaches.

Author Note

This research was supported by the National Science Foundation (NSF DUE-1745442 IIS- 1735785; IIS- 1748739) and the Institute of Educational Sciences (IES R305A170432; R305A130030). Any opinions, findings, and conclusions or recommendations expressed in this material are those of the authors and do not necessarily reflect the views of the funding agencies.

7

International Applications
of Automated Scoring

Mark D. Shermis

CONTENTS

When Page (1966b) prophesied the advent of "automated essay grading" (i.e., *machine scoring* of writing), he viewed the technology as a "teacher's helper" based on his belief that the evaluation of writing was a major impediment for teachers assigning more writing in the classroom; we use the common term *automated essay scoring* (AES) in this chapter. His *Project Essay Grade* (Ajay, Tillet, & Page, 1973) was the first essay scoring engine developed, but it had little practical application since the input mechanisms for computers of the time required either punched cards or magnetic tape. It was not until the late-1990s that an array of machine-scoring engines began to appear, including entries from ETS (Burstein et al., Chapter 18, this handbook; Burstein, Kukich, Wolff, Lu, & Chodorow, 1998; Cahill & Evanini, Chapter 5, this handbook), Pearson Knowledge Technologies (Landauer, Laham, & Foltz, 2001), and Vantage Learning (Elliot, 1999).

In the early 2000s, machines were developed to grade essays in foreign languages, but they used the framework of English grading engines (e.g., JESS; see Ishioka & Kameda, 2004). Vantage Learning's *Intellimetric* was applied to Bahai Malaysian (Vantage Learning, 2002) and Hebrew (Vantage Learning, 2001) with only modest results, though it got much better results with Chinese (Schultz, 2013). In general, the scoring engines that were directly modeled after a parallel machine scoring software package seemed not to have been as effective as the English counterpart.

For the last 10 years, researchers have been developing AES engines that could be used for languages other than English based on native frameworks. In this chapter, we present a survey of AES engines developed for other languages, designed to provide a benchmark of developments in machine scoring. Why is it important to look at work involving machine scoring in other languages? Several reasons emerge as salient. First, the study of machine scoring in other languages may lead to a better specification of conceptual models that take into account discrepancies between English and other languages. Second, work in other languages may help fill in the gaps between what experts say they are looking for in evaluating writing versus what machine-scoring algorithms actually do. Shermis (2018) has referred to this as a "crosswalk" and, based on the specifications of the leading machine scoring vendors, there are domains that are not covered or not covered well. Third, unlike English, there are languages that are governed by academies. It could be the case that there are lessons to be learned for machine scoring with languages that have endorsed rules associated with them. Finally, understanding how machine scoring in foreign languages works may help English-scoring algorithms better serve ESL students.

We review 11 foreign language text-scoring systems, examine the kinds of models that are used, identify whether they employ methods from *natural language processing* (NLP) or not, discuss challenges in parsing and tagging text, and report on evaluation criteria such as reliability and validity information. By having knowledge of work in other languages, the field may be able to address some of the challenges facing machine scoring in English.

Language Differences

In assessing language differences, there are two salient dimensions that may impact how effectively writing in different languages can be evaluated with automated systems. The first dimension pertains to differences between written and spoken language while the second dimension addresses dialectical or regional differences within and across languages; we discuss each of them in the following.

Dimension 1 – Differences between Writing and Speaking

For purposes of this chapter, we use the following definition of 'writing' from *Wikipedia* (https://en.wikipedia.org/wiki/Writing):

> Writing is a medium of human communication that represents language and emotion through the inscription or recording of signs and symbols. In most languages, writing is a complement to speech or spoken language. Writing is not a language but a form of technology that developed as tools developed with human society. Within a language system, writing relies on many of the same structures as speech, such as vocabulary, grammar and semantics, with the added dependency of a system of signs or symbols.

This definition suggests that there is a great overlap between writing and speaking, but that there may be some critical differences. The following is a listing of some of the differences observed between speaking and writing, taken from the *Omniglot* website (http://www.omniglot.com/writing/writingvspeech.htm):

- Writing is usually permanent and written texts cannot usually be changed once they have been printed or written out. Speech is usually transient, unless recorded, and speakers can correct themselves and change their utterances as they go along.

- A written text can communicate across time and space for as long as the particular language and writing system is still understood. Speech is usually used for immediate interactions.

- Written language tends to be more complex and intricate than speech with longer sentences and many subordinate clauses. The punctuation and layout of written texts also have no spoken equivalent. However, some forms of written language, such as instant messages and email, are closer to spoken language. Spoken language tends to be full of repetitions, incomplete sentences, corrections and interruptions, with the exception of formal speeches and other scripted forms of speech such as news reports and scripts for plays and films.

- Writers receive no immediate feedback from their readers, except in computer-based communication. Therefore, they cannot rely on context to clarify things so there is more need to explain things clearly and unambiguously than in speech, except in written correspondence between people who know one another well. Speech is usually a dynamic interaction between two or more people. Context and shared knowledge play a major role, so it is possible to leave much unsaid or indirectly implied.

- Writers can make use of punctuation, headings, layout, colors and other graphical effects in their written texts. Such devices are not available in speech. Speech can use timing, tone, volume, and timbre to add emotional context.

- Written material can be read repeatedly and closely analyzed, and notes can be made on the writing surface. Only recorded speech can be used in this way.

- Some grammatical constructions are only used in writing, as are some kinds of vocabulary such as some complex chemical and legal terms. Some types of vocabulary are used only or mainly in speech such as slang expressions and tags like *y'know, like*, and so on.

Dimension 2 – Differences within and across Languages

Differences within the same language can refer to both spoken and written expressions, which can manifest themselves in a variety of ways. The following are a few select examples:

- **Vocabulary:** Referring to the (North American expression) *hood* of a *station wagon* as the (British expression) *bonnet* of an *estate car*.

- **Pronunciation and Phonology:** If the pronunciation or specific word usage is restricted to a region or social group, it is sometimes referred to as a *dialect*. Examples of regional variations include merging the word *you* with the word *all* to get the expression *y'all* in the State of Texas, which means nothing more than the plural form of *you*. Similarly, in southern Germany, one is likely to be greeted with the expression *Grüß Gott* (may God bless you), whereas in the rest of the country one might be acknowledged by the term *Guten Tag* (Good day!). In the region surrounding Barcelona, Spain, native speakers tend to pronounce the letter "c" (as in Barcelona) with a "th" sound (as in Barthelona).

- **Alphabet:** In German, there are umlauted letters ä, ö, ü, and the double-s, ß. The umlauted letters end up being additional vowels for use in the language. The pronunciation of letters may be different across languages. In Spanish, the vowel "e" sounds like an "a" in English and the letter "i" sounds like an "e" in English.

- **Spelling:** In German, nouns are capitalized whereas in English they are not unless placed at the beginning of a sentence.

- **Grammar:** In German, referring to the past is usually done in the present perfect tense – *I have driven a car* rather than *I drove a car*. One reason German is a difficult language for English speakers to learn is that the verb of the sentence is often at the very end rather than in the middle of sentence.

Parsing and Tagging Text

The differences noted above have significant implications for building AES systems as analyzing text in a foreign language is dependent on being able to parse and tag the text into its various components. Specifically, *parsers* (i.e., algorithms that process written text and segment it into units) have to be able to process the different alphabets used in the language along with any mechanics associated with the language. *Taggers* (i.e., algorithms that assign labels to units) have to be programmed to handle idiomatic expressions. Finally, decisions have to be made about whether dialectical or spelling differences should be corrected before being processed or whether these errors should be incorporated in the model-building process.

Perhaps the best set of statistical parsers resides with the *Stanford Parser* (https://nlp.stanford.edu/software/lex-parser.shtml), which incorporates a series of plug-in modules for a number of different languages. These are freely available through a general public license and could serve as the front-end on scoring engines for those languages; we discuss this option a bit more in the closing section. Most parsers can classify text into over 400 categories (i.e., variables) from which predictions can be made using *statistical models*. Often researchers will use *cluster-analytic methods* to organize the variables to create a representation of a writing domain (e.g., grammar) or use *factor analysis* to similarly identify variables that statistically coalesce. These factors may or may not have a coherency that parallels what the writing community values as important components of the writing process.

For example, Figure 7.1 reflects a mapping of the constructs assessed by *e-rater®* to its component variables. At first glance, the constructs appear to be focused more on the mechanical aspects of writing when compared to common writing traits. However, a closer examination suggests that combinations of the constructs may be good approximations or proxies for underlying valued attributes of writing (Shermis, 2018).

Related to this mapping exercise, some authors have advocated the use of the *Common European Framework of Reference for Languages* (CEFR) (Council of Europe, 2001) as a way to score writing even though it was originally designed for the assessment of spoken language. The CEFR uses range, accuracy, fluency, interaction, and coherence as traits for the assessment of writing communication. If we were to crosswalk these traits with the constructs assessed by e-rater, we might get the following linkages:

- Range: mechanics and style
- Accuracy: grammar, collocations and prepositions, source-use, and vocabulary usage

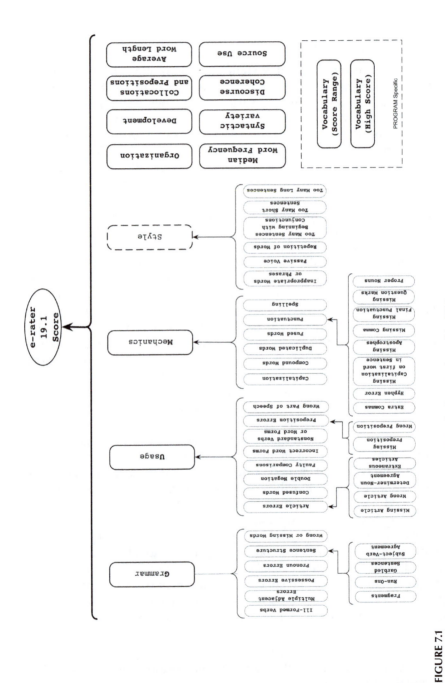

FIGURE 7.1
Organization and construct coverage of e-rater v19.1, which was released in 2019.

- Fluency: syntactic complexity
- Interaction: usage
- Coherence: organization, development, and discourse coherence

Of course, the theoretical constructs do not map perfectly onto the operationalized traits, but the lack of complete overlap suggests either the need for more development on the part of the machine-scoring provider or may lead to further discussion regarding what the traits actually mean.

Automated Scoring Systems for Languages other than English

In the following sections, we summarize a number of machine-scoring engines for languages other than English; note that not all of these are restricted to essay scoring, which is why we do not always use the term 'AES engines'. We have divided this section into two parts: machine-scoring engines that have been modeled after machine-scoring engines in English and those that were developed from scratch based on foreign languages. In this latter category, we further distinguish between scoring engines that do not use NLP as an evaluation component and those that do.

Machine-scoring Engines based on English-scoring Engines

Early efforts to create foreign language machine-scoring engines used intact English versions with a substitution of parsers and taggers. For the most part, these scoring engines showed promise, but the results were not as impressive as they were in their original English versions.

Example 1 – IntelliMetric

The first application of AES using an existing engine was the *IntelliMetric* versions of Chinese, Hebrew, and Bahai Malaysian (Schultz, 2013; Vantage Learning, 2001, 2002). Like other systems, *IntelliMetric* is an intelligent machine-scoring engine that emulates the processes carried out by human scorers. It draws its theoretical underpinnings from a variety of areas, including cognitive processing, artificial intelligence, NLP, and computational linguistics in the process of evaluating the quality of written text (Shermis & Hamner, 2013). The *IntelliMetric* scoring rubric covers five domains (Schultz, 2013):

(1) **Focus and Meaning:** the extent to which the response establishes and maintains a controlling idea (or central idea), an understanding of purpose and audience, and completion of the task;

(2) **Content and Development:** the extent to which the response develops ideas fully and artfully using extensive, specific, accurate, and relevant details (*facts, examples, anecdotes, details, opinions, statistics, reasons,* and/or *explanations*);

(3) **Organization:** the extent to which the response demonstrates a unified structure, direction, and unity, paragraphing and transitional devices;

(4) **Language Use, Voice, and Style:** the extent to which the response demonstrates an awareness of audience and purpose through effective sentence structure, sentence variety, word choice that create tone and voice;

(5) **Mechanics and Conventions:** the extent the response demonstrates control of conventions, including paragraphing, grammar, punctuation, and spelling.

In the references cited above, *IntelliMetric* used a native language parser but employed the standard *IntelliMetric* engine to analyze the essays. The researchers used modestly sized samples of approximately 600 essays, two-thirds of which were randomly selected for modeling and one-third was used for *validation*. These small sample sizes may have restricted the range of human scored essays, may have limited the types of essays that the machine-scoring algorithms were likely to encounter, or may have narrowed the vocabulary that the algorithms encountered.

Chinese essays were modeled best with *Pearson product-moment correlations* between human and machine-ratings at about $r = 0.86$ while Hebrew and Bahasa Malay essays had lower human–machine correlations in the mid-seventies. However, it should be noted that these early studies were conducted over 15 years ago. Vantage Learning, the developer of *IntelliMetric*, currently claims that the technology can be used to score essays in 20 different languages but does not list what these are.

Example 2 – JESS

One of the early non-English systems is the *Japanese Essay Scoring System* (JESS), which was modeled after the *e-rater* engine and incorporates predictors from three major domains (Deerwester, Dumais, Furnas, Landauer, & Harshman, 1990):

(1) **Rhetoric:** ease of reading, diversity of vocabulary, percentage of big words (long, difficult words), and percentage of passive sentences;

(2) **Organization:** routines that determine the logical structure of a document by detecting the occurrence of conjunctive relationships, including additions, explanations, demonstrations, illustrations, transitions, restrictions, concessions, and contrasts;

(3) **Content:** overlap of content in an essay relative to the essay prompts as determined via latent semantic indexing.

Using translated essays that had previously been graded by *e-rater, JESS* was able to obtain acceptable correlations with the *e-rater* score. One big difference from *e-rater* was that the statistical model created for *JESS* was created from the professional writings of the *Mainchi Daily Newspaper.* The developers concluded that *JESS* was able to generate valid score predictions for essays in the range of 800 to 1,600 characters.

Machine-scoring Engines not based on English-Scoring Engines

Machine-scoring engines not based on existing English-scoring engines fall into one of four categories: those that have a statistical component, those that use *latent semantic analysis* (LSA) or some variant of it, those that have an NLP component, and those that are combinations of the above. The most predominant category is the last one, specifically a combination of a statistical model based on the parser's categories (i.e., variables) and LSA (or a variant of LSA). However, most of the foreign language systems are not actually newly built scoring systems per se, but, rather, collections of scoring tools that were pieced together manually.

Example 1 – Chinese

Chang and Lee (2009) proposed a method to automate the scoring of Chinese essays based on conceptual connections among paragraphs. Their argument is that essay grading is dependent on the order of concepts in a paragraph since rhetorical skills are reflected in the order with which the same concepts appear in the essay. They recommend the use of similarity of the connections of paragraph concepts for each document to predict essay scores.

In their system, a test essay is evaluated by searching the training corpus for essays with a similar paragraph structure. The score of the test essay is the average of the given scores of the top *n* similar training essays. Chang and Lee (2009) obtained exact and adjacent rates of agreement of 37% and 84%, respectively. The authors concluded that using paragraph structure was an efficient method for scoring essays based on content.

Example 2 – Filipino (Tagalog)

In this scoring engine, statistical prediction along with *concept indexing,* which is a variant of LSA, is used to analyze and score the content of high-school Filipino essays (Ong, Razon, Perigrino, & Guevara, 2011). For the study cited here, 150 high-school essays were written in Tagalog and scored by an expert for content, grammar, and organization. The authors were able to get acceptable results and found that some common adjustments to their analysis produced improved predictions. For example, correcting spelling errors improved reduced errors of prediction by 8% while removing *stop words* resulted in an improvement of 37%; the exact and exact + adjacent *agreement rates* were 45% and 86%, respectively.

Example 3 – Finnish

Kakkonen and Sutinen (2004) utilized LSA for evaluating Finnish classroom essays. They compared the conceptual similarity between the essays and textbook materials covering the subject matter and applied pre-processing techniques such as *stemming, term weighting,* and the use of the *FINCG Constraint Grammar* parser for Finnish to perform a *morphological analysis* of the essays.

The grading process included four steps. First, a comparison of an essay to the textbook passages covering the assignment-specific knowledge was performed; second, a computation for the *cosine similarity* between the pre-graded essays and the textbook passages was performed; third, values for each score category were then defined after which the cosine similarity between the test essays and textbook passages were computed; and, finally, these values were compared to the cutoffs for the score categories and the corresponding scores were assigned.

The researchers performed this analysis on 143 essays using three evaluation statistics: exact agreement, exact + adjacent agreement, and the *Spearman rank-order correlation coefficient,* which yielded values that ranged from 35.6% to 39.7% for exact agreement, 77.2% to 84.9% for exact + adjacent agreement, and 0.78 to 0.82 for the Spearman rank-order correlation coefficient, respectively. The authors concluded that LSA can be successfully applied to Finnish essays and grading using course materials can provide accurate results.

Example 4 – German

Wild, Stahl, Stermsek, Penya, and Neumann (2005) evaluated responses from a student marketing question written in German to demonstrate that a scoring approach using LSA with a corpus built on a marketing glossary of 302 terms could be effective. The data consisted of 43 responses that were modeled using three "golden answers" (e.g., a model/gold standard based on attributes of writing not human ratings). The document pre-processing steps included the removal of stop words and stemming.

Stop word removal provided the best results although it was found to be slightly further improved with stemming; however, the performance degraded with stemming alone (Zupanc & Bosnic, 2015). To weight the predictors, the authors tested three *local weightings* (i.e., raw-term frequency, logarithm, and binary) and three *global weightings* (i.e., normalization, inverse document frequency, and entropy). Local weightings did not yield significant results; rather, the weighting strategy of using inverse document frequency had the best result for the global strategy, which was improved when combined with raw-term frequency or logarithm normalization.

Dimensionality of the scoring solution was evaluated using four methods to determine the number of reliably extractable factors where methods using normalized and cumulated *singular values* (i.e., summing singular values until a threshold value is reached) provided the best result. Lastly, the authors used Pearson's correlation coefficient, Spearman's rank-order correlation coefficient, and cosine similarity measures to evaluate the performance of the system and found that Spearman's rank-order correlation coefficient provided the best results (Zupanc & Bosnic, 2015).

Example 5 – Thai

An unnamed automated scoring engine exists in Thai, in which a *backpropagation neural network* and LSA were used to assess the quality of Thai-language essays written by high-school students in the subject matter of classic Thai literature (Loraksa & Peachavanish, 2007). Specifically, 40 essays written were each evaluated by high-school teachers and assigned a human score. In the first experiment, raw-term frequency vectors of the essays were used along with their corresponding human scores to train the neural network and obtain the machine scores. In the second experiment, the raw-term frequency vectors were pre-processed using LSA prior to feeding them to the neural network. The first experiment yielded a mean error and standard deviation of errors of 0.80 and 0.83, respectively, while the second experiment yielded values of 0.32 and 0.44, respectively; therefore, the experimental results showed that the addition of LSA technique improved scoring performance.

Foreign Language Engines that Use NLP

Example 1 – French

There is a scoring engine that was developed for the automated assessment of second-language French texts with the potential for applications to be developed for a French *portfolio system* (Parslow, 2015). An analysis was carried out in the error-annotated *FipsOrtho* corpus, which is based on second-language exemplars. The analysis suggested that NLP tools designed principally for use with formal first-language French were surprisingly robust when applied to second-language productions. What was particularly interesting was that the study was predicting to CEFR language categories which range from 'A1' (lowest proficiency) to 'C2' (highest proficiency). The system uses over 50 possible metrics of vocabulary, corrections, verb use, clause use, syntax, and cohesion based on the CEFLE corpus. Confusion matrix (correct classification) scores were reported as 0.67, 0.55, 0.52, 0.74, and 0.86 for levels A1, A2, B1, B2, and a collapsed version of C1 and C2 (referred to as "native" in the study), respectively. Based on this initial study the authors referred to the results as "promising."

Example 2 – Hebrew

The *NiteRater* (2007) system is an AES engine that extracts about 130 quantified linguistic features from a given text, including statistical, morphological, lexical, syntactic, and discourse features. The features are arranged in three main clusters with 61 specific features, 38 combined features, and 15 main factors. The engine builds a prediction model using user determined feature-clusters, training sample size, and characteristics, which is typically based on stepwise multiple linear regression. The features used in the final-scoring model are those that contribute significantly to the prediction with the weights for the features derived empirically. Machine–human agreements rates were in the low mid-70s.

Example 3 – Korean

The *Korean Automatic Scoring System* (KASS) was developed to score short-answer questions in a *National Achievement Test* (NAEA) for public school students (Jang et al., 2014). NLP techniques such as morphological analysis were used to build a token-based scoring template for evaluating the responses. In one study, 21,000 responses across seven prompts from the 2012 NAEA were evaluated using the new system. Results showed about 90% to 95% of the student responses could be processed with an agreement rate of 95% with manual scoring.

In the introduction KASS system, Jang et al. (2014) refer to other work going on in Korea involving statistical and LSA techniques. For example, Jung, Choi, and Shim (2009) used cosine similarity measures between each student response and a given model answer to determine whether responses are correct by matching critical keywords between the response and the question. Similarly, Park and Kang (2011) explored open-ended questions in four categories of responses: short answer, fill-in-the-blank answer, single sentence answer, and multi-sentence answer. The scoring system used classifying questions, exact matching, applying partial credits, critical keywords, and heuristic similarity for making score predictions (Kang, 2011).

Finally, Kang (2011) classified questions into six categories and the system processes them with the targeted methods for each specific category and used morphological analysis, similarity with heuristics, and score calculator. A competing system used techniques from information retrieval with semantic kernels, vector space modeling, and LSA; experiments with actual exam papers showed that accuracy of the system was about 80% (Cho, Oh, Lee, & Kim, 2005; Oh, Cho, Kim, & Lee, 2005).

Example 4 – Swedish

An unnamed system was developed for the machine scoring of high-school essays written in Swedish (Östling & Smolentzov, 2013). The system uses standard text quality indicators and is able to compare vocabulary and grammar to large reference corpora of blog posts and news chapter articles. The system consists of tools to evaluate *simple features* (i.e., text length, average word length, OVIX lexical diversity, parts of speech distribution), *corpus-induced features* (i.e., part-of-speech tag cross-entropy, vocabulary cross-entropy, hybrid *n*-gram cross-entropy), and *language error features* (i.e., spelling errors, split compound errors). The system was evaluated on a corpus of 1,702 essays, each graded independently by the student's own teacher with a blind back-reading by another teacher. The system's performance was fair (machine–human agreement was 58%) and subsequent work using machine learning techniques was done to improve it with modest results (Smolentzov, 2013).

Summary

In this section, we briefly reflect on the state-of-the-art of AES in languages other than English based on our previous discussions and propose some recommendations for practitioners who want to explore this space and learn how to build their own systems; we have summarized key features of all engines in Table 7.1.

Current State-of-the-Art

Many of the foreign-language scoring engines are, at this point, simply proofs-of-concept and are typically a combination of a statistical model used in conjunction with LSA or some variant of it. They usually employ a weighted linear combination of a subset of parser variables in addition to a distance metric derived from the LSA component. In other words, foreign-language scoring engines are generally an assembly of tools rather than a complete system. As such, they typically require manual handling from one step of the process to the next rather than as an embedded set of linked steps. However, they tend to work surprisingly well in many cases (i.e., with *quadratic-weighted Kappas* sometimes in the low 80s) given that training sets are often small or that they employ corpora that are only tangentially related to the essay topic.

Because of the programming-intensive nature of NLP systems, they are often thought to be higher up on the machine scoring evolutionary scale. That is, NLP systems are able to capture well-defined constructs like grammar for which rules can be programmed. Scoring engines with NLP can identify grammatical errors and create a ratio of the number of errors to the amount of text, for instance, which may serve as a proxy for one aspect of good (or poor) writing. If the NLP programming follows a well-described and carefully operationalized construct model such as the one in Figure 7.1, it is possible to combine predictor variables in a way that mirrors a rubric or

TABLE 7.1

Comparison of AES Systems*

AES System	Developer	Technique	Main Focus	Instructional Application	Number of Essays Required for Training	Language
AutoScore	American Institutes for Research	NLP	Style and content	N/A	Unavailable	English
Bookette	CTB McGraw-Hill	NLP	Style and content	Writing Roadmap™	250–500	English
CRASE™	Pacific Metrics	NLP	Style and content	N/A	100 per score point	English
e-rater®	Educational Testing Service	NLP	Style and content	Criterion™	100–1,000	English
Intelligent Essay Assessor (IEA)™	Pearson Knowledge Technologies	LSA	Content	N/A	100–300	English
Intellimetric™	Vantage Learning	NLP	Style and content	MY Access!®	300	English Chinese Hebrew Bahasa Malay
Lexile® Writing Analyzer	MetaMetrics	NLP	Style and content	N/A	None – uses a fixed model	English
LightSIDE	Teledia Laboratory	Statistical	Content	N/A	300	English
Project Essay Grade (PEG)™	Measurement Incorporated	Statistical	Style	N/A	100–400	English
JESS (Japanese Essay Scoring System)	National Center for University Entrance Exams (Japan)	NLP, LSI	Style and content	N/A	External database (Newspaper)	Japanese
NiteRater	National Institute for Testing and Evaluation (Israel)	Statistical, NLP	Style	N/A	~300	Hebrew
Unnamed (Arabic)	Department of Computer Sciences, Yarmouk University (Jordan)	Statistical, Cosine Similarity	Content	N/A	Model answer	Arabic

(Continued)

TABLE 7.1 (CONTINUED)

Comparison of AES Systems*

AES System	Developer	Technique	Main Focus	Instructional Application	Number of Essays Required for Training	Language
Filipino Essay Grader	Department of Computer Science, UP Diliman (The Philippines)	Statistical, Concept Indexing	Content	N/A	150	Tagalog
Automated Essay Grading with Latent Semantic Analysis	Department of Information Systems and New Media, Vienna University (Austria)	Statistical, LSA	Content	N/A	Model answer	German
Unnamed (Finnish)	Department of Computer Science, University of Joensuu (Finland)	Statistical, LSA	Content	N/A	143, External database	Finnish
Unnamed (Thai)		Artificial Neural Network, LSA	Content	N/A	40	Thai
Unnamed (Chinese)	Department of Computer Science and Information Engineering, National Kaohsiung University (Taiwan)	Statistical, R-chain Similarity (paragraph)	Content	N/A	574	Chinese
Unnamed (Chinese)	Digital Content Technology Research Center, Institute of Automation (China)	Latent-semantic Vector Space Models	Content	N/A	970	Chinese
Unnamed (Korean)	Korea Maritime and Ocean University (Korea)	Statistical	Content (short answer)	N/A	Model answer	Korean
KASS (Korean Automatic Scoring System)	School of Computer Science, Kookmin University (Korea)	Statistical	Content (short answer)	N/A	Model answer	Korean

(Continued)

TABLE 7.1 (CONTINUED)

Comparison of AES Systems*

AES System	Developer	Technique	Main Focus	Instructional Application	Number of Essays Required for Training	Language
Unnamed (Swedish)	Institute for Linguistics, University of Stockholm (Sweden)	Statistical, NLP	Content	N/A	External database, 600 essays	Swedish
Unnamed–a collection of tools (French)	University of Paris (France)	Statistical, NLP	Content	N/A	External database	French
Unnamed–a collection of tools (English and French)	Center for Research in Applied Measurement and Evaluation, University of Alberta (Canada)	Statistical	Content	N/A	Not reported	English French
Unnamed (Arabic)	Computer Teacher Preparation Department, Mansoura University (Egypt)	Statistical, LSA	Content	N/A	External database	Arabic
Freeling – a collection of tools (Spanish and Basque)	Development Center of Applications, Technologies and Systems (Cuba)	Statistical	Style	N/A	30	Spanish Basque
Unnamed (Chinese)	National Engineer Research Center for E-Learning, Central China Normal University (China)	Statistical, LSA (Coh-Metrix)	Content and style	N/A	Not reported	Chinese
ReaderBench (Dutch)	University Politehnica of Bucharest (Romania)	Statistical, LSA	Content	NA	External database	Dutch

*Based on Shermis (2014).

a set of articulated writing guidelines in addition to the predictive power of some parser variables that may not have good writing trait proxies.

We reviewed four systems in this chapter that explicitly used NLP. In the unnamed Swedish system, the machine–human agreement percentages were in the high 50s – a marginal result – but as good as what was obtained by human raters for the demonstration set. The unnamed French system performed slightly better, providing machine–human agreement values in the high 60s, and showed promising results. The Hebrew *NiteRater* system had machine–human agreements in the low 70s. The best of the reported systems was with the *IntelliMetric* system for Chinese, which produced machine–human agreement values in the mid-80s. Put differently, when machine–human agreements are used as the sole criterion, it appears as if NLP-only scoring systems fared the worst, followed by statistical models only, followed by statistical models with LSA; the best combination appeared to be statistical models with LSA and NLP for the language considered here.

Only a handful of the systems employed NLP as a component of their scoring models. These typically employed one of three strategies to make predictions:

(1) Complete an error analysis, adjust for text length, and then pattern match against essays with similar error profiles;

(2) Weight components of the NLP system by regressing them against a sample of training essays and then making predications based on a linear weighted multiple regression;

(3) Use the NLP-weighted component in conjunction with parser (meta) variables and an LSA-like component.

This last option generally produced the best results and lends itself to theoretical modeling, which can be compared against traits of good writing.

One of the events that helped propel machine scoring in English was the 2012 *Hewlett Trials* sponsored by the Hewlett Foundation. A total of 159 teams participated in the public competition and most of them used parser variables for modeling writing and some variant of LSA for addressing the content of the essays. None of them employed any NLP programming to address the mechanical aspects of writing. The differences among the competitors' models centered around the learning algorithms used to optimize their models. The conclusion at the time was that NLP was good for helping to create constructs that approximate writing traits and are essential for providing feedback in formative writing, but really did not add much predictive power beyond parser variables and/or metavariables. This seemed also to be the case for studies that examined machine scoring in languages other than English.

In addition, none of the studies in foreign languages made reference to their governing academies or bodies as potential sources of input either for creating NLP components for models or for using parser variables to approximate rules that may govern the language. Since all the languages used in this study were governed by some official body except for English, this result is surprising since deviations from rule-based algorithms can be a rich source of prognostic information.

None of the systems that were developed addressed cross-language problems since they were focused on only one language. Consequently, some of the issues raised at the beginning of this chapter cannot directly be addressed. For example, all the systems used training essays as the sole criterion for model building. By using an empirical corpus, the issue of dialects is normed into the model and its impact cannot be teased out. The same would also be true for the impact of specific vocabulary and unique letters processed by the foreign language parser.

Best Practices

As the examples in this chapter have underscored, if one is interested in developing machine-scoring systems in a foreign language, one needs to start with a good parser and tagger and build a system based on the output from these systems. In the following we describe this work in four steps or layers or increasing complexity.

Step 1 – Parsing and Tagging

There are various effective parsers and taggers nowadays available on the market. As we noted earlier, at the time of this writing, the *Stanford Parser* has modules in Arabic, Chinese, French, German, and Spanish as well as English, which can be used under a general use license. Google has parsers in English, Spanish, Japanese, Chinese (simplified and traditional), French, German, Italian, Korean, and Portuguese, and uses a volume-based pricing scheme for use. Microsoft text analytics also supports parsing and tagging in 18 different languages using again a volume-based licensing pricing scheme.

Step 2 – Systems without NLP Features or Content Scoring

Once the parser has done its work, the raw data can be fed into a program such as *LightSide* through a customized plug-in for modeling. *LightSide* is a public-domain AES engine developed at Carnegie Mellon University, which can create prediction models but neither comes natively with NLP capability nor evaluates content per se (Mayfield & Rosé, 2013). However, it creates prediction models based on the relationship between the parser variables and human-rated essays.

Step 3 – Systems with Content Scoring

If one wants to add variables that capture information about the evaluation of content into the mix of predictors, one can perform a vector space analysis (e.g., LSA or variants) by employing software that will create a vector space based on training samples or publicly available corpora. This can be tricky, however, because most software packages are written for English comparisons but there are select packages that can process other languages as well. For example, *Gallito 2.0* (http://www.elsemantico.es/gallito20/index-eng. html) is a web-based service that will make vector space calculations in multiple languages to generate indices of similarity that can be incorporated in prediction models.

Step 4 – Prediction Models with NLP Features

If one wants to add some basic NLP capability to a prediction model, one can download the *Stanford CoreNLP* package (https://stanfordnlp.github. io/CoreNLP/download.html) and process texts in Arabic, Chinese, English, French, German, and Spanish. Note, however, that some of the features of this package may overlap with the output of the foreign language parser, especially if the Stanford parser was used.

Author Note

Thanks to editors for their careful review of previous drafts of this chapter. Correspondence with the author can be addressed to Mark Shermis, 14 South Braided Branch Drive, The Woodlands, TX 77375. E-mail can be sent to mshermis@gmail.com.

8

Public Perception and Communication around Automated Essay Scoring

Scott W. Wood

CONTENTS

For *automated essay scoring* (AES) professionals concerned with the computational aspects of this work, most daily activities involve the nuts and bolts of the *automated scoring* (AS) engine – working with *data, natural language processing, statistical modeling,* and *machine learning,* as well as writing technical reports that summarize the capabilities of AS engines. Frequently, however, there is a need to adequately communicate to the public how AES works and how AES is used in assessment, which is an area of importance for AES professionals to understand as well.

To illustrate the risks associated with such communicative breakdowns, consider the case of Australia's *National Assessment Program – Literacy and Numeracy* (NAPLAN) Writing assessment in late-2017. Despite evidence presented by the Australian Curriculum, Assessment, and Reporting Authority that supported AES on this assessment, a campaign against the use of AES gained traction within teachers' unions, leading education ministers around the country to halt the use of AES for the May 2018 administration (Koziol, Singhal, & Cook, 2018; Robinson, 2018). Education Minister Simon Birmingham stated, "It's clear more work would need to be done to reassure families about its value," referring to AES (Koziol, Singhal, & Cook, 2018).

Criticism toward educational assessment has been commonplace in recent years, especially in the United States. The Common Core State Standards, which many states had adopted at the start of the 2010s, met political resistance in several states due to their complexity and federal over-reach

(Ujifusa, 2014). Many states left the two major assessment consortia, *Smarter Balanced* and *PARCC*, due to public pressure (Ujifusa, 2015). Parents participated in the "Opt Out" movement, keeping their children home on testing dates (Evans & Saultz, 2015). Grassroot movements stood up to the alleged barrage of year-round testing (Johnson, 2015).

These criticisms remind me of the closing keynote speech from the 2016 meeting of the Association of Test Publishers, a "nonprofit organization representing providers of tests and assessment tools" committed to "advanc[ing] the position of the industry, its technology, and the science that supports it" (Association of Test Publishers, 2017). In this keynote, clips from popular media such as *The Simpsons* (Meyer, 1992) and *Last Week Tonight with John Oliver* (Carvell, Gurewitch, & Oliver, 2015) illustrated how assessment professionals are responsible for "leading the conversation" through good communication with the public. *The Simpsons* used satire to provide a lighthearted look at how assessments are administered, scored, and reported. *Last Week Tonight with John Oliver* used satire to provide a more critical look at how assessments are administered in U.S. schools. The 13.4 million YouTube views of Oliver's program (as of May 1, 2019) represent a large audience receptive to his criticisms. Through our knowledge and expertise as testing professionals, we can address the negative attitudes toward, and reveal the utility of, assessment (Allen, D'Astoflo, Eatchel, Sarathy, & Schoenig, 2016).

Thinking more deeply about that closing keynote and the news of AES being suspended on the NAPLAN writing assessment, it is clear that AES is not immune to criticism and that we, as AES professionals, have our own role in leading the conversation. By understanding past and current concerns about AES, professionals can take steps toward better communication with the public, building trust with assessment stakeholders.

In this chapter we first provide a history of AES criticism, which began to form in the 1990s and continues today. After reviewing the history, we review common concerns from the public around AES before providing suggestions for best practices around communicating AES information to stakeholders. Even though the focus of this chapter is on AES in large-scale *summative assessments*, we include a discussion of AS for *formative assessment* at the end of the chapter.

A History of AES Criticism

To understand public concerns about AES, one must understand the history of such criticism; Figure 8.1 displays a timeline of the major events that helped to shape AES criticism.

Ellis Page invented the first AES engine, *Project Essay Grade®*, in the 1960s (Page, 2003). For several decades thereafter, AES occupied a niche in

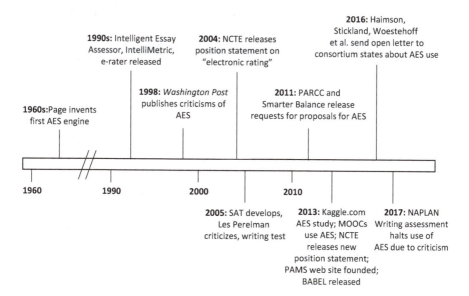

FIGURE 8.1
Timeline of major events that shaped automated essay scoring criticisms.

educational testing. Computing technology was not as ubiquitous as it is today, and laws mandating annual standardized testing were not as prevalent. AES simply was not widespread enough to warrant concern.

Starting in the early 1990s, however, interest in AES began to increase as affordable personal computer technology became more common. Pearson's *Intelligent Essay Assessor*, Vantage Learning's *IntelliMetric®*, and ETS' *e-rater®* were released during the mid- to late-1990s. During this time, the first concerns toward AES were published in *The Washington Post* (Perlstein, 1998; Schwartz, 1998). Editorial writer Amy Schwartz pondered whether AES represented "the cynical abandonment of any pretense that an essay is an exercise in communication" and if AES was a "return to the old ideals of drillwork" (Schwartz, 1998). Perlstein interviewed an English professor, who lamented that AES will keep writers from "weaving thoughtful arguments into compelling prose" (Perlstein, 1998).

In 2004, the *National Council of Teachers of English* (NCTE) released a position statement on "electronic rating" at their *Conference on College Composition and Communication* (CCCC). This is likely the first position statement on AES by an academic organization:

> Because all writing is social, all writing should have human readers, regardless of the purpose of the writing. Assessment of writing that is scored by human readers can take time; machine-reading of placement writing gives quick, almost-instantaneous scoring and thus helps provide the kind of quick assessment that helps facilitate college orientation and registration procedures as well as exit assessments.

The speed of machine-scoring is offset by a number of disadvantages. Writing-to-a-machine violates the essentially social nature of writing: we write to others for social purposes. If a student's first writing-experience at an institution is writing to a machine, for instance, this sends a message: writing at this institution is not valued as human communication – and this in turn reduces the validity of the assessment. Further, since we cannot know the criteria by which the computer scores the writing, we cannot know whether particular kinds of bias may have been built into the scoring. And finally, if high schools see themselves as preparing students for college writing, and if college writing becomes to any degree machine-scored, high schools will begin to prepare their students to write for machines.

We understand that machine-scoring programs are under consideration not just for the scoring of placement tests, but for responding to student writing in writing centers and as exit tests. *We oppose the use of machine-scored writing in the assessment of writing.*

(CCCC, 2004; emphasis mine)

In this statement, NCTE formalizes the concerns first identified by Perlstein and Schwartz. Schwartz's concerns about the loss of the essay as "an exercise in communication" is reflected in NCTE's statements that "all writing is social" and "writing-to-a-machine violates the essentially social nature of writing." Schwartz's concerns about "drillwork" are also captured in the position statement: "[I]f college writing becomes to any degree machine-scored, high schools will begin to prepare their students to write for machines."

Through the first decade of the 2000s, criticism to AES was present though minimal. The few public-facing news articles about AES tended to be promotional, often public-relations stories from assessment companies touting their newest engine features. But in late 2010, the assessment landscape began to shift in ways that opened criticism of educational testing and AES throughout the decade. Federal grants are awarded to two new nationwide testing consortia: *Smarter Balanced* and *PARCC*. In a blog post for *The Huffington Post*, author and former scoring professional Todd Farley predicts the role of AES for the two consortia (2011).

Given the volume and complexity of free-response items from the consortia, Farley sees items being scored by either classroom teachers or automated scoring. Farley recognizes that algorithms can score free-response items "without being able to actually read them" and that the algorithms have been "proven statistically" to score as accurately as temporary human raters. Farley's predictions around AES use were confirmed when requests for proposals were released from both consortia at the end of 2011, and requests for AES appeared in both.

In 2012, the data science competition website Kaggle.com hosted an AS competition sponsored by the Hewlett Foundation (Kaggle, 2012). Called the *Automated Student Assessment Prize*, phase one of this competition allowed the public to develop AES engines by training them on a set of responses from eight essay prompts alongside the established AES vendors. Results from the established vendors were published by Mark Shermis and Ben Hamner (2013), who concluded that proprietary AES engines performed as well on the eight prompts as human raters. The findings were reported on positively by major news outlets, including *USA Today* (Toppo, 2012), the *Financial Times* (Mishkin, 2012), and *Education Week* (Quillen, 2012). The findings were also published on the popular blog *Inside Higher Ed* (Kolowich, 2012), referring to the engines as "robo-graders."

The *Inside Higher Ed* post used quotes from Shermis and from Les Perelman, an academic at the Massachusetts Institute of Technology. Les Perelman has a long and distinguished career in education, having taught collegiate writing courses and directing writing centers over three decades. His name became familiar to psychometricians when he fought changes to the SAT in 2005. Early in 2005, the SAT was revised to include an essay component to be written within a 25-minute testing window. Perelman heavily criticized this, stating that the time period for writing was too short, that the task was too artificial to produce any meaningful writing, and that the scores on the essay were too highly correlated with essay length. David Coleman, the new president of the College Board in 2012, was receptive to Perelman's concerns, and the SAT Writing section was overhauled in early 2016. Coleman credits both Perelman and the stakeholders of the SAT for the changes (Balf, 2014; Weiss, 2014). Given Perelman's fight to eliminate what he perceived as an unfair writing assessment, it is no surprise that his opinions on AES would be elicited as a counterpoint to Shermis.

Also, in 2013, *massive open on-line courses* (MOOCs) became popular, many of which rely on AES to evaluate student assignments. For many people, MOOCs provided their first interaction with an AES engine, increasing visibility of the technology. Eveleth (2013) describes the phenomena of MOOCs and AES for *The Smithsonian*, asking "Can a computer really grade an essay?" NCTE revised their 2004 position statement on AES at the 2013 CCCC, enumerating nine reasons why AES should not be used (NCTE, 2013).

Several of NCTE's criticisms in 2013 match those from their original 2004 position such as the social nature of writing and concerns about "teaching to the machine." Other criticisms update their 2004 positions. For example, concerns about not knowing the criteria for scoring are now replaced with concerns about an engine's inability to handle complex writing features and the use of "surface" features as scoring criteria. New criticisms include the lack of innovative and creative prompt development for writing instruction, the training of human raters to score like computers, the undesired influence of essay length on scoring accuracy, the lack of access to technology, and the possibility of gaming an engine into assigning high marks to a poor essay.

Around this time, a website went on-line which petitioned teachers and academics not to use AES on high-stakes assessments. Called *Professionals Against Machine Scoring of Student Essays in High-Stakes Assessment* (PAMS), the website cites several reasons not to use AES much like the 2013 NCTE position statement (Blalock, 2013). Visitors to the site can electronically sign an online petition to stop government officials, assessment directors, universities, schools, and teachers from using AES. As of February 20, 2019, the petition has over 4,300 signatures.

The PAMS website references criticisms cited in the 2004 and 2013 NCTE position statements and Perelman's concerns about the SAT writing assessment. Common concerns include the suitability of the features used in scoring models for proper construct representation, the lack of score reliability of long essays, short time frames for writing assessments, scores being correlated with length, narrowness of scope for writing prompts used by AES, training human raters to score like machines, score bias, teaching to the machine, the black box nature of scoring engines, and gaming the engine to yield high scores on poor essays. New concerns on the PAMS website include the use of "scoring scales of extreme simplicity," machine scoring bias against specific *subgroups*, machine scores predicting future academic success poorly, and the inability for an engine to measure authentic audience awareness. The PAMS website provides research citations to support each criticism, allowing AES professionals to better understand the evidence behind the criticisms.

In late 2013, Les Perelman illustrated his criticisms about AES in a new way. Along with three undergraduate students, Perelman released the *Basic Automated BS Essay Language Generator* (BABEL), a web-based software program that can produce meaningless, gibberish, counterfactual essays that are syntactically accurate, which the software creators predict will be scored highly by AES engines (Sobel, Beckman, Jiang, & Perelman, n.d.). Freely available on Perelman's personal website, the user enters three topic key words as input and the software returns a poorly written essay that would supposedly receive high scores. One such essay produced in BABEL included the phrases "Mankind will always feign president," "Capitalist economy, frequently at the field of semantics, advocates," and "According to professor of semantics the Reverend Dr. Martin Luther King, Jr., society will always regret patriot." While individual sentences make no sense grammatically or factually, the use of sophisticated words like "mankind," "feign," "semantics," and "advocates" is believed to compensate for the poor grammar in AES. BABEL was reported in media outlets including the *Chronicle of Higher Education* (Kolowich, 2014), *Business Insider* (Eadicicco, 2014), and the *Huffington Post UK* (*Huffington Post UK*, 2014).

Social media allows for the rapid spread of news and information to personal blogs, Twitter feeds, Facebook walls, and more. As a testament to the way that criticisms of AES have spread across the Internet, there are many mainstream blogging sites dedicated to this issue nowadays like Salon.com ("Computer grading will destroy our schools" [Winterhalter, 2013]) and *The Atlantic* ("What Would Mark Twain Have Thought of Common Core Testing?" [Perrin, 2014]), which sometimes include posts by major figures in education like Diane Ravitch ("Why Computers Should Not Grade Student

Essays" [Ravitch, 2014]) and personal educational bloggers with post titles both fair ("What Automated Essay Grading Says to Children" [Hunt, 2012], "The Pros(e) and Cons of Computers Grading Essays" ["Steve", 2013]) and sensationalized ("Beware the Robo-Grader" [Vander Hart, 2014], "The Fraud of Computer Scoring on the Common Core Exams" [Haimson, 2016]).

In April 2016, concerned citizens and members of the *Parent Coalition for Student Privacy, Parents Across America, Network for Public Education, FairTest, Save Our Schools New Jersey,* and other parent groups wrote an open letter to the Education Commissioners of states participating in the *PARCC* and *Smarter Balanced* assessments (Haimson, Stickland, Woestehoff, Burris, & Neill, 2016; Strauss, 2016). In their letter, they explain their concerns about AES use on these assessments and ask the following questions to the commissioners: What percentage of the ELA exams in our state are being scored by machines this year, and how many of these exams will then be re-scored by a human being? What happens if the machine score varies significantly from the score given by the human being? Will parents have the opportunity to learn whether their children's ELA exam was scored by a human being or a machine? Will you provide the "proof of concept" or efficacy studies promised months ago by Pearson in the case of *PARCC*, and AIR (American Institute of Research) in the case of *Smarter Balanced*, and cited in the contracts as attesting to the validity and reliability of the machine-scoring method being used? Will you provide any independent research that provides evidence of the reliability of this method, and preferably studies published in peer-reviewed journals?

A month after the questions were submitted, only four states had responded to their questions directly, none of which were using AES. Two states responded to AS questions indirectly to local citizens, and 17 states and the District of Columbia had not answered their questions at all (Haimson, Stickland, Woestehoff, Burris, & Neill, 2016).

The Public's Concerns

As AES professionals, we have a responsibility to understand and acknowledge the myriad concerns reviewed in the previous section as we craft communications to the public about the use of AES. In the following, we synthesize these concerns thematically; specifically, Table 8.1 contains a list of 14 thematic concerns common across the documents described in the previous section. For each thematic concern, we provide citations from critical documents along with relevant standards from the *Standards of Educational and Psychological Testing* (American Educational Research Association [AERA], American Psychological Association [APA], National Council on Measurement in Education [NCME], 2014). Linking critics' concerns with the *Standards* shows that critics are right in questioning these facets of AES, even if there is disagreement with specific criticisms.

TABLE 8.1

Fourteen Thematic Concerns about Automated Essay Scoring

Theme	Concern	Sample Supporting Quotations	Standards
Standard Assessment Practices	Reliability	"Will you provide any independent research that provides evidence of the reliability of this method, and preferably studies published in peer-reviewed journals?" (Haimson, Stickland, Woestehoff, Burris, & Neill, 2016).	2.0–2.20, 4.0, 5.0, 6.8, 12.2
	Validity	"Machines are not able to approximate human scores for essays that do fit real-world writing conditions." "Machine scores predict future academic success abysmally" (PAMS, 2013).	1.0–1.25, 4.0, 5.0, 12.2
	Bias	"Privileging surface features disproportionately penalizes nonnative speakers of English who may be on a developmental path that machine scoring fails to recognize" (NCTE, 2013). "Machine scoring shows a bias against second-language writers … and minority writers such as Hispanics and African Americans" (PAMS, 2013).	3.0–3.20, 3.8, 4.0, 5.0, 12.2
	Teaching to the machine	"Using computers to 'read' and evaluate students' writing … compels teachers to ignore what is most important in writing instruction in order to teach what is least important" (NCTE, 2013). "Teachers are coerced into teaching the writing traits that they know the machine will count … and into *not* teaching the major traits of successful writing" (PAMS, 2013).	12.7
	Visibility into the scoring process	"often students falsely assume that their writing samples will be read by humans with a human's insightful understanding" (PAMS, 2013). "Will parents have the opportunity to learn whether their children's ELA exam was scored by a human being or a machine?" (Haimson, Stickland, Woestehoff, Burris, & Neill, 2016).	4.18, 6.0, 6.8, 6.9, 7.8, 8.2, 12.6
Unique to AES	"Black box" nature of AS features	"We cannot know the criteria by which the computer scores the writing" (CCCC, 2004). Examinees "are placed in a bind since they cannot know what qualities of writing the machine will react to positively or negatively, the specific algorithms being closely guarded secrets of the testing firms" (PAMS, 2013).	4.19
	Simplicity of engine features	"Computer scoring favors the most objective, 'surface' features of writing" (NCTE, 2013). "To measure important writing skills, machines use algorithms that are so reductive as to be absurd" (PAMS, 2013).	–

(Continued)

TABLE 8.1 (CONTINUED)

Fourteen Thematic Concerns about Automated Essay Scoring

Theme	Concern	Sample Supporting Quotations	Standards
	Essay length	"Machines cannot score writing tasks long and complex enough to represent levels of writing proficiency or performance acceptable in school, college, or the workplace" (PAMS, 2013). "Mere length becomes a major determinant of score by both human and machine graders" (PAMS, 2013).	–
	"Gaming" automated scoring	"Computer scoring systems can be 'gamed' because they are poor at working with human language" (NCTE, 2013). "Students who know that they are writing only for a machine may be tempted to turn their writing into a game, trying to fool the machine into producing a higher score, which is easily done" (PAMS, 2013).	6.6
Beyond the Scope of AES Professionals	Limited construct representation	"Computers are unable to recognize or judge those elements that we most associate with good writing" (NCTE, 2013). "Machines also cannot measure authentic audience awareness" (PAMS, 2013).	1.1, 4.1
	Few social aspects of writing	"Because all writing is social, all writing should have human readers, regardless of the purpose of the writing" (CCCC, 2004). "Computer scoring removes the purpose from written communication – to create human interactions" (NCTE, 2013).	–
	Little variety in writing	"If a student's first writing-experience at an institution is writing to a machine ... this sends a message: writing at this institution is not valued as human communication" (CCCC, 2004). Automated scoring "[dumbs] down student writing" (PAMS, 2013).	–
	Unfavorable assessment conditions	"Computer scoring favors ... 'surface' features of writing ... but problems in these areas are often created by the testing conditions and are the most easily rectified in normal writing conditions when there is time to revise and edit" (NCTE, 2013). "Machines require artificial essays finished within very short time frames (20–45 minutes) on topics of which student writers have no prior knowledge" (PAMS, 2013).	4.2, 4.14
	Limiting human scorer training	"Conclusions that computers can score as well as humans are the result of humans being trained to score like the computers" (NCTE, 2013). "High correlations between human scores and machine scores ... are achieved, in part, when the testing firms train the humans to read like the machines" (PAMS, 2013).	1.9, 12.16

The thematic concerns in Table 8.1 can be organized into three categories: (1) concerns that are related to standard assessment practice, (2) concerns that are unique to AES, and (3) concerns that may be beyond the scope of an AES professional to address.

Concerns Related to Assessment Practice

The first five concerns in Table 8.1 relate to standard assessment practices. Analyses of score reliability, validity, and fairness are important to document and report for any assessment program. Part I of the *Standards for Educational and Psychological Testing*, along with Standards 4.0, 5.0, and 12.2, provide standards for addressing these issues in all assessments and many of these standards apply to AES applications as well.

For example, "teaching to the machine" is a special case of "teaching to the test." Standard 12.7 of the *Standards of Educational and Psychological Testing* is clear on this: "In educational settings, test users should take steps to prevent test preparation activities and distribution of materials to students that may adversely affect the validity of test score inferences" (AERA, APA, & NCME, 2014). Under the comment for Standard 12.7 is the following:

> When inappropriate test preparation activities occur such as excessive teaching of items that are equivalent to those on the test, the validity of test score inferences is adversely affected. The appropriateness of test preparation activities and materials can be evaluated, for example, by determining the extent to which they reflect the specific test items and by considering the extent to which test scores may be artificially raised as a result, without increasing students' level of genuine achievement.
>
> **(AERA, APA & NCME, 2014)**

When teachers excessively teach skills designed to increase automated scores without increasing achievement, that is a violation of teaching to the test. Test owners, users, and stakeholders (including AES professionals) have an obligation to call out that behavior.

Visibility into the scoring process can also be linked backed to several of the standards, including Standard 8.2: "Test takers should be provided in advance with as much information about the test, the testing process, the intended test use, test scoring criteria … as is consistent with obtaining valid responses and making appropriate interpretations of the test scores" (AERA, APA, & NCME, 2014). Thus, scoring procedures that include AES should be publicly documented as part of the entire testing process.

Concerns Unique to AES

The next five concerns in Table 8.1 are concerns unique to AES. These concerns should be acknowledged and addressed by AES staff. Information about an engine's internal "black box" workings, including feature information, may need to be communicated to appropriate audiences. Such communications may be difficult in practice, as the features and methods used in an AES engine are often the intellectual property of the testing organization. However, there are ways to provide visibility into the engine without exposing intellectual property, as will be shown in the next section. The scoring of long essays, the possibilities of gaming the engine, and the ability of an engine to adequately represent the important constructs of interest are also unique issues for AES algorithm developers and data scientists to address through advances in technology.

Concerns beyond Scope of AES Professional

The last four concerns in Table 8.1 may be beyond the scope of the AES professional but are still relevant to others in the test development process. For example, critics of AES have pointed to the *social aspects of writing* and the lack of social aspects around writing to a computer as a fundamental concern. Though AES professionals can contribute to the discussion, such matter may be better served by content experts in writing and *English Language Arts* (ELA). Likewise, concerns around writing assessment conditions, the de-emphasis of variety in writing, decisions about prompt task, prompt type, assessment authenticity, and testing time are often the control of test developers. They are valid concerns in writing assessment, but they require expertise from professionals outside of AES and are often based on external factors such as time allocation and rubric concerns.

Similarly, concerns about human scorers being trained as automated scorers may or may not fall within the responsibilities of AES professionals, especially technical staff. Such staff generally have limited involvement with range finding, hand scoring decisions, human scorer training, or rubric development. Staff in charge of such activities may reach out to AES staff, but rely on their own best practices to make decisions when training human scorers.

Leading the Conversation

As AES professionals, we have an obligation to lead the conversation: to communicate how AES works to alleviate certain criticisms, to make changes to the system to address valid concerns, and to weigh the benefits of AES in time, cost, and scalability against the disadvantages cited by critics. These communications must include both *academic resources* (e.g., technical reports, conference presentations, and peer-reviewed journal articles) as well as *public-facing resources*, which are particularly important. Visibility into the AES process will help the public from spreading disinformation and unfounded criticisms.

To conclude this chapter, we propose the following action items for AES professionals as part of their outreach to the public. These action items are not designed to alleviate all concerns or criticisms about automated scoring, nor do these action items provide a checklist of actions meant to eliminate criticisms. Rather, the purpose is to recommend ideas that AES professionals could take to better communicate with the public and to acknowledge valid criticisms of automated scoring.

1. *To alleviate concerns about validity, reliability, and bias with automated scoring, produce public-facing documentation outlining such evidence in plain English.*

Most testing programs already produce technical information for customers and stakeholders around *validity*, *reliability*, and *bias*. However, this information is not always public-facing and not always in plain English. Sharing this information gives the public an opportunity to review the validity, reliability, and bias evidence needed to evaluate the effectiveness of AES.

Of these three core psychometric properties, the validity of AES scores might offer the widest variety of approaches (Attali, 2013; Bejar, 2011; Bejar, Mislevy, & Zhang, 2017; Bennett, 2004; Bennett & Bejar, 1998; Rupp, 2017). Williamson (2003) distinguishes between the *formal definition of validity* used by psychometricians – validity as defined by Cronbach, Messick, and Kane – and the *informal definition of validity* used by English writing experts meaning "you are measuring what you say you are measuring." Though AES professionals focus on the formal definition, it is important to recognize that the informal definition is on the minds of stakeholders. Yan and Bridgeman (Chapter 16, this handbook) discuss the details of AS validation.

Similarly, *rater reliability* of automated scores is commonly presented in the literature (Attali, 2013; Ricker-Pedley et al., Chapter 10, this handbook; Wolfe, Chapter 4, this handbook). Since the objective of AES engines is to accurately reproduce the score distribution of human raters, rater reliability provides an important way to evaluate the objective. There are a variety of statistics available for evaluating rater reliability with *quadratic-weighted kappa* being a popular choice in the literature; however, *exact agreement* rates may be a better option for communicating results to the general public, especially when coupled with appropriate guidelines for interpretation. When possible, documentation should include agreement statistics for two independent human raters and for a human rater with automated scoring. The public may be surprised if a scoring engine agrees with a human 65% of the time but may not be so surprised if two independent humans agree only 67% of the time on the same prompt.

Bias, fairness, and subgroup analyses of AS are not as common in the literature, but the *Standards for Educational and Psychological Testing* consider such studies so important that two standards have language about them: Standard 3.8 ("When tests require the scoring of constructed responses, test developers and/or users should collect and report evidence of the validity of score interpretation for relevant subgroups in the intended population of test takers for the intended uses of the test scores.") and Standard 4.19 ("collect independent judgments of the extent to which the resulting [automated] scores will accurately implement intended scoring rubrics and be free from bias for intended examinee subpopulations.") (AERA, APA & NCME, 2014).

Most bias studies take the form of either descriptive statistics or *differential feature functioning*. In the descriptive statistics approach, various scoring metrics (e.g., exact agreement, quadratic-weighted kappa, mean scores) are calculated for each subgroup and large differences between subgroups are identified (Bridgeman, Trapani, & Attali, 2012; Chen, Fife, Bejar, & Rupp, 2016; Ramineni, Trapani, & Williamson, 2015; Ramineni, Trapani, Williamson, Davey, & Bridgeman, 2012a). The differential feature functioning approach (Penfield, 2016; Zhang, Dorans, Li, & Rupp, 2017), which is relatively new, uses differential item functioning techniques on the engine features individually or in aggregate.

When providing information to the public, AES professionals may reach more people if alternative forms of documentation are provided. Besides reports and one-page summaries, leading the conversation could include social media strategies, engaging with stakeholders in public forums, and utilizing video-sharing services. Addressing validity, reliability, and fairness in a variety of mediums allows for more opportunities to connect with test stakeholders.

2. *To alleviate concerns about visibility into the scoring process, produce different types of documentation for different public audiences: parents, students, and teachers.*

Many testing programs are responsible for producing information about their assessments that is accessible to parents, teachers, and students, describing how the assessment works and how to interpret scores. The *Standards of Educational and Psychological Testing* reflect this via Standard 7.0 ("test documents should be complete, accurate, and clearly written so that the intended audience can readily understand the content") and Standard 6.10 ("when test score information is released, those responsible for testing programs should provide interpretations appropriate to the audience. The interpretations should describe in simple language what the test covers, what scores represent, the precision/reliability of the scores, and how scores are intended to be used") (AERA, APA & NCME, 2014). These standards suggest that similar documentation around AES is also important.

The five questions reported in the open letter of Harrison, Stickland, Woestehoff, Burris, and Neill (2016) are a good foundation for such documentation. Parents, students, and teachers may want to know what percentage of responses or items will go to AES, what percentage will go to human scoring, and what percentage will go to both. How reliable are automated scores for the kinds of items being assessed? Is the exact agreement between an AES engine and a human rater similar to the exact agreement between two human raters? If the response gets very different scores from a human and an AES engine, how is the final score determined?

Additionally, it may be useful to provide a high-level overview of how AS works. There is no need to dig deep into the details of features and statistical modeling, but the public may be relieved to know that AES models are based on actual responses to the items, that the scores input into the system come from human markers, and that two independent human markers can disagree over scores as often as a human scorer can disagree with an AES engine. If such documentation is produced for an assessment program, it should be made easily available on the website for the assessment along with other public-facing documentation about the test. Once again, information should not be limited to writing reports. Social media and online videos offer opportunities to connect with a wider audience about AES.

3. *To alleviate concerns about the features of an AS engine, carefully reveal the computational design of selected features via public-facing documentation.*

This action item may be difficult for some testing companies as components within their AS engines are considered intellectual property. Engines are expensive to develop, implement, and maintain, requiring significant investments. A new feature that can improve scoring performance enough to gain a significant market edge can be valuable. With careful writing, though, we can discuss the operation at a level of abstraction appropriate to the level of knowledge of an average user – alleviating the "black box" concerns that critics raise – while not being so overly detailed as to compromise intellectual property concerns. For example, consider the public-facing work produced by Ramineni, Trapani, and Williamson (2015), which includes a feature-space diagram showing how the features group together in major categories (see also Shermis, Chapter 7, this handbook). This diagram reveals much about e-rater's dozens of features without revealing any of the proprietary algorithms, underlying corpora, or statistical methodologies

4. *To alleviate concerns about "gaming" the AS engine, publish research on engine robustness.*

An examinee can receive a high score on a writing assessment via proper mechanics, good style, and the right vocabulary. However, if the examinee attempts to increase his or her score by leveraging the engine in "artificial" ways, then the examinee is "gaming" the engine. The idea of gaming an AES engine is frequently cited by Perelman as a concern. Common methods for gaming an engine include artificially adding length to the essay by copying and pasting one's response repeatedly, arbitrarily adding sophisticated vocabulary to the essay, or adding HTML/XML tags to the response to cause unpredictable behavior in the engine (Lochbaum, Rosenstein, Foltz, & Derr, 2013). As part of the research and development of an AES engine, it is important to investigate the effects on scoring when gaming strategies are used, and all AES systems should have anti-gaming algorithms as a core component (Henderson & Andrade, 2019).

5. *To alleviate concerns about the representation of the construct with AS engines, involve writing professionals such as teachers and didactic experts through development consultation and written commentary geared to the public.*

When critics opine that AES results in the de-socialization or de-valuation of writing, the issue goes beyond scoring and becomes an issue for writing assessment in general. Nevertheless, AES professionals can gain buy-in and trust if they initiate conversations with both writing assessment professionals and writing teachers and gain better understanding on how best to conduct writing assessment.

In the case of large-scale summative writing assessment, it may not be possible to satisfy the needs and expectations of all stakeholders, and tradeoffs must be considered to ensure the assessment is standardized for examinees. For example, it would be ideal for examinees to be able to obtain the prompt, then complete multiple rounds of pre-writing and draft writing, perhaps with a peer, to create their final essays – just like one might go about writing such documents in practice. This time investment is likely impractical for a traditional standardized testing session however.

Similarly, it would be ideal for human raters to be able to read the essays thoroughly, reflecting on what was written and annotating comments throughout, but scoring costs and volumes prevent raters from spending more than a minute or two on each response. For the social nature of writing, it would also be ideal for a human to always read each essay, but scoring volume, costs, and turnaround time may prevent a testing program from doing so. Such tradeoffs may be easier to address in formative writing assessments, though issues around testing time and scoring resources still apply.

It may also be useful to provide to the public some explanation about why AS was considered in the first place (e.g., AS may provide formative diagnostic feedback in addition to scores) or why a writing assessment was administered in a certain way. Highlighting factors like costs, time required to complete scoring, or formative feedback may help to convey the advantages of AES to stakeholders. By explaining the factors that go into the length of time for the writing assessment, test development professionals convey the decisions considered to balance effective writing assessment with testing time and resources. In other words, it may help the public to consider the gains of AES and not just the limitations.

6. *To alleviate concerns about AS usage more broadly, consider the operational use of hybrid human-computer scoring in writing assessment.*

This action item has less to do with communication and more to do with acceptance and buy-in of AS for large-stake summative assessments. Some believe that one must either use only AS or only human scoring. While there may be a reason to use only one method of scoring, there are benefits to using both automated and human scoring together (see Lottridge & Hoefer, Chapter 11, this handbook; Lottridge, Winter, & Mugan, 2013). One approach to hybrid scoring is to have an essay scored twice: once by the AES engine and once by a human rater. Then, the scores are aggregated in some way to determine a final score. Another approach is to have an essay scored by an AES engine, but if the resulting score is very low or very high, the essay goes to a human rater for evaluation. A third approach is to have an engine and a human scorer score all the essays, but the engine's scores are only used for quality control; the final scores are always assigned by a human rater.

In these three approaches, the benefits of both AS and human scoring are leveraged. Both the engine and the human rater are scoring according to the rubric, but the engine will not fatigue and will produce scores in a fraction of the time. Further, one can get the benefits of human scoring while reducing costs via the replacements of a human rater with an automated scorer. Hybrid scoring may not be appropriate or possible in all large-scale situations, but it is an option that should be considered by test developers and stakeholders.

7. *To alleviate concerns about general best practices in the profession, develop a set of standards around automated scoring.*

The development of standards for educational assessment is common, with documents ranging from the *International Test Commission Guidelines* to the *Standards for Educational and Psychological Testing*. Most of these standards documents are not focused directly on automated scoring, though some standards in each document would be applicable. In practice, AES professionals may also rely on highly cited journal articles (Williamson, Xi, & Breyer, 2012) or major technical reports (McGraw-Hill Education CTB, 2014; Pearson & ETS, 2015) for standards around good practice although there is an inherent risk in overgeneralizing cited recommendations to use cases that are broader than the original scope of the authors.

In this vein, Haisfield, Yao, and Wood (2017) and Haisfield and Yao (2018) reviewed the automated scoring, data science, and assessment literature to compile standards appropriate for AES engines and applications; their list of ten core standards appears in List 8.1. This list is neither exhaustive nor does it address all of the components of AS engines (such as standards for the underlying hardware and software); it does, however, provide a foundation for further development of AES standards. Standards show that AES

LIST 8.1

Ten Core Standards of Best Practice for Automated Essay Scoring Professionals

1. Automated scoring scores should meet absolute and relative thresholds in the industry when compared with human rater scores.
2. How automated scoring engines and procedures operate and satisfy construct coverage should be transparently communicated.
3. Scores produced by automated scoring should demonstrate fairness and equity for all populations.
4. Criterion-related validity studies establishing logical relationships in how AS relates to other constructs should be conducted.
5. The stakes of the test, item types for informing scoring decisions, and scoring approaches for the integration of human and automated scoring should be considered.
6. Accuracy and reliability of scoring via process monitoring should be made available to the client during testing.
7. The quality of inputs into an automated scoring engine (response, human scoring, item structure) should be carefully evaluated.
8. The impact or consequences of automated scoring on the reported score or test should be considered and defined.
9. Procedures should be in place to identify alert papers (e.g., cheating, disturbing content).
10. Measures around how and when to appropriately recalibrate or retrain the engine should be established.

professionals adhere to a high level of professionalism and care and, consequently, can help to engender trust with stakeholders.

Automated Scoring and Formative Assessment

Throughout this chapter, we have viewed AES through the lens of large-scale, high-stakes essay scoring. Before concluding, let us consider how AS in the formative assessment space may help to alleviate concerns from critics.

Many AES engines now provide feedback to examinees in formative applications. In these low-stakes environments, students can utilize the pre-writing, writing, and revision processes that are standard in real-world writing situations. Students can submit their drafts to an AES engine that can give instantaneous feedback on a variety of essay characteristics. Writing teachers are free to devote their resources to selected elements within the writing process or to certain students that need additional instruction (e.g., Burstein et al., Chapter 18, this handbook).

Research into automated feedback in formative writing assessment has shown promising results. Kellogg, Whiteford, and Quinlan (2010) showed that continuous automated feedback may help to "reduce errors of mechanics, usage, grammar, and style" among college writing students. Wilson, Olinghouse, and Andrada (2014) showed that automated feedback led to improvements in writing quality across revisions on a benchmark writing assessment for students in grades 4 through 8. Wilson and Czik (2016) found that teachers incorporating automated feedback were able to save time while focusing their feedback on "higher-level writing skills."

AS use in formative contexts allows for its strengths to be put on display. By showing the utility of automated feedback, students and teachers can use such technology in a low-risk space, yet reap the rewards such feedback can bring, including reduced wait times to receive feedback and the allowance of teachers to focus their feedback on higher-level skills. As teachers and students become comfortable with AS in these applications, they may become more amenable to using AS in summative situations.

Conclusion

AES has existed since the 1960s, but critics of AES have become more commonplace since 1998. Led by NCTE, Les Perelman, and the creators of the "Professionals Against Machine Scoring" website, several position statements have been published outlining the reasons why AES should be avoided in high stakes assessment. Criticisms range from a lack of visibility into the scoring process to psychometric concerns to concerns about the social value of writing. The authorities behind these criticisms have allowed the spread of these concerns on major and personal blogs.

As AES professionals, we have an obligation to communicate the intended uses of AS through documentation in plain English, social media, stakeholder forums, and public videos – mediums designed for a broad audience. Through these materials, we can provide more visibility into the scoring process, address basic information about validity, reliability, bias, and engine "gaming," and set standards for good practice. In doing so, trust can be built with the public about the use of this new technology. How will you lead the conversation around AES?

9

An Evidentiary-Reasoning Perspective on Automated Scoring: Commentary on Part I

Robert J. Mislevy

CONTENTS

Late last century, Bennett and Bejar (1998) argued that truly understanding the role of *automated scoring* (AS) in an assessment involves not just the scoring procedure itself, but its relationship with *construct definition* and *test design*, through *validity argumentation*. They were right then, and they still are; it's not just about the scoring. I too have found an *argumentation framework* helpful in studying traditional assessment topics like validity, fairness, and task design, and for more recent developments in simulation, process data, and *situative, cognitive*, and *sociocultural psychology* (Mislevy, 2018). As AS draws in all these topics, I offer my comments on the chapters of Part I from this standpoint.

I begin by reviewing a basic argumentation schema that locates AS, its connections with other *assessment elements*, and the contributions of the chapter authors. The argumentation concepts and tools lie in the so-called domain analysis layer of *evidence-centered design* (ECD) (e.g., Mislevy, Steinberg, & Almond, 2003), a *principled assessment design* approach that DiCerbo, Lai, and Ventura employ in Chapter 3. They aptly argue the role of principled assessment design more generally and describe and illustrate design considerations for AS in the *conceptual assessment framework* and *implementation layers* of ECD. I may therefore focus on argumentation.

Assessment Arguments

Evidentiary Reasoning

Assessment design, interpretation, and use arguments build on a schema from Stephen Toulmin (1958), shown in Figure 9.1. Specifically, **Data** support a **Claim**, which is some inference we are interested in, by virtue of a **Warrant** that has some empirical or theoretic **Backing**. The claim may not hold due to some **Alternative Explanation(s)**, which may be bolstered or weakened by **Rebuttal Data**. In assessment, data arise from learners' performances in tasks and support claims in the form of interpretations and decisions. Warrants, backing, and alternative explanations come from theory, research, and experience from psychology, assessment practice, and the domain of application.

Roughly speaking, warrants in assessment are about the kinds of things people with certain kinds of capabilities say, do, or make, in situations with certain kinds of characteristics. That is, we observe learners say, do, or make particular things in particular situations, and reason back through the warrant to make inferences about their capabilities and what they might do in other situations. In assessment terms, claims are typically about *constructs*, which are characterizations of aspects of performances, tasks, criterion situations, and peoples' capabilities. Characterizations concerning peoples'

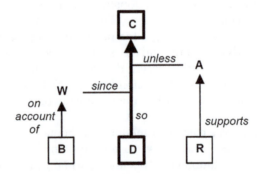

Reasoning flows from *data* (D) to *claim* (C) by justification of a *warrant* (W), which in turn is supported by *backing* (B). The inference may need to be qualified by *alternative explanations* (A), which may have *rebuttal evidence* (R) to support them.

FIGURE 9.1
Basic Toulmin argument schema.

capabilities appear most frequently in the assessment literature and are most strongly associated with this term; after all, inference about them is the raison d'être of assessment. However, Chapter 2 by Bauer and Zapata shows that constructs concerning performances are pivotal in AS as well, because constructs are assessors' operationalizations of philosophers' categories, which, in turn, provide the semantics for the task and test scores in assessment, as well as the corresponding variables in measurement models.

This has always been the state of affairs in assessment, though mostly implicit in familiar and established practices – AS forces us to be explicit. Even though we can technically implement AS without a foundational understanding of these elements of assessment arguments, we cannot establish their meanings or investigate their qualities without it. Chapter 2 by Bauer and Zapata is therefore an appropriate opening to this handbook, as it illuminates the cognitive grounding of category formation and discernment as it pertains first to human scoring and extends to AS. We begin here to see how AS is fundamentally rooted in constructs yet introduces novel warrants and demands backing into sub-arguments for evaluating learners' performances and gives rise to corresponding alternative arguments for the quality of the inferences it is meant to support.

Basic Assessment Arguments

Figure 9.2 begins to expand Toulmin's schema on assessment arguments, addressing a single task and a single use; later in this commentary I deal with some elaborations that AS further provokes. There are actually two connected main arguments, each embedding sub-arguments that distinguish three kinds of data – all of which are to be understood in terms of constructs (i.e., categories). Specifically, the lower half of Figure 9.2 capturing the **Design/Interpretation Argument** is, as the name implies, about designing an assessment and interpreting results,* while the upper half of Figure 9.2 capturing the **Use Argument** is about using the results for inferences such as predicting performance or providing feedback. In the center of Figure 9.2, the design and use arguments share a claim about the learner – often expressed as scores that are interpreted as measures of a construct – that is both the output of the argument interpreting the performance and an input to the use argument.

The **Cloud** at the bottom represents a learner's performance – what they say, do, or create in the task context such as an essay, a sequence of troubleshooting moves, or utterances in a conversation with an avatar. A performance is not itself data for inferring the learner's status on the construct, but

---

* We reason through the same structure for both design and interpretation. We use it *prospectively* when we design an assessment, make decisions for its elements, look toward alternative explanations, and generally anticipate how to interpret learners' performances. We use it *retrospectively* once the elements are in place and we observe performances and additional information about individuals.

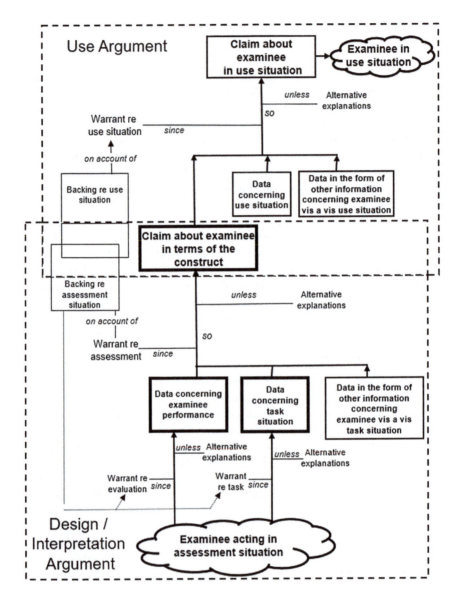

FIGURE 9.2
Assessment design/interpretation and use arguments. Examinee = learner.

rather evaluations of it. The **Arrow** from the cloud representing the performance to the **Box** named **Data Concerning Examinee Performance** represents this evaluation, which may include a human rating, an identification of a multiple-choice response, an automated score of an essay, or a real-time interpretation of a sequence of actions. These evaluations depend on a conception of the construct (see Chapter 2 by Bauer and Zapata) and what we think is evidence of it (see Chapter 3 by DiCerbo et al.), the core aspect of the warrant that runs through the reasoning for design, interpretation, and score use alike. The evaluations are grounded in sub-arguments with their own additional warrants, backing, and alternative explanations, in some cases for multiple strands or stages of processing (see Chapter 5 by Cahill and Evanini, on AS for speaking and writing).

Features of learners' performances are the most familiar one of three essential kinds of data in the interpretation argument. But the second kind, features of the task situation, is equally important, which is represented by the **Box** named **Data Concerning Task Situation** and the associated **Arrow** from the **Cloud** in Figure 9.2. Actions make sense only in light of context, content, and purpose. Task features contribute to the several meanings of the task situation – meanings as intended by the designers and meanings as perceived by learners – and shape their actions. With multiple-choice items, designers construct their features so they are likely to evoke responses that are both construct relevant and straightforward to evaluate – this reasoning is part of the warrant. With more complex performances, identifying and interpreting evidence often requires more complex processing; specifically, with interactive performances both task features and performance features emerge as performance unfolds.

Also critical, though usually tacit, is a third kind of data: information that the assessment user knows about the relationship between the learner and the assessment situation, which is represented by the **Box** named **Data in the Form of other Information Concerning Examinee vis a vis Task Situation** to the far right at the bottom of Figure 9.2. Such information can rule in or rule out various alternative explanations that threaten the validity and the fairness of uses. Simply knowing the testing population can assure that learners are sufficiently equipped with knowledge that is necessary but not construct relevant. In complex tasks, learners' unfamiliarity with interfaces, expectations, background knowledge, and standards of evaluation are all alternative explanations that seriously degrade score interpretations and uses. All of these take particular forms in AS that need to be mitigated by learner preparation, averted in design, or detected when they are encountered.

The central message of Chapter 3 by DiCerbo et al. is that design decisions concerning all three types of data – what is discerned in performances, what is built into tasks, and what else will be known about learners in operation – shape the meaning of a learner construct as it is operationalized in ways that interact with the performance construct(s) enacted in AS. These considerations substantially impact the measurement properties of scores for both interpretations and various uses. The correspondence between these assessment design/interpretation elements and corresponding elements of the score-use situations can render scores more or less informative for various subsequent uses, quite beyond the construct labels assigned to the scores of performances and the assessment as a whole.

The **Assessment Use Argument** in the upper rectangle of Figure 9.2 has essentially the same structure as a **Design/Interpretation Argument** in the lower rectangle. Specifically, the **Claim** at the top in the **Box** named **Claim about Examinee in Use Situation** addresses a learner's state or performance in a different situation such as improvement after feedback, prospective learning outcomes in an instructional treatment, or job functioning. This claim is supported through three types of **Data**. The first kind of data for this claim is the records (e.g., scores, ability estimates, evaluations) coming from the assessment, often omitting the details of all the three kinds of data from which they were derived. The second kind of data for this claim is features of the assessment use situation, which may be known specifically or only vaguely. In fact, how well they are known and how well they match up with the assessment task situations are integral consideration that undergird validity, which I discuss further below. The third kind of data for this claim is what else is known about the learner with respect to these use situations.

As with design/interpretation arguments, this information is usually tacit – built into the use situation with some features recognized but others not – but it also is critical to validity and fairness in assessment use. Chapter 5 by Cahill and Evanini shows how the intended testing population (e.g., language learners versus advanced students) and use of AS of written responses (e.g., control of mechanics versus coherence of argumentation) can require substantially different methods from *natural language processing* at certain stages of AS for written essays.

Finally, note that the warrant in the use argument overlaps the warrant of the design argument since it provides the rationale as to why we might expect features of performance in these tasks to provide information about performance in use situations. It may be the similarities in the situations, similarities in the capabilities they draw on, or some mixture, depending on the nature of the construct at issue. These inevitable differences generate alternative explanations, some of which may be problems in some learner populations, as with stylistic differences across cultures. This can become a fairness issue that can arise with both human ratings and AS, a particular issue discussed in Chapter 4 by Wolfe.

Scoring Sub-Arguments

The simplified form of the scoring sub-argument in Figure 9.2 can be detailed further. In the following, I sketch some scoring sub-arguments that arise in AS of completed task products such as final essays, artifact designs, and logs from written responses or science investigations; I mention extensions to real-time evaluation of interactive tasks only briefly (see Chapter 16 of Mislevy, 2018, for further thoughts along these lines).

Figure 9.3 addresses the usual way of thinking about scoring processes for multiple-choice items and human ratings of performances as having one

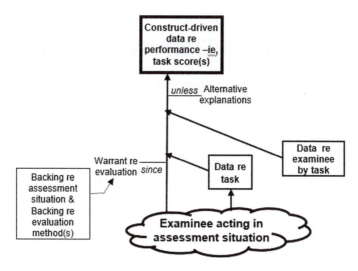

FIGURE 9.3
Single-stage evaluation argument. Examinee = learner.

stage.* Note that the claim is about a construct anchored in performances, one or more features of the performance expressed in numbers, categories, or functions – fully articulated or not – with respect to the universe of possible performances on this task – possibly enough cast generally to apply to other tasks. With multiple-choice items, the warrant is that students with higher proficiency are more likely to select the correct response as the features of the item require a learner to reason through construct-relevant knowledge or skill. Backing includes the substantive content of the task, item statistics from pretesting, expert panels' bias reviews, and learners' familiarity with the item and response type.

Human Scoring Characteristics

As discussed in detail in Chapter 4 by Wolfe, during human scoring raters assign scores along one or more dimensions. The warrants posit that the scores characterize targeted qualities of a performance and that the raters are providing those scores on the basis of those qualities and with sufficient accuracy. Backing comes from various sources in different applications. The simplest type of backing is the presumption that the raters are experts, which is a claim that is backed in turn by credentials or experience. A more substantial type of backing comes from the design and use of scoring rubrics, although their nature ranges from rubrics that include only vague descriptors

---

* The engineers who build machines that score paper answer sheets know better, having learned to deal with marks of various shapes and weight, erasures, and the errors that humid conditions can cause.

of qualities to rubrics that include more detailed characterizations of proto-typical features at each score point to rubrics that include highly proscrip-tive scoring rules to abstract evaluations that call for broader judgments and interpretations. Additional kinds of backing include the principled use of training procedures and example materials, examinations of the *cognitive processes* by which raters produce their scores, and statistical analyses of rat-ers' agreement with one another or with "anchor" performances. These – and other – forms of backing and their particulars in a given application can all be charted using *graphical argumentation techniques* (Schum, 1994) and related to analytic methods such as *generalizability studies* regarding tasks as well as raters and fairness investigations with respect to construct-irrelevant variance related to raters, learners, or their interaction.

Chapter 4 by Wolfe brings out the copious methodological questions that underlie human ratings. There are several reasons why such a chapter is criti-cal in a handbook on AS. In early days, agreement with human raters was the sole standard of AS. However, the fact that an automated score agrees well with human scores is certainly part of a multifaceted warrant to be backed by empirical results, but it is not sufficient. The validity of the human scores on which the automated score is based must also be established. Wolfe shows us that much more is going on in human scoring than meets the eye and how to investigate the many issues of evidentiary reasoning involved, how to articulate alternative explanations that arise, and how to identify sources of backing that can be marshaled. Of particular importance is his discussion of the values of psychometric modeling, including *item response theory* (IRT) and *generalizability theory*, which are valuable tools for investigating human ratings when, as he notes, human ratings are to be the target criterion for AS. These techniques can be used not only to improve human ratings, but also to provide information about aspects of performance that are hard to identify or confounded. Moreover, the same models can be pressed into service in analyzing the properties of scoring systems that incorporate both human and AS scores.

Comparison of Human and Automated Scoring

Even when automated scores in an application agree well with human scores and the human scores are demonstrably grounded in both construct-relevance and rater cognition, there is a separate issue of whether the same aspects of performances are employed in the two processes. The validity argument for construct-based score interpretation is weaker if this "under-the-hood equivalence" is not established. Indeed, in Chapter 5, Cahill and Evanini describe how a major focus of research for AS in writing and speak-ing has been extending techniques from natural language processing to bet-ter align with higher-level constructs such as event coherence and sentiment in narrative writing. This may not be critical in some score-use applications that only use final scores for research, evaluations, or policy decisions – the key issues in the score-use argument is whether differences in the two processes unfavorably affect either the intended use or the future behavior of students, educators, or policy makers.

There are two cases, however, in which establishing the relationship between human and AS is necessary. The first is when the intended use depends not just on overall performance but on more detailed aspects of knowledge or skill, as in feedback for learning. The second is when a deeper understanding of human ratings, in terms of both cognitive processes and *substantive grounding*, helps AS developers create more strongly grounded processes. The cognitive processes that have for years resided in raters' heads hold insight into better AS processing. It is surprising how far one can go with even a basic "bag of words" scoring process using *latent semantic analysis*, a suitable *training corpus*, and a good set of human-rated performances to calibrate to. I note here that the web of claims, warrants, and backing beneath these could also be explicitly charted.

Sociocognitive Perspective

But we see that further improvements do not come just from better analytical techniques, but also from techniques in concert with a deeper understanding of the raters' cognitive processes as well as the *sociocognitive* processes from which peoples' capabilities arise. DiCerbo et al., in Chapter 3, argue for the importance of a coordinated construct-based perspective throughout assessment design and implementation. Later, in Chapter 6 by D'Mello on inference from performance data that are both *multimodal* (i.e., that capture multiple aspects of performance such as speech, bodily actions, eye gaze, and gesture) and *temporal* (i.e., unfolding over time), his reliance on constructs at multiple levels of performance seems indispensable. He considers constructs concerning aspects of persons, task, and performance, all derived from situative and sociocognitive psychological perspectives, supplementing or supplanting constructs from trait and behavioral perspectives.

A couple of key ideas from situative and sociocognitive strands of psychology and related fields suffice to highlight these emerging developments for AS, and the implications that follow for task design, validity argumentation, and psychometric modeling. As noted above, a sociocognitive perspective focuses on the interplay of cross-person *linguistic, cultural, and substantive* (LCS) patterns that emerge over the myriad of individuals' experiences and events (this is the *socio-* part of sociocognitive) and the within-person resources that individuals develop to engage productively in those experiences (this is the *-cognitive* part); see Gee (1992) and Sperber (1996) for an introduction on a sociocognitive perspective.

There are many such LCS patterns, at all levels of granularity, that are mixed and matched, combined and recombined, extended and adapted in every unique interaction with the particulars of that event, and there are many kinds resources at varying levels of granularity that individuals draw upon to understand and act therein. *Situative psychology* focuses on the ways that meaning arises in the moment jointly in an evolving interaction in the external world and an agent's internal cognition in the situation. Both the situations we act in and the mental structures we soft-assemble to act through are hierarchically organized. Consequently, so too, then, can be the person and performance constructs, the task designs, and the analytic techniques around which we construct and use assessments in general and AS in particular.

Game-Based Example

In Figure 9.4, I sketch one such argument structure while in Table 9.1 I give an example from the *SimCityEDU* game-based assessment for thinking about levels of meaning in performance in evidence identification (Mislevy et al., 2014). LCS patterns of various types are involved at each level; evidence identification processes can work both bottom-up and top-down, identifying and characterizing likely meanings at a higher level in patterns at lower levels, using various processes like the ones discussed throughout this handbook. Processing may address fewer levels, or add some, and there may be many distinct LCS patterns at a given level. Different processes can be used at different stages of analysis such as *data-mining detectors* at lower levels, *Bayes nets* or *Markov processes* at middle layers, or *deep-learning neural nets* that operate at multiple levels in a common layered model.

Note that each level in the argument that is labeled describes the (perhaps highly dimensional) output at each level as characterizing information at that level, as found in a given realized performance. Each successive level is a further inference about a feature that is constructed around lower levels and becomes meaningful as an element that takes a particular situated meaning in concert with other elements at the same level or levels above or below. These evaluations are often interpreted semantically in terms of a construct about performances. This is generally the case when a process identifies a feature from lower-level input on the basis of previous human expert identifications of features in training examples. However, it need not be, as when functions of lower-level features are obtained as data-reduction variables from exploratory analyses (see Chapter 3 by DiCerbo et al.).

Similarly, in Chapter 6 by D'Mello on multimodal analytics, we see that intermediate levels of modeling are posited to govern multiple manifestations of phenomena such as mental states at the time (i.e., constructs about persons' specific realized states). In assessments of collaboration, they can be patterns of moves and actions that constitute a response to a partner in which the interpretation depends as well on the partner's previous actions and the state of the problem. As a result, the evaluation of the learner's action depends on the evaluations of the partner's previous actions and on the problem state, perhaps both as the team members view it and as the assessment system evaluates it.

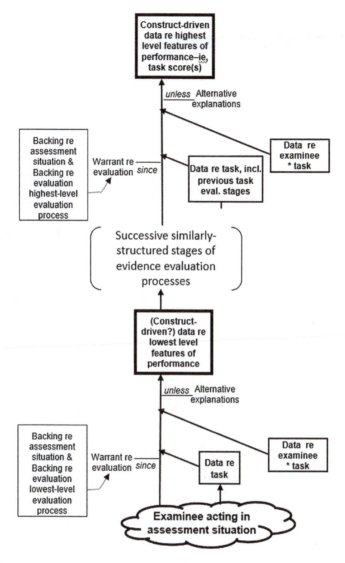

FIGURE 9.4
Multistage evaluation argument. Examinee = learner.

Reusability Considerations

The issue of reusability of AS processes arises Chapter 3 by DiCerbo et al. on design, with respect to cost, Chapter 5 by Cahill and Evanini with respect to both lower-level feature capture and higher-level genre qualities, and Chapter 7 by Shermis on international applications of essay scoring. There is an important conceptual point here that the layers of ECD bring out: the meanings of real-world human activities always involve a conjunction of

TABLE 9.1

Levels of Reasoning from Performance in a Game-Based Assessment

Level of Reasoning	Explanation and Example
Proficiency structures	Learner-level constructs that correspond to student-model variables in ECD (e.g., aspects of students' knowledge, skills, propensities, and strategy availabilities). In psychometrics, they are modeled as giving rise to pragmatic sequences of semantic actions, traces of which are contained in work products.
Pragmatic structures	Sequences of semantically meaningful actions generally arise from reasons, motivated by a player's goals and their understanding of the current situation, and knowledge and skills. They help to articulate reasons for why a learner was carrying out various semantic actions in order to differentiate whether they represent an attempt to check the effect of a move, cause a change in the game state to move toward a goal, or to re-play a thorny situation for example.
Semantic structures	Actions and products of certain kinds following the grammatical rules that have meaning with respect to the game (e.g., pull up help menu, ask a nonhuman game character a question), which are the essential building blocks from which play is fashioned. These actions can arise from different combinations of raw movements and the same raw movements can have different interpretations in different game contexts.
Grammatical structures	Rules of organization and combination that are needed in order to understand if epiphenomena can be used as evidence (e.g., screen shots may contain lines and edges, mouse clicks on a keyboard can produce words and sentences, mouse actions can be radio button selections or drag-and-drop to hot spots, and eye movements can signal fixations and jumps).
Epiphenomena	Rules of organization and combination that are needed in order to understand if epiphenomena can be used as evidence (e.g., screen shots may contain lines and edges, mouse clicks on a keyboard can produce words and sentences, mouse actions can be radio button selections or drag-and-drop to hot spots, and eye movements can signal fixations and jumps).

Note: ECD = evidence-centered design.

persons doing things in situations, and assessment arguments therefore involve inference about what persons will do in other situations. However, once an argument is fleshed out, it is often possible to design elements in student, evidence, and task models and then construct, implement, use, and re-use them distinctly as pieces of machinery to instantiate arguments in operational assessments. Working out such arguments is essential to design and implement assessments, all the more so with complex ones like those involving AS. This may be done implicitly, by adapting existing assessment pieces and systems that seem to work, or by clever creation, careful thought, and iterative improvement. Or it may be done explicitly, with the aid of argumentation concepts and structures. The newer, the more complex, the costlier, and the greater the value of component re-use in a proposed assessment, the better the "explicitly" looks.

Therefore, as discussed in Chapter 7 by Shermis, we may note that in AS systems that comprise multiple layers, some of the processes may apply across tasks, topics, or languages much as is; others may build on the same higher-level principles but require more extensive tailoring; and still others may need to be constructed almost from scratch. As an example, consider the scoring of textual responses in either English or in Mandarin, concerning either persuasive essays or critical incident reports in nursing. The lowest level of feature identification necessarily differs in English and Chinese responses, as do strings and constructions. However, conventions above these levels for essays may be more similar across language, and effectiveness of critical incident reports as to how to include and how to present it may be more standard in metropolitan hospital settings, so that adaptations of a common procedure may suffice. On the other hand, lower-level processes for essays and incident reports may be virtually identical in English and in Chinese, while higher-level evaluation processes differ more in accordance with the conventions of the respective genres. Note that these issues are not simply technical matters, but issues of high-level person constructs, sociolinguistic considerations, and linguistic grammars and conventions, as they inform the hierarchies through which of the meanings of responses are built.

Latent-Variable Models

The chains of AS evaluations discussed above are about performances, not about learners. Moving from claims about performances to claims about learners is where concepts from psychometrics become useful for reasoning about performance across tasks, not just within specific realized performances. It is positing *latent variables* that characterize tendencies of learners across some domain of relevant tasks and beyond them in interpretations and uses, which of course needs to be investigated and warranted. Modeling concerns not just descriptors of what learners have done in particular tasks, but also capabilities, propensities, and potentials that are relevant more broadly in other similar or dissimilar tasks (*reliability* and *generalizability*) or in real-world criterion or educational situations (*validity of extrapolations* in score uses). Latent variables are vehicles for framing such claims and synthesizing evidence coherently for such purposes and it is in their terms that the claims about learners in Figure 9.2, the literal center of the argument, are expressed in probability models. This framing further provides tools to characterize the weight and direction of evidence and for investigating certain important classes of alternative explanations as well as introducing new ones, because model adequacy is an additional part of the warrant.

Figure 9.5a shows the location of this step of reasoning and modeling in the frequent circumstance of assessments consisting of multiple tasks, with AS or some other type of scoring used to evaluate performance within each and with a latent-variable psychometric model being used to synthesize information across tasks. In ECD terminology, the strands of argumentation within tasks are specified in terms of assessment elements and processes as the *evidence identification* component of an evidence model and the cross-task argumentation are specified in terms of the psychometric model as the *evidence accumulation* component.

As an example, in an assessment with ten short responses on ten different topics scored by human raters, an IRT model could be used to synthesize evidence in each rating in the form of evidence about a learner's θ, which is a parameter for a person. A *rating-scale model* combines θ through the IRT model along with parameter(s) for prompts' difficulties and raters' severities to give probabilities of performances at each possible rating value, from a given person, task, and rater. Figure 9.5 adds an argument step that underlies such psychometric modeling across multiple tasks. Warrants now include the presumption of *model fit, exchangeability of raters* with the same severity parameters, and differences in *task difficulties*. Alternative explanations include model misfit as well as concerns around systematic interactions between rater and learner characteristics or interactions between raters and rubrics that would constitute unfairness as quantified via differential functioning and bias indices. With such models, generalizability analyses, or further hybrids and extensions of the ideas, we can investigate reliability, validity, and fairness not as measurement issues but as *social values*, as extended and adapted to new forms of data, analytics, contexts, uses, and psychological underpinnings (Messick, 1994).

As DiCerbo et al. note in Chapter 3, it is also possible to use psychometric models to synthesize evidence across tasks at lower levels for feedback to learners. Figure 9.5b suggests this usage, which could be implemented in an integrated model with higher-level person constructs. This cross-task, model-based, and finer-grained feedback contrasts with specific feedback on specific aspects of given performances. An example of the latter might be "Consider signaling your change in topic at this point in your essay, to improve the flow of your argument," which can be useful in and of itself for learning. An example of the former might be "You often miss opportunities to indicate the flow of your arguments," which can be useful for a decision to offer a targeted strategy review. In either case, feedback requires a construct-relevant interpretable feature of performance at this level of evaluation, whether to talk about a specific performance in terms of a task-based construct or about a tendency to exhibit such behavior as a corresponding person-based construct.

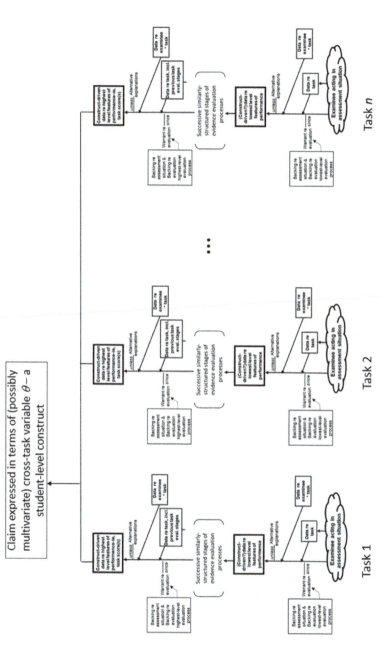

(a) Student-based score at highest task-level scoring level

FIGURE 9.5

Arguments for synthesizing evidence across tasks. Examinee = learner. (a) Student-based score at highest task-level scoring level. (b) Student-based scores at highest task-level scoring level and a finer-grained score at an intermediate level.

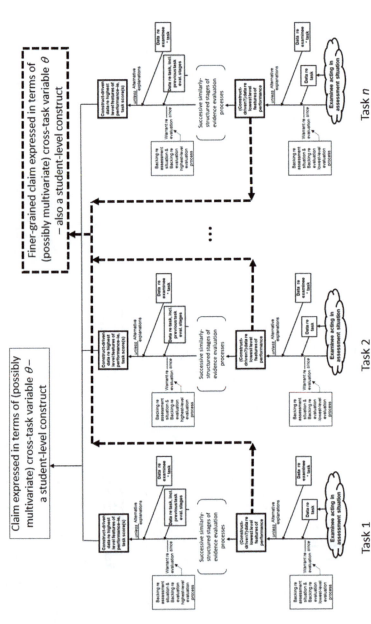

FIGURE 9.5
(Continued)

Many discussions of AS focus on within-performance evaluation, which is not surprising given the many challenges of new forms and greater volumes of data as well as new analytic techniques being developed to make sense of them. But evaluating qualities of performances takes us only halfway there – it is by synthesizing this evidence in terms of learners' capabilities that we arrive at the student-model data that ground interpretations and subsequent uses.

Closing Comments

'Scoring' is a term inherited from familiar processes of comparing multiple-choice responses with keys and eliciting evaluative ratings from human raters. The title *Handbook of Automated Scoring* is thus a good name for this volume for the purpose of allowing potential readers to quickly recognize its subject at a glance. However, 'scoring' is far from a sufficient label for the purposes of design and analysis, as well as of validation and implementation, that we face in assessments that capitalize on AS. The wide-ranging discussions in the chapters in the first part of this handbook concern valid uses of AS whose characteristics go far beyond just swapping computational algorithms for human raters in an isolatable process of an otherwise unexamined system; see related criticisms of some AS systems in Chapter 8 by Wood. Concepts and representations from evidentiary argumentation help us think through the elements and relationships that are involved while structures and terminology from principled assessment design help us to effectively implement such reasoning in operational systems.

Part II: Operational Methodologies

10

Operational Human Scoring at Scale

Kathryn L. Ricker-Pedley, Susan Hines, and Carolyn Connelly

CONTENTS

With the rapidly expanding use of *automated scoring* (AS) in many assessment contexts, it might be tempting to assume that AS may eliminate the complexities and costs associated with *human scoring*. However, it is important to remember that even if the intention is for all constructed responses to be 100% automatically scored there is typically always some degree of human scoring necessary. For example, accurate and reliable human scores are often needed for determining reported scores, for collecting data to support updating or developing new AS models, or for quality-control purposes.

When discussing evaluation of constructed responses, a term that is commonly used is 'scoring,' which suggests a sole focus on the resulting numerical output of a process. However, a more systemic view that focuses on the entirety of the *ratings collection and resolution design* as well as its procedural implementation is likely more productive in this regard. Specifically, in such a system, *constructed responses* are collected, *ratings* are assigned to the responses by human raters, potentially divergent ratings are resolved or *adjudicated* using *ratings rules*, and the resulting ratings are used in conjunction within *scoring rules* to create individual or composite *reported scores*.

In this chapter we address some of the key considerations of designing and operating such a large-scale human scoring effort. When many hundreds of thousands or even millions of ratings are assigned by thousands of raters for high-stakes testing, the sheer scale of the enterprise requires that the systems and processes are thoughtfully designed to help manage the work. To ensure consistently accurate scoring, there are many decisions that must be made about how to recruit and hire raters, collect the ratings, and monitor scoring quality.

As practitioners with many years of experience in large-scale constructed-response assessment, the considerations and recommendations that we describe here come from the joint perspectives of *human resources, assessment design,* and *modern psychometrics,* and are informed by work on programs including those of summative assessment in K–12, assessments of English language learners, and professional licensure, to name but a few contexts. The scope of these assessments ranges from single essays or short content-focused answers on a mixed-format assessment (i.e., an assessment with selected response and constructed responses) to assessments with spoken responses or submitted pieces of art, as well as videos of complex performances.

We designed this chapter to help raise awareness of the kinds of consideration that must be made by interdisciplinary teams as well as to offer some advice for what kinds of efforts are needed when the intent is to eventually transition to partial or complete AS. The general principles described in this chapter apply to any of these response or assessment types, though each type of assessment has its own special considerations and conditions that require additional fine-tuning or tweaking of the specific procedures and practices. For research foundations in this area see, for example, Chapter 4 by Wolfe (this handbook).

We have organized the chapter into four main sections. In the first section, we describe key requirements for an electronic ratings collection / scoring system. In the second section, we discuss key elements of designing a scoring process and planning a scoring event with such a system. In the third section, we review important considerations around rater training. Finally, in the fourth section, we present key operational considerations for monitoring raters.

Scoring System Design

If a testing program intends to use AS to evaluate responses in real time, then these responses must be collected in electronic form. It seems logical, then, that an *electronic score collection system* would be used, in which responses are evaluated by human raters via a computer, as opposed to a system where raters evaluate responses on paper-based *scanning sheets* for instance. Beyond significantly simplifying ratings capture, there are numerous advantages to using an electronic system, especially if the system is built to optimize the ratings collection process.

The biggest advantage of an electronic ratings collection system is how much easier it is to manage scoring logistics, particularly as the scale of the scoring moves from hundreds to hundreds of thousands or even millions of test-taker responses. But another significant advantage of these systems is they help to implement procedures that detect and/or minimize any *systematic errors* or *biases* in rating procedures that are present. By reducing the undue influence of any individual rater, other responses, or other factors that are not related to the quality of the response and the construct that is to be captured, the system helps reduce to *mitigate risks* around *negative unintended consequences* of score interpretations. Of course, great care must be taken to train and monitor raters as well as to supervise scoring to ensure that scoring is conducted accurately and fairly; this includes training that is focused on how raters can identify and try to minimize their own biases. Put differently, carefully designing a scoring system with a *systemic mindset* acknowledges that a risk of errors and biases exists throughout the system and helps the scoring organizers take precautions to minimize the potential impact on scoring, including any costly corrective actions required.

While it is technically possible to orchestrate a scoring effort on paper with some of these properties / features, it is significantly more difficult and time consuming; typically, it is simply impractical. Similarly, trying to mimic paper-based scoring process directly via an online system does not automatically result in scoring processes and *quality-control* properties that are desirable; these must be specifically designed to take advantage of the online environment in which it operates. In the following subsections we describe some key recommended features of the electronic system that can help in this regard.

Feature 1: Ability to Design Rater Assignment

One of the greatest benefits of using an electronic scoring system is the ability to mix rater assignments to responses, typically via *randomization*. In non-digital paper-based scoring efforts, responses are often carried from rater to rater in file folders for ease of tracking, organization, and movement. In such scoring, folders are a necessity to keep the testing materials organized and ensure all scoring is completed. This organization means that responses move together, so any two responses tend to be scored in succession each time they were scored by a new rater. Therefore, if a given response follows a particularly strong or particularly weak response, it might look relatively weaker or stronger, respectively, when compared with that previous response. This effect could multiply across all the responses to items on an assessment if there is more than one constructed-response item.

Furthermore, if all the responses from one testing center, school, or school district are scored at the same time, we might expect that those responses could look more like each other than like other responses in the entire pool. For example, if a particularly strong school district had all its responses scored in sequence, then raters might be tempted to make differentiations among responses with regard to neighboring score points where no such differences should exist by design of the scoring rubrics. That is, some responses might get lower scores by relative comparison because raters were expecting to see some weaker responses at some point. Having a system that can mix and randomize assignment of responses to raters helps prevent such effects and minimizes the impact of any potential bias on individual test takers.

In some cases, it is also advantageous to use a *stratified random assignment* of responses to raters. For example, if a representative sample of the responses is needed for preliminary *psychometric analyses* (e.g., *item analysis* or *test score equating*), then some sample of responses could be selected and prioritized to be scored first with all other responses scored subsequently. Higher- and lower-priority responses can still be randomized within the priority strata, thus retaining the benefit of randomization.

Another beneficial feature of an electronic-scoring system is that information about a rater can be stored in a *database*, which is beneficial in helping minimize the risk of introducing rater biases into scoring because it can be used to allow for the selective *matching* of raters to responses. For example, if students are offered topic choice in a response, the system could match a response in the domain of economics to a rater with economics expertise or a response in the domain of history to a rater with history expertise. It can also prevent a rater from scoring a response of someone they might know personally (e.g., by avoiding assigning responses from test takers with matching institution or zip codes).

Feature 2: Routing Materials to Support Desired Scoring Rules

One of the most important – yet typically overlooked – parts of constructed-response scoring is setting up the scoring rules mentioned at the outset of the chapter. These are the rules that govern how ratings are translated into scores. The ratings collection system must be able to support and enforce desired scoring rules; below are some examples of the types of scoring rules that might benefit a scoring design.

One of the most basic scoring rules is how many human or automated ratings will be applied to each response (i.e., one, two, at least two). When most responses will receive only one rating, it is desirable to assign a second rating to some sample of the responses for computation of *rater agreement statistics* or to assign additional ratings to responses where the score puts the test taker near a *cutscore* for the test to increase precision about the classification. For example, all responses for which the total test score is within one *standard error* of the *passing score* might receive an additional rating or multiple additional ratings to allow analysts to have greater confidence in the final score.

Whenever two or more ratings are used to calculate a final item score, it is important to have additional *score resolution rules* that can resolve any discrepancies that arise among the ratings. The simplest rule is to not resolve discrepancies and to take the average of the original ratings. Empirical studies have looked at the value of discarding ratings in score resolution; some consider this practice as removing an error while others think of it as losing information from a rater. For two very interesting discussions of score resolution rules, please refer to Cohen (2015) as well as Johnson, Penny, Gordon, Shumate, and Fisher (2009).

Apart from this simplistic rule, there is a large range of response routing options for scoring to consider. For example, what rules should be used if one rater assigns a response a '1' and another a '2'? What about a '1' and a '3'? Are the two ratings under consideration "close enough" to signal agreement about the demonstrated skill sets? The allowable difference that is acceptable before adjudication ratings are required depends on the use of context and associated stakes of the assessment. If such adjudication ratings are required, should they come from another rater in the pool or should they come from a scoring supervisor or some other rater with a more senior designation? Should certain types of ratings only be assigned by more senior/experienced raters? If a test taker has entered a response in the wrong section of the test, should there be an easy way to route those responses to scoring supervisors, who can determine whether the response can be scored? Regardless of whether certain responses or score adjudications are assigned to more senior raters or not, it is important that the system always allows raters to defer responses that they are unable to score with certainty, rather than forcing them to assign an uncertain rating.

Similarly, if test takers provide *unusual / aberrant responses* (e.g., off-topic responses, responses signaling emotional distress, otherwise nonscorable responses) it might be desirable for raters to send these responses directly to more experienced raters. In this context, it is becoming increasingly important to be able to quickly handle what are known as "crisis" responses. These are responses where a test taker – particularly minor children – disclose an issue such as likely abuse or endangerment or where a test taker issues or refers to a threat that should be investigated. These types of responses need to be identified and routed quickly so that appropriate actions can be taken in a timely manner.

Unusual responses typically receive customized *advisory flags* or *condition codes*, which are often nonnumeric *classification labels* (e.g., 'B' for "blank," 'OT' for "off-topic"); these are sometimes subsequently converted to numerical scores (e.g., an off-topic response may eventually be scored as '0' since it shows a lack of understanding of the task requirements). This scoring flow is significantly easier in a system that can handle the appropriate routing automatically. Related to this point, it is also important that scoring rules specify how condition codes are handled when calculating agreement statistics; for example, the two ratings may be considered discrepant even if the numeric conversion of the condition code classification is a '0' or a '1.' The discrepancy between a condition code and numeric scores is sometimes enforced (even if the resulting scoring is '0') if assigning condition codes is considered an important differentiation from the type of response that warrants a '0.'

Feature 3: Ensuring Rating Independence

It is commonly desirable to use *quality-control statistics* (e.g., rater agreement indices such as various *kappa* variants) for monitoring scoring quality, which often make the assumption that ratings are collected independently of each other. However, if two raters discuss a response before assigning ratings or if a single rater assigns multiple ratings to a response, then the ratings are not independent and, therefore, are more likely to agree than if they were truly independent. For assessments for which AS models will be eventually developed, these human–human and human–machine agreement statistics are an important part of the evaluation of whether AS models are appropriate for use and, thus, being aware of their robustness and sensitivity to violations of key assumptions for proper inference is important.

The concept of independence of ratings should also be extended to rating multiple responses of a given test taker, as mentioned earlier. For example, if a test has multiple constructed-response items, ideally different raters will score each one, especially if the scores for the constructed responses make up a large proportion of the overall test score. Further, a rater should not be aware of what ratings the test-taker received for the other constructed responses since this induces similar dependency effects that technically violate the assumptions of key quality-control statistics.

Another important function of the electronic ratings collections system is to assure that all raters in the pool are *randomly paired* with other raters in the pool whenever multiple ratings are assigned to a response, to the extent that this is possible within the overall design. Specifically, if two raters were only paired with each other, we would only know the extent to which they agreed with each other. However, if raters were randomly assigned across the entire pool of raters, then we would have a better idea of the extent to which all raters agree with each other. The latter is a more appropriate evaluation of the entire rater pool, which is typically the focus of the rater-monitoring process. Moreover, if discrepancies between raters are a mechanism for identifying rating errors, then, by triggering additional ratings for resolution, pairing raters runs the risk of missing errors made my pairs of raters who tend to score like each other and requires more score resolutions than would be necessary with rater pairs who tended to disagree.

Yet, there are some instances where nonindependent rating is appropriate, or at least necessary. In a process known as *consensus scoring*, raters – typically, *expert raters* or *scoring supervisors* – will score responses for the purposes of identifying *exemplars* (i.e., sample responses) that are used to train and monitor the raters. Exemplars are then used as *benchmarks* (i.e., responses that exemplify the meaning of different score points along the rubric) for training and quality-assurance purposes.

Rather than providing independent ratings, those involved in consensus scoring discuss the responses openly and articulate rationales to explain the appropriate score points to the raters. Such discussion is necessary to assure that raters are aligned on the scores for these responses, because they are treated as canonical scores. Another condition where consensus scoring is appropriate is when an assessment has such a low volume (e.g., is newly deployed, targets a small population, is utilized for very specialized purposes) that few (or no) exemplar materials are available to train and calibrate the raters. In these situations, raters may rely on discussions with supervisors and each other to ensure that the rubric is applied properly. That being said, the data derived from consensus scoring should not be used to calculate rater agreement statistics for the reasons mentioned previously.

Feature 4: Enabling Rater Monitoring

A well-designed scoring system should have capabilities that allow supervisors to monitor rater-scoring accuracy in real- or near-real time with appropriate quantitative tools and in ways that are actionable. On the one hand, supervisors should be able to see objective and reliable measures of scoring accuracy such as scoring statistics derived from responses that have received a *consensus score* and have been inserted into the operational rating stream for quality-control measures (i.e., *validity responses*). To collect and present these statistics, the system must track and report statistics computed on operational responses and separately report statistics computed on responses from validity papers used for quality-control monitoring.

On the other hand, scoring supervisors can also use more subjective measures of scoring accuracy by, for example, *back-rating* (i.e., "read-" or "listen-behind") a rater, which requires a rater and their supervisor(s) to be able to access a response at the same time for discussion. Ideally, supervisors can conduct both "informed" back-rating (i.e., knowing the rater's rating on a response prior to assigning their own) and "blind" back-rating (i.e., not knowing the rater's rating on a response prior to assigning their own), because both are useful for different types of monitoring.

Supervisors must also have a means to provide timely and meaningful feedback and guidance to raters. That is, they need to see *performance statistics* for the rater pool, including statistics that help identify areas of difficulty with scoring that the entire pool needs to be trained or re-trained on and identify exemplars and samples that might be problematic. This may be necessary if, for example, there is a tendency of the pool to confuse scores in the middle of the score scale based on evidence from validity paper scoring. They also need to be able to identify individual raters who are struggling and in need of *remediation*. It is especially helpful to have the system be able to automatically restrict scoring for those raters and redirect them automatically into remediation training. It can also be beneficial for the supervisors to be able to easily invalidate the scores these raters have assigned over a time window, if desired, and for the system to re-assign the responses to other raters for re-scoring.

As a general principle, information that is provided to raters and to supervisors needs to be *meaningful, specific*, and *actionable*. That is, supervisors need to use the information to decide how to best help raters learn how to score more accurately, and raters need to use the information to learn more about how to appropriately apply the scoring rubric to responses. However, amidst all the activities of a *scoring session / shift*, it is also entirely possible to overwhelm raters with too much information, and it can be extremely distracting for raters to see statistics constantly updating while trying to focus on scoring. Thus, a careful balance needs to be struck and the effectiveness of any electronic-scoring collection system for supervisors and raters needs to be evaluated through on-going research as part of the operational enterprise.

Generally, useful feedback must address those parts of the scoring rubric with which an individual rater or a team of raters is struggling and be contextualized in the specifics of a response or type of response. One of the reasons back-rating can be so effective is that it provides an opportunity for a scoring supervisor and a rater to discuss the features of a specific response to help the rater understand the rationale for why the response warrants one score level but not another.

Feature 5: Facilitating Communication between Scoring Supervisors and Raters

There are many reasons for supervisors to communicate with raters. While scoring can occur face-to-face, with an electronic-scoring system raters can also work in a distributed manner across many geographical locales, which is consequently often referred to as *distributed scoring*. While *phones* can be used in a distributed scoring model, internal *chat functions* or other text-based means of communication are often most helpful. These functions should be enabled through a user-friendly *graphical user interface* (e.g., with clear nesting of conversational threads, the ability to easily include IDs for responses) within a *secure system* to ensure that no test materials or *personally identifiable information* about raters or test takers are accidentally leaked. Beyond the core rating functions, supervisors and raters also need to know whom to contact for help with any logistical questions they have around system issues, scheduling inquiries, performance expectations, or payroll questions.

Feature 6: Ensuring Sufficient Capacity and System Performance

For operational scoring at a large scale to be successful, a digital-scoring system must be able to accommodate the expected number of *simultaneous users* (e.g., raters, scoring supervisors) without slowing down significantly. While sounding somewhat trivial, a slow system or a system that temporarily *freezes* or *crashes* can have devastating effects on scoring resulting in cognitive distraction of users due to the disruption, frustration with the process that might affect rating quality, and potential *loss of data*. It will also most certainly slow the rate of scoring, which is costly. In this regard, a system that works well during field test conditions may nevertheless have difficulty performing under operational scale, which makes pre-testing under operational conditions via operational trials or simulations essential.

Scoring Event Design

In this section, we discuss key design features for large-scale scoring events in more detail. Specifically, we discuss in-person vs. distributed scoring as well as *workforce estimation*.

In-Person vs. Distributed Scoring

A key decision to be made very early in the planning / scoping process is about whether scoring will occur in-person or in a distributed network where raters score from a secure location at either home or office. Both types of scoring have different advantages and disadvantages, which are shown in Table 10.1.

In short, in-person scoring may be preferred for perceived ease of face-to-face training and communication in the work environment but can significantly add to cost and logistical considerations and can be limiting in terms of where very large scoring events can be held. It can also be extremely difficult to adjust and is not as *robust* to issues such as unexpected scoring schedule changes or disruptions caused by issues such as large-scale *weather events* (e.g., snowstorms, tornadoes, floods) that can cause disruptions to test delivery or scoring (e.g., via resulting power outages, delayed flights, transportation bottlenecks). Additionally, Wolfe, Matthews, and Vickers (2010) compared face-to-face and distributed rater training and, while they found

TABLE 10.1

Advantages and Disadvantages of In-Person vs. Distributed Scoring

	In-Person Scoring	Distributed Scoring
Advantages	• Easiest opportunity for exchange and communication between rater and scoring supervisor during training and monitoring • Easiest to know/see exactly what raters are doing	• Easier to find raters with appropriate qualifications and availability • Scoring times do not need to be limited by when a scoring center can be open • Raters provide their own technology infrastructure (but will still need IT support) • Easier to add or reduce scheduled raters as needs change
Disadvantages	• Requires rental of large spaces if scoring requires hundreds of raters, with associated costs • Requires extensive equipment and on-site technical support for sound systems as well as IT infrastructure (computers and networks) with associated costs • Limits geographical area for recruiting raters and may require payment for travel and accommodations if raters with required qualifications within that area cannot be found • Can be limiting in terms of how much scoring can be done overall unless scoring centers are open for very long hours	• System and training must be carefully designed to maximize the ability of raters to train effectively and communicate well with their scoring supervisors and other scoring staff (extra time, effort, and cost upfront) • Perception that raters who cannot be seen cannot be trusted with secure materials, requiring various security mechanisms and legal documents • Raters may be more likely to not report for their shifts for distributed scoring • Need to manage hiring and paying raters across jurisdictions, which can create a lot of complexity because of differences in labor laws and regulations

no difference between the methods in rater-scoring quality, distributed training completed faster, which translates to cost- and time-savings.

As a result, distributed scoring is often preferable. Apart from costs-savings and less susceptibility to disruptions, it also makes recruitment from a larger geographical area easier and less costly, which subsequently allows continual scoring *across regions*. It makes it easier to recruit raters from a much larger area, which becomes increasingly important if raters require specialized credentials that limit the potential rater pool. It also makes it easier to react to unexpected changes to scoring volumes or time schedules. Furthermore, technological solutions such as the use of a *learning management systems* (LMS) and chat functions can overcome many of the perceived disadvantages for training and communication that distributed scoring has versus face-to-face scoring. On the downside, distributed scoring makes scoring more dependent on the ability of raters to use the electronic-scoring system, it requires significant upfront investments in design and maintenance, and leads to legal complexities with rater renumeration due to varying laws across states.

In the end, which type of scoring setup is more appropriate depends on the needs and purposes of an assessment as well as existing resources. In both setups, care needs to be taken to ensure that sensitive materials are handled and secured properly during the scoring process and afterward. However, given the significant advantages of an electronic score collection system, using one where possible is preferable.

Workforce Estimation

One of the most important parts of planning a scoring event is to determine how many raters will need to be recruited; we discuss this process in the following.

General Principles

In large-scale assessment contexts, especially those for which AS is used for efficiency, the number of responses requiring scores will likely be in the tens of thousands, hundreds of thousands, or even millions. The larger the number of scores required, the more human raters are needed, and the more opportunity there is for error in the estimation of how many raters are needed to complete scoring accurately, reliably, and, on time. Pilot or field test data can prove helpful for making sure the key assumptions about the process are realistic, provided these tests were administered under similar conditions to what is planned operationally. Importantly, most operational large-scale assessments have hard deadlines for reporting scores to stakeholders, which makes it critical that scoring is completed on time. The time between the testing date and when scores are operationally reported, which we refer to as the *scoring window*, involves many activities beyond scoring however.

For example, in addition to time allotted for scoring itself, time may be required for pre-scoring activities such as having scoring experts select or supplement existing exemplars for use as benchmarks in training, monitoring, and feedback. Other considerations include how many hours per day raters will be allowed to score and whether lengthening scoring shifts can be employed to extend the amount of scoring completed each day. Similarly, an important consideration is whether each rater will be required to work in every shift (e.g., in a model where raters score for an eight-hour shift every day during the overall scoring window), whether raters will be scheduled for specific shifts only, or whether raters can choose to participate in some, but not all, shifts within the scoring window. Moreover, if test takers produce responses at different times (e.g., days or even weeks apart), it might also be important to consider if more raters might be needed at certain times during the scoring window to accommodate fluctuations in volumes.

As such, it is generally easiest to consider the number of scoring shifts that are needed, rather than the number of raters at each shift, with the understanding that some shifts might have more raters scoring than others. This number can be computed in a three-step process based on estimating first the overall number of hours of *productive scoring time* and then the hours of *nonproductive scoring time*; see Table 10.2 for a brief overview.

Step 1 – Estimating Overall Hours of Productive Scoring Time

Two important factors are used in determining the overall amount of scoring time required; which we discuss in the following.

Total Number of Human Ratings Required

First, the number of human ratings that need to be assigned during the scoring window will depend on the *assessment design* (i.e., how many items/

TABLE 10.2

Key Facets for Workforce Estimation

Facet	Definition	# of Shifts Increases for...
Number of productive scoring hours	All hours that are used for scoring actual operational responses	More overall hours
Number of overall ratings	Total number of scores that need to be assigned by human raters	More overall ratings
Scoring rate	Typical number of ratings that can be assigned for each item/task	Slower average score rate
Number of nonproductive scoring hours	Total number of working hours that are not used for scoring	More nonproductive time
Break time	Time for lunches and resting	More break time
Training and retraining time	Time for training raters before scoring starts or retraining them while scoring is in progress	More training time
Rater evaluation time	Time for evaluating the readiness of raters for scoring before scoring starts or while scoring is in progress	More rater evaluation time
Rater proficiency	Number of raters that do not qualify for operational scoring despite training	Fewer proficient raters
Rater attrition	Number of raters who drop out of the scoring window for unexpected personal reasons	More rater attrition
Assessment design	Key characteristics of the assessment that affect variation in scoring workload	Depends on interplay of factors

tasks need to be scored), the *volume of test takers* expected, and the *scoring design* (i.e., how many ratings are required per response). For example, if all responses need to be *double-scored* by raters, then at least twice as many ratings are needed than if only single ratings were assigned. If single ratings are used, then only a sample of the responses is typically assigned to get a second independent rating for quality-assurance purposes (i.e., to compute *inter-rater agreement* statistics).

If double-scoring is used, whether by two humans or one human and an AS engine, then score resolution rules must be considered. If a third rater from the pool or a scoring supervisor is used to resolve discrepancies, then those ratings will need to be added to the total expected count. For example, if a difference of more than one score point between ratings is set to require a third rating and these discrepancies occur in 2% of responses based on historical data, then an additional 2% of human ratings would be required. Likewise, if validity responses will be used – which we highly recommend – then the validity response *insertion rate* must be factored in as well, which depends on many factors (see Ricker-Pedley, 2019).

Average Scoring Rate

Once the total number of human ratings for a scoring window is determined, it is critical to be able to estimate how much rater and scoring supervisor time is needed based on reasonable estimates of *average scoring time* for a given type of response. For example, if raters are scoring short answers (e.g., a few sentences or a paragraph), then the average scoring rate for the rater pool might be well over 50 responses per rater per hour, whereas raters scoring long and complex performance assessments with multiple traits (e.g., extended essays, portfolios) might score only a few responses per hour – perhaps even just one or less than one. The number of score points in the rubric, the grade level of the test takers, and the content domain of the item / task can all have an influence on the scoring rate, regardless of the length of the responses themselves. Thus, it is sometimes most accurate to calculate rater pool needs separately by grade, item / task type, or item / prompt. Once the number of required human ratings and the scoring rate have been estimated, then the number of scoring hours required can be calculated by the following simple formula:

Scoring hours = total number of required human ratings/scoring rate

Step 2 – Estimating Overall Hours of Nonscoring Time

Once productive scoring time is estimated, nonproductive scoring time must be estimated as well. This time includes time for activities like breaks, training, rater evaluation, and any other time incorporated into the scoring window. Nonscoring timing is considered separately from scoring time because, unlike scoring time, it cannot be divided among additional raters to lessen the load. For example, if every rater gets a 15-minute break, then four raters each get a 15-minute break leading to a loss of one hour of productive scoring time.

Break Time

Breaks are important to factor into scoring times for ensuring raters do not become overly fatigued in addition to adhering to labor laws. For example, if raters score for an eight-hour shift, we might assume two 15-minute breaks as well as a 30- to 60-minute midday meal break. That meal break might be unpaid but must still be factored into scoring scheduling time. As might be imagined, if scoring is occurring in a distributed setting, several scheduling options might be considered (e.g., some four-hour shifts, some eight-hour shifts). This introduces additional complications for calculating how much scoring time is needed, particularly if raters have choices as to the scoring shift length, and for managing the rater assignments.

Training and Retraining Time

Training time is an important consideration in planning a scoring event. Training involves orientations on how to use the scoring system and understanding the overall scoring procedures as well as more substantive training on the general features of the scoring rubric and how to apply it consistently. Training might be possible to do in advance of the actual scoring session, particularly if using online training via an LMS or *webinar technology*; if an in-person setup is used, it might be more practical to wait until the scoring window begins. Training based on secure materials such as item-specific scoring guides, notes, benchmarks, and sample responses may have to wait until the operational scoring window begins no matter what setup is used for test security purposes. In addition, it may also happen that some materials (e.g., sample responses) might simply not exist until immediately prior to the scoring session when they are selected from incoming operational responses.

General orientation and training on an item / task might take anywhere from a few hours to a full day or more, depending on the complexity and difficulty of what the raters will be scoring. Additional time should be factored in for any possible retraining that might be needed during the scoring session if issues with scoring accuracy are uncovered. When factoring in training time, the fundamentals of the scoring design mentioned above will also come into play. For example, if raters will score only one item/ task during the scoring window, then training time will be less than if raters will switch items / tasks throughout the scoring window and will need to re-train.

For the purposes of considering nonscoring time, it is helpful to calculate the *average training time per shift*. For example, if a total of six hours of initial training time is assumed, along with three hours of retraining time and 10 scoring shifts, that results in an average of 54 minutes (or 0.9 hours) of training time per shift.

Rater Evaluation Time

Rater evaluation procedures, sometimes known as *calibration* or *qualification* processes, are frequently used prior to beginning a scoring shift to ensure that raters can produce consistent and accurate scores. The frequency of rater evaluation during operational scoring can have a significant impact on total nonproductive scoring time. For example, raters may be evaluated at the beginning of each shift, every few days, or every few shifts, with the amount of time spent on rater evaluation varying significantly as a result.

Understanding the average scoring rate is also important for estimating time needed for rater evaluation. However, as a note of caution, experience has shown that raters often score much more slowly during evaluation sessions than during regular scoring – often up to half the rate – likely because they are very aware of being evaluated and the stakes are high (i.e., if they fail calibration too often they may not be assigned to score and may not get paid for a particular shift). Like training and retraining time, it is easiest to calculate the average time per shift. For example, if the rating design prescribes that a rater must calibrate every third shift and this evaluation is assumed to take 15 minutes, then an average of five minutes of evaluation time per shift is effectively assumed.

Step 3 – Calculating the Required Number of Raters

Once the number of scoring hours and nonscoring hours is known, the *number of raters* can be calculated. As a simple illustrative example, consider a large-scale assessment with two items / tasks that yield extended responses. For both items there are responses from 500,000 test takers, which is comparable to data from one grade level in a reasonably large statewide K–12 assessment. Furthermore, assume that Item 1 has an estimated scoring rate of 25 responses per hour per rater and that Item 2 has an estimated scoring rate of 20. Both items will be double-scored; the first item will be scored by two human raters while the second item will be scored by one human rater and an AS engine.

The scoring rules for both items allow for a difference between ratings of one point without adjudication. Any discrepancies that are either greater than one point between two human raters or 1.50 points or greater between the human rating and the automated score will require a third rating by a human rater from the same pool (i.e., not a scoring supervisor). Both items have an estimated discrepancy rate of about 2% based on field test data. The second item has approximately 0.5% of ratings that cannot be scored by AS for reasons specific to the response (e.g., too short, too long, too many grammatical errors), which must be human-scored. Both items have a validity response insertion rate of 10% for rater-monitoring purposes and the second item will have 1,000 additional responses randomly selected for a second human rating for the purposes of collecting inter-rater agreement data and to update AS models in the future. Note that the assumption here is that the 1,000 responses are distinct from the responses requiring additional human ratings from insufficient human–AS agreement, but 1,000 is effectively the upper limit as there is likely some overlap between the two sets of responses.

Finally, assume that raters will work in eight-hour shifts with two 15-minute breaks and a 30-minute meal break for a total of one hour of nonproductive time due to breaks per shift. Raters are required to complete a six-hour training session prior to the scoring window and, during the scoring window, an average of 20 minutes of retraining/remediation per shift is assumed as well as an average of 10 minutes of rater evaluation per shift. Therefore, the overall total is 1.5 hours of nonproductive time per shift leaving 6.5 hours of productive scoring time.

With these data at hand, the overall number of *required human ratings* is computed as follows:

$$\text{Total \# of ratings} = \left(\text{Total \# of responses } x \text{ \# of human ratings per response}\right)$$
$$+ \text{Total \# of adjudication ratings}$$
$$+ \text{Total \# of second human ratings}$$
$$+ \text{Total \# of validity ratings}$$

This yields the following for our scenario:

Item 1 total ratings: (500,000 responses × 2 human ratings) + 10,000 adjudication ratings + 101,000 validity ratings = **1.111 million ratings**

Item 2 total ratings: (500,000 responses ×1 human rating) + 10,000 adjudication ratings + 3,500 second human ratings (1,000 maintenance + 2,500 where AS fails) + 51,250 validity ratings (10% of the sum of all other ratings) = **564,850 ratings**

Total ratings: 1.111 million + 564,850 = **1,675,850 million ratings**

Next, the overall number of *required scoring hours* is computed as follows:

Total # of scoring hours = Total # of ratings/average read rate per hour

This yields the following for our scenario:

Item 1 scoring hours: 1.11 million ratings/25 ratings per hour = **44,400 scoring hours**

Item 2 scoring hours: 564,850 ratings/20 ratings per hour = **28,242.5 scoring hours**

Total scoring hours: 44,400 + 28,242.5 = **72,642.5 scoring hours**

Finally, the number of *required shifts* is determined as follows:

Total # of shifts = total # of scoring hours/average # of scoring hours per shift

This yields the following for our scenario:

Item 1 rater shifts: 44,400 hours/6.5 scoring hours per shift = **6,837 shifts**

Item 2 rater shifts: 28,242.5 hours/6.5 scoring hours per shift = **4,345 shifts**

Total rater shifts: 6,387 + 4,345 = **10,732 shifts**

The final number of raters will now depend on the length of the overall scoring window during which the work needs to be completed, how many eight-hour shift periods are available for scoring overall, and how much flexibility raters have in selecting scoring shifts.

Additional Considerations

The above considerations relate to design factors that are largely under the control of the organization that develops and oversees the rating procedures; the following two factors are under less direct control.

Rater Proficiency

The proficiency of raters in the overall rater population from which the rater pool is created is a significant factor in estimating the size of the needed rater pool and the overall number of hours expended for a rating project. For example, it is reasonable to assume that some proportion of selected raters will not successfully pass rater evaluation and, therefore, may not be able to score.

Rater Attrition

With a large rater pool, it is also reasonable to expect that some raters might drop out of the project unexpectedly or might cancel / miss selected shifts because of personal issues, so it is prudent to assume a small number of raters (e.g., about 5% based on experience) will not be able to work their scheduled shifts at any given time; this phenomenon is known as *attrition*.

Assessment Design

In the example above, all test takers responded to the same items / tasks. In a more realistic complex test design, there may be multiple prompts for each item / task, each associated with potentially different read rates; representative field testing of items / tasks / prompts can help provide data for developing read rate assumptions. Further complications may include the use of *multi-stage* or *adaptive testing*, which makes it difficult to predict the exact test-taker volumes for any item / task, or long testing windows during which it is difficult to predict when exactly test-taker responses will arrive for scoring.

A Note on Scoring Supervisors

Depending on the model of supervision that is used, scoring supervisors can serve many different functions including monitoring rater quality, providing guidance to raters, assigning ratings to responses that are more difficult to score (including adjudication ratings), and identifying potential additional exemplars. They might also be involved in managing rater work such as reaching out to raters who are not logged into the system when scheduled and managing time for breaks. When organizing a scoring event, the organizers must consider which roles the scoring supervisors will play to determine how many are needed at any time / shift and what characteristics they will need; at a minimum, being able to provide accurate ratings and effective feedback. The nature of the work expectations will also determine how many scoring supervisors are needed. Shortages of scoring supervisors can become a bottleneck in scoring even if adequate raters are available to score, so careful attention to scoring supervisor workforce needs is important for success.

Rater Training

General Principles

One of the biggest organizational challenges in a large-scale scoring event is how to coordinate and deliver training. If all raters are physically co-located and score at the same time with an in-person setup, then face-to-face training led by a rater supervisor might be most efficient. However, if scoring is distributed, then either training will need to be held live via an online meeting application or raters will need to be trained asynchronously using digitally delivered materials.

The advantage of live training is the opportunity for raters to directly interact with a supervisor / trainer and ask questions face-to-face. With very large groups of raters, however, those opportunities might be limited, especially in a distributed presentation where raters may feel less inclined to be engaged. It might also be difficult to have all raters scheduled at the same time, especially if the scoring window will be a long time and raters might not all start scoring at the same time. This would require multiple live training sessions, which could lead to inconsistencies in how raters are trained.

When raters self-train, the obvious advantage is convenience for the rater and consistency in what materials are provided. It might also be advantageous to raters who benefit from the ability to revisit or repeat materials when needed, or who require a different pace than what might be used with a larger group of people. To ensure that all raters complete all materials, using an LMS is recommended, which can track what a rater reads / watches and can issue automatic certificates upon the satisfactory completion of embed-ded evaluations. The ability for raters to be able to access and refer to train-ing materials on demand during operational scoring is also recommended.

Initial Training

Regardless of the choice between in-person and distributed training, it is important that initial training provides all the information that raters need to score accurately using the correct processes and procedures. First, time should be spent with raters thinking about potential sources of bias in scor-ing generally and what their own biases might be to help reduce their influ-ence. Providing information about the prompt and the rubric is of course key but training also needs to involve a significant amount of time with exem-plars responses to help contextualize the rubric in the responses.

Examples of responses at all ratings levels, with different response char-acteristics, should be presented. This includes variation in *length* (i.e., short and long answers), *linguistic structures* (e.g., poor and sophisticated gram-mar/spelling whenever they are not relevant to scoring rubric), and types of *construct-irrelevant variance* to help the rater have a fuller understanding of all the ways that a response might fit into a score category. *Annotations* should explain why a response warrants a score and why it does not warrant a neighboring score, especially if the response is representative of responses near the conceptual boundary between two score points. Raters should move from more structured / guided training on clear-cut samples at each score point to less structured / independent scoring with immediate feedback on less clear examples across the score points. Raters may need to complete sev-eral training sets before they are ready to calibrate.

Supplemental Training and Remediation

Even with excellent initial training, monitoring scoring quality may reveal previously undiscovered issues with scoring accuracy that can arise over time. For example, the rater pool may have unexpected difficulty with dif-ferentiating between two neighboring score points. Once such issues have been detected and diagnosed, supplemental training sets, usually targeting specific score points or types of responses, can be created and deployed to individual raters or the rater pool overall. Struggling raters can be required either to repeat initial training or to complete a new training set. Afterward, it is common practice to require that a rater successfully complete either an evaluation set or other short assessments before being allowed to return to operational scoring. Another option during operational scoring is to employ *seeded feedback*, which inserts an exemplar response with a canonical con-sensus score and annotations amongst the operational responses-basically, a validity exemplar that provides instantaneous feedback to raters once they enter a rating. This type of training is particularly effective in cases where new types of responses are uncovered during operational scoring that were not represented in the original training set.

Rater Monitoring

General Principles

Monitoring human scoring in high-stakes testing is always important to minimize risks of scoring errors that lead to reductions in score reliability as well as associated threats to the interpretation and use of test scores. Importantly, it is also critical whenever the intention is to deploy AS in the short or long term since human scores are commonly used to train and evaluate AS models as well as to monitor their continued performance after implementation. When trying to decide on the statistical criteria for *flagging* raters or rater pools, care must be taken to consider several factors, including the purposes, stakes, and potential negative consequences associated with the scores if they are incorrect, the design of the items / tasks, and the contribution of the constructed-response scores to the total calculation of the final section or overall assessment scores.

For example, statistical criteria and associated resolution actions are often more stringent for items / tasks whose scores contribute a significant portion to the total test score, especially in high-stakes contexts (e.g., on a certification/ licensure exam or a graduate admissions exam). Complex *holistic rubrics* are going to be more difficult to apply consistently in scoring than more straightforward or *analytic rubrics*. As a result, if constructed responses fail to consistently produce human ratings that meet required standards of score quality, despite interventions to enhance rater training and remediation, then changes to the design of the items/tasks and associated scoring rubrics may be necessary, especially if the desire is to be able to eventually deploy AS models.

Amount of Monitoring

Quality assurance in scoring can be a significant cost driver of constructed-response scoring. Therefore, it makes sense to try to monitor only as much as is needed to ensure that quality is appropriate for the stakes of the test. If second ratings are collected purely for evaluating consistency / accuracy of the rater pool, then it would make sense to only apply second ratings to a (stratified) random sample of the total response pool rather than all responses. For example, double-scoring of a random sample of 10% of responses up to some maximum number (e.g., 5,000) could yield a sufficient sample size for evaluating the rating pool performance, but a higher second rating percentage might be needed to be able to evaluate individual raters with sufficient data if inter-rater agreement is used as a monitoring criterion.

Likewise, the number of validity responses depends on the purposes that the validity data are used for as well as the stakes of the test. For example, it may be common practice to use an approximately 10% validity insertion rate within a given use context (i.e., roughly every 10th response is a validity response), but lower percentages might be sufficient if the rater pool is well trained and few raters are expected to require remediation. Using lower insertion rates means that it will take longer to accumulate enough data to evaluate a rater or raters however. It might therefore be desirable to seed in validity responses at a higher rate when raters begin scoring and then reduce the percentage once enough validity data are collected.

Furthermore, if the data are used to help guide scoring supervisors in deciding which raters to prioritize for back-reading and feedback, then a lower validity seeding rate might be acceptable. In contrast, if the intention is to detect raters with poor accuracy and invalidate the ratings for the responses they have scored, the additional cost of a higher validity insertion rate might be worthwhile because it will reduce the amount of re-scoring that is needed (see Pedley, in preparation, for more discussion).

Timing of Monitoring

When and how often to monitor scoring is driven by a few things, but the overall driver of monitoring is how quickly the team working on the scoring project can act on the results of the monitoring. Simply monitoring scoring quality is not enough to ensure good quality. In other words, more monitoring is not always better and, in fact, may waste resources that might be better used to act on less frequent monitoring. For example, rater supervisors continuously monitor the raters assigned to them, because they can take immediate action on any information they have by intervening with the raters.

In contrast, when someone monitoring the overall quality of scoring during scoring identifies an issue with the scoring of an item / task, it might first take several hours for a team to investigate and determine a course of action. As a result, materials might need to be prepared and deployed (e.g., new training sets or a communication to raters / scoring supervisors about an issue), and then several additional hours / days might have to pass to evaluate whether the intervention has had the desired effect. Thus, if this cycle takes several hours or even a few days, then hourly monitoring is probably not helpful or a good use of resources.

Time should also be considered a resource. Identifying issues early in scoring is critical, particularly if scoring windows are short. Otherwise, there might not be time to take meaningful action or significant re-scoring may be needed later if warranted by the stakes of the test. Having a well-coordinated team with a thorough quality monitoring plan (i.e., clearly defined statistical criteria and sequences of actions that should be taken when any issues are identified) will help make reaction to issues in scoring faster, which will then be more likely to have time to be effective.

Conclusion

Constructed-response scoring for high-stakes standardized assessments is always complex and becomes increasingly complex and resource-intensive as the scale of the scoring enterprise increases. What might be a simple process at a classroom, school, or district level becomes an enormous project at the state or national level. While it is difficult to make specific recommendations because the circumstances for each assessment will call for different requirements, there are certain general recommendations that we can reiterate at this point:

1. It is critical that the design of the overall scoring process and the supporting electronic ratings collection systems support the requirements of the assessment in terms of the reliability, validity, and fairness. Proper planning of the entire scoring process is critical.

2. The scoring process design must be carefully customized for each assessment context to determine the appropriate number of ratings, shifts, and raters. Not carefully planning all aspects of scoring can lead to problems with managing production deadlines, poor score consistency, or scores whose validity is questioned.

3. The larger the scale of the scoring event, the more care must be taken to assure that training of raters is comprehensive and consistent among all raters.

4. Once trained, special care must be taken to ensure appropriate quality-assurance processes to allow for an effective monitoring of the performance of all individual raters as well as the overall rater pool. A plan should be put into place to identify how different scoring issues will be identified and what actions will occur so that they can be addressed in a timely manner. Without timely action, quality monitoring is just expensive documentation.

5. Feedback during scoring should be prompt, explanatory, and actionable to allow raters to improve performance with minimal additional training and to avoid remediation and re-scoring events.

6. Even when the goal is to eventually use AS for most of the scoring process, accurate and consistent human ratings will always be needed to either build AS models and/or to evaluate their performance during deployment.

7. Key assumptions required for psychometric analyses to evaluate rater performance (e.g., independence of ratings for inter-rater agreement statistics) should ideally be met or, alternatively, the robustness of these analyses to common violations of these assumptions must be clearly understood.

All these aspects require nontrivial amounts of resources for the design, evaluation, and implementation of systems. This, in turn, requires the interdisciplinary collaboration of a team of experts from assessment development, psychometrics, learning sciences, human–computer interaction, and user-interface design, to name but a few.

11

System Architecture Design for Scoring and Delivery

Sue Lottridge and Nick Hoefer

CONTENTS

In this chapter we describe the fundamentals around building *automated scoring* (AS) software for assessment programs. The scores produced by AS software are part of a rigorous *psychometric and reporting pipeline* and must adhere to the common *professional standards* for score quality and fairness outlined in the *Standards for Educational and Psychological Testing* (APA/ NCME/AERA, 2014). As such, the *architectural design* must consider the *validity, reliability,* and *fairness* of scores delivered. It must also be able to train models quickly and process responses at speeds and volumes that meet the assessment needs at low cost. Finally, it must also enable monitoring of scoring performance and the ability to conduct research to improve the system.

Because architectural design is covered in many books (e.g., Bass, Clements, & Kazman, 2013; Evans, 2003), our focus in this chapter is on the unique problem of designing AS software. Although AS software can produce *qualitative feedback,* we focus on the production of *scores* (holistic, trait-based, or otherwise) modeled on *rubric-based human scoring* and applied to examinee-typed *responses.* We also assume the use of the scores in other aspects of the psychometric and assessment operations – such as combining engine scores with human-applied scores or with other item scores to produce an overall set of test scores – will be managed by other systems. Still, we believe the ideas and procedures covered in this chapter support broader applications. These applications include the provision of *non-rubric scores* (e.g., specific feedback on the variation in sentence types) and other *modalities* (e.g., spoken responses).

We have organized this chapter as follows. In the first section, we discuss stakeholder considerations. In the second section, we discuss use cases such as summative, interim, and formative implementations. In the third section, we discuss software considerations arising out of the stakeholder and use case sections. In the fourth section, we provide more detailed recommendations around architecture components. We close the chapter with a few recommendations for team composition when developing software.

Stakeholder Considerations

Stakeholder considerations are the foundation for software requirements that describe the features of a system and drive the architecture of AS software (IEEE, 1998, 2009). Specifically, the primary drivers for using automated scoring are the reduction of hand-scoring costs, faster score turn-around and reporting, and the provision of accurate, consistent, and fair scores (Boyer, 2019).

The design, implementation, and monitoring of software architecture is influenced by myriad factors that can be ascribed to one of five stakeholder contexts. These contexts are highlighted in Figure 11.1 and are described in this section; at a high level, these are (1) psychometrics, (2) natural language processing / data science, (3) test operations, (4) assessment applications, and (5) software.

Psychometrics

Psychometrics is interpreted broadly (APA/NCME/AERA, 2014) to include *foundations* (validity, reliability, and fairness), *operations* (test design and development, score production, and test administration), and *documentation* (technical reports in support of the foundations). In particular, scores produced by AS software are analyzed using psychometric methods to evaluate their validity, reliability, and fairness when replacing human scorers and when combined with other information to produce test scores such as scaled scores and achievement levels. The ability of AS software to properly model human scoring is dependent upon the *item development plan, forms administration procedures, sampling plans,* and *hand-scoring requirements* defined by psychometric best practices. In this way, then, the AS software architecture must consider the choices and needs of the psychometric team.

NLP/Data Science

Natural language processing (NLP) and *data science* provide analytic frameworks complementary to psychometrics when building AS software. NLP is concerned with representing and modeling language and, in current

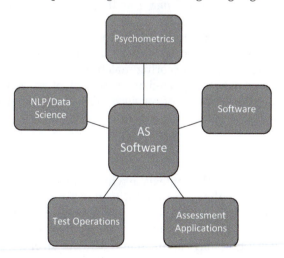

FIGURE 11.1
Key areas affecting the architecture of AS software.

practice, many of the expert-designed features used in AS software are built by NLP experts or using NLP tools (Lottridge, 2018; Rupp, 2018). Data science is a larger analytic framework concerned with all aspects of the analytic process including data capture, data processing, prediction of values, and maintenance of data (Berkeley, 2019). The architecture of AS software relies on the methods used in each of these areas. NLP methods operate at the response level, modeling aspects of language hypothesized to be related to the rubric. The validity and reliability of the NLP methods has overlap with psychometrics because the NLP features support the construct validation processes (Williamson, Xi, & Breyer, 2012). The data science methods drive the statistical modeling, overall model training, deployment, and monitoring processes (Géron, 2017; Zinkevich, 2019).

Test Operations

Test operations refers to the actual administration of tests and the reporting of scores during and after the assessment of examinees throughout which AS software is one element related to the flow of data (CCSSO & ATP, 2013). Typically, a test is administered to examinees via a *test delivery platform*. The AS software receives responses from the test delivery platform and returns scores that are then used by downstream systems to produce *aggregate scores* and *score reports*. In terms of architecture, care must be taken in defining specifications for *incoming requests* and *outgoing responses* (i.e., scores and score meta-data) to ensure that unexpected changes are identified and handled appropriately.

For instance, changes to the examinee input field or storage of responses can impact how the software processes and scores responses. In terms of outgoing responses, *downstream programs* that use data to route responses to hand-scoring, to provide reports to users, or to calculate total scores may be impacted by changes in the meaning and calculation of these outputs from the engine. In addition, care must be taken to ensure that the *latency* and *volume expectations* around scoring match those in the larger flow; item scores not returned when expected may disrupt other activities such as item administration in the case of *computer-adaptive testing* or hand-scoring activities.

Assessment Applications

Assessment applications are the contexts in which scoring is conducted, including workplace, credentialing, and educational testing (APA/NCME/AERA, 2014). In our experience, automated scoring has been used primarily in educational testing in a summative, interim, or practice / formative context. Specifically, *summative testing* can include K–12 state-level testing conducted as part of federal law – currently, the *Every Student Succeeds Act* (ESSA) – or for college or graduate school entrance. AS has also been used in *interim testing*, which is typically offered by states or districts to measure student progress toward proficiency. Finally, AS has been used in *practice / formative contexts* in which students can take tests to measure their own understanding and skill with or without guidance from an instructor. The type of assessment application affects architecture because it can impact requirements around *reproducibility* of scores within and across *testing windows*, the degree of *system availability*, the level of *audit data* stored, and the frequency in which models are *calibrated*.

Software

Software development principles and practices are closely related to software design or architecture (Scaled Agile, Inc., 2018). Like any other software, AS software must be built, tested, deployed, and maintained. AS software development typically uses *prototype methods* whereby a prototype is built by NLP experts and data scientists, which is then evaluated for its performance by key stakeholders, including psychometricians. A successful prototype is then refactored or rewritten to create production-ready software that meets organizational standards around testing, security, privacy, stability, latency, and volume. Testing of AS software is heavily data-driven and reliant on trained model data and thresholds. Finally, examinee data may contain *personally identifiable information* because examinees may provide their name, phone number, or address in their response. Thus, access and storage of such data must adhere to the laws and principles where AS is used.

Architectural design of the AS software is affected by all of the above factors. Of course, organizations will prioritize some contexts over others, which will affect software design. For example, portions of the test operations may involve *third-party systems*, which typically involve already-fixed *interfaces* for data collection and management that AS software must deal with. Finally, it is helpful to remember that change is constant in each of these contexts; an effective design must account for the evolution of the *stakeholder contexts* over time.

Illustrative Use Cases

In this section, we first describe a high-level use case for AS software that was designed to represent characteristics that are common to most implementations. While the flow of steps we describe is common across specific implementations, details around the flow such as speed of testing will vary by implementation. To this end, we wrap up this section with a brief description of these details for the common assessment applications: *low-stakes* (e.g., formative or practice assessment), *medium-stakes* (e.g., interim assessment), and *high-stakes* (e.g., summative assessment). The stakeholder considerations described in the previous section should be kept in mind while reviewing these use cases.

High-Level Use Case

The high-level use case for AS software involves training, evaluating, and implementing an AS model based on human-rater scores that serve as the *target of prediction*. The different steps involved in the AS work include sampling, training, validation, model approval, model packaging, system deployment and testing, and scoring, which are shown in Figure 11.2. These steps are typical in the training and deployment of *machine-learning models* (James, Witten, Hastie, & Tibshirani, 2013; Géron, 2017; Zinkevich, 2019). In the following, we describe each step and note the roles of key parties for each step, which we summarized in Table 11.1.

Data

AS software is typically designed to model human scores by capturing key response characteristics of human raters. This core purpose means that the process of training rests on access to responses and human-applied scores. The *hand-scored responses* are typically generated under the leadership of psychometricians and conducted by hand-scoring teams. In some cases, meta-data about the item are included in the model-training process. Examples of these data include associated stimulus material, rubrics, or presentation condition meta-data (e.g., mode of input). Once handscored and loaded into a system, access to these data are managed by data experts in the organization. Psychometricians are typically accountable for the quality of the data, defining specifications around sampling and hand-scoring.

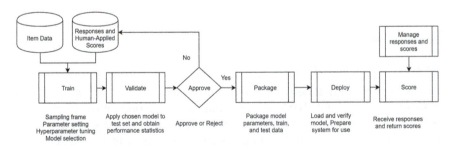

FIGURE 11.2
Overview of AS model development and deployment.

TABLE 11.1

Team Roles for Developing AS Software

Role	Role Description
Business Owner	This person understands the business concerns deeply and is presumably well versed in scoring, psychometrics, data science, and technology. Their role is to lead the team in terms of high-level goals and direction.
Software Developer	This person has generalized software skills, but with knowledge of testing, automation, deployment, and the software used in the training. It is preferable for this person to have a strong data and mathematics background with a machine-learning background a plus. This also includes front-end developers for interfaces.
Deployment Engineer	This person is responsible for packaging the software for deployment and defining the specifications for the appropriate hardware given the business needs and software capabilities.
Data Scientist/Statistician	This person provides knowledge of the training pipeline, including sampling methodology, text normalization, feature extraction, and particularly the appropriate methods for estimation. This person also provides test data for verifying engine performance.
Psychometrician	This person provides metrics and standards for model evaluation and use, specifications for data access and sampling, analyzes performance of models, and writes reports on results. This person, combined with the data scientists, may provide descriptions and documentation on the engine high-level design for clients.
Natural Language Processing Expert	This person provides insight and input into training pipeline, including the development of sound methods for text normalization and feature extraction, and the interpretation and justification of those choices.
Content/Hand-Scoring Expert	This person or persons provides insight and input into the intent of the item and item rubrics, as well as the interpretation of the rubric as used in range-finding and hand-scoring. This information is critical in determining the types of features to develop to align to the rubrics and items the engine is intended to score, and in providing guidance on how human raters interpret responses and rubrics that can be operationalized in the software.
Business Analyst	This person analyzes business and user needs and translates user requirements to specifications interpretable and testable by software engineers.
Technical Writer	This person writes technical documentation including documentation, training materials, and user guides.
Technical Program Manager	This person manages the software development process, resourcing, and timelines.
Quality-Assurance Manager	This person evaluates whether software meets the criteria outlined in the requirements.

Training

Once training materials are obtained and stored in a data repository, the *training step* begins. In this step, rules for sampling the data are defined, models are chosen, and modeling parameters determined. If relevant, *hyperparameters* – that is, parameters which govern the machine-learning model functioning – are examined and tuned. The outcome of this step is a set of models that are evaluated with one model chosen for moving forward to the validation step. This step is typically undertaken by specialists who are familiar with the features and methods, are familiar with response data characteristics, and have statistical and psychometric knowledge. Psychometricians are typically accountable for the training specifications and models generated.

Validation and Approval

Once the final scoring model is chosen based on training results, the *validation step* is undertaken. In this step, this model is used to score the validation data. A report on the performance of the model relative to the human-applied scores is produced out of this process. Once produced, the performance of the engine is evaluated. If the engine results meet performance criteria, then the model is approved for production use. If not, then more work is conducted to improve model scoring (e.g., more representative data are obtained, engine feature sets are improved, modeling approaches are changed). The approve-or-reject decision is determined by the person(s) – typically psychometricians – with responsibility for scoring in the program that includes the assessment for which the item under question is used.

Packaging

Once a scoring model is approved, then the model is *packaged* and deployed for production use. This packaging includes all information needed to score the model (e.g., parameter estimates) and typically also contains supporting information regarding the training and validation data sets. This package can also contain self-describing information to alert the *calling program* to the model characteristics and versioning. In essence, all information necessary to retrain, validate, and use the model is stored in the model package. Unlike the earlier steps, this procedure is primarily a software engineering task.

Deployment

Once the model is packaged, it is then loaded into a *secure repository* accessible to the AS software deployed for scoring. During this step, a number of tests are enacted. First, a representative set of responses is submitted to the model to ensure that the correct scores and score meta-data are returned. Second, the software is tested for *latency, stability,* and *accuracy* under various testing loads. Finally, the integrations of the engine with upstream and downstream systems are tested to ensure the flow of responses and scores functions according to expectation. The functions performed at this step are fully dependent on data and require software engineers and quality-assurance personnel with strong data skills. These tests are repeated as the software makes its way through the steps in the release process. These include quality assurance, user acceptance testing, and production releases. Once the engine passes user acceptance testing, it is deployed for production use.

Scoring and Monitoring

Once the engine is deployed for production use, the software is part of test operations, receiving scores from the test delivery system and returning scores. While the scoring process is occurring, systems are used to manage and monitor engine scoring. For example, systems may monitor the latency of the scoring by sending targeted responses at regular intervals and computing return times. *Infrastructure health* is watched (e.g., memory usage, CPU usage) by software engineers and scoring functions are tracked (e.g., percentage of responses at each score point) by psychometricians.

Specific Use Cases

As we noted in the first section, AS software is used in the context of assessment applications that have different stakes for examinees or other stakeholders; this has architectural implementations. In this section, we describe three use cases:

1. a *low-stakes use case* whereby examinees use the scores and score meta-data to improve their own learning,
2. a *medium-stakes use case* whereby stakeholders (e.g., teachers) use the scores to monitor learning and improve instruction, and
3. a *high-stakes use case* where stakeholders (e.g., state education entities) use the scores to assess learning.

Each of these use cases has design elements that impact architectural design in critical ways. Considerations include tolerance for system down-time, duration of access needed throughout the year, requirements around score reproducibility, requirements around the speed in which scores are returned (latency), and data that may accompany scores (meta-data); we discuss these in more detail in the next subsections.

Low-Stakes Use Case

In low-stakes implementations, the AS software is integrated with assessment systems in which an examinee enters a response and expects to see results immediately. Typically, the examinee can use the system at any time.

The purpose of the formative system is to provide the examinee the opportunity to practice, receive immediate feedback, and use that information to learn. This use case has several architectural implications. First, the software must always be available in a day across a *broad time window* (e.g., full year or academic year) and return results immediately. Because this is a practice system with few immediate consequential decisions, the scores produced by the software need not be *identical over time* for a given associated response.

In fact, in such systems it may be preferable that the software improve over time (i.e., learn) as more data are gathered, which means that underlying scoring models and associated predicted scores will change. Moreover, these systems typically provide more detailed feedback to the examinee. Thus, the AS software may return expert-feature *qualitative output* (e.g., thesis sentence identification, citation quality) in addition to a score. Given these requirements, the process outlined in the previous subsection typically needs to be implemented at regular intervals (or in real-time) to continually update the AS model. This has to be balanced with deployment and testing consideration so that these processes do not interfere with the reporting of results.

Medium-Stakes Use Case

In medium-stakes implementations, the AS software is integrated with assessment systems that are part of a larger suite of items, tasks, or activities. The goal is to give instructors or other stakeholders (e.g., teachers, school administrators, district administrators) information on student performance in the aggregate for the purposes of improving instruction and curriculum and as professional development. Results from these systems are typically expected to be available within a work-day period within a defined *testing window* (e.g., a semester). Clients expect no or minimal delays in the return of scores to ensure that the data can be used effectively. The scores provided by the AS software for a given response should be *reproducible* to ensure fairness in scoring to all examinees across the window. Because the item score data in which the AS software is used are combined with other item data, only scores (versus qualitative feedback) are typically provided in these systems. Given these requirements, the process outlined in the previous subsection is typically implemented prior to the testing window while the system is only maintained during the testing window, with updates conducted as needed during off-hours during the day.

High-Stakes Use Case

In high-stakes implementations, the AS software is integrated with assessment systems that are part of a summative test with a goal of providing the highest-quality assessment information. The stakeholders can include schools, certification agencies, and governments. Results from these systems are typically expected to be available within a work-day period within a narrowly defined testing window (e.g., two weeks). Clients expect delays in the return of scores to ensure that the data that are reported are accurate and adhere to strong psychometric principles around reliability, validity, and fairness. The score data may be combined with ratings from professional human-scoring efforts as well.

Just as in the medium-stakes use case, the score provided by the AS software for a given examinee response should be reproducible to ensure fairness in scoring to all examinees across the window. Because the item score data are combined with other item data, only scores are typically provided by these systems. Because the primary goal is the accuracy of scores, more frequent oversight, specifications, and testing of the process outlined in the previous subsection are conducted. Great effort is made to ensure the scores coming out of each step meets professional quality standards and expectations.

Taking the time to outline the particular design constraints for a given use case – or for sets of use cases – when developing AS software can offset confusion and issues during development. AS software has many stakeholders because there are multiple stages in modeling and software deployment that involve different roles and because the team developing the software is often multidisciplinary. Having a common understanding of the key needs and requirements for AS software (e.g., reproducibility of scores during a test window) can prevent day-to-day choices that may contradict those requirements when coded (e.g., ignoring randomness in model training). These seemingly small decisions can have negative consequences when identified too late.

Software Context

This section builds upon the prior sections to further illustrate the software context that impact architectural design of AS software. We chose to highlight software context because it drives AS software development and architecture. It is known that software systems transform with advances in technology, evolving stakeholder contexts, and hardware improvements; AS software is no different. As an example, K-12 assessment evolves slowly but has indeed evolved over time as illustrated by the relatively new ESSA regulations that allow for more flexibility in assessing students (CCSSO, 2018).

The field is also recommending "high-quality" assessments that measure *higher-order cognitive skills*, are *instructionally sensitive*, and have *high fidelity* (Darling-Hammond et al., 2013). Finally, states are increasingly using *online testing* (EdTech Strategies, 2015), which enables the use of AS with the digital collection of responses. At the same time, the technological advances in *artificial intelligence* have exploded (Economist, 2016) and available *processing power* has grown quickly (Simonite, 2016). The teams building software systems should thus recognize that change is constant and they should limit over-investment in a specific technology.

In our experience, one can generally expect to significantly rework AS software every six to eight years. The process of reworking may involve recoding parts of the system but preserving interfaces, in which case the process of reworking would involve recoding selected feature sets, statistical models, or deployment scripts. Rework is best done in a *modular* fashion, which can reduce cost and ensure that models and data can be rigorously compared when such changes are made.

We have overseen the rework of two AS software engines, each of which needed to be rebuilt after six years from the initial deployment. In these two situations, the choice to rebuild occurred for two different but related reasons. First, the *core engine codes* were rewritten in *Python* (from *Java*) to enable the data scientists to more easily read, test, and contribute to the code base, which enabled faster prototyping-to-deploy cycles. Second, advances in *computational linguistic* and *machine-learning libraries* in *Python* allowed for quicker use of new methods to gain scoring accuracy. At the time of writing, *Python* and its associated libraries are the most commonly used software for *machine learning* and *computational linguistics* (TIOBE, 2019; Kaggle, 2017; Robinson, 2017).

In the past 5–10 years, the state-of-the-art in AS software has evolved in its use of data and associated architectural complexity in order to achieve gains in *scoring accuracy* (Burrows, Gurevych, & Stein, 2015; Chen, Fife, Bejar, & Rupp, 2016; Taghipur & Ng, 2016). This evolution can be divided into three areas: (1) *feature-based approaches* whereby experts identify features relevant to scoring a type of item and models produce scores using these features; (2) *feature-free approaches* whereby features are learned alongside the predictive model; and (3) *ensembling approaches* whereby scores from competing models are combined to produce a score. The architecture of systems needs to accommodate these changes, particularly in the areas of prediction methods, data management, and model packaging.

The prediction method can vary from relatively straightforward (e.g., *rule-based methods* such as *decision trees*) to very complex (e.g., *deep learning models* with many layers such as *neural networks*). As prediction methods increase in complexity, the software that is needed to manage that complexity will increase for both engine training and scoring. Not surprisingly, data management complexity increases with model complexity. For example, all prediction methods need a training sample for model training and a validation sample for model evaluation.

However, more complex methods require the training sample to be partitioned into different types of training and validation sets. For instance, training sets can be divided into training and intermediate validation samples using a single split, multiple splits, or sampling methods such as *bootstrapping, k-fold cross-validation*, or *batching*. If *ensembling* is used with statistical methods, then the training sample needs to be further divided into a set for model training and testing as well as a set for ensemble model validation (see Yan & Bridgeman, Chapter 16, this handbook). The creation and storage of these data sets is influenced by the AS software approach.

Finally, the creation, packaging, and use of the models themselves is bound to the approach used as well. Model packages will vary in complexity in terms of their model architecture, model parameters, and format for parameter storage. The *predictive model markup language* (Grossman, Bailey, Ramu, Mahli, Hallstrom, Pulleyn, & Qin, 1999) serves as an example of principled thinking and implementation around model packaging and storage.

Additionally, there has been an explosion of *open-source* software, corpora, and libraries that have enabled the faster development of machine-learning systems, including AS software (Thakker, Schieson, & Nguyen-Huu, 2017). Open-source software is developed as a public collaboration where the *source code* is made freely available under a license governing its use. The increasing use of open-source software, corpora, and libraries has architectural impacts in a few important ways (Cruz, Wieland, & Ziegler, 2006).

First, AS software is likely to require increased *modularization* to manage multiple open-source systems, particularly if they use different programming languages. Second, AS software must explicitly manage *version control* in order to deal with updates from the various open-source software packages. Third, the *licenses* of the open-source software components should support the intended use (e.g., commercial use). Fourth, care must be taken to understand how open-source software operates *"under the hood"* to be certain that it works as intended. While these systems may be easy to use and it may be tempting to use them without understanding how they work, a hasty, ill-informed choice of a particular set of algorithms without proper empirical vetting will most certainly cause some unexpected surprises in development and operations.

Architectural Recommendations

In this section, we discuss specific architectural recommendations in light of what we discussed in the previous sections organized into three areas of considerations: (1) *functional components*, (2) *response handling, training, and deployment*, and (3) *software development life cycles*.

Functional Components

There are various ways to architect the components of the AS software. No singular exemplar architecture exists because many approaches can be used successfully when building software systems. As we discussed in the previous section, the details of these systems depend upon the prioritization of concerns, skills of the team, tolerance for complexity, resource availability, and time allocated to development. A common approach that is aligned with our experience is to divide the system into three main components, each with a primary function as shown in Figure 11.3: (1) *model training*, (2) *model deployment*, and (3) a communications *application programming interface* (API).

These components are aligned with the core functions surrounding automation scoring: training and validating models, scoring responses, and managing traffic to the software. People in different roles typically use each component: (1) data scientists and psychometricians conduct engine training and validation; (2) software engineers with a data science specialty manage the software that packages and scores responses; and (3) engineers manage the API calls to and from the system. We describe the three high-level components and their functions in more detail in the following.

Training

The training component handles model training and packaging for use by the scoring component. This component connects to *response data stores* and *item data stores*. We expect fast, iterative, and multiple model builds here – potentially automated – resulting in a final model choice. The system should have export capability of models and results; NLP experts, data scientists, and psychometricians are typically responsible for this phase as shown in Table 11.1.

Scoring

The scoring component receives the model package, verifies results, and then expects to score incoming responses using the model. The system needs to be optimized for high throughput and low latency and so must manage storage around model usage depending upon incoming requests. We note that

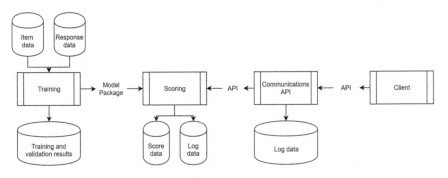

FIGURE 11.3
High-level architectural components for AS software.

assessment data are typically "bursty"; that is, examinees often test within a brief testing window and, within that window, at around the same times in the day. In the event that many hundreds – or even thousands – of models need to be stored, the system must have quick access to each model for each scoring request. Data from the *scoring flow process* (e.g., response variations, features) need to be saved for auditability purposes and the system needs to manage the security of the data around that storage. The system will also need to store *log data* that records key steps in the program, which are useful for *debugging*. It is expected that no changes to the model are made once the system is live. Software engineers who specialize in data science manage this process as shown in Table 11.1.

Communications API

The API insulates the engine from the systems it communicates with and helps to protect the core scoring functionality from external systems; consequently, it is optimized to handle aspects around this concern. The core concerns with this method are model / request identification and formatting, user authentication and security of access, managing the flow of requests, logging critical actions, and the production of alerts. Software engineers typically manage this process as shown in Table 11.1.

Response Handling, Training, and Deployment

In this section, we highlight in more detail key design elements in the architectural design of AS software that we have found to be critically important to consider when designing AS software. These design elements are (1) *response handling*, (2) *training*, and (3) *deployment*.

Response Handling

Response handling is a vitally critical element of AS software in that the incoming response should reflect how the examinee responded and that transformations of the response should be auditable. In Table 11.2, we cover the design elements around response handling including *response formatting, response audit trail, quality management*, and *corpora management*.

AS software typically predicts scores of responses to open-ended items that place few restrictions on the examinee input. As a result, responses can vary in almost an infinite number of ways. Defining assumptions, standards, and protocols to standardize some elements of the examinee response can prevent scoring issues, which help the software to create consistent rules when processing responses (e.g., consistent identification of newline characters).

Auditing occurs when it is necessary to trace exactly what the system did in response to a particular input at a point in time. As an example, intermediate data might be stored securely and outputted so that scores can be reproduced

TABLE 11.2

Architectural Design Elements around Response Handling

Design Element	Considerations	Example
Response Formatting	Assumptions	• Language • Item design • Entry requirements
	Standards	• Item Management Standards • Unicode
	Protocols	• Markup • Higher-ascii characters
Audit Trail	Response, Feature, and Score Output	• Original response • HTML-cleaned response • Versions of processed response • Feature values associated with response • Engine score, confidences, codes
Managing Quality	Data Quality and Failure Tests / Messaging	• Non-escaped special characters • Higher-ascii values • Empty or blank responses • No sentence- or paragraph-demarcations • Very long sentences or responses • Gibberish, many characters, repetitions
Corpora Type Management	Generic	• Versioning and updates • Storage • Security • Part of software versus model

exactly if needed. For data quality, most AS software is modular with components drawn from various open-source libraries. It is important that components return results in a responsible manner with error messaging for out-of-bounds conditions that can occur for a variety of unusual responses.

Finally, AS software generally uses both generic (i.e., cross-item or system-level) corpora and item-specific corpora, where *corpora* are defined as "any body of text that is used during the processing of the response", including word lists, dictionaries, and collections of documents. Generic or system-level corpora are often stored as part of the software package and consequently versioned; item-specific corpora – typically treated as more secure – are packaged with the item model itself.

Training

Training is concerned with the statistical estimating and packaging of models. In Table 11.3, we outline design elements around model training that affect AS software architecture including *code-base management, user interface,* and *model packaging and storage.*

Specifically, within this process, code-base management can be divided into *training*, which is optimized for training speed, multiple model builds,

TABLE 11.3

Architectural Design Elements around Training

Design Element	Considerations	Example
Code-Base Management	Training	• Training speed • Multiple model builds/comparisons • Response, feature, model details impacts • Packaging
	Scoring	• Model use • Scoring speed • Audit trail
User Interfaces	Training	• Training design • Validation results • Model choice • Model packaging • Deployment
Model Packaging and Storage	Storage Protocol	• Level of sensitivity to versions • Binary versus text • Size
	Data	• Enable model retraining • Ensure reproduction of scores
	Model Architecture and Parameters	• Rule-based vs statistical model • Feature-based vs feature-free • Ensemble method vs single method
	Versioning	• Corpora • Software • Libraries
	Randomness	• Random seed scope • Parameter initialization • Sampling (feature, response, model) • Spell correction

inspection of model features contributing to scoring and model packaging, and *scoring*, which is optimized for fast model load and use, scoring speed and auditing production of scores. User interfaces are typically created that enable a data analyst to conduct all aspects of training including model packaging and deployment. User interfaces also enable this person to conduct all aspects of scoring including model loading and verification and scoring.

The last design element outlines the considerations around model packaging. As outlined earlier, model packaging can be a complex process if the scoring models themselves are complex. A key consideration in model packaging is the ability to load the package into the scoring system and reproduce training results exactly. As models grow more complex, controlling randomness in the software and ensuring version control becomes particularly thorny when ensuring the reproduction of scores. The size of the individual packaged models can also be quite large (e.g., 50 megabytes) and can be expected to grow more as models increase in complexity.

Deployment

There are numerous key architectural AS software design elements relevant for deployment, which include code-base management around the API, deployment scripts, security handling, and monitoring interfaces around infrastructure. As discussed earlier, a communications API is optimized to protect the core scoring functionality from external clients. Because the full suite of software needs to be deployed on multiple systems (e.g., quality assurance, user acceptance testing, production), the ability to quickly build and deploy systems is critical. Related to deployment, handling of communications and any stored data need to be secure and auditable. Finally, interfaces which monitor the infrastructure and functional performance should be considered part of the architecture as well.

Software Development Life Cycle

We briefly discuss elements of the *software development life cycle* process that are unique to AS software. While there are many approaches (e.g., Agile, Waterfall, Lean, XP), here we are less concerned with the specific method of software development and more concerned with the unique elements that distinguish these systems from traditional software development projects.

Prototyping

The development of features and statistical engines are usually *prototyped* by NLP / data science staff and then converted to production by software engineers. Once the prototype system is considered adequate for production use, it is then documented and refactored by software engineers with attention to maintainability, testing, security, stability, and speed. This sequence is necessary because the prototype code is usually written in such a way that it offers *functional value* but is not appropriate for production use. This means that the traditional path of writing specifications (or "user stories") involves a fairly exhaustive first step of prototyping. It also means that the documented prototype code and output serves as truth when evaluating the functional performance and accuracy for the production code.

Multiple Engine Versions

The interplay between the prototype and the production engine can complicate the development or revision of features and statistical models. In some cases, a prototype is built and maintained so that it parallels the production code. This approach requires the maintenance of two separate *code bases* but allows the data scientists free rein for experimentation. In other cases, a version of the production code is used by the prototyping staff for improvements. This can complicate the software development process because data scientists may not be familiar with standard software versioning methods such as *gitflow* (Driessen, n.d.) and because multiple versions require more tracking and coordination.

Score Reproduction

The complexity of AS software and the variation in inputs mean that the principal method for testing is the reproduction of scores from a *verified model* (prototype or otherwise). The testing procedures include examinee responses and scores, and each tested element is expected to exactly reproduce scores given the examinee input. While common testing to requirements may offer some value (i.e., assurance of boundary cases, violations of expected behavior which are well known), such testing is of limited value in AS software. Particularly for high-stakes uses, the exact reproduction – or reproduction within well-defined *tolerance levels* – of feature values, scores, and score probabilities or confidences is necessary. Reproducibility of each element in the software can be a great help when diagnosing places where errors appear. In these complex systems, identifying the source of differences can be painstaking and quite difficult; it is best to create data-based tests at each level of the software to support this type of debugging.

Performance Monitoring

Fourth, monitoring of AS software can be tricky because: (1) responses are typically unconstrained and changes in response characteristics over time may be difficult to detect; (2) scores produced by the engine are not readily discernable as accurate and require analysis by hand-scoring experts; and, (3) scores have enough variability in them that errors due to software bugs may not be distinguished from random error. To illustrate these challenges, we provide a brief real-life scenario to suggest that monitoring systems should provide a suite of timely data and reports to provide a coherent picture of all elements of scoring.

In this scenario, a series of AS models were originally calibrated on responses which were stored in HTML format. At some later point in time, the item interface design was changed so that the responses were stored in plain text format. The owners of downstream systems were not notified of the change because the change was not considered critical. Because all testing for deployment was conducted on the original HTML-formatted training and validation set, the software and model was deemed to be performing correctly. In live testing, however, those monitoring the system noted a distributional change in score confidences while the score distributions were similar to expectation. The software was revised to consider the "new" response format and all incoming responses were rescored. With the revisions, about 8% of the responses received a different, typically adjacent score.

The choice in the methods used to build and maintain AS software is likely influenced by organizational practices. Regardless of the methods used, it is wise to recognize that the uniqueness of AS software production described in this section is likely to impact existing processes. Organizations with a strong foundation in building technical data-driven software may undergo less of a shift in perspective but the transition to machine-learning approaches still may be a challenge to organizations unfamiliar with those methods (Zinkevich, 2019).

Conclusion

We have presented a range of issues surrounding the architecture and design of AS software. These issues involved considerations around the stakeholder context, various use cases, software considerations, architecture components, design elements, software development lifecycles, and team composition.

As we have underscored throughout this chapter, the development of AS software is challenging because of the range of stakeholders involved in the use of the software, the difficulty in evaluating the "truth" of the system when monitoring AS performance, and the emphasis on strong data and statistical methodology during development, validation, and testing of the software and models. Given the multiple stakeholders involved in building, evaluating, deploying, and monitoring AS systems across varying use cases, this work clearly requires a multidisciplinary team.

Having led multiple teams across organizations and time, we have found that the leader of these teams should have a deep knowledge of assessment and psychometrics in order to understand use cases and context, should have a willingness to actively identify and limit bias in interpretation, should be appropriately skeptical of empirical results (Silver, 2012), and should have a strong belief in the value of dissent to encourage the refinement of best approaches (Epstein, 2019). This set of core beliefs should translate to a leader that can communicate the underlying rationale for use cases and requirements, dedicate ample time for full-group and ad hoc discussion that centers on open communication and debates of methods and results, and tolerate multiple failures when prototyping new features or methods.

The key roles for a development team were presented earlier in Table 11.1. People with a multidisciplinary focus can be cross-functional and people can be partial or periodic members of the team depending on the phase of development. Thus, the roles below do not necessarily map one to one with people on a team; software can be built with as few as three to four people and as many as 12 or more.

When it comes to developing technical software like AS software, it is our belief that a team comprising *fewer but highly technical* staff is preferable to a large team. In this sense, we very much follow the *Google* approach in which they use software developers with an aptitude for software testing rather than software testers (Whittaker, Arbon, & Carollo, 2012). It is our experience that small teams with key functional roles help maintain engagement and ownership in the software development process and reduce confusion (Hoff, 2019). We have also found that staff in some roles outlined in Table 11.1 – for example business analyst, technical writer, quality-assurance staff, and technical program manager – often lack the technical or statistical background to effectively contribute to the development of software and it is difficult to teach these skills.

All of these management challenges, while daunting, also suggest a path forward. We should expect to routinely improve the software by evaluating the accuracy of all elements that impact score prediction, adding or revising features, and improving statistical modeling. We should recognize that these systems live within multiple contexts including psychometrics, test operations, and assessment applications. This means that the directors and team members should be communicating with these stakeholders as necessary to ensure coherence to each contextual goal. Most importantly, in our view, is that directors talk often with clients – be they state departments of education, teachers, curriculum coordinators, or other entities – and listen carefully to their expressed concerns as it is precisely these concerns that help identify scoring issues and help improve AS software.

12

Design and Implementation for Automated Scoring Systems

Christina Schneider and Michelle Boyer

CONTENTS

There is increased interest and pressure for test developers to pursue operational uses of *automated scoring* (AS) with large-scale educational assessment. Traditional *human scoring* processes are time consuming and state departments of education desire rapid score reporting to better inform instructional actions (Conley, 2014; Shermis, 2015; Shermis & Hammer, 2013). Historically, operational use of AS has been in licensing and college admissions testing (Bejar, 2010; Trapani, Bridgeman, & Breyer, 2011), with statewide operational AS use being sparse (e.g., Rich, Schneider, & D'Brot, 2013). As a result, most research regarding AS has been published outside of grades 3–12 educational assessment setting. We specifically consider assessments in grades 3–12 in this chapter based on our own extensive experience in this area.

From 2012 to 2015, AS researchers conducted several high-profile studies to assess the capabilities of AS in grades 3–12 (Shermis & Hammer, 2013; Shermis, 2015; McGraw-Hill Education CTB, 2014; Pearson & ETS, 2015). First, two worldwide competitions were implemented to identify the "state-of-the-art" in AS of essays and short-answer items at the time (Shermis & Hammer, 2013; Shermis, 2015). The competitions created a highly visible incentive for advancing the development of valid and reliable AS models.

Second, the *Smarter Balanced Assessment Consortium* (SBAC) commissioned studies that involved training and validating multiple AS raters for their potential use in scoring the SBAC short- and extended constructed-response items (McGraw-Hill Education CTB, 2014). Third, the *Partnership for Assessment of Readiness for College and Careers* (PARCC) commissioned research regarding the potential use of AS for a prose constructed-response item type (Pearson & ETS, 2015). These studies moved state departments of education forward with using AS operationally (e.g., AIR, 2017; Strauss, 2016).

Broadly, these studies found that automated essay scoring capabilities were often comparable to, and in some cases better than, human scoring; however, constructed-response scoring capabilities for students in grades 3–12 were lower in quality compared to human scoring. The studies have served as a basis to increase interest in, and to explore, the potential AS uses in statewide *accountability assessments*, prompting continued research and development of constructed-response scoring capabilities. Furthermore, analyses of these studies have provided useful guardrails for test developers who are planning for AS implementation.

Concurrent with the study of AS there was also a policy shift from measuring students' achievement using a status measure to an increasing interest in measuring *student growth*. Under the *No Child Left Behind Act* (NCLB, 2002), statewide accountability assessment programs were mandated to measure year-end achievement through the use of individual student scores and *achievement level classifications*. Measures of change in student knowledge over time, interpreted as *student growth*, were privileged for students who moved over the 'proficient' cut score into the proficient category.

The use of student growth indicators regardless of status has increased in *accountability weighting* and *teacher evaluation systems* under the *Every Student Succeeds Act* (ESSA) (ESSA, 2015). This trend has direct implications for how score *validity claims* are structured and supported for both human scoring and AS. Where score use is centered on achievement level classifications via cut points, the focus is on claims related to specific regions on a score scale. However, where growth measures are used, the focus is on offering strong support for validity and *reliability claims* at every point on the scale.

Test score uses and their *interpretive arguments* have implications for the implementation of AS from a variety of perspectives. AS and the resulting scale scores should be evaluated for their score quality from a *classification consistency* perspective and for their ability to maintain *longitudinal trends* along the full test scale. That is, AS should not jeopardize score comparability over time. Conceptually, AS is expected to produce equally reliable and valid score uses as human scores. In practice, however, there are conditions that may present challenges to such an expectation.

First, enhancements to AS models are ongoing. The implication of these enhancements is that, over time, we can expect that automated raters will score student responses with changing levels of rater error. Ideally, automated score quality is expected to improve, but such changes require that psychometricians attend to their potential impact on downstream measurement scales. Second, states may transfer contracts to a different vendor. A change in vendor could also mean a change in AS models and features extracted from an essay, which introduces the risk that levels of automated rater accuracy will change from one automated rater to another.

Third, student population characteristics often change over time due to a variety of reasons. For example, growth in *English language learner* (ELL) populations and an increasing percentage of students identifying as two or more races have been noted throughout the U.S. across many testing programs. Changes in any one of these conditions of scoring may present a threat to *score comparability* within items and over test forms and time. Consequently, careful attention to the item-level quality of automated scores, automated and human rater characteristics, and student demographics is needed because of their potential influence on the psychometric properties of the test scale.

In this chapter, we highlight important considerations for state or vendor psychometricians as they (a) monitor the design of items for AS, (b) conduct or approve AS model training and evaluation, and (c) operationally deploy AS systems. Such psychometricians would typically work as part of a larger AS team or receive scores from such a team either internally within her own company or as a part of a subcontract.

We use the term *AS model* during the training and validation phase of AS work and the term *automated rater* for a scoring model that has moved into production and is scoring student responses operationally in real time. We use the term *extended constructed-response item* to refer to an essay and *short constructed-response item* to refer to a response that is a paragraph in length or shorter. As part of our discussion we give special attention to the issue of test score validity and to the longitudinal use of automated scores where measurement of student growth is central to an assessment program's interpretive argument and intended score use.

We have divided this chapter into three sections. In the first section, we review general design-related considerations in conceptualizing and planning for AS. In the second section, we provide a more detailed review of common *quality assurance* procedures used in the construction and validation of AS systems. In the third section, we discuss a variety of considerations around the operational deployment of AS systems. We conclude the chapter with a discussion of important considerations for implementing AS in a new or existing large-scale testing program.

General Design Considerations for Automated Scoring

State-developed *requests for proposals* (RFPs) for grades 3–12 assessment programs are increasingly specifying interest in AS. State stakeholders issue RFP requirements that range from (a) the intent to use hand-scoring for operational testing while piloting the use of AS to (b) the use of automated scores as the score of record. Regardless of the timeline for operational use, it is important for test developers to consider the constructs to be measured by automatically-scored items and how the attributes of those items and student response features increase or decrease the likelihood of accurate scoring. In the following we discuss several key design considerations organized in five different subsections: (1) scorability, (2) validity, (3) special populations, (4) score use, and (5) model building.

Scorability

In this chapter, we define the *scorability* of an item as the degree to which it can be accepted for operational deployment using an AS model after a thorough evaluation by a psychometrician. Determining potential scorability is often an interdisciplinary activity. A team with expertise in item development, hand-scoring, AS, and psychometrics is recommended to investigate the language used in scoring rubrics and the item demand characteristics to predict the likelihood of AS success with a particular set of items. We have summarized key considerations via a scorability rubric in Table 12.1, which may be used by assessment experts for this purpose.

TABLE 12.1

Sample Rubric for Determining Scorability for Automated Raters

Scorability Criterion	0	1	2	Score
Rubric Language	The rubric describes knowledge and skill using generic, holistic language such as "sufficient evidence" and "limited evidence."	The rubric describes increases in knowledge and skill across higher score levels but uses the lack of attributes at others (e.g., language such as "many errors" verses "numerous errors").	The rubric describes increases in knowledge and skill using concrete, explicit language describing the observed attributes that are present for each level.	
Depth of Knowledge Level (CR Items)	Level 3	Level 2	Level 1	
Complexity (CR Items)	Response can have more than one correct claim or inference with the text, providing five or more possible text-based supporting details the student can use to justify the claim.	Response requires a low-level inference with the text, providing two to four possible text-based supporting details the student can use to justify the claim.	Response requires students to create a response from explicit information with few varied supporting details to choose from.	
Item Demand	Students must perform five or more steps with inferred information and make decisions on how to access item.	Students must perform more than one step with inferred information and make decisions on how to access item.	Students must perform more than one step with explicit information or memorized procedures.	
Item Score Quality	Response distributions are highly skewed so that there are few responses at higher or lower score point levels or sample sizes for combined training and evaluation sets are less than 1,500 for essays and 500 for constructed-response items.	Randomly selected response distributions are slightly skewed so that there are few responses at the highest point level, but human ratings are sufficiently high (e.g., QWK > 0.75) with moderately robust sample sizes for training and evaluation sets.	Randomly selected response distributions are symmetrical with robust counts for training and evaluation sets, and human ratings are sufficiently high (e.g., QWK > 0.75).	

As seen in Table 12.1, scorability is generally determined by a set of complex interactions between the cognitive task for students and the implications of that complexity for both human and automated raters. In the case of well-established state-level summative assessment programs, states will often seek to introduce AS into an existing testing program where robust item parameters have been established previously. In such cases, the items will have already been developed, *field tested*, and scored by humans and will consequently have item parameters demonstrating that the item functions with the student population.

Where year-to-year *test equating* is embedded in the test design, such scoring changes likely preclude the use of the automatically-scored items as *anchors*. Moreover, in a *pre-equating* scenario, rubrics should stay constant across time to ensure the scoring of the item maintains its connection to the item parameters. Decisions to implement AS need to explicitly consider the tradeoff between maximizing AS uses and minimizing threats to *score comparability* across time and forms. If these items are not to be used as anchors or to support pre-equating, there is an opportunity to re-examine *item development specifications* and *rubrics* as new items are scaled to determine which items are good candidates for AS as well as possibly *retrofitting* scoring rubric language to increase the likelihood of automated scorability. Regardless of whether rubrics are adjusted, responses from existing human scored, operational, or field test items should be shown to be scorable with an automated rater prior to their operational deployment. If responses were originally collected in *paper-and-pencil format*, the items should be field tested again with responses collected digitally.

Rubric Attributes

A focus on *rubric quality* is anchored in findings that some attributes of rubrics make item responses more amenable for AS than others. Generally, rubrics often describe a lack of attributes in a student's response versus increases in knowledge and skills across all score levels. Because many AS systems typically rely on counting and then normalizing *lower-level proxy features* of student responses, rubrics that explicitly describe the attributes required in student responses will have a greater chance of being accurately scored by a machine and, incidentally, a human (Leacock, Gonzalez, & Conarroe, 2014).

Therefore, investigating the specificity of the rubric language used to differentiate levels of student thinking and content knowledge is critical for investigating the likelihood of AS success. For example, Leacock et al. (2014) and McGraw-Hill Education CTB (2014) found that when rubrics differentiate among levels of student understanding with holistic language such as "sufficient evidence" and "limited evidence," human raters had a more difficult time making distinctions between levels of performance than when concrete, explicit language or exemplar responses are provided. In an operational setting, calculating the percentage of items that have rubrics using this type of language is a quick way to identify items that may need additional attention when considering a move to AS.

Item Attributes

In addition to considering rubric attributes, item attributes must also be considered. Leacock, Messineo, and Zhang (2013) found that human rater agreements declined in short constructed-response items when (a) there were more than five possible text-based supporting details, (b) there was more than one possible correct answer, and (c) the correct answer required students to make an inference. Similarly, McGraw-Hill Education CTB (2014) found that human rater agreements declined in short constructed-response items when (a) there were four or more text-based supporting details or (b) the *depth of knowledge* (Webb, Alt, Ely, & Vesperman, 2005) for the items was identified as Level 3, which indicates items that require reasoning, planning, and using evidence to justify a response.

There is a complex interaction between the *cognitive task* for students and for raters, which, by extension, affects whether an automated rater can be trained to successfully score items. Harder items for students become harder items for humans to score. As most automated raters are trained on human scores, these conditions are likely to affect the quality of automated rater scores. As a result, the assessment development team should review the item demand characteristics that indicate what students have to do to respond successfully to items in terms of processing steps. The increased number of concepts and processes that students must integrate to respond to an item correctly can make items more difficult for students (Ferrara & Steedle, 2015). Because such items are likely to be more difficult for humans to score as well, AS indexes may not meet operational implementation criteria (Leacock et al., 2014; McGraw-Hill Education CTB, 2014).

Rater Error

In addition to reviewing rubric and item attributes, the consistency of human scoring for existing item types should be examined. Traditionally, score quality by human raters is evaluated using *inter-rater reliability* statistics, including the percent of responses with *exact and adjacent agreement, correlations, kappa,* and *quadratic-weighted kappa* (QWK). If human agreement rates are not sufficiently high (e.g., if a QWK is less than 0.75), there may not be sufficient *precision* in the human scores to adequately develop an AS model.

In addition, if the distribution of item scores is skewed so that there are few responses at higher score point levels or if many students earn a score of '0' on an item, the item is likely not a candidate for an automated rater

(Pearson & ETS, 2015). While most testing programs train automated raters to score at the trait level of a rubric, there have been occasions where training and validation of responses have been at a *sum-score level* (e.g., Shermis & Hammer, 2013). Such configurations of an AS system mean the *score reliability* will be lower (Shermis, Burstein, Brew, Higgins, & Zechner, 2015), which is not recommended when growth interpretations are a desired inference.

Evaluating Scorability

Because the *interactions* between items, rubrics, examinees, and human raters can be complex, *scorability judgments* can be equally complex. Lottridge, Wood, and Shaw (2018) found that assessment experts had a relatively more difficult time predicting whether automated raters could score reading items than mathematics or science items when reviewing items holistically. By paying attention to and better understanding the complexities of item scorability, test developers can increase the validity of score interpretations with both automated and human raters. *Principled assessment design* approaches that involve steps such as using *achievement level descriptors* to define the content interpretation differences across achievement levels to support item development (e.g., DiCerbo et al., this volume; Egan, Schneider, & Ferrara, 2012) may hold promise for supporting automated rater scorability in addition to supporting a *construct-centered validation approach* for the use of AS in grades 3–12.

Scorability rates can change depending on the *business requirements* from one testing program to another and from one automated rater to another. For example, not all automated essay raters measure the content of an essay. If a state has a business requirement that content coverage must be measured in the essay, this may produce a lower scorability rate compared to a state that allows for the provision of credit for a well-written essay that is off-topic.

Clearly, though, not all items are machine scorable. For example, in the SBAC study of scoring capabilities for several automated raters, roughly one-third to one-half of constructed-response items were determined to be unscorable, and English language arts / literacy constructed-response items tended to be more difficult to score than mathematics constructed-response items (McGraw-Hill Education CTB, 2014). The proportion of essays that can be automatically scored tended to be higher than that of constructed-response items (McGraw-Hill Education CTB, 2014; Pearson & ETS, 2015; Shermis, 2015).

Validity

Whether building a new test or transitioning an existing testing program to AS for constructed- and extended-response items, evidence of the *construct validity* of those scores is an important element to consider and expand in research planning and technical documentation.

Constructs Representation

Constructs are broadly defined as the "qualities that a test measures" (Messick, 1989, p. 16). A construct is theory-based, regardless of whether it is defined as 'mathematics achievement' or 'college and career readiness in mathematics.' Examinee responses and their corresponding scores can be conceptualized as indicators against a theoretically defined construct whose measurement is operationalized through the test questions (Cronbach & Meehl, 1955; Loevinger, 1957; Messick, 1989; Kane, 2006; Shepard, 1993; AERA, APA, & NCME, 2014).

In the grade 3–12 context that is the focus of this chapter, constructs largely center on students being on track for *college and career readiness* in content domains such as mathematics and English language arts, with growing attention to new and potentially *multidimensional constructs* such as those covered in the *Next Generation Science Standards* (NGSS; NGSS Lead States, 2013). This kind of added dimensionality can add a layer of complexity to AS considerations and how scores on separate dimensions are best handled in downstream psychometric analyses. Decisions about how to combine and calibrate scores on separate dimensions is beyond the scope of this chapter, but the presence of multiple test dimensions does have important implications for how score scales are constructed.

Whether scoring is done by human or automated raters, the appropriateness of the *correspondence* between the features of an examinee's response and the targeted construct of measurement is a central component of any strong *validity argument* (see Yan and Bridgeman, Chapter 16, this handbook). Consequently, construct validity arguments can be thought of as arising from both *theoretical* (construct definitions) and *empirical* (true-score variation) perspectives. Between the theoretical construct definition and the assignment of scores, raters essentially make a determination that links an examinee's response to a theory of the construct. The stronger the evidence is that a rater has assigned the most accurate score that represents the true score for an examinee's response for a given an item with a given rubric, the stronger the validity claims may be about the usefulness of item and test scores for their defined purpose.

Validity theorists typically agree that test content should represent the construct domain and should not contain *construct-irrelevant variance* (Cronbach & Meehl, 1955; Messick, 1989; AERA, APA, & NCME, 2014). According to Kolen (2011), the three broad characteristics of an assessment that define the construct represented by a test score are (a) the *content of an assessment*, (b) the *conditions of measurement*, and (c) the *examinee population*. AS interfaces with two of these elements as it is one facet of the conditions of measurement, and it is one facet of test content when the intended inference of a score is referenced through the evidence described by a rubric and extracted and summarized via examinee response features. The types of evidence that are appropriate for supporting construct-centered score interpretation validity claim is similar to, and ideally extended from, that of human scoring.

Rater Error

An underlying challenge in gathering support for construct validity claims for an automated rater trained on human scores is that it is not fully understood how humans score examinee responses. It is generally accepted that *human rater error* is likely present in examinee scores, but such error is not typically accounted for in AS models (Bennett & Zhang, 2016). However, without mitigation procedures, any rater error in the human scores can be propagated into the AS models and impact the psychometric properties of the score scale.

Although many models have been developed that can be used to examine rater error (e.g., Wilson & Hoskens, 2001; Patz, Junker, Johnson, & Mariano, 2002; Wang, Song, Wang, & Wolfe, 2016), they tend to be computationally complex and more time consuming to implement because they require multiple raters per item, which also increases both operational scoring and psychometric costs. Moreover, they tend to be used during post-hoc quality assurance processes which makes them of limited value if the desire is to monitor scoring in real time. Consequently, they are not often used assessment settings for grades 3–12. Instead, more descriptive statistical measures are often employed to assess human rater quality and the presence of some level of rater error is generally accepted in practice. Mitigation occurs primarily through detailed human rater training and score monitoring, often with interventions in real time through comparisons with expert scores.

Consequently, a significant challenge in demonstrating the construct validity of automated scores is inherent in the use of potentially fallible human rater scores as the criterion against which an automated rater is built and evaluated. If the human scores used to train an automated rater do not accurately reflect the ideal scores for given examinees for given responses to a given set of items, those scores may not accurately reflect the variance that can be explained by the given construct. In this sense, construct validity is threatened before the automated rater is ever trained, so it is fundamentally important that test developers and users fully understand and document the quality of the human scores used in automated rater models.

Feature Design

Another challenge in demonstrating construct validity is in how automated raters relate scoring rubrics to examinee responses during scoring. In traditional human scoring, raters match features of the examinee's response to the scoring rubric to produce a score. The current state-of-the-art for many AS raters, however, is to extract *proxies of rubric features* from examinee responses and use those proxies in *statistical prediction models* provide the best prediction of human assigned scores (e.g., Attali & Burstein, 2006).

This difference in approach is important because human and automated raters use different processes to assign scores to examinee responses. Humans align features of the examinee responses directly with the rubrics whereas automated raters *parse* responses into a set of *proxy features* that are *statistical representations* used to model human scores. Ideally, the features identified by human and automated raters are the same, but in practice they can be different, as they often are. Therefore, if the human approach is assumed to be more valid because of the direct link between responses and scoring rubrics, construct validity arguments for automated rater scores could be threatened by their reliance on arguably fewer direct links.

Methods of Documentation

Whenever certain features used in the AS system are limited from a construct-representation perspective and can reasonably be identified, one way to address this potential threat to construct validity arguments is to provide a *crosswalk* between the rubric criteria and the features used by the automated rater. Shermis, Burstein, Brew, Higgins, and Zechner (2015) recommended documenting the relationship between the features scored by humans and automated raters using such a crosswalk.

In addition, Bennett and Zhang (2016) advocated for providing the types of evidence traditionally used to support the validity of scores produced by human raters, including the following: (a) evidence that examinee response processes, scoring rules, and rater processes are those intended; (b) the appropriateness of treatment of unusual examinee responses; (c) accurate prediction of scores on other tasks from the universe of tasks; (d) the relationship of scores with external variables; and (e) the population invariance of scores. Similarly, Rupp (2018) discussed a wide variety of design decisions that are made in the development, evaluation, and deployment of AS systems and noted various challenges for creating a coherent construct-representation argument on the basis of these. Yan and Bridgeman (Chapter 16, this handbook) discuss details of AS validation.

These difficulties are often compounded by a lack of data external to the automated scores that might be used to validate the relationship of the automated scores with other measures of the intended construct. For a psychometrician charged with creating a technical report, determining the types of evidence to collect early on in the process (e.g., during *proposal costing*) is useful, with the caveat that the complexity and limited transparency of some automated raters can make building the construct validity argument more challenging.

A few extensions of traditional construct validation methods hold potential for both understanding and documenting the validity evidence of automated rater scores in terms of the internal structure of the tests in which they are used. One such approach would be to demonstrate the amount of *construct-relevant variance* of the automated rater scores through *dimensionality analyses* such as a comparison of the fit of *confirmatory factor analysis* models using human and automated scores. Another approach might be to implement an *analysis-of-variance* (ANOVA) across human and automated rater conditions or a fully-crossed *Facets analysis* between the two to examine the invariance of examinee scores across raters (see Linacre, 1989).

In addition to the potential contribution of features as a component of the construct validity argument, the investigation of *feature invariance* across student *subgroups* can be used to provide evidence of *measurement equity*. Given the dearth of research regarding the functioning of AS for students with disabilities and ELLs in the grades 3–12 context, it is essential to understand how and if features are weighted in producing scores that may create unintended separation between human and automated scores for some populations.

Psychometricians for state testing programs often present automated rater agreement statistics compared to humans as the sole evidence source to support the use of automated raters. Table 12.2 presents our recommendations for planning the documentation of construct validity evidence for automated raters that we believe should be minimally considered for inclusion in a technical report, especially in cases in which the automated rater is the sole rater of record.

Special Populations

A fast-growing policy trend for grades 3–12 is to screen early elementary students for *dyslexia markers* (see, e.g., policies in Oregon, Arkansas, Missouri, and South Carolina). This suggests that with more equitable *early identification*, attending to the fair scoring of examinees with a growing number of recognized dyslexia may be on the horizon. For these students, the most notable deficit over time is *spelling* (Eide & Eide, 2006; Habib, 2000; Parrila, Georgiou, & Corkett, 2007) with successful adults often relying on technology to assist them in compensating (Schneider, Egan, & Gong, 2017).

As Gregg, Coleman, Davis, and Chalk (2007) reported, in the context of high-stakes essays with trained raters, spelling errors, and varied and sophisticated vocabulary were found to strongly influence raters' perceptions of dyslexic writing. There is a potential of spelling errors and purposefully chosen simple word choices that dyslexics sometime use as a strategy to mitigate potential spelling errors, introducing negative *human rater bias*. Buzick, Oliveri, Attali, and Flor (2016) raised the concern that spelling errors could have an influence in features producing a larger, negative influence on scores for dyslexic examinees than intended.

Buzick et al. (2016) found that college-age test takers with learning disabilities made the highest percentage of spelling errors, but their overall

TABLE 12.2

Recommendations and Sources of Validity Evidence

Recommendation	Source of Evidence
Document the construct humans use while scoring	Training materials and scoring rubrics for human raters, including item and response rationales for score assignments on training papers
Examine features used by humans while scoring	Cognitive labs with raters during scoring during field test analyses
Document the types of features (and their relative weights) used by automated rater in scoring	Documentation produced during AS system development
Crosswalk features used by automated rater with features used by humans based on rubric criteria	Cognitive labs with human raters and feature weights used in automated rater
Provide evidence of the appropriateness of treatment of unusual examinee responses	Flagging criteria, regression-testing process, and studies investigating system functionality
Provide evidence of the generalizability of automated rater results across items/tasks for generic automated raters	Prediction of scores on other tasks from the universe of equivalent tasks designed for a particular score interpretation
Provide evidence of the population invariance of scores	Standardized mean difference of feature scores across student subgroups and differential functioning analyses on item-level scores comparing human and automated rater
Provide evidence of the construct-relevant variance of the automated rater scores	Dimensionality analyses such as a comparison of confirmatory factor analysis model fit over humans and machines; analysis-of-variance across human and automated scorer conditions; fully-crossed Facets analysis between humans and machines to examine the invariance of examinee scores across raters

scores tended to be higher than the general population. Conversely, students with disabilities in grades 3–12 score generally lower than the general student population. Including spelling in a scoring model may over emphasize lower-level writing skills at the expense of the more sophisticated elements of writing associated with college readiness. From a *universal design for learning* perspective around AS, it may be advisable in certain situations to not score spelling errors and, instead, allow spell correction for all examinees during testing or correct spelling before scoring as a *pre-processing* step.

As a real-life example, the College Board does not count spelling during essay scoring (College Board, n.d.) and some automated raters have already moved to correcting spelling for all students (S. Lottridge, personal communication, August 27, 2018). In short, the role of spelling correction as a component of the construct and in the pre-processing steps for scoring may need more investigation and consideration in the industry. This is especially important given the anecdotal evidence that the mis-spellings of phonetic writers or other English variations can get flagged as non-English as well as other unusual use of word patterns that may be common to ELLs or students with disabilities (S. Lottridge, personal communication, August 27, 2018).

Furthermore, AS requires the use of *digital technology* in the collection of responses, but there are populations of students for whom digital technology is not the optimal way to collect evidence regarding what they know and can do. States typically allow small numbers of students to take a paper-and-pencil form as an accommodation, which means human ratings are required unless an adult representative then types the responses for AS. It is possible that paper-and-pencil accommodations may introduce *modality effects* in a strict sense, but it is critical that students be allowed to test in the mode most appropriate for them.

Scoring Approaches

In high-stakes testing programs for grades 3–12, it is historically most common to have two human raters score short- and extended constructed-response items with *resolution reads* (i.e., *adjudications*) occurring for discrepant responses. In operational practice, this is generally considered to be the method most likely to identify the ideal ratings for all item responses. However, this process is expensive in labor and time. Consequently, in times of economic recession, states have often reduced hand-scoring to a single read with a 5 to 15 % human back-read as a cost-savings measure, placing higher demands on scores to be reliable in single rater scenarios.

This scenario has also been observed for AS (AIR, 2017; Rich, Schneider, & D'Brot, 2013) but contrasts with more conservative uses of automated raters in other high-stakes contexts (e.g., Breyer, Rupp, & Bridgeman, 2017; Yan & Bridgeman, Chapter 16, this handbook). More conservative scoring approaches that combine human and automated scores for reporting, for instance, avoid certain risks around legal challenges to construct representation, validation, and fair treatment of examinees. In any practical scenario, AS teams have to balance a portfolio of competing considerations related to bounding risks, associated constructed-response scoring requirements from peer-review processes, limited funding, and state politics driving a rapid return of scores.

In the grades 3–12 context, AS approaches for operational reporting that have been evaluated include using the automated scorer as the *single rater* of record (AIR, 2017; ETS, 2018; McGraw-Hill Education CTB, 2014; Rich, Schneider, & D'Brot, 2013), as a *second rater* (Lottridge et al., 2018; McGraw-Hill Education CTB, 2014) and as both the *first and second rater* (McGraw-Hill Education CTB, 2014). When an automated rater is used as the sole rater, the quality of the scores are typically evaluated through validation studies, a small percentage of human *read-behinds*, and *check reads*. This reduces the number of human reads, which lowers costs and allows responses to be scored in real time or quickly thereafter. However, it does not allow sufficient time and sample sizes of human reads to investigate human rater drift or other issues that can be found during operational implementation.

Bennett (2011) further suggested independently training and deploying two separate automated raters for each item and routing papers to a human when the automated raters were *discrepant*. However, McGraw-Hill Education CTB (2014) found that roughly 20% to 50% of papers may need to undergo a *resolution read* under this scenario. This approach was used by combining the two top-scoring engines and specific outcomes in other contexts likely depend on both the scorability of the items and the reliability of the automated raters. In other words, non-exact agreement resolution, while desirable, may not yet be practically feasible in all situations for technical reasons. Moreover, the critical *policy issue* with such an approach would be to gain buy-in from stakeholders that some students' scores will and should arrive earlier than others because of the need to ensure accurate scoring.

It is critical to prioritize what is valued most (e.g., speed, cost, accuracy) and make trade-offs related to those values. For example, if measuring growth were of high value, this additional time could be essential. If measurement accuracy were the prioritized value, then the most robust method of scoring short-text constructed-response items and essays is using an AS system with a human rater (McGraw-Hill Education CTB, 2014) as a second read. If an organization is focused on reducing potential sources of measurement error across time to support longitudinal analyses, resolving non-exact agreements between the two automated raters or an automated rater and human rater could certainly be accomplished. If speed of scoring is the central consideration, a sole rater is likely the best option, although human back-reads should be sufficiently robust with daily monitoring so that the psychometrician is able to troubleshoot should human-automated rater separation be discovered and return of responses should be held for a short amount of time to ensure score quality.

Automated Scoring Model Development

As we noted in the previous section, when beginning the process of training an automated rater, it is prudent to examine the *human rating quality* and *rater qualifications*. It is also important that the AS team understands the score resolution processes stipulated for a program and for determining the score of record.

Sample Size

Although different vendors quote various sample sizes for training and evaluating AS models, McGraw-Hill Education CTB (2014) found that, in the grades 3–12 context across a subset of vendors, 1,000 on-grade responses was a reasonable minimum for a *training set* and 500 on-grade responses was a reasonable minimum for a *validation set* for short-text constructed-response items. Similarly, their research established that roughly 1,500 on-grade responses was a reasonable minimum for a training set and 500 on-grade responses was a reasonable minimum for a validation set for essay items. These are only very general *rules-of-thumb*, however, and required sample sizes can vary based on the item type, item attributes, rubric attributes, type of scoring model, and distribution of scores.

Pre-Processing

The first step in creating models is generally pre-processing of sampled responses. According to Shermis et al. (2016), responses for training and evaluation are generally randomly selected. However, according to Pearson and ETS (2015), it may be beneficial to *oversample* when responses are sparse in the tails.

During the pre-processing stage, digitally collected responses may be cleaned as they are moved from an *HTML format* to a format that can be processed by the computational routines of the AS system. It can be important to visually *spot-check* a random set of responses after cleaning to ensure that new anomalies introduced by students are not present. Once responses have been *parsed* and *tagged* and features have been *normalized*, there should be *output checks* to ensure that features are behaving as expected in terms of distributions. Documentation of the statistical attributes for the specific features for each item can be collected for each model as it is approved as a comparison point for future administrations.

Bejar (2010) as well as Rich, Schneider, and D'Brot (2013) described performing quality assurance on *advisories* as a component of the validation process. Routing responses through *advisory processing* prior to model building results in a more accurate scoring process overall (Zechner & Loukina, this volume; Ramineni, Trapani, Williamson, Davey, & Bridgeman, 2012a, b) and can allow humans to begin additional *condition coding* of responses, if necessary.

Because most testing programs for grades 3–12 are high-stakes for administrators and teachers rather than students, some students will, at times, work to *game* the automated rater. This is especially true when students also have multiple opportunities to work with an automated rater for formative purposes. In-state programs, advisories are often assigned to papers that are off-topic, repeat the prompt, have repetitive text, or have other non-scorable issues such as key banging or too many spelling errors.

Model Type

An additional consideration during AS model construction is whether *item-specific* or *generic automated raters* should be constructed. This decision is often based on intended uses of scores, the level of precision a state may require as a component of a contract, and the purpose of the automated rater. In our operational experience, automated raters used in the grades 3–12 context for high-stakes uses are typically item-specific (i.e., trained on human scores for a single item). Item-specific automated raters tend to perform at higher agreement levels than generic raters due to the additional usage of item-specific information in the predictive model although many automated raters in the industry intended for formative purposes are often generic. That is, the same generic automated rater is used to score the same type of prompt (e.g., informational) across essays in a single grade, which are sometimes considered *prompt families* similar to *task types* or *task families* (Breyer et al., 2017).

Generic automated raters are generally trained on a large sample of examinee responses from a prompt family that is appropriately representative of the range of writing quality defined by the targeted construct. An advantage of this type of automated rater is that it is cost-effective under conditions in which a large number of similar prompt types need to be scored on the fly. Generic raters are often used in scenarios in which students can submit revisions of an essay for scoring to check progress. A drawback of such an approach is that generic raters are not particularly useful for determining if students are writing on topic because only features of writing such as grammar, usage, mechanics, and structure are included in the model. Even for more formative contexts, periodic human back-reads are needed to ensure that human-automated rater separation is not occurring across time.

Model Evaluation

In the grades 3–12 context, the practical use of AS model evaluation criteria has been strongly influenced by the *scoring framework* articulated in Williamson, Xi, and Breyer (2012) (see, e.g., Rich, Schneider, & D'Brot, 2013; McGraw-Hill Education CTB, 2014; Pearson & ETS, 2015; Lottridge et al., 2018; Yan & Bridgeman, Chapter 16, this handbook). The framework is influential because of its multifaceted review into automated rater functioning as well as areas in which human-automated rater separation can occur.

When implementing this framework, a series of *statistical flags* are created by the psychometric team that are to be considered through a human judgmental process when approving AS models. For example, McGraw-Hill Education CTB (2014) created statistical flags for three areas of hierarchical import to better automate scoring model acceptance, review, and reject criteria:

a) automated score compared to final human score of record,

b) inter-rater agreement between an automated score and human score compared to inter-rater agreement between two human raters, and

c) *differential functioning* of automated scores for subgroups compared to human scores.

Beyond single item-model evaluation procedures, Kieftenbeld and Boyer (2017) proposed an automated rater evaluation approach based on Demsar (2006) and Garcia and Herrera (2008) for evaluating automated raters over two or more items. Unlike the more typical *descriptive methods* that are applied one item at a time to an automated rater evaluation, the methods investigated by Kieftenbeld and Boyer (2017) relied on *inferential approaches* to examine rater quality over sets of items. Specifically, the approach examines the appropriateness of both *parametric* and *non-parametric tests* for differences between raters and between automated raters and a human rater criterion, which can be conducted via a *repeated-measures ANOVA* on *ranked QWKs* without much loss of information.

In addition to the above, psychometricians should scrutinize items to identify items with *highly skewed distributions* or *sparse observations* at particular score points. For example, high agreement between two raters (human vs. human, human vs. automated) may occur artificially when rubric score ranges are *truncated* due to limited rater use of the range of scores (Bridgeman, 2013) or when those scores are not observable in the population (e.g., the use of high scores as evidence in writing in the elementary grades is rarely seen). This can lead engines to be highly reliable for the task, on average, while being unreliable at a particular score point. There is also a potential to introduce error in *trend analyses* as student abilities increase and more students begin populating a higher category or raters exhibit *rater drift*. Therefore, it is prudent to monitor accuracy at both the item and score point level, with nonadjacent, maximally discrepant human-automated score papers being reviewed by humans for anomalies as a quality control.

Post-Processing

Once AS models have been selected for particular items and placed into production, depending on the setup of the AS system, an important next step is to process the selected model validation sets once again through the scoring system as a quality assurance measure. Psychometricians need to ensure that the appropriate models have been captured for production and to make sure that the system can handle all kinds of expected aberrant response. Thus, it is generally also advisable to send check papers through the system to ensure that advisory flags are working as intended.

Implementing an Automated Scoring System

Once AS models are placed into production, it is important to continue monitoring and documenting scoring quality. The timing of such evaluations is one of several critical design decisions (see Lottridge & Hoefer, Chapter 11,

TABLE 12.3

Quality Assurance Steps

Process Step	Considerations
Before Model Building	
Perform program-level regression testing	Do you have baseline year papers to support evidence that scoring systems enhanced over time provide comparable student scores to the baseline year, provide the same accuracy or better, and comparable or reduced SMD differences for subpopulations?
Clean student responses	Has a random visual inspection of a subset of responses occurred prior to parsing and tagging to ensure no unexpected anomalies?
During Model Building and Validation	
Examine normalized feature distributions	Do normalized feature distributions fall into a reasonable range based on previous experience?
Examine maximally discrepant papers	Does an expert review of maximally discrepant papers identify systematic scoring issues during model building and implementation?
Rescore responses with approved models to ensure correct models are moved to the production environment	Are final approved scoring models accurately moved into production for operational scoring?
Route responses to advisories	Are advisories working as intended and accurately coded for the state?
During Operational Scoring	
Examine normalized feature distributions	Do normalized feature distributions fall into a reasonable range based on previous experience?
Examine maximally discrepant papers	Does an expert review of maximally discrepant papers identify systematic scoring issues during model building and implementation?
Examine Inter-rater reliability Statistics	Are human scores accurate on trend papers? Is evidence showing no indication of automated rater-human score separation?
Randomly select responses with advisories for review	Are advisories working as intended or are other anomalies detected?

this handbook; Rupp, 2018) and likely depends on the score processing timing patterns (i.e., whether responses are processed continuously, in batches, or at the end of the administration), which influence when evaluations are best accomplished (e.g., continuously, weekly, daily, at the administration end). We have summarized our recommendations for this process in Table 12.3.

Changes in Scoring Conditions

Changes in any one or more of the conditions of scoring (e.g., new feature enhancements, updated statistical models, requests for representing student achievement trends, and changes in student demographics) can create technical challenges in AS because the corresponding distribution of student response characteristics is very likely to change. These distributional changes may affect both the quality and comparability of scores derived from automated rater when compared to their human counterparts and it can be at odds with the goal of measuring student growth on comparable metrics. Implementing the collection of a program-level set of responses from the baseline year as a regression-testing set is advisable.

In addition to trend scoring analyses, it is important to consider if the AS systems will be trained using *historic examples* of the essays (e.g., field test papers or operational papers from the previous administration) or whether *early return samples* will be human scored and then used to train the engines during an operational administration. Psychometricians should document and monitor features for the same items from different administrations, which is critical should unexpected results occur.

However, in the context of assessment in grades 3–12, such inputs will often change because students are expected to increase in ability as teachers shape curriculum to be target state standards. This may be especially problematic whenever items are reused as *test equating anchors* across administrations or raw score outputs are used to calibrate items parameters in an *item response theory* framework; in these cases, additional evidence of rater quality will be helpful to ensure score comparability across scoring modes.

Aberrant Responses

An important consideration in the implementation of an AS system is the detection of unusual or *aberrant responses*. As we noted above, one common undesirable response type is typically referred to as *gaming* and indicates that an examinee makes deliberate response choices that are unrelated to the intended item demands and lead to construct-irrelevant variance (see Higgins and Heilman, 2014). A common example is when examinees write very long responses to capitalize on tendencies for AS models to weight *response length* in scoring. This is one reason why comparing the performance of AS models with and without response length during development is important in order to support the validity of scores.

More generally, Higgins and Heilman (2014) proposed a framework for how to best handle aberrant responses, which involves simulating data for the hypothesized conditions and computing a *gameability metric*, which can be computed longitudinally. Should construct-irrelevant examinee responses become more pervasive over time, this would alter the range of scores and associated ability distributions and, as a result, the prediction models used in the AS process would tend to lose true-score precision over time. This would affect their operational performance and could be observed even after the relatively short time that elapses between field testing and operational implementations (V. Kieftenbeld, personal communication, 2017). As a result, it may become necessary to train or *recalibrate* new models in successive operational administrations where indications of changing examinee behaviors exist.

Rater Error

Research investigating the psychometric impact of *automated rater bias* on the reliability of reported scores is still relatively early and incomplete. Rater bias in general is relevant in situations where (a) different automated raters are used to score the same items over time, (b) improvements are made to automated raters that score the same items over time, and (c) examinee behavior and distributions changes over time.

On the human side, human scoring as back-reads can play an important role in quality assurance processes, but this means that qualities of human scores must also be monitored closely. Shifts in human rater expertise and/or performance contribute to shifts in human rating quality over time (Schneider & Osleson, 2013). For example, when human rater agreement trends differ across time, it becomes difficult to determine if human-automated rater differences are due to an AS issue or due to human rater drift (McClellan, 2010; Schneider, Waters, & Wright, 2012).

As human raters are not perfectly consistent, we do not expect absolute comparability of raw scores across different human raters. However, AS has the potential to consistently produce the same score for a given examinee response, no matter how many times the response is rated. As AS methods continue to advance and produce scores that are increasingly consistent with – or exceed – the score quality of the best human raters, the amount of automated rater error in the scores for constructed-response items may be reduced.

On the automated side, scoring consistency can be evaluated by reviewing feature distributions and summary statistics from the production environment at regular intervals and comparing them to those from benchmark sets. For example, Trapani, Bridgeman, and Breyer (2011) and Yan and Bridgeman (Chapter 16, this handbook) recommend calculating the *standardized mean score difference* for each feature to understand if *distributional shifts* in features may have occurred. They recommend this step as part of a pre-processing feature quality assurance phase to provide an early alert regarding the potential for systematic human-automated score differences (i.e., *score separation*).

To disentangle sources of potential differences in scoring between human and automated raters, responses that have been scored by expert human raters should be included in operational scoring for both human raters and automated raters. This process is often referred to as *trend scoring, validity scoring*, or as embedding *check sets*. Schneider and Osleson (2013) found that check sets are best designed to be common across raters and should include the entire range of rubric score points.

Discrepancies with the expert scores can help detect the presence of human rater bias and drift in an operational setting can serve to correct such bias through rescoring procedures when used for retraining raters (human or automated). This becomes especially important when using AS in measures designed to support longitudinal interpretations of changes in student ability as this procedure facilitates isolation of changes due to student growth from changes due to rater bias or drift.

Equating

Recent studies on the effects of differing levels of rater bias on the psychometric properties of tests indicate *degradation* in the accuracy of parameter estimates for item discrimination and difficulty from item response theory (Boyer, Patz, & Keller, 2018; Boyer, 2018; Boyer & Patz, 2019). These studies demonstrate that increasing levels of rater error may degrade test equating functions and produce substantive bias in examinee scores in cases where higher levels of bias are present in rater scores. In both *single-group* and *nonequivalent groups anchor test* designs, higher levels of rater bias and unreliability result in *equating bias*, particularly at the tails of the score scale. The higher the bias, the greater the range of the scale that is impacted. The implication for examinee scores is that score comparability across test forms may be overtly threatened at different points along the scale. Under such conditions, growth interpretations on a *vertical scale* would be impaired.

The equating bias noted in the results of these studies appears to be influenced by the proportion of constructed-response items used in the equating (Boyer, 2018; Boyer, Patz, & Keller, 2018; Boyer & Patz, 2019). The greater the number of constructed-response items used in equating, the more notable the equating bias regardless of rater type. This is consistent with rater drift studies that find that the use of constructed-response items in equating may result in greater equating error, particularly where the correlations between multiple-choice and constructed-response items are not strongly correlated (Dorans, 2004; Hagge & Kolen, 2012; Kim & Walker, 2009).

Consequently, depending upon test content and equating designs, different testing programs might draw different conclusions about the level of rater error that is acceptable in either a human or automated rater and how much change in automated rater error across testing occasions might be tolerated. These results also suggest that a conservative approach to adding automatically-scored constructed-response items to a high-stakes testing program may be to limit the number of such items included in current test designs.

Conclusion

In this chapter, we reviewed several quality-control processes that state or vendor psychometricians can utilize in order to ensure reliable, valid, and fair reported scores whenever AS systems are operationally deployed.

One important focus has been on improving score accuracy through a careful evaluation of the item demands and scoring rubric qualities. These evaluations can be used to modify existing items and rubrics as well as to develop new items and rubrics for use in AS programs to make them amenable to more accurate scoring for both human and automated raters. Improving human scoring consistency and construct representation improves AS, primarily because human scores most often serve as the quality criterion in the construction and evaluation of AS models. In turn, improved human scoring also increases the likelihood that more constructed-response items are scorable with an automated rater, which reduces costs.

A second focus was on systematic evaluations of AS systems that go beyond simple inter-rater reliability indices to an evaluation of the underlying feature space and the impact of automated scores on the psychometric properties of tests, including the consequences of their possible use in anchor sets. Evidence that rater bias does not adversely impact examinee scores overall or for particular subgroups is fundamental to supporting validity claims for a test with automatically-scored constructed-response items. For example, studies on the impact of using automated rater scores on test reliability, measurement error, model parameters, and equated scores may contribute to a more complete understanding of the "degree to which evidence and theory support the interpretations of test scores for proposed uses of tests" (AERA, APA, & NCME, 2014, p. 11).

Such studies are important because the purpose of including an essay or constructed-response item type on an assessment is not to determine how well the examinee performs on one particular item but to use that score as a proxy for an ability that would generalize to other measures in the same general domain and on items on different topics written at different times (Bridgeman & Trapani, 2011). Critical to this work is investigating how such pieces of evidence support a score interpretation centered in what students can do. Validity of the score interpretations for automated raters, which are increasingly attended to under principled assessment design approaches, need to center on content interpretations that support equitable and comparable interpretations of student abilities across different subpopulations, on different occasions, and across all ranges of the score scale.

13

Quality Control for Automated Scoring in Large-Scale Assessment

Dan Shaw, Brad Bolender, and Rick Meisner

CONTENTS

In this chapter, we discuss best practices in *quality control* (QC) for an *automated scoring* (AS) system from a *construct-centered* and *cross-system* perspective; see also Lottridge and Hoefer (Chapter 11, this handbook) as well as Schneider and Boyer (Chapter 12, this handbook). In general, when something goes wrong in an AS pipeline, it can be quite maddening to determine the *root cause* of the breakdown. There are many potential points of failure between the various *interdependent systems* that make up an AS modeling pipeline, and the symptoms from different points of failure can manifest themselves in frustratingly similar ways. In order to be successful in a root-cause analysis to correct errors, one must develop a full understanding of the component systems and how their interdependencies can compromise the effectiveness of neighboring systems.

A Holistic Cross-System Perspective

Technical systems include systems for test delivery, data storage and transfer, distributed human scoring, score processing, feature extraction, model-building algorithms, and model evaluation and results analysis. *Organizational systems* include systems for test development, rater training and certification, rater monitoring and remediation, human resource management, scientific research and development, customer relations and business units, and sample recruitment for pilot testing and data collection. Thus, we would encourage practitioners to embrace a broad definition of systems when thinking about QC for AS and urge them to consider a variety of organizational processes and staff that are closely related to the production and management of assessment score data. In this broader view, a system includes the totality of computational architectures, human team members, work flows, artifacts, and documentation that work in concert to accomplish defined objectives in support of the AS modeling pipeline.

The process of test development, for example, is instantiated through an intricate system of complicated human processes that can directly impact the quality of scores fed into an AS pipeline, whether it be reflected in the quality of the test items or the underlying support processes that ensure quality scores. For instance, the system for training human raters and producing scores for responses bears directly on the quality and reliability of one's modeling data. This includes the *human resources system* that hires and retains staff for each of these units given the requisite need for sufficient scoring expertise and experience, which can profoundly impact the quality of the resulting scores. Alternatively, consider the team responsible for recruiting participants for the pilot testing that generates examinee responses, which needs to ensure that no significant sampling issues affect AS.

Put differently, in order to trace problematic results, one must have a *holistic cross-system perspective* on all of the human and technological systems working in concert and how different sources of variance or error can potentially come from multiple possible systems. In this regard, Avi Allalouf (2007, 2017) has written compellingly about the rigorous QC procedures and controls that must accompany standardized testing. He rightly asserts that "QC must be implemented separately for each stage of the assessment process, beginning with test development, right through test administration and test analysis, and ending with test scoring, reporting, and validation" (2017, p. 59).

While it is true that each system needs its own process-specific QC steps to ensure quality inputs and outputs, it is equally critical that a single *"watch dog" owner* – or a small team of owners in close contact – have insight into the functioning and practices of each component system in the AS chain. It is at the points of intersection between systems in an AS pipeline – the *hand-offs, transformations*, and *post-handoff processing steps* – that built-in *assumptions* and crucial context for the data as well as critical information about its collection or computation are most likely to get lost in the shuffle. Acting every bit as much as an interpreter as well as an auditor, this person (or team) needs to be able to drive root-cause analysis to its most efficient and effective resolution by putting the various systems into dialogue with each other and analyzing the potential points of breakdown between them. This level of oversight is only truly possible when a deep understanding of each system and its potential effects on interconnected systems is achieved.

Allalouf (2017) further recommends that a trained professional be assigned the task of routinely monitoring QC in testing procedures and that this person should operate independently of those involved in routine test scoring, analysis, and reporting. We recommend that this be taken one step further in dealing with the intricate web of technical processes that feed into AS by assigning the task to someone who is steeped in those processes but who can also assume the mantle for oversight across all the relevant systems in the AS modeling pipeline. This not only centralizes the investigation into root cause analysis and process improvement, but it furthermore encourages *transparency* from each contributing system, as they must remain in open communication with "central command" about the technical details of their work. Consequently, we have written this chapter for this kind of person (or team), but we hope that it can also be instructive for those who work closely with this person (or team) or those who may be charged with hiring such an individual (or team).

Quality Control in Large-Scale Assessment

Nowhere is a systems-level approach to QC more important than in the context of a large-scale operational implementation of AS. Deploying an AS system in such a context changes the nature of the attendant QC processes because of the high stakes for the stakeholders, which trickle down into high stakes around QC for each component part. For *examinees*, the impact of the test score on their lives is often significant (e.g., placement, admission). For *decision-makers*, using these test scores to make life-altering decisions positions them as gatekeepers to opportunity. For *testing organizations*, it is imperative to deliver high-quality AS score information to their customers and to prevent such information from deteriorating over time. It is therefore incumbent upon the organizations who provide such services to demonstrate that their AS processes and score interpretations are *valid* (i.e., the scores mean what we say they mean), *defensible* (i.e., effective procedures and standards have been followed), *accurate* (i.e., both human and automated scores reliably measure ability or mastery), and *interpretable* (i.e., processes are sufficiently transparent such that stakeholders can grasp how the scores were derived).

We have organized this chapter according to these four criteria. In each section, we discuss QC principles from both a system-specific and cross-system / organizational perspective while also detailing the critical role that documentation plays in the support of both perspectives. Specifically, we examine how component systems should work together in an integrated fashion to produce AS scores in large-scale assessment and describe how QC procedures and close scrutiny of each of these systems and their interconnected dependencies can help to build an important *validity argument* for our stakeholders. That is, we seek to demonstrate the advantages of a holistic cross-system perspective for understanding and evaluating the impact of any particular design flaw, error, variance, or other shortcoming to the whole system.

Validity

In the simplest of terms, a key goal in assessment is to make claims to stakeholders that assessment scores represent a high-fidelity measurement of a specific target construct of interest. Let us take a closer look at one key process in building our AS validity argument – *feature validation* – in order to see how the interconnected systems can impact each other in subtle, but significant, ways. By *features* we mean the *statistical variables* that are created through the *automated textual analysis* of each essay, which are subsequently used as inputs for an AS model. Such features can cover a wide range of general types including, but not limited to: *descriptive features* (e.g., raw word count, average characters per word, average syllables per word); *word frequency measures* (e.g., measures based on how frequently words occur in a pre-defined large corpus); *cohesion measures; syntactic complexity measures; n-grams* (e.g., the 1-, 2-, or 3-word strings used in the essay); *wordlist-based measures* (e.g., measures of the use of particular types of connectives), and so on; see also Cahill and Evanini (Chapter 5, this handbook); Zechner and Loukina (Chapter 20, this handbook).

Feature-Specific Analyses

Engineering an effective *feature set* helps to ensure the *fidelity of construct measurement*, which requires an important interaction between the test development and AS systems, specifically those scientists charged with *feature engineering*.

Qualitative Approaches

Construct experts from test development can optimally contribute to this important dialogue when they have been familiarized with the technical details of how the extracted features are computed and can understand what element each feature is trying to capture within the construct of interest. This allows them to work together with the feature engineers in a careful process of mapping the proposed feature set onto the assessment's scoring *rubric* and *evidence model* such that areas of strong coverage or glaring omissions can be identified. It is critical that the process of *mapping* the theoretical construct to the computational feature set be a direct conversation, with plenty of back and forth and mutual education about the meaning and intent of key terms and concepts from both teams. When the two groups of experts involved in test development and feature engineering attempt to perform this analysis in isolation, there is too much opportunity for slippage in understanding, missed assumptions, and misinterpretations about each group's conceptions of the feature set and rubric elements.

Quantitative Approaches

There are also a number of quantitative methods for ensuring that a given feature is contributing productively to score prediction in AS models. At the most basic level, a common practice is to look at *correlations* between feature variable values and AS scores to determine that some level of *signal* is contributing to score prediction. In the case of *parametric statistical models* such as *linear* or *logistic regression*, we might further look at the model weights that the algorithm generated for the features to determine *relative feature importance*. When more advanced *machine learning algorithms* such as *gradient boosted decision trees* or *linear support vector machines* are used, the validation methods typically benefit from an additional level of computational complexity.

For example, gradient boosted decision trees are able to produce feature importance tables that can calculate the amount that each feature in the model has contributed to the reduction in score prediction error, helping the AS researcher to confirm that a feature is carrying its weight in determining the appropriate score for the responses in the model. One important related aspect of QC work is the systematic and rigorous documentation of key design decisions with their underlying rationales and evidence so that processes can be understood and potentially be replicated (see Rupp, 2018).

Feature Selection

Careful consideration of feature inclusion criteria is especially important for QC in large-scale high-stakes contexts, helping to inform and guide a disciplined, systematic, fair, and effective approach to AS. An adequate minimal list of feature inclusion criteria address issues of relevance, potential bias, legality, and practicality of implementation, and should include:

- *Construct relevance.* The theoretical basis for inclusion of the feature may include reference to writing construct experts' judgment and/or published research studies and should be thoroughly investigated and documented.

- *Relationship to other features.* Though construct-relevant, a feature's effect may be highly correlated with that of another feature in the set, making it redundant and relatively unhelpful for score predictions, even though its computation may help with the design of advisory flags that detect undesirable response behavior.

- *Accuracy improvement.* Even features that are construct-relevant and partially independent may fail to provide a discernable improvement in predictive accuracy. This may be due to misalignment with the ability level of the targeted learner population for example as a feature may measure something too simple or too sophisticated for it to sufficiently discriminate between low- and high-ability examinees in the targeted population.

- *Bias.* Differential feature effects for subgroups may be difficult to predict and should be a matter of empirical study, requiring sufficiently large and diverse sample populations in both preliminary non-operational studies and by auditing results throughout ongoing operational scoring.

- *Ease and speed of extraction.* A feature that may be ideal by every other criterion may need to be excluded from consideration if its extraction is too computationally slow or difficult in the context of real-time AS. Application of this criterion requires determination of both the maximum allowable time and the actual time of extraction per essay for a given considered feature.

- *Robustness to gaming.* Gaming is a serious issue in a high-stakes context as it constitutes an examinee's attempt to deliberately misrepresent his or her ability. The gaming of AS systems has been a popular media topic with outspoken critics having demonstrated limitations in certain AS systems by submitting absurd essays that receive high scores; see Wood (Chapter 8, this handbook).

Feature Engineering

Thorough documentation of the feature engineering process will include documentation of how each feature meets each of the feature inclusion criteria described in the previous subsection. However, additional technical documentation that pertains to the actual computation of the feature typically includes the following:

- *Dictionary.* This should include feature name, format (e.g., real number at 2 decimal places), and a short description for each feature.
- *Licensing.* This should include the databases, lists, software libraries, and tools used for the extraction of various types of textual features, including their licenses for use, which cover a broad spectrum from 'none' to 'highly restrictive.'
- *Pre-processing.* This should include a detailed, replicable description of the steps involved in pre-processing the essays is an essential step prior to feature extraction to address how flagging and/or processing occurs for issues such as (a) blank essays and/or score fields, (b) off-topic essays, (c) essays consisting of gibberish such as random character strings, (d) presence of incorrect/unrecognized Unicode characters, (e) tokenization into words and sentences, and (f) spelling correction, to name but a few key ones.
- *Parameterization.* This should include detailed, replicable description of the computational requirements and steps involved in extracting each feature. Examples include (a) what code or tools is used and where it is stored, (b) the parameterization of said tools and any recommended best practices, and (c) the accompanying word lists, lexical databases, or corpora that were called in the computation of the feature.

In short, the lifeblood of effective QC around feature engineering – and other procedures – is systematic and rigorous documentation of *robust procedures* so that team members and owners of the various intersecting component systems can understand what protocols and procedures have been followed throughout the AS pipeline. Such documentation enables contributors to pin down assumptions about what the data and information they are working with represent and supports the creation of effective interdisciplinary workflows.

System-Level Interactions

In some instances, one might notice that certain AS scores are failing to correlate with scores from other *external validation measures* such as grades in courses related to the construct of interest or scores from other related assessments. One may also find that AS scores are not agreeing well with the human scores for the responses in the data set across the board. Alternatively, as is routinely the case in direct writing assessment for instance, one may find that the AS models do well at predicting scores for a certain range of the score scale – typically, the lower or middle portion of the ability spectrum – but that the accuracy starts to deteriorate at the upper end of the rubric scale where the construct gets more complex and involves more nuanced writing and rhetorical abilities; see Yan and Bridgeman (Chapter 16, this handbook).

This is where a review of the computational *feature extraction system* and the construct experts' articulation of the construct in rubric and framework documentation help to diagnose the potential source of the breakdown. Was a proper mapping of the feature set to the scoring rubric and task elements completed such that the textual evidence of writing ability has been effectively captured by the AS feature set? If so, did the staff involved have the proper qualifications and take the time to understand the intricacies and technical details of the other systems involved? Did the construct experts grasp the feature computations devised by feature engineers and document their own theoretical basis for including said features? Did the feature engineers have a full picture of the key elements of the construct and scoring rubric as they implemented the feature set? It is incumbent upon the person (or team) charged with QC for the AS pipeline to help the organization architect sound procedures that will forestall such preventable problems before they occur.

Other potential systemic sources of unwanted *construct-irrelevant variance* of scores may be indistinguishable at first glance. For example, the human scoring system that produced the scores may have some deficiencies associated with it. Were human raters trained as effectively in the upper range of the scoring rubric as they were in the middle and lower range of responses? It is not at all uncommon that raters disagree most about the more complex, top-level responses in a performance assessment, so investigating disagreement among human raters at specific score points can reveal whether one might have a scoring issue in the upper-range scores of the score data. The best feature set in the world cannot effectively model data that contains too much noise from rater disagreement.

It may make sense to look even further back for more fundamental issues within the organizational human resources system. The system that staffs, schedules, and maintains the human scoring operation can have a surprisingly outsized impact on the quality of the data that is produced by the scoring group. Is the staffing group following best practices and QC for ensuring that capable and experienced staff with the appropriate content expertise are being recruited and retained for organizational continuity? One important principle in this regard is to not be afraid to ask hard questions about their protocols and requirements for hiring and retaining scoring staff, should one find that scoring problems are not isolated, but ongoing and persistent. The key lesson here is to be on the lookout for those dependencies among systems and to be alert for signs that the issues one is observing in one set of AS model results may be more systemic and, indeed, may have their roots in other foundational systems.

Defensibility

Recall that by 'defensibility' we mean specifically whether effective procedures and standards have been followed so that processes and score interpretations are considered sound by key stakeholders.

Professional Standards

Ultimately, specialists reference *industry standards* and the available authoritative literature from the supporting disciplines when they look for guidelines of correct procedures and practices. Yet, in a rapidly developing field like AS, it is equally important to develop *internal handbooks* and *internal guidelines* for best practice drawn from an organization's own historical work and empirical findings. Many colleagues in AS have one iron in the practitioner fire and one iron in the researcher fire at all times, constantly seeking improvements and advances for operational activity. It is important that everyone remembers to curate that working knowledge and document it for both posterity and for our collaborating colleagues.

Internal Documentation

For human scoring practices in particular, a rigorous adherence to scoring protocols and QC-monitoring procedures is essential to producing consistent, high-quality human score data from a human scoring system (see Ricker-Pedley et al., Chapter 10, this handbook; Wolfe, Chapter 4, this handbook). Nowhere is the old maxim "garbage in, garbage out" truer than in building and evaluating AS models. For example, to verify that raters were trained with procedures that ensure effective scoring across the full range of scores, we can review documentation from rater training and recalibration. We can use the documented artifacts of those processes (e.g., the results of rater certification and recalibration tests) to quantify the reliability of the human rater pool at various score points.

More fine-grained aspects of the score-quality monitoring protocols should also be documented and adhered to religiously. For example, it is best industry practice to ask experienced *expert readers* to focus their enhanced QC efforts on those scores that are deemed most likely to be troublesome. These QC efforts include *back-reading* and *front-reading* responses from particular raters who are struggling with a task, evaluating responses at score points that are more challenging to score in general, or review scores from tasks with concepts that are more difficult to score reliably. The important thing is to establish QC procedures with an eye firmly trained on the *score data integrity* needed for effective AS modeling.

Data Deterioration

Allalouf (2007) warned the testing industry of the perils of giving in to *time pressures* or *resource constraints* when QC is at stake, especially when compromising in the name of time or resources will push the integrity of a process past its breaking point. We specifically advise to not carry over the priorities from day-to-day operational human scoring – large-scale production output with lean efficiency in score monitoring and calibration – into the scoring environment where the responses being used to train AS engines are scored by human raters. *Unassailable human score quality* is paramount here, and the size of the pool of responses needed to calibrate an AS engine is relatively small in comparison to typical large-scale assessment operational response volume.

One particularly telling example of this data quality deterioration can occur when writing assessment programs attempt to use operational scoring data from single-rater human scoring systems for building AS models. With no blind *second read* and no *expert adjudication* of *discrepant scores* between two human raters, many instances of inappropriate / incorrect scores will likely remain in the single-rater data set. One simply has no idea if the data from the score points at the ends of the scale are accurate and for prototypical ability distributions that approximate normal distributions there are often few of these tail-end datapoints to begin with, leading to discrepant AS score predictions in these score ranges.

Additional investments in increased QC measures implemented during AS engine *calibration-pool scoring* are likely to pay off handsomely in the form of more *robust* and *efficient* AS scoring models down the line. Those models will themselves realize substantial operational efficiencies due to a reduction in human scoring labor needed for score resolution and monitoring. In other words, *calibration-pool scoring* is no place for a bottom-line mentality.

Expert Involvement

One of the most important intersections is ensuring that construct experts from the test development team are also fully involved in QC throughout the rater training, score production, and monitoring procedures similar to what we mentioned above about the feature engineering process. The most effective AS operations are those that have a tight *feedback loop* between these teams, with test developers participating fully in rater training and score monitoring and with scoring leaders participating in test development review and task feedback.

The goal is to avoid the kind of missed assumptions and partial understanding that can occur when tasks and responses are just "thrown over the wall" to be scored separately from the processes and expertise that generated them. Integrating these two systems into a fully functioning, cohesive unit goes much more smoothly when you have detailed and comprehensive documentation outlining the roles and responsibilities of both parties. Formalizing the arrangement in scoring protocols also ensures the buy-in of each group's leadership to this important organizational model.

Software Coding

It is also critical to establish bulletproof QC for *software coding* by adhering to established best practices for development and maintenance of code is strongly recommended. This includes thorough documentation of code with comments, adherence to code style guidelines (e.g., *PEP 0008 Style Guide for Python Code*), use of standard version control tools (e.g., *GitHub* or *Bitbucket*), documentation of code testing, including any initial parameter settings and results. There are many other important software development processes across the AS-relevant systems that require a similar level of QC scrutiny in order to ensure the defensibility of the overall modeling pipeline, but these are beyond the scope of this chapter; see Lottridge & Hoefer (Chapter 11, this handbook).

Accuracy

In the previous section, we sought to establish that data quality from human scoring is the guiding beacon of QC for AS systems, especially in large-scale assessment contexts. In practice, we try to balance complex conceptual considerations about score accuracy with simpler practical considerations to develop a scalable system for which QC at scale can be put into place.

Evaluation Standards

Fundamental to knowing whether we are getting the scores "right" is having the appropriate evaluation standards and metrics in place to judge results. Fortunately, there is no shortage of industry standards to reference, from the broad principles of the *Standards for Educational and Psychological Testing* (2014) to more AS-specific guidelines like the Williamson, Xi, and Breyer (2012) review of AS industry standards, which has caught on as the go-to reference in the AS literature and industry discussions of score quality. Bennett and Zhang (2016) have also described a host of other evaluative measures and sources of evidence which might be used to establish the validity of scores.

Evaluation Metrics

A key question commonly arises of what to do when the standard QC statistics (e.g., *percentage agreement rates*, correlations like *Pearson's r* or *quadratic-weighted kappa*, *root mean squared error*) do not provide sufficient information to ensure the accuracy of scores at the level needed for AS modeling.

Score Accuracy

On the one hand, many researchers and psychometricians lean toward defining accuracy or consistency across the entire dataset, whether it be between human and automated scores or between various human raters. On the other hand, staff on the floor of a scoring room might alternatively define accuracy in terms of the *target ("true") score* that would be given to an essay by the most expert and experienced human raters, with the deciding vote often cast by the lead trainer or test developer. Indeed, this very process – having an expert group decide the true score for an essay – lies at the heart of the oft-implemented *range-finding procedure* used to identify the optimal training exemplar essays used for human rater training. The goal is to get as close to the true score as is knowable within the constraints and processes of standardized scoring.

We must attach ourselves at any given point in the AS pipeline to a particular definition of accuracy that is appropriate to our QC purpose at that time and then choose *evaluation statistics* and *data-collection schemes* that provide us with evidence of that accuracy so that we can make an effective QC judgment about whether our score data is sufficiently accurate. Quantitative measures of overall *score reliability* across an entire scoring sample are excellent for identifying broad trends and assessing the overall scoring calibration, particularly at the outset of a scoring project. However, even satisfactory statistical results at the level of the overall data set can still allow a surprising amount of inaccuracy at the level of the individual responses used in AS model training and evaluation.

Specialized Metrics

We encourage practitioners to research or devise suitable metrics in the spirit of continuous improvement to help shed greater light on the precision of the scores in one's data sets. For example, some of the most innovative assessment tasks, which might use different scales and rubrics than the traditional four- or six-point scales of direct writing assessment, require alternative metrics and standards to deal with their atypical score outputs. Scores produced on a *dichotomous scale* ('0' vs. '1'), for example, can benefit significantly from the application of evaluation metrics drawn from the broader fields of machine learning and medical science where binary diagnostics outcomes have a rich tradition of refined statistical analysis techniques. For example, the *diagnostic odds ratio* and *Matthews correlation coefficient* each provide additional useful information about the ratios of true positives / negatives to false positives / negatives, which is crucial information in dichotomous score outcomes.

In our work, we developed a new statistic to help us evaluate AS-human agreement across the entire score scale. It is essentially a simple averaging of the *conditional agreement by each score point* (CABS) across all score points, which is known as the *equally-weighted conditional agreement* (EWCA). This statistic is particularly useful for comparing multiple iterations of AS models using slightly varying methods and feature sets. It has been instrumental in helping us fine-tune our modeling methods and establish whether certain feature sets were more effective at modeling the stubbornly resistant high-end and low-end score points in direct writing assessment.

Differences between scoring models during evaluation, selection, and tuning can sometimes be quite subtle, so additional meaningful statistical information about how the model might be expected to generalize to the wider population of novel operational responses, for example, can be quite helpful. For example, establishing CABS / EWCA expectations for consistent score accuracy across all score points in a writing assessment can ensure that even those essays that are most challenging to score with AS will be scored with a level of accuracy that is comparable to essays from the middle of the score scale.

Expert Scores

The CABS and EWCA statistics are one example of taking the existing model results data outputs (e.g., a *confusion matrix / contingency table* between human reference scores and AS score predictions) to derive additional insight. On other occasions, one may need to seek out additional data points in order to evaluate score accuracy. For example, comparing individual human raters to expert scores, which typically result from resolutions of non-adjacent scores and randomized back-reading, can help better identify individual raters who were severely out of step with the group's scoring calibration, which might necessitate a rescore or excluding their compromised score data. Similarly, comparing individuals or the entire group to expert third raters by score point can help to understand the pervasiveness or degree of human scoring issues at specific score points. If one finds, for example, that rater agreement viz-a-viz expert scores falls off dramatically at the top two score points, one would expect the AS modeling of the top of the score scale to consequently suffer from accuracy issues.

System-Level Interactions

The previous example illustrates just how important it is to use all the information at one's disposal to obtain the most accurate possible scores for AS model training and validation. Approaches that simply rely on standard rater scores as a reference score or that, on occasion, may pick a single-rater score (e.g., the score from Rater 1) for AS model building or evaluation may have the advantage of producing more straightforward comparisons between two scores on the same scale (i.e., a Rater 1 score and a predicted AS score, both on a scale from '1' to '6') but there is no way to quantify *rater error*. Our internal analyses of third-rater scores (i.e., non-adjacent score resolutions as well as randomized or focused back-reading) in our use contexts have shown that as much as 20% to 30% of standard rater scores may be overturned by a more expertly trained and experienced third-rater scores, even in highly effective scoring operations. Consequently, the choice for AS model building and evaluation seems obvious, namely to use all available information, particularly expert scores.

Typically, Rater 1 scores are chosen for training or validation reference scores because of the simplicity of using them to calculate various metrics. *Exact agreement* and *quadratic-weighted Kappa*, in particular, are designed to be calculated using two scores on the same scale such as a Rater 1 score and a machine score. However, this does not solve the question of what to do with all of the Rater 2 scores (when available) that disagree with Rater 1 scores – often, all of the Rater 2 information is unnecessarily lost for modeling when a single-rater score scale is selected for training and evaluation.

The greater issue here is that the reported scores from high-stakes large-scale assessments usually need to retain the increased level of score accuracy made possible with blind second-rater scoring and targeted third-rater scoring in cases of adjudication. This can lead to scores that are based on different amounts of information across the score scale. That is, the reported scores for most responses are a combination of scores from Rater 1 and Rater 2 while the reported scores for backread essays on which they disagree are often based on just scores from Rater 3.

Thus, if QC metrics are designed to be calculated on a '1' to '6' scale, but the most accurate dataset for engine training and model results evaluation are on a '2' to '12' scale, what are the options for analysis and reporting? One simple option is to create AS engine predictions on the '2' to '12' scale, which may work if the AS engine is designed to be the single source of reported scores. In high-stakes large-scale assessment, however, most customers expect to see some human scoring contribute to the final reported scores, at least with the current state of the art in AS. Therefore, the AS score prediction will have to be on the same '1' to '6' scale as the human ratings.

A more appealing solution is to transform the data or quality metric calculations to accommodate a one-rater vs. two-rater scale comparison. For example, for overall exact agreement, we can credit the AS engine with agreement whenever it successfully matches one or both final resolved scores. For quadratic-weighted kappa, agreement can be calculated using a decimal-based formula or the table of associated quadratic weights can be modified to include the additional 'split score' score points on the '2' to '12' dimension of the quadratic-weights table. For the CABS / EWCA statistics, calculations can be performed for every class of response score from '2' to '12' even though the *even-numbered score points* (at which both rater scores agree) are of more interest for accuracy evaluation than the *odd-numbered score points* (at which the two scores are adjacent) since the AS engine has greater odds of matching one of the two reference scores in the split score.

Aside from providing us with reference scores for engine training and model evaluation, data from the human scoring system can also provide us with some useful *reference thresholds* for use with our quality metrics. It is common for analysts to utilize both *absolute criteria* as well as *relative criteria*, with associated thresholds that are determined based on human scorers' accuracy in scoring the same dataset. For example, we might expect an overall exact agreement of 70% or more or no more than 10% less than the agreement of two human raters on the same data.

We can extend this concept to the CABS / EWCA statistics as well. For example, we can stipulate that the AS model agreement with human rater scores by score point must be 50% or higher at all score points or, failing that, that it must not be lower than the human rater agreement at that same score point. Taking our cues about performance from the empirical results of the human scoring system gives us confidence that the AS scoring system is performing in comparable ways, an important prong of our validity argument for AS.

We also sometimes find that our quality metrics reveal deficiencies in the AS model calibration that can be remedied with the continued use of human scoring in operational practice.

For example, the composition of our testing populations frequently leaves us with imbalanced subsets in our dataset. Most often, as we noted earlier, this is seen at the tail ends of the score scale, where it is not unusual to find only 1% to 2% of the overall population at these score points. If the AS model has not been trained with sufficient data – or with possibly insufficient feature coverage – to effectively score the top or bottom score points, there will be performance problems.

Yet, if QC procedures are precise enough to confirm that the model works well for the remaining score points, as demonstrated by the CABS / EWCA statistics, for example, one can implement a combined scoring approach, sometimes also known as *split scoring*. Under this approach, one identifies the AS decimal score above (or below) which the model results fail to meet *performance standards*, and one then routes all responses that receive an AS score outside those boundaries to fully human scoring, dropping the AS score from the reported score calculation. Subsequent analysis can then help confirm that the operational scores from the combined AS/human and fully human scoring meet acceptance criteria. In many cases, one can still automatically score 90% to 95% of operational responses with this scoring approach while ensuring that the more difficult-to-model score points are receiving the additional human scores needed to ensure optimal accuracy across the full human score scale.

Interpretability

Nonlinear and multilayer machine learning models that are being implemented for AS today can offer substantial performance improvements over more traditional linear regression models. A downside, however, is that the scoring mechanisms from such complex prediction models can be more difficult to interpret directly, often earning them the nickname "black box models." However, methods have recently been developed that continue to improve our ability to interpret the effects of features on score predictions. Practitioners and external stakeholders want assurance that a scoring model implemented in production will use construct-relevant writing features to generate scores, and that the features will affect scores in plausible ways.

In the validity section, we discussed selection criteria that can be used to determine which computational features should be included in an AS model, arguing for the crucial role that feature set analysis plays in building an argument for the validity of scores. In this section we discuss additional, more precise, *feature analysis methods* that help us communicate with key stakeholders in interpreting the meaning of scores. Specifically, we review the additional information offered by three such analyses: (1) *feature importance tables*, (2) *partial dependence plots*, and (3) individual *conditional expectation plots*.

Feature Importance Tables

One method for interpreting machine learning models is to determine the relative influence of each predictor variable in predicting the response and reducing model error, often referred to as relative feature importance as we noted earlier. Feature importance values should be collected in *feature importance tables* and be reviewed by construct experts during the model validation phase to confirm that the relative values are plausible from a theoretical perspective. Features related to core writing competencies, such as measures of verbal sophistication, should appear near the top of the list, while other features that are beneficial but have less predictive value should appear further down.

Methods for measuring feature importance vary between AS modeling algorithms. For example, for a gradient boosted regression tree model, the importance of a variable in an individual tree reflects the reduction in error achieved by splitting observations based on a certain value of that variable. Feature importance values for the full boosted tree model are then calculated by averaging values across all trees. Finally, relative values are then calculated either by assigning a value of 100 to the largest feature importance value and scaling the others accordingly or by computing feature importance values as proportions of the sum of overall error reduction between 0 and 1. Feature importance tables with this information can be exceptionally useful for communicating to non-experts the degree of influence that each feature in your model has exerted on the final score outcomes.

Another aspect that feature importance tables can capture is the relative proportion of error reduction that is attributed to individual features. For example, if a single feature accounts for an 85% reduction of the overall prediction error, then the model may be at risk of generating poor predictions in the presence of atypical variations along this feature. This may be the case when a gaming strategy is used by examinees that avoid response characteristics that this feature can capture. When low-level features are used, a robust AS model with a relatively even distribution of feature importance may be more resistant to crude gaming attempts like examinees writing long, nonsense-filled essays.

Partial Dependence Plots

A visual method for feature interpretation is the *partial dependence plot* (PDP) proposed by Friedman (2001), which visualizes the change in a generated prediction as a specified predictor is varied across observed values; see Figure 13.1. An advantage of the PDP is that it is *model-agnostic* because the visualization technique only relies on predictions and not directly on any internal computational algorithm. Thus, a PDP may be generated for any complex model, including *support vector machines, gradient boosted trees*, and *deep neural networks*.

It is again useful to have construct experts review PDPs for the most important features in a model during the validation phase. They should confirm that the directionality of the PDP makes sense given an understanding of the operationalization of the construct of interest through the assessment and AS algorithm. For example, many features used as inputs to an AS model for essay responses have a positive cumulative effect on predictions (e.g., the 'Academic Words Used' feature in the left panel of Figure 13.1), but there

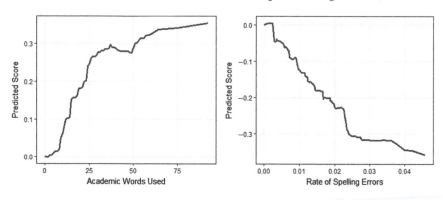

FIGURE 13.1

Partial dependence plots (PDP) for the number of words used from *Coxhead's Academic Word List* (2000) and rate of spelling errors.

are a few that tend to have a negative cumulative effect (e.g., the 'Rate of Spelling Errors' feature in the right panel of Figure 13.1). Confirming the directions of the effects of important variables can support scoring model validity arguments.

Individual Conditional Expectation Plots

Similarly, *individual conditional expectation* (ICE) plots (Goldstein et al., 2015) refine the PDP by adding functional relationships between a predictor and target for individual observations; see Figure 13.2. Beneath a prominent PDP on the top layer, ICE plots contain points representing predictions and real values of the specified predictor for a sample of observations *stratified* by the target variable. Individual ICE curves are drawn by generating predictions for all observations as the value of the specified predictor is varied across all observed values. Like PDPs, ICE plots can be generated for any complex model including support vector machines, gradient boosted trees, and deep neural networks.

ICE plots may be centered by adjusting all ICE curves to achieve a y-intercept of 0. When all ICE curves start at 0, the cumulative effect of a feature may be plotted as the value of the feature is increased along the x-axis. This eases interpretation by compressing ICE curves into a smaller range, making them somewhat more intuitive for non-experts to grasp. Reviewing centered ICE plots is particularly useful for evaluating AS model effectiveness, because ICE plots indicate the degree to which the effects of particular features interact with the effects of other features. If the cumulative effect of a feature varies widely for different responses, then it indicates interactions with other features. In some cases, ICE plots may even illuminate complex interactions where the cumulate effect of a single predictor could be positive for some observations and negative for others, depending on the values of interacting features.

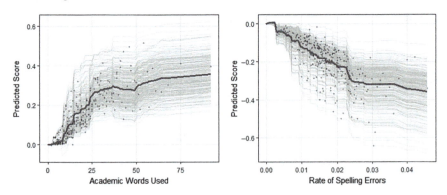

FIGURE 13.2
Individual conditional expectation (ICE) plot for the number of words used from *Coxhead's* (2000) *Academic Word List* and rate of spelling errors.

ICE plots can also provide more fine-grained information about the potential effects of features on predicted scores. While the PDP indicates the potential effect of a feature on predicted scores on average, the ICE plot additionally indicates the potential effect on individual responses. This may be a better representation of the potential size of the effect and non-expert stakeholders generally feel more conceptually grounded in discussing documentation that displays information aligned to the rubric scale. At the same time, some caution is warranted since individual responses could be atypical along other dimensions in any variety of ways, so the cumulative effect could be due, in part, to strong interactions with extreme values of other features. Nevertheless, understanding and communicating the potential effects of important features on predicted rubric scores, and describing how this aligns with writing theory, provides additional support to validity arguments for QC purposes.

Final Thoughts

We were not able to exhaustively describe all of the many QC processes and procedures required to successfully implement AS in a large-scale assessment context in a single chapter. Hopefully, though, by highlighting some key processes with our various case studies, we have been able to convey a feel for the kind of strategies and analyses that will help readers explore and understand the interdependencies between the different systems that comprise an AS pipeline and its supporting network of connected resources. Above all, this type of holistic, cross-system QC perspective encourages adroit solutioning that allows us to balance competing needs in a complex environment where many factors are brought to bear.

To recap, here is a list highlighting the "Top 10" best practices that are key takeaways from this chapter:

1. **Develop a comprehensive holistic understanding** of the component technical systems and human systems in the AS pipeline and get a feel for how their interdependencies can either bolster or compromise the effectiveness of neighboring systems.

2. **Appoint a single "watch dog" owner** who has insight into the functioning and practices of each component system in the AS chain. Both interpreter and auditor, this person can drive root cause analysis most effectively by putting the various systems into dialogue with each other and analyzing the potential points of breakdown between them.

3. **Document design decisions** as a tool for communicating data context to other interconnected parties in the AS pipeline. Systematic and rigorous documentation of key design decisions, with their

underlying rationales and evidence, allows processes to be understood and potentially be replicated.

4. **Document feature selection** methods using criteria that can address issues of relevance, potential bias, legality, and practicality of implementation, which helps establish the validity of scores and adds to score interpretability with key stakeholders.

5. **Invite construct experts** to work together with the feature engineers in a careful process of mapping the proposed feature set onto the assessment's scoring rubric and evidence model, such that areas of strong coverage or glaring holes can be identified. This should be a direct conversation, with plenty of back and forth and mutual education about the meaning and intent of key terms and concepts from both teams.

6. **Integrate construct experts with AS experts** throughout the rater training, score production, and QC procedures to enhance the quality of human scores and subsequent AS system predictions. The most effective AS operations are those that have a tight feedback loop between these teams, with test developers participating fully in rater training and score monitoring and with scoring leaders participating in test development review and task feedback.

7. **Do not carry over operational scoring priorities** – large-scale production output with lean efficiency in score monitoring and calibration – into the scoring environment where the relatively small number of responses being used to train AS engines are scored. Unassailable score quality is paramount here: "garbage in, garbage out".

8. **Develop internal handbooks and guidelines** for best practice drawn from your organization's own work and findings in your particular use cases that supplement standard industry metrics and procedures. In particular, we encourage practitioners to experiment and constantly seek new evaluation methods and statistics that can shed greater light on the accuracy of the scores in your data sets on a more fine-grained level, not just in the aggregate, so we can trust their use in AS modeling. Discover or craft statistics to fit your use case while still retaining their essential statistical function.

9. **Use all the human score data at your disposal** to determine the most accurate possible scores for AS model training and validation datasets. For example, use second-rater scores and extant score corrections from third-rater resolutions to evaluate score accuracy and AS model performance with greater precision.

10. **Use sophisticated feature analysis methods** such as feature importance tables, partial dependence plots, and individual conditional

expectation plots to help communicate with key stakeholders in interpreting the meaning of scores, regarding the specific ways that each feature is impacting score predictions.

As AS moves forward into the next generation of capabilities, it is likely that we will see powerful new algorithms emerging from *deep learning* or perhaps advanced capabilities possible through *reinforcement learning*; see D'Mello (Chapter 6, this handbook) as well as Khan and Huang (Chapter 15, this handbook). In these approaches, AS engines can *self-refine*, learning from their mistakes and improving their predictions as more data accumulate. The broader collection of diverse data points made possible by 'Big Data' approaches to educational measurement may finally afford us an external validation measure that is more robust and precise than standard human rater scores, changing the measuring stick for AS engine success from "as good as human raters" to something that is potentially much closer to the "true measure" of student ability. This will of course challenge our current collection of best practices and compel us to continue to devise new QC methodologies that can regulate these powerful new methods.

Nevertheless, these computational methods will still require the same close scrutiny of the overall process as we have described above and will likely necessitate an even greater level of sophistication in QC controls. There is a continued importance of a broad view that looks for systemic dependencies across organizational and computational systems and traces root causes throughout these systemic connections. If the devil is hiding in the details, it will take not just vigilance and persistence to root him out, but also creativity and holistic understanding to devise effective monitoring methods that will keep him at bay.

14

A Seamless Integration of Human and Automated Scoring

Kyle Habermehl, Aditya Nagarajan, and Scott Dooley

CONTENTS

Increases in the use of digital technology in educational markets (Darling-Hammond, 2017) and an increased demand for low-cost assessments of *higher-order skills* (Stosich et al., 2018) have been driving a rise in demand for *automated scoring* (AS) of essay and open-ended, constructed-response items. This has added pressure on AS providers to find ways to develop significantly more AS models, at lower costs, with reduced times for development and implementation cycles.

At Pearson, *automated essay scoring* (AES) systems in particular were first developed for large-scale assessments. In those assessments, the focus is on customization, high volumes, and more recently, automation. *Human scoring* and AS groups historically had been separate groups at Pearson. *Conway's Law* (Conway, n.d.), an oft-quoted law of computer system architecture, states that an organization that designs a system will produce a design whose structure is a copy of the organization's communication structure. Prior to the introduction of this new architecture, Pearson systems and processes had evolved in accordance with Conway's Law and exhibited a *linear waterfall structure*.

In spring 2015, *innovation workshops* were held with participants across Pearson – including colleagues from content development, scoring operations, technology, psychometrics, and business development. During *brainstorming sessions*, a key framing question emerged – "what-if data 'flowed continuously' between organizations, systems, and people?" This deceptively simple question promoted creative ideas involving changes across business processes, organizational structures, software architecture, and operations.

It ultimately drove evolution of a strategy to inject *artificial intelligence* (AI) and *deep learning* across the enterprise. It fundamentally changed how humans and software systems interact to provide better scoring products to our customers. A new *software component* was developed to implement and explore ideas envisioned during the innovation sessions, which quickly became a core application in Pearson's *machine learning* and *scoring suite*. It became known within the company as the *Continuous Flow* (CF) system and has now been in operation at Pearson for several years.

In this chapter we discuss the changes and challenges that arose during conceptualization and development of the CF system, the accompanying lessons learned, and the ensuing results of putting the ideas into practice. In the first section, we provide the background and limitations of traditional approaches used in the industry. In the second section, we describe the architecture of the CF system at a high level and provide sample workflows that illustrate the implications of the architecture for how interdisciplinary teams engage in their work. In the final section, we discuss additional implications of the CF system on business and processes.

Background

To understand the context of the genesis, development, and use of the CF system it is helpful to have some background in both machine learning and in the traditional methods for developing and deploying AS models. We address each of these in the next two sections.

Machine Learning

In this section we provide a high-level overview of machine learning for AES; we note that many of these concepts also apply to short-answer item scoring. A good introduction to machine learning can be found in Murphy (2012) while a good introduction to *natural language processing* (NLP) can be found in Manning and Schütze (1999). Specifically, *computer programming* is a process of explicitly developing instructions for a computer to perform some task. However, it is extremely difficult and very expensive to develop explicit instructions for a computer to perform the task of scoring / grading essays. Machine learning is a process of providing data to a computer algorithm that can then "learn on its own" to perform a task without explicit instruction in how to do it. Machine learning is what has made it cost-effective to instruct computers in how to score essays.

In the case of machine learning for AES, the key ingredients are *ratings / scores* that humans have assigned to each student *response* (i.e., essay). In traditional machine learning terminology, these scores are called *labels*. Next, *model training* refers to the process where the machine attempts to learn to grade like the humans, resulting in a statistical *prediction model*. Conceptually, a model *encodes* the rules, instructions, and knowledge for scoring. Once trained, a model can be deployed for operational scoring. We use the term *scoring model* in this chapter to refer to all the artifacts needed to enable a computer to accept a student essay and compute a *holistic score* and/or multiple *analytic / trait scores* according to a defined *scoring rubric*.

Human scores are a key ingredient for training and evaluating AES models. The rigor applied to the human scoring processes for model training depends upon the *stakes of the assessment* for which AES is used since the quality of data used for model training can greatly affect the quality of the trained model. Notably, the costs and effort to train and certify scorers for a high-stakes assessment will be much different from those in which a teacher or teaching assistant fulfills the human scoring function to train a scoring model. Adding more human scores typically improves the models but usually with *diminishing returns* at some inflection point, and not all models reach sufficient quality standards for operational scoring. See Ricker-Pedley et al. (Chapter 10, this handbook) as well as Wolfe (Chapter 4, this handbook) for more details.

Traditional "Waterfall" Approach for AS

Developing AS models has traditionally involved separate groups performing discrete sets of tasks, with each group delivering the results of their work to the next; we refer to this as the *linear waterfall approach* as mentioned at the outset. In this section we provide an overview of the waterfall approach as it had been in place at Pearson before the development of the CF system. Not included in the scope of our description here are the component tasks for *item authoring* and *test form creation*, whose goal is to create and identify items that can serve as the best possible tool for evaluating a student's knowledge and understanding regardless of the background of an individual.

A traditional linear waterfall approach to the creation of an operational AS model is shown in Figure 14.1. The waterfall approach begins with *data collection*, which is the process of obtaining written student responses (e.g., essays in the case of AES). The cost of this step can vary widely between high- versus low-stakes assessments and their embedded test items. For high-stakes assessment, data is usually collected from *field test items*. For low-stakes items, data may be collected in a variety of ways. For example, a formative writing product may be used to collect responses as students practice writing for a classroom setting.

Once student responses have been collected, the next step is selection of an appropriate *subset of responses* for human scoring. Costs and procedures for human scoring are a function of whether the test item for which AS is to be used is envisioned for high-stakes or low-stakes assessment. Items on high-stakes assessments require multiple human scores from highly trained scoring experts to ensure sufficient *score reliability*. The quality of machine scoring models is judged by comparisons between human-to-human and human-to-machine agreements. For item development in low-stakes assessment contexts, the number of human scores per student response and the training and skill requirements around human scoring are not as stringent. The important point here is that the selection of the responses is done once and this set of responses is sent *downstream* for human scoring in what can be referred to as a *batch-processing* step (i.e., processing a set of information in a bundle or packet).

The *model training step* takes student responses with human scores and uses machine learning techniques to construct AS models; we provide an explanation of how machine learning models for AS are trained later in this

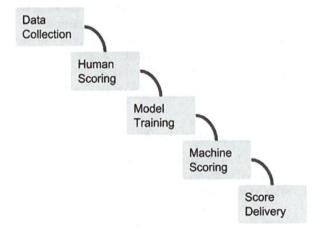

FIGURE 14.1
Key steps to operational scoring prior to CF.

chapter. As we noted above, it is important that accurate and reliable human scores are available for model training because the machine is learning how to score the responses based on the patterns in the human scores. Once scoring models have been created and judged to be of sufficient operational quality, they are deployed for scoring.

In high-stakes scoring contexts, deployment implies "freezing the model" from all future changes. In other words, high-stakes models are typically deployed once. During operational use, new student responses are submitted, scores are computed, and results are returned to a *test management system*. It is often the case that additional human scores may be received once a model is in operation. As we shall see later, the CF system supports multiple operational models where this additional data can be used to automatically build improved scoring models.

In this traditional process, from the moment that students first interact with a test item, the discrete-step processes of the waterfall begin. One team (and system) processes the data and hands it off to the next. In addition to being inefficient, each *handoff of data* introduces risks that the data format and content may not be fully understood by the receiving team and that mistakes will be made in manipulating the data. Any such mistakes will negatively affect the quality of the scoring model.

Process Issues and Implications

At Pearson, the previously described process worked satisfactorily for *large-scale summative high-stakes assessment* for many years until it became too cumbersome to manage and too costly to execute. Specifically, key factors that were driving the high costs under this approach included the number of (a) field test items, (b) scoring models to create and deploy for operation, (c) student responses collected per field test item, (d) human scorers needed overall and the cost of hiring and training them, (e) human scorers needed for operational scoring, (f) human scores needed to train a machine model, and (g) student responses to be scored during operational scoring.

Planning for a large-scale assessment often occurs far in advance of its actual delivery and staffing a *human workforce* for scoring large volumes in short time windows is challenging (see Ricker-Pedley et al., Chapter 10, this handbook). One of the driving motivations for AS is to reduce cost, so whenever the cost to build an AES model is high and an acceptable model cannot be created, the initial investment benefits are quickly offset since the item may need to be 100% human scored. Therefore, a key concern is mitigation of the risk of building an unacceptable AS model for which several strategies can be employed, which we have listed in Table 14.1.

The batch nature of the procedures that are needed to apply the strategies described above makes it difficult to adjust them during implementation. For example, during data collection from a field test, if the number of student responses collected results in a small number of scores at the upper

TABLE 14.1

Risk Mitigation Approaches in Large-Scale Assessments

Risk	Mitigation Strategy	Impacts
Too little time to develop acceptable scoring models	Collect field test data far in advance	Increases time-to-market to create items
Coverage of score points in student data insufficient to train acceptable scoring models	Collect more student responses than may be needed	Increases costs; adds to student testing fatigue
Human scorer agreement insufficient for training acceptable scoring models	Human-score more than may be needed for model training	Increases costs and time
Models only acceptable for some score points	Partial automated scoring	Adds complexity to planning

end of the score scale, it may be impossible to go back and collect more responses. In this case, a common approach is to plan for the worst-case scenario and collect more student responses than would be needed under ideal conditions. Similarly, if the human raters for an item cannot sufficiently agree leading to low *inter-rater agreement* for the item, it is expensive and time consuming to either retrain or hire more scorers. Consequently, more scorers than should be necessary are often hired pre-emptively and they then tend to score more responses than should be necessary for creating an acceptable AS model.

Unbalanced score distributions can result in inefficiently consuming the budget for human scoring as well. For example, using a *simple random sampling* approach to obtain responses for human scoring may result in too many or too few scores at particular score points of the score scale than are needed to train an acceptable AS model. Unfortunately, this deficiency might not be discovered until later in the workflow when it will be costly to re-engage scorers to go back and score additional responses. Similarly, *partial scoring* – also called *split scoring* – uses AS and human scoring at different regions across the score range to mitigate cases when AS works well only in specific score regions. Partial scoring prevents a total loss of investment by allowing the scoring model to be used for a subset of the student responses.

Business Drivers

Related to the cost drivers for AS there are several broader business drivers that drove the development of the CF architecture at Pearson. The company wanted to achieve reuse across formative / interim / summative scoring, reduce barriers for innovation across groups using AS, support a higher number of smaller-volume customers, achieve more flexibility in combining human and AS, and use machine learning for non-content scoring. That is, Pearson had historically split formative, interim, and summative assessments across different organizational units but wanted a system and architecture that could be leveraged across business units and customer segments.

Clayton Christensen's *disruptive innovation theory* (Christensen et al., 2015) describes a process by which a product or service takes root initially in simple applications at the bottom of a market and relentlessly moves up-market, eventually displacing established competitors. In our context, a good example is new techniques in machine learning, which can be perfected in low-stakes markets and then move up-market into high-stakes markets. Large-scale assessment, by definition, assumes a smaller number of high-volume assessment customers. One of our internal business drivers was to better support smaller-scale assessments, however, and, thus, to acquire a larger number of smaller-volume customers. The batch approach in the linear waterfall architecture breaks down as the batch size decreases and the CF automation was considered key to achieving this goal.

In addition, different customers require different mixes of human scoring and AS. For example, rather than scoring every student response with either a human or a machine, it may be advisable to use a *confidence-in-scoring* signal to route responses to human scoring only when the machine is not sufficiently precise in its score prediction. This allows human scorers to focus on boundary and unusual writing cases. Related to the idea of student response routing, Pearson notifies school district customers if a student's writing indicates a case of *learner-in-danger* (LID) such as when a student indicates that he or she is being bullied. A solution must be provided for customers when the scoring mix is mostly oriented toward AS because human raters are not available to catch these responses through careful reading. The solution approach for this difficult problem is discussed as part of the next section.

Continuous Flow

What-If?

We started our discussions that guided the development of the CF system with the key question "what-if data continuously flowed to all available parties?" We later added an additional related qualification to this question that spurred us to explore more creative uses of automated scoring: "and, what if we could drive machine learning/modeling time and costs closer to zero?" Specifically, this 'what-if' exploration question led to the following ideas:

- Use machine learning and active learning to rank student responses for human scoring
- Train initial scoring models in real-time using humans to best train the machine

- Train or retrain AS models to search for improved models
- Adjust confidence-in-scoring using real-time evaluations of human-machine agreement
- Detect LID responses in real-time and continuously retrain underlying AS models

These questions started inter-divisional dialogs that ultimately reframed how Pearson viewed and now uses machine learning in its scoring business. The framing of the discussions changed from a perspective of "human-versus-machine" to one of "human-plus-machine." Essentially, it became clear that the desire was to develop the process represented conceptually in Figure 14.2. That is, rather than stopping after the first sequence of student response collection, human scoring, model training and AS, the process continues with directed human scoring, model re-training, and improved AS.

System Architecture

High-Level Architecture

As suggested through Figure 14.3, the CF system is central to the training and evaluation of AS models as well as the use and improvement of those models in operational scoring; all configurations are done through well-defined *application programming interfaces* (APIs). Specifically, the CF system, which is

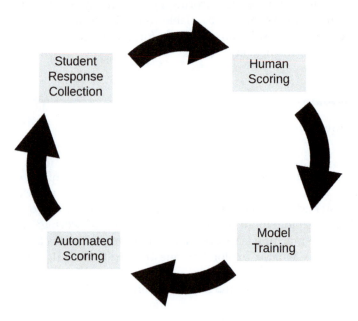

FIGURE 14.2
The AS development cycle in CF.

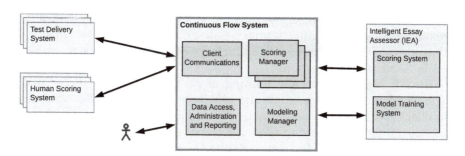

FIGURE 14.3
High-level architecture of the CF and surrounding systems.

shown in the center, communicates with other systems responsible for test delivery and human scoring, which are shown on the left, as well as with systems responsible for AS scoring and model training, which are shown on the right, through its APIs; at Pearson, the systems responsible for AS scoring and model training are known as the *Intelligent Essay Assessor* (IEA).

On the left-hand side, it is the responsibility of the 'Test Delivery System' to manage the collection of student responses while it is the responsibility of the 'Human Scoring System' to collect human ratings for these responses. Through a 'Client Communications' API, this information is then received by the CF system in the center. Via the 'Scoring Manager,' the CF system takes these student responses and human ratings and passes them on to the 'Scoring System' within the IEA, which is designed to evaluate them to produce appropriate content scores (e.g., holistic scores, trait scores, feedback) and flags for aberrant responses, including LID responses. The CF system dynamically prioritizes human or automated scoring of particular student responses within the 'Scoring Manager' component based on the characteristics of each response.

AS Model Training and Evaluation

The CF system utilizes *active learning* techniques within its 'Modeling Manager' that optimize the flow of information based on empirical data properties and human-configured parameters that control the overall process, rather than direct human intervention at specific steps. This leads to a more *efficient, effective,* and *scalable* process for model management than systems that are human-controlled.

Training an AS model requires scored or labeled responses from human raters, even within the CF system. The CF system directs the solicitation of these by implementing targeted policies for the selection of specific *subsets of student responses* for human scoring. Without these policies, it would passively receive human scores from the 'Human Scoring System,' most likely such that only a subset of the score range would be sufficiently represented, thereby leading to inefficiencies in AS model training. Instead, the CF system uses the active learning approaches described above to drive labeling of the training set to be able to create a high-quality AS model sooner and with fewer human-labeled responses.

The CF system directs the training of AS models within IEA's 'Model Training System' and performs the evaluation of those models according to client-specific evaluation criteria within the 'Modeling Manager.' After model training, if a scoring model does not meet the evaluation requirements, the 'Modeling Manager' puts the test item back into the candidate pool for future data collection and subsequent model training. When the collection of student responses and human scores meet re-training requirements, it initiates a new round of data loading and training.

After deploying an acceptable AS model for operational scoring via the 'Modeling Manager,' the CF system immediately begins requesting scores from the 'Scoring System' within the IEA for all student responses received for the specific test item. As those scores become available, it sends them to the 'Client Communications' API, which passes them on for operational reporting through the 'Data Access, Administration, and Reporting' component.

For some student responses, however, the 'Scoring System' within the IEA may indicate that the response cannot be assessed with a high degree of confidence or that the system has detected an *advisory condition* while scoring. In those cases, the CF system creates an alert for the 'Human Scoring System' that the student response needs a human scorer's attention, which allows the system to route the student response for appropriate human review. Typically, the 'Human Scoring System' also collects an independent set of human ratings for a small portion of the student responses that have been machine-scored to support human-machine reliability analyses.

For high-stakes assessments, it is generally not acceptable to change the AS model during test administration. However, for lower-stakes assessments – or across test administrations for high-stakes assessments – it is often acceptable and desirable to make system improvements. In these cases, the CF system can be configured so that even after an acceptable AS model has been trained and deployed for operational scoring, it will attempt to improve upon that model. While preparing for and training a potential replacement AS model, the CF system continues to communicate with IEA's 'Scoring System' to generate scores using an existing scoring model. If a more accurate AS model can be trained for a test item, scoring for that item is paused while the new model is deployed and then resumed so that new student responses are assessed using the improved model.

Assessment Components and Scoring Management

For each test item there is typically a single AS model that receives most the emphasis in training and whose results are used for the vast majority of the associated student responses. However, there may also supplemental scoring models. Some of these may modify or override the evaluation of the primary model or they may provide an additional evaluation type of the student's responses. These *supplementary models* often can be continuously improved during a test administration, even for high-stakes assessments. Some of these supplemental scoring models are used to detect: potential LID instances, off-topic responses, non-English responses, refusals to answer, and plagiarism of source material or other student responses.

Each supplementary model plays an important part in creating a comprehensive evaluation of each student response. The CF system manages the retrieval of both the primary and supplementary evaluations, combines them to produce a final outcome, and communicates that outcome to the requesting client system. It also reviews the auxiliary evaluations, determines whether the response warrants additional communication with the client system, and takes action as appropriate.

Administrative and Data Display Capabilities

As outlined above, the CF system is making many automated decisions during the preparation, evaluation, deployment, and use of AS models for each test item. These decisions are based on human-controlled *configurations* at the customer, test administration, and test item level. Specifically, the CF system's 'Administrative API' provides functions for these configurations as well as model training strategies and model evaluation requirements. Some configuration is accomplished by direct communication from the customer system while the remainder is directed by an administrator using either the 'Admin User Interface' or the 'Administrative API.' The goal is to strive for a high degree of direct configuring by the 'Customer System' because it enables much quicker setup and testing for a new test administration than was traditionally possible.

After configuration, data can begin flowing to the CF system for each test item. These data include student responses and human scores for use in training new AS models, requests to score responses for which scoring models are already available, human scores to use for model reliability verification, human-detected LID cases, and so on. The variety and quantity of data makes is especially important for the assessment team to be able to visualize the data flow, monitor system performance, and make intervention decisions. The CF system provides a 'Data Access API' along with various 'Dashboards' and 'Report Generation' user interfaces for monitoring data collection, model training, and operational scoring. They are used by scoring managers and developers during test administrations for monitoring and by data scientists during and after test administrations for model research and development.

We provide screenshots of two sample-monitoring interfaces used during operational scoring of test items to illustrate these capabilities. The first screenshot in Figure 14.4 shows an interface with charts that help monitor

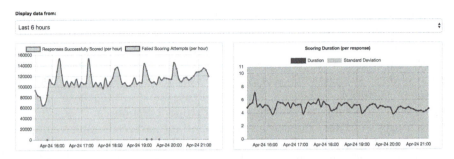

FIGURE 14.4
Scoring velocity and time cost dashboard.

the scoring rates in responses scored per hour and the average time required to score each response. The second screenshot in Figure 14.5 shows an interface with a table of per-test-item data, including the number of student responses and requests to score received to date, the number for which outcomes have been generated ('Processed'), the number returned to the client system ('Sent'), and so on.

AS Model Management

The key design driver for the CF system was to enable *seamless integration* of human and machine scoring to obtain the best possible AS model at a lower cost. As we noted before, AES systems use various techniques from machine learning and NLP to score different aspects of the response ranging from detecting bad-faith responses to assessing the grammar usage and mechanics trait of the response.

In particular, *supervised machine learning* and *unsupervised machine learning* are two broad categories of machine learning widely used in AES systems. Supervised machine learning requires a collection of *human-labeled data*; in AES this is a set of essays that are scored by expert human scorers' for multiple traits. Unsupervised machine learning is a branch of machine learning that deals with *extracting patterns* from large sets of *unstructured data* that do not have human labels. Unsupervised machine learning techniques are

Export ALL Scores	Export ALL Scoring Distributions							
Prompt		Score Requests	Processed	Unprocessed	In Review	Sent	Status OK	Status WARN
4_B_1423 (review) ▾		75216	57703	17513	57703	0	56968	735
4_H_37496 ▾		3974	3729	245	0	3729	3427	299
4_V_42359 ▾		90570	86647	3923	0	86647	83687	2959
5_E_07459 ▾		9505	6805	2700	0	6805	6793	12
5_C_98474 ▾		106199	99999	6200	0	99999	97822	2177

FIGURE 14.5
Scoring counts dashboard.

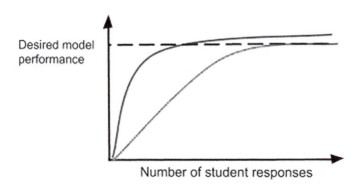

FIGURE 14.6
Model performance curves as a function of sample size.

important because unlabeled data are available in abundance in the form of textbooks and online reference material such as *Wikipedia* and news articles.

Put differently, unsupervised techniques identify patterns that become *signals* for supervised learning while supervised learning techniques enable prediction. A third category, *semi-supervised learning,* combines supervised and unsupervised techniques, typically using a large dataset without human labels for unsupervised learning and smaller labeled supervised dataset with human labels. An advantage of semi-supervised learning is that it requires less labeled data than supervised techniques.

Finally, active learning is a semi-supervised learning approach whereby a machine algorithm can query (or request) human scores so as to maximize the diversity of examples at each score point in the training set. This allows the algorithm to control the data distribution and prevent label imbalances during AS model training. Several active learning approaches have shown to work well for AES (Dronen et al., 2014) and short-answer scoring (Horbach & Palmer, 2016).

Figure 14.6 shows a well-known result of applying active learning to train an AS model vs. not using active learning. Specifically, the graph shows scoring accuracy from a model trained iteratively by adding samples of randomly selected human-labeled examples from the target population (bottom curve) and a scoring accuracy from a model trained by requesting human scores on select examples using active learning (top curve). The top curve shows that the active learning model achieves a higher accuracy at a faster rate than the model without active learning, thus supporting the use of active learning to reduce the number of required human scores for AS model training.

IEA Engine

At the core of the CF system is the IEA training and scoring engine that we mentioned earlier, which implements various supervised, unsupervised, and semi-supervised machine learning algorithms (Foltz et al., 2013). Similar to other AES systems such as ETS's *e-rater*® (e.g., Attali & Burstein, 2006), IEA uses NLP techniques to derive linguistic and lexical features to assess writing quality.

The IEA scoring engine enables data scientists and AS engineers to tailor models to address *prompt-specific* rubrics and writing criteria. Furthermore, in constructed-response assessments it is common to give multiple scores to a response in the form of analytic / trait scores, which represent differences on the targeted writing construct. These traits are modeled independently in IEA using a gamut of features that are extracted from the text response such as grammar and mechanics, idea development, and overall quality (see Foltz et al., 2013). Figure 14.7 shows an example of an IEA model that is used for AS. The box on the top left a text transformation operation, the other boxes on the left are feature extraction stages, and the boxes on the right are machine learning scoring models, typically regression models.

Text Pre-Processing

The *pre-processing stage* for textual responses within an NLP pipeline consists of *transforming* or *normalizing* these responses so that they can be adequately prepared for the subsequent *feature extraction stage*. For example, in short-answer scoring, punctuations and capitalization are typically not important to determine the score of a response and, thus, the text may be transformed into all-lower-case before the feature extraction stage. *Stemming* the text which is the process of reducing inflected or derived words to their word stem, base, or root form, may also prove to be useful in short-answer scoring. For example, the root 'increase' may appear in the forms 'increased' and

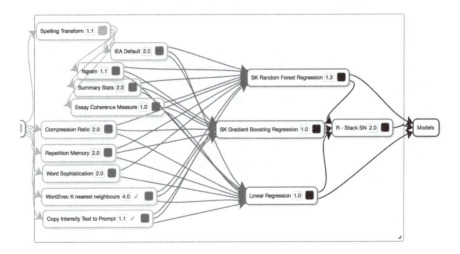

FIGURE 14.7
Connection between different components within the IEA.

'increases' and transforming these variants back to the root can be beneficial to simplify the computation of a diverse array of features.

Feature Extractors

The feature extraction stage of the IEA scoring pipeline involves extracting statistical measures of writing quality and content across various dimensions. The different writing measures extracted from the text are represented by a list of statistical variables (i.e., *vectors*), which become the representation of the text response for supervised machine learning (see Cahill & Evanini, Chapter 5, this handbook; Zechner & Loukina, Chapter 20, this handbook). In the following, we briefly summarize some of the features in IEA:

- **Content-based features.** Central to the IEA scoring engine is the *latent semantic analysis* (LSA) approach (Deerwester et al., 1990; Landauer & Dumais, 1997). LSA is a statistical model that projects the semantic information of text documents into a multidimensional vector space based on large corpora of documents. Once a model is trained on a corpus, a new document or essay can be represented in this vector space and *vector algebra* can then be performed to derive features from these documents. The vectors can then be used to measure similarity between other documents, where responses that convey the same meaning are placed closer together in this space via techniques such as *cosine similarity* or variants of the *word2vec technique* (e.g., Bojanowski et al., 2017; Mikolov et al., 2013).

- **Discourse coherence.** Discourse coherence is a subjective characteristic of an essay used to measure the flow of ideas and understandability of the writing (Foltz, 2007; Burstein et al., 2013). The IEA scoring engine generates a set of essay *coherence measures* based on *semantic similarity* of units of texts in a response. These units include words, phrases, sentences, and paragraphs. The units are first converted to the LSA vector space, the cosine between these units are computed, and statistical measures such as means and standard deviations become features that model coherence. Furthermore, the standard deviation of the cosine measure across sentences gives a quantitative measure of sentence variety.

- **Grammatical errors.** Detecting grammar errors has a rich history in AS and most of the automated essay scoring engines has a variant of grammar detecting feature that are based on two techniques (Gamon et al., 2013). In IEA there are two types of techniques that detect grammar errors: *rule-based* and *statistical*. The rule-based grammar features use *part-of-speech representation* of the text response to derive error counts for each type of grammar error such as preposition and article errors, sentence run-ons, punctuations, and subject-verb agreement; weighted error counts become input to the AS model.

IEA also uses statistical-based techniques to detect grammar errors that are based on *n-gram language models* (see Jurafsky & Martin, 2019) and *neural language models* (Bengio et al., 2003). Language models are unsupervised machine learning techniques that learn the syntactic structure of a language from large corpus of data while *n*-gram language models learn the *co-occurrence statistics* or *likelihood* of a set sequence of words occurring together.

- **Lexical sophistication.** Word maturity is a measure of learners understanding the meaning of various forms of a word and how it is used by a learner at the various stages of the student's life (Landaur et al., 2011).

Scoring Models

The features extracted above are quantitate descriptions of writing quality across the various dimensions of writing targeted in the scoring rubric. Although these features may not capture all aspects of a student's writing that an expert human scorer will be able to identify, they typically capture enough information for machine learning algorithms to learn how to grade constructed responses in an accurate and consistent manner or provide trait scores. Commonly used machine learning algorithms are *linear regression, support vector machines,* and *decision trees* (Hastie et al., 2009). To achieve higher accuracy and better generalization performance, *ensemble techniques* such as *random forests* (Breiman, 2001) and *gradient boosting* (Chen & Guestrin, 2016) are often used where multiple algorithms are trained on subsamples of the training data.

Model Deployment

The CF system *packages* all information about AS models trained by IEA to ensure reproducibility during training and scoring. This includes the following *metadata* about the trained prompt:

- Prompt text
- Passage readings
- Trait information such as score point ranges of each trait
- Scoring rubric information
- External corpora used for modeling
- Scoring model for each trait

Scoring Pipeline

As AS systems are increasingly used as the first and only score in high-stakes assessments, it is important to ensure high reliability of scores for any response. The CF system accomplishes this by using scores from multiple AS models that were trained on different tasks (e.g., detecting off-topic essays for a prompt, predicting trait scores, identifying LIDs). The final score returned to the customer is determined by a post-processing logic configured in the CF system once all model scores are gathered. The final outcome determined by the post-processing logic in CF can be a numerical score or a customer-specified *advisory code*. Furthermore, this outcome can be predicted with high or low confidence. In practice, *high-confidence predictions* do not require a human review whereas *low-confidence predictions* require a human review.

Illustration

In this subsection, we provide a brief case study example to illustrate and clarify how the CF system works in practice. The context for the case study, LID alerts, was chosen because it demonstrates concepts of CF and also demonstrates the power of combining human intelligence with machine intelligence. As testing becomes more dependent upon AS to achieve cost savings, it becomes more important for AS systems to detect such responses.

As we noted in previous sections, student responses occasionally contain disturbing content which may warrant alerting school officials such as content reflecting: extreme depression, extreme violence or intent to harm others, rape, sexual, or physical abuse/bullying, suicide or self-harm, neglect, and criminal, alcohol, or drug use.

Detection of these LID responses is an extremely challenging technical problem. Social media companies such as *Facebook* and *Twitter* are facing similar issues; however, social media companies have access to deeper and richer data sets (e.g., entire history of posts, demographic profile, network associations) whereas for assessment and learning companies the only data typically available to a detection model is the student response and, potentially, prior student responses to other questions.

Similar to the general scoring setup for essays, the LID detection problem was originally looked at as an "either-or" problem. That is, technology teams were asked to "build a black box that flags learner alerts as well or better than humans." This sequential solution approach neither seemed adequate to changing language nor provided guarantees of finding all alerts. The solution approach combines human and machine intelligence, utilizing the machinery of the CF system. In particular, the approach consists of the following steps:

- Using machine learning models to flag responses for human review/ labeling
- Feeding labels and annotations back to machine learning systems in near real-time
- Training and deploying new models as new information is received

- Automatically scaling the system to score large sets of student responses
- Adjusting thresholds to account for item-specific language

The mechanism starts with a basic machine learning model trained using available examples, of which there are typically few available. In essence, this model behaves the same as any content-scoring model except the score it returns is interpreted as the likelihood / probability that a response warrants an LID alert. Next, LID alerts that are created by human scorers during normal scoring are immediately sent to the CF system and are also forwarded to a scoring supervisor for review and escalation. The supervisor's score is also sent to the CF system allowing the system to learn from the supervisor by training a new LID alert model. At the same time, the CF system also uses its own LID score for each student response to decide whether to request a human review and then learns from the requested human reviews. The result of these two processes is a *self-reinforcing* and *self-adjusting loop* leading to continuously improve the LID detector.

Although no approach can guarantee to find every LID alert, the above approach has shown great promise. Of course, humans are not perfect either. In a *beta-test* of the above, the automated system caught nearly all of the alerts that humans detected but also caught some alerts that humans had missed. In both cases, the human scorers learned from the automated system, the automated system learned from the humans, and the system continually learns, adjusts, and adapts over time.

Conclusion

The acceleration of AI is potentially disruptive to many business models because the threat of AI is causing fear, uncertainty, and doubt to businesses and work forces. Within the domain of educational assessment, discussions about "robot scorers" have been a topic for many years (see Wood, Chapter 8, this handbook). Framing the discussion somewhat simplistically as "human-versus-machine" has fueled this debate because scoring robots are seen as a threat when viewed as plug-in replacements for human scorers.

Research by Dietvorst (2016) shows people quickly lose confidence in algorithmic forecasts even when they seem to outperform human forecasters. Prior to the development of the CF system, the relationship between automated and human scoring groups at Pearson was distant. An inherent bias and adversarial posture precluded proactive discussions about better ways to work together. Framing the overall scoring problem as "human-plus-machine" instead of "human-versus-machine" has been a key to success.

This reframing has changed internal conversations toward optimizing how to operationally combine human and machine intelligence rather than choosing one over the other. Daugherty and Wilson (2018) recently described a third wave of business transformation in which technology's greatest power is *complementing / augmenting* human capabilities in an area they called the "missing middle."

Nevertheless, change is difficult. A *Gallup Business Journal* poll indicated that more than 70% of change initiatives fail because decisions that threaten status quo create stress in staff (Leonard & Coltea, 2013). Reimagining new processes and process re-engineering is even more challenging. Ready (2016) discusses issues facing change leaders and paradoxes around digitization of value chains and points out there are no easy answers. In addition to organization and cultural issues, the development of the CF system also involved pushing the boundaries of technology. It required new innovations in machine learning fueled by the operational demands for scaling and training machine learning models in near real-time. The effort required changes to the core information technology systems used for human scoring and AS. It also significantly changed existing processes for logistics, planning, costing, and resource management. In short, this amounted to a large and risky change.

Given the scope and extent of the changes, Pearson executive management made organizational changes to more closely align the human scoring and automated scoring organizations. This involved uniting teams together with shared – instead of competing – incentives. Most importantly, they empowered the teams with the freedom to carry out the work and make the project successful.

The CF system architecture started as a thought experiment but, over time, has resulted in changes in organization structure, business processes, systems, technology, and culture. The CF system is agnostic to which human scoring platform or which student management system is being used. It supports various degrees of automated/human scoring mixes and provides benefits even with 100% human scoring. Most importantly, it allows Pearson to provide a richer mix of products and services for its customers across various market segments as it supports formative, interim, and summative testing in high and low-stakes settings, allowing Pearson to be nimbler and more flexible as a result.

15

Deep Learning Networks for Automated Scoring Applications

Saad M. Khan and Yuchi Huang

CONTENTS

To be successful in today's rapidly evolving, technology-mediated world, students must not only possess strong skills in areas such as reading, math, and science, but they must also be adept at *21st-century skills* such as critical thinking, communication, problem-solving, trouble-shooting, and collaboration (Farrington et al., 2012). However, the assessment of skills such as collaboration and communication is difficult because often it involves understanding the process used to arrive at a conclusion rather than simply the end product (Bejar, 1984; Romero & Ventura, 2007).

In the past, digital interactions that were used to make inferences about learners' *knowledge, skills, and attributes* were constrained because *automated scoring* could only process very limited types of information. That is, the technology did not exist to process large amounts of data, extract particular features, and apply complex scoring rules in an automated fashion. This meant that the primary mode of data collection was paper-and-pencil tests with the form of these tests highly oriented toward fixed responses. Although complex *interactive performances* were employed in assessments, these usually depended on *human raters* to recognize the often-subtle features of the interactions that constituted evidence about examinees' capabilities. This meant that the subtleties were all handled in the raters' internal thought processes, explicated – if at all – only at high levels in part due to the prohibitive costs of scaling up.

Advances in fields like *machine learning, educational data mining, computer vision*, and *affective computing* have made it possible to score a much wider range of responses. This is enabling the expansion of the types of activities that can be used to characterize the knowledge, skills, and attributes of learners and, hence, an expansion of the forms of assessment we can use (von Davier, 2018). In this chapter, we present a *computational model* designed for the assessment of a variety of *complex constructs* and competencies that are not readily amenable to more traditional item-response type analyses. These competencies may include a wide variety of cognitive, non-cognitive, 21st-century, and workforce-readiness skills.

The measurement of these constructs entails the recognition and extraction of patterns of behaviors and activities from *time-series data* that capture the interaction of learners with the simulation task as well as with one another. These data are very rich and *multimodal*, which means that they encompass a multitude of *sensory modalities* such as mouse clicks, keystrokes, audio, and video collected from these interactions. Inexpensive and ubiquitous *sensors* like webcams and microphones enable comprehensive tracing of the learners' behaviors with real-time capture and recognition of the learners' gaze, facial expressions, gestures, body poses, spoken responses, and speech tone. These different modalities can then be integrated to model the learner's cognitive thought processes and non-cognitive (i.e., affective/emotional) states like engagement, motivation, and persistence (see D'Mello, Chapter 6, this handbook).

Realizing this presents a number of technical challenges. The raw time-series multimodal data captured in digital learning environments may not have any direct *semantic meaning* and may not be interpretable by humans as such. It is typically constituted of simulation *log files* that contain audio and visual data, which cannot be analyzed for meaningful information without sophisticated computational models. Consequently, the inferences and corresponding interpretations from the raw multimodal data may contain information at vastly different levels of semantic meaning and abstraction (e.g., specific facial expressions vs. level of engagement, turn-taking sequences). Moreover, even if one can extract a large number of features that are indicative of a specific competency, it is often unclear how they should be combined to obtain a *quantitative representation* of standing on a particular skill.

Recent advances in *artificial intelligence* and machine learning are making it possible for us to address these challenges. In particular, *deep hierarchical computational frameworks* are being developed that can bridge the gap between the raw, low-level multimodal data and the measurement of high-level constructs. In the following we provide some details on these approaches and briefly introduce the inner working mechanism of various *deep learning* techniques. We have divided this chapter into two main sections. In the first section, we review recent advances this area, which include advances in *deep neural networks, deep sequential networks,* and *deep generative networks.* In the second section, we illustrate how these advances are leveraged to provide meaningful information to learners via an extended case study centered around facial expressions in a virtual tutor.

Deep Learning

Deep learning is part of a large family of data-driven machine learning methods focused on learning implicit representations of complex data by discovering underlying – sometimes unobserved (i.e., *latent*) – patterns. Originally known as *artificial neural networks* (ANN) in the 1980s, these models were loosely based on neurons in animal brains. Their initial form only consisted of two to three *layers* of connected units or *nodes* called *artificial neurons*. While ANNs showed early promise, unfortunately they did not scale to larger tasks due to limitations in training and computing power. Moreover, during the 1990s advances in alternative methods of *supervised machine learning* such as *support vector machines* and *random forests* meant ANNs went out of favor in the machine learning community. Around 2006, ANNs were rebranded by Hinton as deep learning, in which the idea of *unsupervised pre-training* and *deep belief nets* was introduced (Hinton, 2007). A major feature of this new type of networks was that they could have many more layers of artificial neurons than their predecessors.

A major breakthrough happened in 2012 with the publication of a highly influential paper (Krizhevsky et al., 2012), in which the authors achieved state-of-the-art results on the *ImageNet* object recognition challenge by halving the previous best error rate. More importantly, their work was the first to apply massive *labeled data sets* and *graphics processing units* for model training. From then on, deep learning architectures yielded numerous state-of-the-art results in a wide range of current artificial intelligence applications such as *speech recognition, image recognition,* and *language translation.* In a number of cases, they have produced results comparable to – or even superior to – human experts. An attractive quality of deep learning is that it can perform well without external human-designed resources or time-intensive *feature engineering.* Despite these advantages there have been limited applications of these models to solving some of the most challenging technical problems in educational assessment and learning.

Deep Neural Networks

Deep neural networks (DNN) are typically *feedforward ANNs* in which data flows from the *input layer* to the *output layer* without *looping back.* In contrast to ANNs, however, they contain more *hidden layers* between the input and output layers, which allow DNNs to model *non-linear relationships* in highly complex data.

The building component of DNNs (as well as ANNs) is the single-hidden-layer *multi-layer perceptron* (MLP) (Rosenblatt, 1961). An MLP can be viewed as a *logistic regression classifier* where, technically speaking, the input is first transformed using a learned *non-linear transformation*, which projects the input data into space where it becomes *linearly separable.* An MLP with a single-hidden-layer can be represented mathematically as shown in Equation 15.1:

$$f(x) = G\left(b^{(2)} + W^{(2)}\left(s\left(b^{(1)} + W^{(1)}x\right)\right)\right) \qquad (15.1)$$

where $b^{(1)}$ and $b^{(2)}$ are bias vectors, $W^{(1)}$ and $W^{(2)}$ are weight matrices, and G and s are non-linear activation functions such as *tanh*, *sigmoid*, or *rectified linear*.

Figure 15.1 illustrates the relationship between the hidden intermediate layers and the observable input and output layers. A single-hidden-layer or the composition of two to three hidden layers is typically sufficient for MLPs to perform well. However, the number of layers the MLPs can contain is the major difference between ANNs and DNNs. Specifically, DNNs usually use many such hidden layers together to form deep learning architectures where the extra layers of DNNs enable abstraction of features, potentially modeling complex data with fewer units than a similarly performing but more shallow network. *Backpropagation* (Rumelhart, Hinton, & Williams, 1986) is used to learn the parameters from model training in all layers, in which *error* is calculated at the output layer and, as the name implies, distributed back through all the network layers.

As illustrated in Figure 15.1, in DNNs network nodes are *fully connected* across pairs of layers. Although this structure can be used to learn features as well as classify data, it is not practical to apply them to high-dimensional data such as images. This is because a very large number of nodes have to be applied due to the very large input sizes associated with images where each element of the input variable represents an image pixel. For instance, a fully connected layer for a small image of size 100 × 100 has 10,000 weights for each node in the second layer.

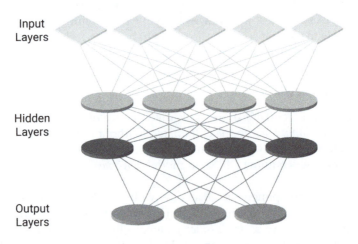

FIGURE 15.1
Illustration of a DNN with a two-layer MLP.

In this way, fully-connected DNNs with a large number of *free parameters* (i.e., the consequence of additions of multiple layers) are prone to *overfitting* and *higher sensitivity* to rare dependencies (e.g., noise) in the training data. *Regularization methods* such as *weight decay* (Krogh & Hertz, 1992) or the *drop-out technique* (Hinton et al., 2012) can be applied during training to combat overfitting. Data can also be augmented via methods such as *cropping* and *rotating* such that smaller training sets can be increased in size to reduce the chances of overfitting. However, the effects of these methods are limited.

The alternative solution is to create a new neural network architecture with more straightforward structures and convergent training algorithms. *Convolutional neural networks* (CNN) are one such deep neural network, which are biologically-inspired variants of MLPs. The basic components of CNNs are *convolutional layers*, which imitate the functions of the visual cortex in mammalian brains. In nature, cells in the human visual cortex are sensitive to small sub-regions of the visual field, which are tiled to cover the entire visual field. These cells act as *local filters* over the input space and are well suited to exploit the strong *spatially-local correlations* present in input signals.

Inspired by this biological structure, the convolutional layers consist of a set of learnable *kernels*, which have a small *receptive field*, but extend through the full depth of the input volume. Technically speaking, during the *forward pass* of input signals, each kernel slides through the width and height of the input volume, computes the *dot product* between the entries of the filter and the input to produce a two-dimensional *feature map* with respect to that kernel. In this way, the number of parameters in convolutional layers is largely decreased because "repeated" kernel parameters replace weight parameters in a fully connected structure. Similar to DNNs, the kernels – along with parameters in other layers of CNNs such as pooling and full connection layers – are learned by backpropagation.

Historically, *AlexNet* (Krizhevsky et al., 2012) was the first – and one of the most influential – CNNs and won the first place of the *ImageNet Large Scale Visual Recognition Challenge* (ILSVRC) in 2012. Two years later, the champion of the ILSVRC 2014 competition was *GoogleNet* (Szegedy et al., 2014), which consisted of a CNN that was 22 layers deep but reduced the number of parameters from *AlexNet*'s 60 million to four million. The most recent and major improvement of CNNs is a novel architecture called *residual neural network* (ResNet) (He et al., 2016a, b), which introduced techniques such as *skip connections* and *heavy batch normalization* and won the first place of ILSVRC 2015; this framework contains 152 layers and beat human-level performance on the *ImageNet* dataset.

Based on the above discussion, it is clear to see that CNNs are especially useful in image and video recognition. They are also widely utilized in various sub-domains such as speech processing, natural language processing, and recommender systems. For example, CNNs could be utilized in classroom settings to detect and analyze *interactional dynamics* among students and teachers, in video analysis and recommendation system for *online educational platforms*, and on educational text data to perform *semantic parsing*, *search query retrieval*, and *sentence modeling*, to name but a few possibilities.

Deep Sequential Networks

In a traditional ANN, it is assumed that all inputs and outputs are independent of each other; however, this is not the case for many tasks using sequential or *time-series data*. If we want to predict the next word in a sentence, for example, it is beneficial to know the sequence of words that came before it. This is particularly relevant for *process data* in digital educational and learning environments that captures a variety of time-series data such as simulation log files, sequences of student actions, and responses (Hao, Shu, & von Davier, 2015). Unfortunately, feedforward networks such as DNNs discussed earlier are not ideally suited to model such time-series data.

These limitations can be overcome with a framework of models called *deep sequential networks* (DSNs). In particular, *recurrent neural networks* (RNNs) (Lipton, Berkowitz, & Elkan, 2015) and their popular variant *long-short-term memory unit* (LSTM) (Hochreiter & Schmidhuber, 1997) are ANNs designed to recognize patterns in sequential process data such as text, speech, handwriting, or numerical time-series data emanating from educational learning systems. These algorithms have a temporal dimension, which takes time and sequence into account.

Different from feedforward networks such as DNNs, DSNs such as LSTMs take as their input not just the current input example they see, but also from what they have perceived previously. They perform the same task for every element of a sequence with the output being dependent upon previous computations. Another way to think about an LSTM is that they have a "memory" that captures information about what has been calculated so far. Basic LSTMs are a network of neuron-like nodes, each with a one-way connection to every other node. Each node has a time-varying real-valued activation and each connection has a modifiable real-valued weight. Similar to previously discussed networks, nodes are either input nodes, output nodes, or hidden nodes that modify the data on route from input to output.

Technically speaking, as shown in Figure 15.2, a cell state vector C_t is created to pass the information across different time steps. The LSTM has the ability to remove or add information to the cell state, carefully regulated by structures called *gates*, which are composed of a *sigmoid neural activation* (σ) and a *point-wise multiplication operation*. Specifically, σ activation outputs numbers between 0 and 1 describing how much information should be let through.

From left to right in Figure 15.2, the LSTM has three such gates. The first gate ('forget gate') decides what information to keep or remove from the previous cell state; the second gate ('input gate') decides what new information is being stored in the current cell state; the last gate ('output gate') decides what is being output. *Gated recurrent units* (Cho et al., 2014) are another gating mechanism whose performance on different applications was found to

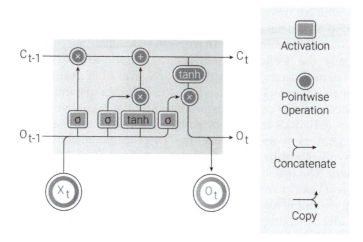

FIGURE 15.2
A typical LSTM unit that replaces the activation function in RNNs.

be similar to that of the LSTM but have fewer parameters as they lack an output gate.

Applications of DSNs like LSTMs include robot control, time-series prediction, speech recognition, grammar learning, human action recognition, business process management, and prediction in medical care pathways, among others. LSTMs are especially useful in various problems in natural language processing such as text classification / categorization / tagging, semantic parsing, question answering, multi-document summarization, machine translation, which are all important tasks in educational learning and assessment systems.

Deep Generative Networks

Deep generative networks are an effective way to learn data distribution using *unsupervised machine learning techniques.* These models are designed to learn the underlying data distribution from a training set in order to generate new data points with some perturbations; however, in practice, it is not always possible to implicitly or explicitly predict the exact distribution of our data. To solve this problem, the power of ANNs has been leveraged in recent years to learn a function that can approximate the true data distribution via a model-generated, rather than pre-specified, distribution.

One of the most commonly used efficient instance within this model family is *generative adversarial networks* (GANs) (Goodfellow et al., 2014). These are themselves a family of architectures that can learn the underlying data distribution of a real-world process from training sample sets, which can then be used to generate new data samples. Importantly, a GAN does not work with any explicit density estimation; instead, it is based on a *game theory* approach with an objective to find what is known as the *Nash equilibrium* between two neural networks, the 'Generator' and 'Discriminator' networks. These two networks compete with each other in a *zero-sum game theory* setup.

Specifically, the 'Generator' network $G(z)$ learns to map random inputs z from a latent space to a particular data distribution of interest while the 'Discriminator' network $D(x)$ discriminates between instances from the true data distribution and candidates produced by the 'Generator' network. The 'Generator' network's training objective is to increase the error rate of the 'Discriminator' network (i.e., "fool" the 'Discriminator' network by producing fake synthesized instances that appear to have come from the true data distribution). The task of 'Discriminator' network is to take input x – either from the real data or from the 'Generator' network $G(z)$ – and try to predict whether the input is real ('1') or generated ('0').

Technically speaking, this can be thought of as a *min-max two-player game* where the performance of both the networks improves over time, which can be represented mathematically via an *objective function* shown in Equation 15.2:

$$\min_G \max_D \left(\mathbb{E}_{x \sim p_{\text{data}}(x)} \left[\log D(x) \right] + \mathbb{E}_{z \sim p_z(z)} \left[\log \left(1 - D(G(z)) \right) \right] \right) \quad (15.2)$$

where the first term of the equation is the *entropy* (i.e., informational value) that the real-data distribution passes through the 'Discriminator' network, which tries to maximize $D(x)$ to 1. The second term is the entropy from the 'Generator' network, which generates a fake sample of noisy data passing through the 'Discriminator' network. In this term, the 'Discriminator' network tries to minimize $D(G(z))$ to 0 while the 'Generator' network tries to maximize $D(G(z))$ to 1.

In a basic or *Vanilla GAN* (Goodfellow et al., 2014), there is no control on modes of the data being generated. However, by conditioning the model on additional information such as class labels, some part of data, or even on data from a different modality, it is possible to direct the data generation process as shown in Figure 15.3. For example, in a *Conditional GAN* (Mirza & Osindero, 2014), both the 'Generator' and 'Discriminator' networks are conditioned on some extra information y. This kind of conditioning can be performed by feeding y into both the 'Discriminator' and 'Generator' networks as additional input layers.

The objective function of a two-player minimax setup is shown in Equation 15.3, which is structurally identical to Equation 15.2 with the exception of the conditioning components on x and z:

$$\min_G \max_D \left(\mathbb{E}_{x \sim p_{\text{data}}(x)} \left[\log D(x \mid y) \right] + \mathbb{E}_{z \sim p_z(z)} \left[\log \left(1 - D(G(z \mid y)) \right) \right] \right) \quad (15.3)$$

An early GAN utilizing the CNN was the *deep convolutional GAN* (DCGAN) (Radford, Metz, & Chintala, 2015), which took as input 100 random numbers drawn from a uniform distribution and output an image of desired shape.

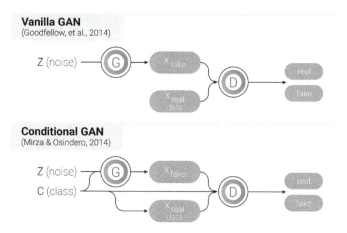

FIGURE 15.3
Architecture of vanilla GAN and conditional GAN.

It consisted of many convolutional, deconvolutional, and fully-connected layers where deconvolutional layers were used to map the input noise to the desired output image and batch normalization was used to stabilize the training of the network.

Previous work based on GANs, CGANs, and DCGANs has shown promise in a number of applications such as future-state prediction of time-series data (Mathieu et al., 2015), video generation (Vondrick et al., 2016), image manipulation (Zhu et al., 2016), style transfer (Li & Wand, 2016), text-to-image (Isola et al., 2016), image-to-image translation (Reed et al., 2016), and 3D shape modeling (Wu et al., 2017). Table 15.1 summarizes the performance of the key deep learning networks we discussed in this section along with popular software packages and libraries that offer efficient implementations for quick reference.

Case Study

In this section, we provide an overview of a GAN approach to creating a human-like *avatar* to embody a *virtual tutor* in educational applications. Generally speaking, human communication involves both verbal and non-verbal ways of making sure messages are heard. A simple smile can indicate approval of a message while a scowl might signal displeasure or disagreement. Moreover, the sight of a human face expressing fear typically elicits fearful responses in the observer as indexed by increases in autonomic markers of arousal and increased activity in the *amygdala* (Morris et al., 1996). This *mirroring process* – whereby an observer tends to unconsciously mimic the

TABLE 15.1

Characteristics and Software Packages for Deep Learning Networks

Network Types	Classic Models	Architecture Characteristics	Data to Process	Typical Applications	Software Packages/ Libraries
Deep Neural Networks (DNN)	Convolutional neural network (CNN), Residual neural networks (ResNet)	Feedforward, supervised	Images/videos, acoustic signals, documents	Object/speech recognition, object detection/segmentation	*Tensorflow* (Abadi et al., 2015), *Torch*/ *Pytorch* (Paszke et al., 2017), *MxNets* (Chen, 2016), *Keras* (Chollet, 2016)
Deep Sequential Networks (DSN)	Vanilla recurrent neural network (RNN), Long short-term memory (LSTM)	Recurrent, sequential, supervised	Various types of sequential or time-series data	Time-series prediction, speech recognition, grammar learning, human action recognition, business process management	
Deep Generative Networks (DGN)	Generative adversarial networks (GAN)	Adversarial (two competing DNNs), unsupervised or supervised (conditional)	Used to generate new data samples	Image/video/text/speech generation, manipulation and style transfer, text-to-image and image-to-image translation	

behavior of the person being observed – has been shown to impact a variety of interpersonal activities such as collaboration, interviews, and negotiations among others (Tawfik et al., 2014).

A classic study by Word et al. (1974) demonstrated that interviewees fared worse when they mirrored less friendly body language of the interviewer compared to what they did in the friendly condition. In parallel with the unconscious face-processing route there is a conscious route that is engaged, for example, when volunteers were explicitly asked either to identify facial expressions or to consciously use facial expression as communicative signals in *closed-loop interactions*. In many situations, an additional cue (i.e., an ostensive signal such as briefly raised eyebrows during eye contact) is produced to indicate that the signaling is deliberate (Sperber & Wilson, 1995).

Recent research in *autonomous avatars* (Devault et al., 2014) has aimed to develop powerful *human–agent interfaces* that mimic such abilities. Not only do these avatar systems sense human behavior holistically using a multitude of sensory modalities, but they also aim to embody *ecologically-valid* human gestures, paralinguistics, and facial expressions. However, producing realistic facial expressions in virtual characters such as embodied automated tutors that are appropriately contextualized and responsive to the interacting human remains a significant challenge. Early work on *facial expression synthesis* (Bui et al., 2001) often relied on rule-based 2D or 3D face models. Such knowledge-based systems have traditionally utilized *facial action coding systems* (Ekman & Friesen, 1978), which delineate a relationship between facial muscle contractions and human emotional states.

Later, statistical tools such as *principal component analysis* were introduced to model face shapes as a linear combination of prototypical expression basis. By varying the *base coefficients* of a shape model, the model is optimized to fit existing images or create new facial expressions (Blanz & Vetter, 1999). A key challenge for such approaches is that the full range of appearance variations required for convincing facial expression is far greater than the variation captured by a limited set of rules and base shapes. Advanced *motion-capture* techniques have also been used to track facial movement of actors and transfer them to avatars (Thies et al., 2016) recreating highly realistic facial expressions. However, these solutions are not scalable to autonomous systems as they require a human actor in the loop to "puppeteer" avatar behavior.

In this case study, we study human dyadic interactions to tackle the problem of generating video sequences of dynamically varying facial expression in human–agent interactions using conditional GANs. As described in the previous section, CGANs are generative models that learn a mapping from random noise vector z to output image y conditioned on auxiliary information x (i.e., $G: \{x, z\} \rightarrow y$). Our work uses GANs to model the influence of the learner's facial expressions who is interacting with the virtual agent to generate an appropriate response of the virtual agent (Huang & Khan, 2018a, b).

Specifically, we developed a multi-stage optimization of GANs that enable modeling of complex human behavior as well as compelling *photorealism* in the generated agent; see Figure 15.4 for an example. At the first stage, termed *Affective-sketch GAN* (AS-GAN), we designed a single layer encoder that takes a sequence of facial expression descriptors of the interacting human as input and output conditional features, which are in turn used by a GAN to generate an intermediate representation. This intermediate representation consists of affective shape descriptors which are then used to reconstruct affective face sketches. The resulting affective face sketches are a *contextually valid* (i.e., they are conditioned on the interacting learner's behavior) and consist of a *temporally smooth* sequence of generated agent's face sketches. They have the advantage of being rich enough to capture valid human-like emotions and generic enough to be mapped onto a number of different agent face identities.

The second and final stage termed *Photo GAN* (P-GAN) fleshes out agent affective face sketches into high-quality photorealistic videos. At this stage, a new *L1-regularization term* computed from layer features of the 'Discriminator' network is employed to enhance the image quality of video frames while novel session constraints are proposed to ensure appearance consistency between consecutive frames. In a recent publication that details this study (Huang & Khan, 2018a, b) we demonstrated that our approach is effective at generating videos that depict visually compelling facial expressions. Moreover, we quantitatively showed that facial expressions by the agent in the generated video clips reflected valid emotional reactions relative to the behavior of the human partner.

Even though this example focused on facial expressions, various complementary modalities such as speech, paralinguistics, and gestures can be readily integrated to more comprehensively model behavior of the human partner. For example, our model can be enhanced by fusing a multitude of modalities to more comprehensively model behavior of the human partner; these modalities include speech, paralinguistics, interaction context, and past behavior of the virtual agent itself. This is a significant result that points to a new way to approach building interactive agents, particularly in educational contexts like *automated tutors*.

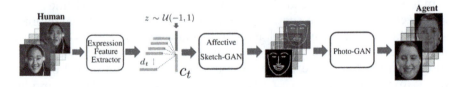

FIGURE 15.4
Two-stage GAN for generating virtual agent expressions that mimic learner expressions.

Conclusion

Deep learning has enabled breakthrough results on artificial intelligence challenges such as automated speech recognition, medical imagery analysis, drug discovery, and genomics. This has been possible due to deep learning's ability to build computational models that are composed of multiple processing layers to learn representations of highly complex data with multiple levels of abstraction. These advances also have significant implications for the design of educational learning and assessment systems.

Traditionally, digital educational interactions that were used to make inferences about learners' knowledge, skills, and attributes were constrained due to limitations in the richness and types of data that could be processed automatically. This has particularly been the case for complex social–emotional learning skills as well as complex competencies such as collaborative problem-solving and communication ability. In this chapter, we have highlighted advances in deep learning technology that open up new possibilities for measuring and remediating such complex competencies in an ecologically valid manner.

The ability of deep networks to develop intermediate representations of data allows us to fuse information at vastly different levels of abstraction and distill from it actionable insights on learner's cognitive thought processes and social–emotional behavior. With models such as CNNs, LSTMs, and GANs we are able to model temporal dynamics and integrate highly unstructured, multimodal learner data that may include gaze, facial expressions, gestures, postures, speech and tone among others.

Due to the hierarchical design of these models they are ideally suited to exploit concept hierarchies that reflect the latent structure of learner interaction data and goals of the assessment. However, as any other data-driven machine learning solution, deep learning models are also susceptible to *biases* in data that can lead to unintended negative outcomes. Best practices such as conducting statistical analyses of training data to identify and remediate biases and skewness along demographic features (e.g., gender and age) and conducting rigorous *cross-validations analyses* have proven to be effective mitigating factors.

Deep learning is quickly gaining popularity in education and is already being utilized in the design of automated tutoring, personalized learning, and automated scoring systems. As the technology matures and gets more widely adopted we anticipate a plethora of new educational applications including automated content generation for learning and assessment, content recommendation systems, and performance outcomes prediction, among others.

16

Validation of Automated Scoring Systems

Duanli Yan and Brent Bridgeman

CONTENTS

A comprehensive *validity argument* should contain an integrated set of logic and data-based statements supporting a particular use of the scores from an assessment; it is the uses, not the assessment itself, that must be validated (Kane, 2006; Messick, 1989). The essence of a validity argument is providing support for specific test use claims and evaluating arguments that might support or undermine those claims.

Overview of Validity Considerations

Validity arguments related to *automated scoring* (AS) initially focused on the extent to which the automated system could replicate scores provided by human raters who were evaluating the same essays as the automated system (Page, 1966a, 1968). Although such evidence may be relevant for claims related to the extent to which the automated system replicates scores from human raters, it could not support all desired claims (Bennett & Bejar, 1998; Williamson, Bejar, & Hone, 1999).

For example, suppose that a test publisher wanted to claim, explicitly or implicitly, that an automated score on a 30-minute essay on a university admissions test was valid for contributing to admissions decisions. Simply showing that human scores and scores from the AS system were related would be insufficient. It would be necessary to show that writing skill is related to success at the university, that a 30-minute test can be a reasonable (though certainly incomplete) measure of this skill, and that raters, whether human or machine, can assign scores with sufficient reliability to be useful.

Note that if the assessment is to use both human and machine ratings of the essay it would not be necessary to show that the human and machine scores are highly correlated. Indeed, if the humans and machine were assessing different relevant components of the writing construct, a zero correlation theoretically could be acceptable. The evidence needed to support a validity claim depends on exactly what the claim is. If the claim is simply that the automated score assesses some relevant aspect of the writing construct, demonstrating a correlation with a human score would have little value, but if the claim is that the machine can substitute for a human rater, then the human–machine correlation becomes relevant.

Human and Machine Scores

The notion that a score from an automated essay grading system can be validated by demonstrating the relationship to a human score on the same essay goes back more than 50 years to the research on *Project Essay Grade* (Page, 1966b). More recent studies have similarly based validity arguments largely on the correlation of human and machine scores (e.g., Alpaydin, 2010; Attali, Bejar, 2011; Bridgeman & Trapani, 2011; Attali & Burstein, 2006, Breyer, Rupp, & Bridgeman, 2017; Dikli, 2006). A contest to identify the best system for the AS of essays defined best in terms of correlations with the scores assigned by human raters (Shermis, 2014). To the extent that the machine can predict the scores by human raters, the claim that the machine score can be used as a substitute for a human score is supported.

Powers, Escoffery, and Duchnowski (2015) proposed a modest refinement of this gold standard by suggesting that the qualifications of the human raters were also relevant. Specifically, they observed that machine–human correlations were higher when the raters were highly experienced compared to relatively untrained raters. Two books (Shermis & Burstein, 2013; Williamson, Mislevy, & Bejar, 2006) further explored recent development of AS for complex tasks.

AS systems can assess some of the same essay features that are also attended to by human raters. Among other things, such systems can evaluate certain grammatical features, sophistication of vocabulary, and usage errors (Attali & Burstein, 2006). The *e-rater®* scoring engine, which is used for high-stakes tests at ETS, also has features to evaluate 'organization' and 'development,' but the way these features are defined (e.g., via the number of lexical units and number of words in a lexical unit) are not the same as the definitions that would be applied by human raters. Computers cannot yet identify exceptional, inspirational essays or even essays that make serious logical errors in argumentation. Critics of *automated essay scoring* have noted that with the current state-of-the-art, the machine can evaluate only a small fraction of a comprehensive writing construct (Condon, 2013; Perelman, 2014; Wresch, 1993; see Wood, Chapter 8, this handbook).

Despite these limitations, the correlation of a machine score with the score from a human rater is often as high or higher than the correlation between two human raters (e.g., Attali, Bridgeman, & Trapani, 2010; Ramineni, Trapani, Williamson, Davey, & Bridgeman 2012a, b). The high correlation does not imply that the human and machine are focusing on the same features in the essay however. All that is necessary for a high correlation is that essay writers who score high on the features that the machine can evaluate also score high on the features that humans evaluate; these could be the same features, but also could be unique for each type of rater (human or machine). Similarly, AS of speech measures only some of the aspects of spoken language that can be evaluated by human raters (Xi, Higgins, Zechner, & Williamson, 2008; see Cahill & Evanini, Chapter 4, this handbook, and Zechner & Loukina, Chapter 20, this handbook). But to the extent that what the machine can evaluate is related to what humans can evaluate, it is still possible to get reasonable correlations between human and machine scores.

Population and Subgroup Differences

If an AS system is to be used for making decisions in diverse population subgroups, then it might be argued that the automated system should show equal agreement with human raters in all subgroups. But this is true only if the claim to be supported is that the machine can function as a *proxy* for a human scorer. If the human and machine are assessing different aspects of the writing construct, then it would not be unreasonable for some subgroups to do better on what the machine is best able to assess while other groups do better on what humans can assess. The existing research in this area assumes that the machine is functioning as a proxy for a human rater so that differences between human and machine scores ideally would be identical in all subgroups.

For example, Burstein and Chodorow (1999) evaluated differences between human and machine (an early version of *e-rater*) scores for several subgroups on the *Test of Written English*®. They found that Arabic and Spanish speakers tended to get relatively higher scores from humans than from *e-rater* while Chinese speakers received higher scores from *e-rater* by about half of a standard deviation. Later research with larger samples from the TOEFL® and GRE® programs and with more advanced versions of *e-rater* also noted significant differences across different subgroups, but the finer-grained analyses suggested an even more complicated picture (Bridgeman, Trapani, & Attali, 2012).

Differences for Chinese speakers, for example, might at first seem to be related to the Mandarin language being quite different from English, but in Mainland China test takers get relatively higher scores from *e-rater* while in Taiwan (which is also a Mandarin speaking area), human rater scores are slightly higher (Bridgeman et al., 2012). Reasons for these discrepancies are not entirely clear. Attali et al. (2010) and Bridgeman et al. (2012) noted that essays from Mainland China tended to be relatively long and, while both humans and *e-rater* reward the verbal fluency demonstrated in longer essays, the length reward is especially salient in the *e-rater* scores. Ramineni and Williamson (2013) argued that different approaches to rating could account for some of the differences with humans using conditional logic and a *rule-based approach* while *e-rater* uses a linear weighting of all features via a *multiple linear regression* approach.

External Criteria

If the only validity claim is that human scores can be a proxy for machine scores on the same essays, then correlations of human and machine scores can be used to support this claim. Broader claims related to predicting external criteria require different evidence. For example, Weigel (2011) evaluated claims that automated scores from the TOEFL iBT essays are related to non-test indicators of writing ability. Specifically, she showed that *e-rater* scores were related to both self and instructor ratings of writing ability; the correlation with instructor ratings of writing ability was 0.29 for a human rater and 0.28 for *e-rater*.

Bridgeman, Ramineni, Deane, and Li (2014) evaluated the ability of human and *e-rater* scores on a two-essay placement test to predict scores on a *portfolio assessment*. Portfolio scores were obtained from a sample of 352 freshmen at the *New Jersey Institute of Technology* (NJIT). The portfolios contained a sampling of major writing assignments completed during the freshman year ranging from expository, memoir-style essays to a final research paper with at least one revision for assignment presented for comparison purposes. The portfolios were scored by NJIT faculty on a 1–6 holistic scale. Human scores on the placement tests correlated 0.26 with this criterion while for *e-rater* scores the correlation was 0.27.

Bridgeman and Ramineni (2017) evaluated the ability of human and machine scores from two 30-minute *GRE* essays to predict scores on actual writing assignments completed by graduate students. A sample of 194 graduate students supplied two examples of their writing from required graduate school coursework across a variety of graduate disciplines (i.e., these were not just assignments in English courses). Human scores on the *GRE* Issue essay correlated 0.29 with this criterion and *e-rater* scores correlated 0.31 while for the *GRE* Argument essay the correlations were 0.27 and 0.34 for humans and *e-rater*, respectively.

Bridgeman, Powers, Stone, and Mollaun (2011) evaluated the ability of an automated speech scoring system (ETS's *SpeechRater*™) to assess measures of oral communicative language proficiency; see Zechner and Loukina, Chapter 20, this handbook, for an overview. A sample of 555 undergraduate students listened to speech samples from the TOEFL® iBT. Evidence of what they were able to understand was evaluated in two ways; they responded to multiple-choice questions that could only be answered if they understood what they had heard and they were asked to use 5-point rating scales to evaluate various aspects of what they had heard (e.g., the effort required to understand the speaker and how confident they were that they understood what the speaker was trying to say).

Note that the emphasis was on understanding the message that the speaker was trying to convey and not on how smoothly or fluently the speaker spoke, with an expectation that *SpeechRater* might be much better at evaluating fluency than the message. Averaged across two test forms, human correlations with the multiple-choice total score and the rating scale total score were both 0.73 while the correlations of *SpeechRater* scores were 0.40 and 0.37 for the multiple-choice and rating scales, respectively. It is not clear if this represents a "glass-is-half-empty" or "glass-is-half-full" scenario however. *SpeechRater* scores were also significantly related to real-world indicators of communicative competence, but the correlations were not nearly as strong as they were for human raters. It is also important to note that this study used a very early version of *SpeechRater* and that the current version contains many improvements. Bernstein, Van Moere, and Cheng (2010) discussed validating a speaking test with a criterion score on oral proficiency interviews. Compared to the 50+ years of validating AS systems by their ability to replicate human scores on the same tasks, predictive validation based on relationships to external criteria is relatively new. More work in this area is clearly needed.

Gaming the System

A threat to the validity of any scoring system is the extent to which scores can be obtained by methods and "tricks" that are not related to the construct that the test is intended to measure. In the case of automated writing evaluation systems, test takers may know that the AS system rewards long essays and so might add words and sentences that are not really needed to fulfill the assigned task.

To investigate this, Powers, Burstein, Chodorow, Fowles, and Kukich (2001) invited writing experts to write essays that trick *e-rater* into assigning scores that were too high or too low in response to *GRE* prompts. Writers were more successful in writing essays that scored too high. The clear winner was a professor who wrote a reasonable paragraph and then repeated it 37 times. As a result of this type of research, AS systems have evolved and now include filters that will flag repeated text and other strategies intended to beat the system. An additional protection against the use of such strategies is to include a human as one of the raters as a human is unlikely to assign a high score with the same paragraph repeated 37 times. A similar approach was used when study participants were asked to fool an engine that was designed to evaluate the content of written responses (*c-rater*™). In that study, writing prompts came from the *PRAXIS*® assessment and the intent of this research was to identify filters that could be built into *c-rater* that would catch these gaming strategies.

To consider the above validity concerns and ensure the quality performance of AS systems, Williamson, Xi, and Breyer (2012) discussed a framework for evaluation and use of AS consisting of five areas of emphasis: construct relevance and representation; accuracy of scores in relation to a standard (human scores); generalizability of a resultant scores; relationship with other independent variables; and impact on decision-making and consequences of using AS. In the following sections, we focus on the accuracy of automated/machine scores and broader AS system evaluations. Specifically, we give illustrations on scoring system models, scoring system evaluations, and evaluation criteria using examples from *e-rater*, *SpeechRater*, and *Inq-ITS* (Gobert, 2005; see also Mislevy et al., Chapter 22, this handbook).

Automated Scoring System Models

While acknowledging that replicating human scores is only one of many possible validity claims, we recognize that this is currently the dominant claim related to the validity of AS engines. Therefore, we focus on this claim when discussing automated engine scoring models. These models match various characteristics of responses with the scores assigned by expert raters and use these relations to predict scores for new responses. They often use statistical models and/or *machine learning* to determine how features of responses relate to scores assigned by expert raters. They can score a variety of response types, including short answers of a few words, extended essays, spoken tasks, and behavioral traces from *intelligent tutoring systems* (ITSs).

An AS system evaluates aspects of the writing construct with a set of *natural language processing* features. For example, Figure 7.1 depicts the construct coverage of the automated engine used at ETS in the *e-rater*® engine. As shown in the figure, *e-rater* uses a fixed set of 11 features with each feature consisting of multiple microfeatures (see Burstein, Tetreault, & Madnani, 2013). Under the classical automated writing evaluation approach, the 11 feature scores are entered into a *multiple linear regression* model to predict a single human score or the average of two human scores on the task. Once the model weights are determined on a representative sample, the scoring model is applied to all operational responses to derive scores. Any time a new model is trained, its performance is evaluated on a separate cross-validation sample at the overall level, individual prompt level, and subgroup level using recommended evaluation criteria (Williamson et al., 2012).

Statistical Models

There are many statistical models that are used for AS engines, and these models are also called AS models. We describe some commonly used models implemented for operational AS engines (e.g., *PEG*®, *e-rater*®, *SpeechRater*™, *c-rater*™, *IEA*®, ITSs), and some models that are in research and development for operational implementation.

Multiple Linear Regression

A multiple linear regression model is a traditional *ordinary least-squares regression* (OLR) model with m predictor variables $X_1, X_2, ..., X_m$, and a response Y. It can be written as

$$Y = \beta_0 + \beta_1 X_1 + \beta_2 X_2 ....\beta_m X_m + \varepsilon$$

where $\varepsilon \sim normal(0,1)$ are the residual terms of the model and $\beta_0, \beta_1, \beta_2...\beta_m$ are regression coefficients. In some cases, if a β_j is less than or equal to 0, it may be dropped from the model.

Lasso Regression

Lasso regression is a type of linear regression in which data values are shrunk toward a central point such as the mean (Tibshirani, 1996). This type of regression is well-suited for models showing high levels of multicollinearity or when there is a desire to automate certain parts of model selection, such as variable selection/parameter elimination

$$\sum_{i=1}^{N} (y_i - \sum_{m} x_{ij}\beta_j)^2 + \lambda \sum_{m=1}^{M} |\beta_j|$$

where λ is a tuning parameter and is basically the amount of shrinkage.

Nonnegative Least-Squares Regression

Nonnegative least-squares regression is a type of constrained least-squares solution in which the coefficients are not allowed to become negative (Lawson & Hanson, 1995).

$$\sum_{i=1}^{N} (y_i - \sum_{m} x_{ij}\beta_j)^2$$

where y is a response vector, and $\beta_j \geq 0$.

Support Vector Machine

A *support vector machine* is a supervised machine learning algorithm that finds a *hyperplane* in an N-dimensional space (N- the number of features) that distinctly classifies the data points (Cortes & Vapnik, 1995).

Artificial Neural Network

Artificial neural networks are considered nonlinear statistical data modeling tools in which the complex relationships between inputs and outputs are modeled or patterns are found (Hastie, Tibshirani, & Friedman, 2001).

Deep Learning

Deep learning is an artificial intelligence function in processing data and creating patterns for use in decision-making. It is a subset of machine learning in artificial intelligence that has networks capable of unsupervised learning from data that are unstructured or unlabeled. This often works with artificial neural networks and is also known as *deep neural learning* or *deep neural network* (Goodfellow, Bengio, & Courville, 2016; see Khan & Huang, Chapter 15, this handbook).

Multimodal Learning

Multimodal learning utilizes and triangulates among non-traditional as well as traditional forms of data in order to characterize or model student learning in complex learning environments (Worsley, Abrahamson, Blikstein, Grover, Schneider, & Tissenbaum, 2016; see D'Mello, Chapter 6, this handbook). Multimodal learning with graphical probability models including Bayesian networks, and neural network and deep learning can be very useful for modern complex simulation-based assessment.

Best Linear Prediction

Best linear prediction is used to predict true test scores from observed item scores and ancillary data (Haberman, Yao, & Sinharay, 2015; Yao, Haberman, & Zhang, 2019b). The linear function $\hat{v} = \alpha + \beta'X$ can be obtained by minimizing

$$S(a,b) = E\left(\left[v - a - b'X \right]^2 \right).$$

with a and b as estimate of α and β, over a real constant α and a M-dimensional vector constant β with $\beta_m, 1 \le m \le M$, if $\beta = \left[\text{Cov}(X) \right]^{-1} \text{Cov}(\tau)c$ and $\alpha = (c - \beta)' E(X)$, where $X = \tau + e$ (X has M dimensions), composite observed score $O = c'X$, and the composite true Score $v = c'\tau$.

For all the models discussed above, when they are implemented for operational scoring, it is necessary to examine model assumptions and abnormalities. For example, for multiple linear regression, one may have to investigate if the relative weights for certain feature change more than a preset threshold under different model configurations or if the variance inflation factor (VIF) > 10/statistical tolerance < 0.10, indicating that there may be a dominance of certain features.

Model Impacts

In practical model selections, various AS models are built, and the statistical performance of these scoring models on the cross-validation sample is evaluated. The impact of the model weights from different models on the total group and subgroup are evaluated, the machine–human score differences for the subgroups are evaluated, and the weighted composite score of human and machine is evaluated. Impact analyses results can be evaluated for the scale score, average deviations, and root mean square deviations, which can indicate which models perform well and which scores have major impacts (Williamson et al., 2012; Bridgeman et al., 2012; Ramineni et al., 2012b). The commonly used evaluation criteria are listed in the following sections.

AS System Evaluation

System Architecture Evaluation

AS engines are typically evaluated and upgraded at regular intervals. These upgrades may include advances in technology, natural language processing, machine learning, and new methodologies in psychology and statistics. Impacts of population changes on the functioning of engines also need to be regularly evaluated.

For example, the *e-rater* engine has been upgraded regularly at ETS. The main reasons for the engine upgrade include: system platform changes, such as the changes in the system and codes; main additions and modifications to the *e-rater* automated essay scoring engine, such as feature revisions and/or addition of new features; procedures used to build and evaluate the scoring models change; and application population changes (Bejar, Mislevy, Zhang, & Rupp, 2017; Bennett & Zhang, 2016; Rupp, 2018; Williamson et al., 2012; Zhang, Bridgeman, & Davis, in press).

When an AS system architecture changes, it is necessary to re-evaluate the system's performance. For example, when *e-rater* engine architecture was completely rewritten using state-of-the-art coding techniques and quality-control methods, software was modernized and modularized for ease of development, testing and quality control, and some of the *e-rater* modules were completely rewritten. Also, when modifications to the macrofeatures of *e-rater* are incorporated into a proposed engine, the *e-rater* engine needs to be upgraded and evaluated. In Figure 7.1, the macrofeatures consist of many microfeatures. For example, the *Grammar* feature includes many microfeatures such as *Pronoun Errors, Sentence Structure*, and *Wrong or Missing Words*; *Usage* feature includes *Article Errors* and *Confused words*; *Mechanics* feature includes *Spelling* and *Capitalization*; *Style* feature includes *Inappropriate Words* or *Phrase and Repetition of Words*. Any changes made to even one microfeature result in a re-evaluation of the functioning of the revised scoring engine. Changes could include: module was updated to use an updated preposition corpus, dictionary updated and added functionality to detect some multi-word spelling errors, component was updated to use a more accurate computation method, corpus changes, and score scale changes. Addition of features such as *Syntactic Variety, Discourse Coherence*, and *Source Use* is a change of *e-rater* infrastructure, which requires a full engine evaluation on all the operational programs using *e-rater*. Note that the feature labels in Figure 7.1 provide a useful shorthand for describing collections of microfeatures, but these labels are not necessarily consistent with how an English teacher might use the label. In particular, *Organization* and *Development* in *e-rater* may not reflect the more nuanced understanding of these terms by English teachers.

When an AS engine's evaluation methodology and processes change, such as the statistical and machine learning model, and/or when the application samples change, it is necessary to re-evaluate the system's performance. The re-evaluation of an AS engine involves feature training sample and evaluation sample refreshment, feature evaluations, comparison of the proposed engine vs the current operational engine, and evaluations of engine performance on all operational programs and for subpopulations (Williamson et al., 2012).

Linguistic Feature Evaluations for Essay and Speech

An AS engine for essay and speech tasks consists of many linguistic features such as those for *e-rater* shown in Figure 7.1. These use Natural Language Processing (NLP) techniques to extract the features for each essay or speech response. Once the features are extracted, the AS algorithms often mimic human rating and support the human scoring rubric to evaluate the essay features at each score level to produce a machine score for each essay. AS engines can perform evaluations for essays, speeches, and conversations, and identify behaviors and actions to help humans to increase scoring and evaluation efficiency, accuracy, and reliability on those tasks. The scoring engine attempts to mimic the human scores by assigning weights to a small number of macrofeatures that were created from a large number of microfeatures that the machine can assess (see Figure 7.1). New methodologies such as best linear predictor are implemented to predict not only human scores but also external criteria related to writing and speaking skills.

Feature Preprocessing/Filtering

Before evaluating the task features (e.g., essay, speech and conversations, mathematics, ITS), these features in an AS system need to be processed before they are included in the scoring engine. The preprocessing/filtering process identifies the nonscorable responses, including (1) system-initiated nonscorable responses such as responses from the recording stage when a computer, or microphone, or other equipment malfunctions; (2) user-initiated nonscorable responses such as responses that are entirely not relevant to the task, memorized responses from external sources, or responses in a language other than the target language. AS systems may fail to produce a valid output during feature generation resulting in feature values that are not representative of test-taker proficiency, or not configured to detect such responses thus to assign an inappropriate score to test takers (Yoon and Zechner, 2017; Yoon et al., 2018). Zechner et al. (Chapter 20, this handbook) discuss different rule-based or statistical models for *SpeechRater* preprocessing and provide an overview of scenarios and solutions.

It is also a common practice that an AS engine assigns *advisory flags* for problematic responses. If a response contains serious problems such as an empty response, it is flagged by a *fatal advisory flag*. Less serious problems that do not prevent the assignment of a score, but indicate noteworthy unusual performance in some respect, result in *nonfatal advisory flags* that are provided as supplementary information. It is common practice at ETS to exclude responses with both types of advisory flags for model training and building purposes. However, essays with nonfatal and fatal flags are retained for advisory evaluation purposes and only essays with fatal flags are excluded for model evaluations.

Feature Evaluations

After the features pass the filtering models, feature-level statistical analyses are performed to see how the features are performing in the scoring engine (Cahill, 2012; Williamson et al., 2012). The feature-level analyses are performed for each operational program, including analysis for (1) microfeatures, (2) macrofeatures, and (3) advisories. Commonly used statistical analysis are feature descriptive statistics such as N, mean, SD, skewness, kurtosis; feature distributions such as box plots, histograms, frequency distribution plots, and prompt-level histograms; proportion of advisories with large discrepancies on human vs machine scores; percentage of responses flagged as advisory by both current and proposed engines.

The feature evaluation also includes consistency analyses such as inter-rater reliability, cross-task correlations, and cross-task macrofeature correlation analyses, interfeature correlations, human–feature correlations, gender, and other differential feature functions (DFFs). Cross-engine classifications flag distributions conditional on human scores (at overall model level and prompt level). For example, the examination can include: (1) If any feature has only one or two frequency bars; (2) If any feature has negative or positive or both shift; (3) If any feature has negative human–feature correlation; (4) If Delta in correlations is bigger than a preset threshold, say $|0.2|$, 5) If $|DFF| >$ preset threshold (e.g., 0.05), and relative weight > preset threshold (e.g., 10%); and 6) If any feature or flags for any prompts cross-engine has high distribution changes. The goal is to identify any abnormality (Zhang, Dorans, Li, & Rupp, 2017).

Sampling for the Training and Evaluation Samples

Representative samples are drawn from the testing populations, including subgroups such as gender, ethnic groups, and native country. These samples must be large enough and cover the whole range of prompts. They include training samples for NLP to train the scoring features for AS engines, evaluation samples including model building sample and model evaluation sample. There may need to be samples for a specific set of analyses, such as reliability sample, repeater sample, and the special accommodation sample, if available. (Williamson et al., 2012).

Random and Stratified Sampling for Different Assessments

For most operational programs, it is important to perform new random sampling and random splitting procedures for AS engine evaluation's training and evaluation data samples. This is to prepare the data refreshment for the development and evaluation of features at the beginning of the upgrade process, as well as the model building and model evaluation processes. The data are refreshed to maintain a current representative sample of the test-taker populations for different programs in terms of proficiency and background characteristics as well as prompt pools. Refreshing data is highly desirable for all programs but is especially critical for high-stakes programs. Various random sampling approaches are used for different programs in accordance with the delivery and score reporting characteristics of each program. For example, for the *GRE Issue* and *GRE Argument* tasks, refreshed data were randomly selected from all data from administrations with minimally acceptable volumes for a past few years as engine upgrade samples.

Often, stratified samples need to be selected due to data availability or requirements of the operational program. For example, *PRAXIS* data samples were randomly selected across prompts meeting minimum volume requirements for each prompt from selected administrations, and sometimes with different weights applied to the sample selection for specific modeling needs. Each of the engine upgrade datasets are then split randomly into an independent *model building* and *model evaluation* sample. Often a *reliability sample* and a *repeater sample* are selected excluding test takers in the *engine upgrade* samples.

These samples need to be examined for abnormality, for example, if (1) Difference between sample and population is greater than preset threshold, say 2%; (2) Difference of feature distribution between model building sample and model evaluation sample > preset threshold (e.g., 0.001).

Engine Training Sample

When developing NLP features, the macrofeatures need to be trained before they can be used. The training sample is used to train the macrofeatures from the extracted microfeatures using machine learning approaches. For example, *e-rater* extracts a count of grammatical errors, number of words, fragments, run-ons, and so on. From the hundreds of microfeatures extracted from the responses, NLP computes and iteratively evaluates the macrofeatures using traditional and modern machine learning technics including regression methodologies, support vector machine, artificial neural networks, and deep learning. These features are the basis for an automated soring engine (Williamson et al., 2012).

Engine Evaluation Samples

Engine evaluation samples include model building samples and model evaluation samples. For example, a large data sample of TOEFL essays is randomly selected across prompts meeting minimum volume requirements, then randomly split into model building and model evaluation samples. The model building sample is used to build a scoring model, then this model is applied to the model evaluation sample to see if the model performs as expected before the model is deployed for operational use. The evaluation of the samples scans for abnormality, for example, if the difference between scores assigned in the model building sample and model evaluation sample is greater than the preset threshold.

Scoring Engine Evaluations

Scoring engine evaluations include evaluations due to sample or population changes and engine architecture changes. When the application population changes, even if an AS engine performs well, it is important to select new representative samples, such as rolling samples from past years to re-evaluate the scoring engine with new model building and model evaluations. If there is any engine architecture or enhancement change, it is necessary to re-evaluate the scoring engine with new model building and model evaluations, and to compare the scoring models between the two engines. We usually call the scoring engine in use as the current operational engine or current engine, and the engine with any changes the proposed engine.

Scoring engine evaluation involves many analyses such as macrofeature analyses that are used to examine the zero-order Pearson correlations of each macrofeature from the current operational engine to the proposed engine for selected model groups. The typical evaluation criteria include agreement statistics between human and machine scores such as quadratic-weighted kappa (*QWK*), Pearson correlations, and standardized mean score differences (Williamson et al., 2012), see the next section.

For feature weights in final models, *proportional standardized feature weights* for selected model groups in the current operational engine and the proposed engine are examined. These particular weights are computed by dividing each standardized regression coefficient by the sum of all such coefficients in a model. The computation of these weights is merely a heuristic and assumes that the predictors are uncorrelated (Johnson, 2000). For example, when *e-rater* features are correlated, proportional standardized feature weights are not optimal for making decisions about feature importance. An alternative approach to quantifying feature importance that does take into account the multicollinearity of the predictors can be obtained through a principal component analysis to create *relative importance weights* (Johnson, 2000). These weights can be expressed as percentages such that they represent the independent contribution of features to the model (Yoon and Zechner, 2017).

In a practical AS engine upgrade, human–machine agreement analysis summarizes the core human–machine agreement statistics in the engine evaluation samples for each operational program. The statistics include the mean and standard deviation of the first human score, the mean and standard deviation of the machine score, the human–machine agreement statistics of *QWK*, *r*, and Cohen's *d* for current operational engine and the mean and standard deviation of the machine and the agreement statistics for the proposed engine.

The scoring models for all the operational programs exhibit satisfactory statistical performance overall when most human–machine agreement statistics are within the quality-control limits for acceptable performance. The key is to evaluate the agreement statistics overall and at the prompt level. For example, the exact agreement rates should be greater than 85%, the adjacent agreement rate should be greater than 95%, and the QWK value should be greater than 0.95 (Ramineni et al., 2012b).

In addition to writing and speaking tasks, AS engines or algorithms are used in many other complex task applications including *intelligent tutoring system* (ITS) (e.g., Mislevy et al., Chapter 22, this handbook). In the case for an ITS evaluation or upgrade, it is similar to essay or speech scoring engine evaluation. That is, it is also based on specific feature improvement, such as learning from teachers/students about ways to improve or modify the system as a whole. For example, *Inq-ITS®* has specific low-level features (i.e., ways to assign lab assignments, ways to present information, how to organize reports). System improvements are such as addition of facets of Next Generation of Science Standards (NGSS) practices and how the practices overlap/interplay with each other. *Inq-ITS* has the same kind of evaluation process as essay or speech: human scoring data, comparing models' performance on old and new data (Sao Pedro, Baker, & Gobert, 2012). *Inq-ITS* also uses an automated checking in place such that if there is a tweak for algorithms to support new cases (i.e., new aspects to assessing competencies) that the old cases still work as intended. *Inq-ITS* performance is evaluated on a variety of usability, usefulness, fidelity of implementation, and efficacy metrics. Theses includes human–human, human–machine reliability of machine scores; construct validity in terms of making sure the reporting out to teachers metrics are both meaningful and actionable; the efficacy, measuring performance across inquiry tasks with and without scaffolding support (Mislevy et al., Chapter 22, this handbook).

The key evaluation for any scoring engines include: (1) for cross-engine agreement guidance, it may suggest biases if exact agreement < 85%; exact plus adjacent agreement < 95%; QWK < 0.95; if % of case in upper/lower off-diagonals are large and/or with uneven values. The overall percentage of cases in the upper and lower off-diagonal should be about equal. If the machine–machine comparison is not about equal, there is a bias that must be addressed. Look for any prompt that shows a lot of bias. (2) If Delta > |2%|, look at really large differences, say larger than 2%. It's important to examine if there is either a large discrepancy at one score or no scores at the top of the scale, or a flagged % change for the mean. Other key analyses include consistency analyses, interrater reliability, cross-task correlations, cross-task macrofeature correlations, cross-engine classifications, and flagging distributions conditional on human scores (at overall model level and prompt level). Sometimes, a feature or flag analysis for a prompt cross-engine shows high distribution changes or low agreement, there is a possibility that the feature or flag analysis for the prompt cross-engine may show the same very low human–human agreement.

Population Subgroup Evaluations

Evaluations of current operational and proposed engine upgrades include an analysis of the engine's performance on each of several specified subpopulations, such as gender, ethnic, country, and language groups. Although country and language groups overlap, they need to be evaluated separately; some countries have different language groups, and language groups can include several different countries. For example, although Mandarin Chinese is spoken in both Mainland China and in Taiwan, the human–machine discrepancies are different in these two locations – Mainland Chinese students tend to get higher scores from *e-rater* than from human raters while the opposite is true in Taiwan (Bridgeman et al., 2012). The evaluation of subgroup differences should consider the correlation between human and machine scores in each subpopulation as well as the mean difference of standardized scores between human and machine scores. These analyses are typically run for only relatively large subgroups ($N \geq 500$).

Subgroup differences between machine and human scores are typically evaluated at the individual task level (e.g., separately for the TOEFL iBT Independent and Integrated tasks) and at the total score level once the human and machine scores are combined and adjusted, if necessary, according to the program's scoring rules. Note that human–machine discrepancies for a particular subgroup that appear to be substantial at the individual task level may have a relatively small impact at the reported score level especially when the final reported score is based on both human and machine scores (Breyer et al., 2017).

System Evaluation Criteria

The commonly used evaluation criteria include descriptive statistics, percentage exact agreement, percentage exact and adjacent agreement, Pearson correlation, Cohen's kappa, QWK, standardized differences for total population and for subgroups, mean squared differences, mean squared error (*MSE*), proportion reduction of mean squared errors (*PRMSEs*) (Yao, Haberman, & Zhang, 2019a, b; Williamson et al., 2012; ETS AS TAC discussions).

Suppose N is the total number of test takers, X_1 and X_2 represent item-level ratings 1 and 2, respectively. Depending on the cases in which combined scores are used, for example, at the item level, X_1 may represent one of the human rating and X_2 may represent a second rating (e.g., human rating 2, adjudication rating, AS rating), and N_1 is the number of test takers with 1 human score. N_2 is the number of test takers with 2 human scores. $\bar{H}_i$ is the average observed human score for test-taker i. At the test level, X_1 may represent total-level scores using a combination of two human scores and X_2 may represent total-level scores using machine scores or a weighted combination of human and machine scores. Also, P is the number of prompts, N_p is the sample size of each prompt.

Basic Statistics

Basic statistics include means, standard deviations, kurtosis, skewness, and nonparametric percentiles of the human and machine score frequency distributions. This information is illustrated by plots of distributions such as frequency plots or box-and-whisker plots whenever appropriate. For example,

$$\bar{X}_1 = \sum_1^N \frac{X_1}{N}; \bar{X}_2 = \sum_1^N \frac{X_2}{N}; SD(X_1) = \sqrt{Var(X_1)}; SD(X_2) = \sqrt{Var(X_2)}$$

$$Var(X_1) = \frac{\sum_1^N (X_1 - \bar{X}_1)^2}{N-1}; \quad Var(X_2) = \frac{\sum_1^N (X_2 - \bar{X}_2)^2}{N-1};$$

$$Cov(X_1, X_2) = \frac{\sum_1^N (X_1 - \bar{X}_1)(X_2 - \bar{X}_2)}{N-1}$$

Agreement Statistics

An exact agreement rate is the percentage of score pairs that perfectly agree with one another while an adjacent agreement rate is the percentage of score pairs which agree with one another by up to one score point.

$$\text{Percentage Agreement} = 1 - \sum_{k=1}^K \sum_{j=1}^K w_{jk} \frac{n_{jk}}{n}$$

where K is the number of item score categories, n_{jk} is the number of times rater 1 assigned a score of j and rater 2 assigned score of k, n is the total number of responses scored. For *Percentage of Exact Agreement*, $w_{jk} = 0$ when $j = k$; and $w_{jk} = 1$ when $j \neq k$. For *Percentage of Exact + Adjacent Agreement*, $w_{jk} = 0$ when $|j - k| \leq 1$; and $w_{jk} = 1$ otherwise.

Pearson Correlation r is the standard correlation coefficient for two continuous score variables that is not affected by mean differences of score distributions.

Unweighted Cohen's Kappa (K) and Quadratic-Weighted Kappa (QWK) for score pairs characterize the degree of agreement after correcting for chance agreement and are thus more robust measures than raw percentage agreement. The *QWK* statistic includes differential weights for different amounts of disagreement (i.e., larger penalties for larger disagreements) to provide a more fine-tuned measure of agreement. However, K and QWK are affected in complex ways by differences in the distributions of cases for disagreement (bias) as well as the distribution of cases for agreement (prevalence). Other statistics including *bias-adjusted kappa (BAK)*, *prevalence-adjusted kappa (PAK)*, and *prevalence-adjusted and bias-adjusted kappa (PABAK)* are statistics with quadratic weighting schemes. For example, the threshold for agreement between human and machine measured by QWK is 0.7. As it is widely accepted the agreement between human raters, measured either by QWK or the Pearson correlation coefficient, ranges from 0.7 to 0.8 (Power et al., 2015; Williamson et al., 2012). The computation formulas are:

Cohen's Kappa

$$K = 1 - \frac{\sum_{k=1}^{K} \sum_{j=1}^{K} w_{jk} X_{jk}}{\sum_{k=1}^{K} \sum_{j=1}^{K} w_{jk} m_{jk}}$$

when $k = j$, $w_{jk} = 0$; when $k \neq j$, $w_{jk} = 1$, and $m_{jk} = (n_{j+})(n_{+k})/N$ is the present of chance agreement.

Quadratic-Weighted Kappa (QWK)

$$QWK_{real} = 1 - \frac{E[X_1 - X_2]^2}{\text{Var}(X_1) + \text{Var}(X_2) + (E[X_1] - E[X_2])^2}$$

Degradation Measures include *Signed difference* in QWK and r between human–machine and human–human scoring conditions. These difference statistics capture the performance of the AS models relative to the human–human baseline for interrater reliability. Various quality-control criteria/thresholds have been developed as guidelines to identify models and/or prompts that appear to have poor performance characteristics at the rating, task score, or total score level (Williamson et al., 2012).

Standardized Difference

Standardized differences are used to evaluate the scoring engine performances in several situations. Standardized mean difference (d)/Cohen's d for two score distributions is computed as the difference between the two distributional means divided by the pooled standard deviation of the two sets of scores.

Mean Differences

$$\text{Mean Difference for Total} = MD/Hsd = \frac{\bar{X}_1 - \bar{X}_2}{Hsd}$$

$$\text{Mean Difference for Prompt Level} = MD/PHsd = \frac{\bar{X}_1 - \bar{X}_2}{PHsd}$$

where *Hsd* is the *SD* for human scores and *PHsd* is the Pooled *SD* for human scores

$$PHsd_{pooled} = \sqrt{\frac{\sum_{p=1}^{P}(N_p - 1)\text{Var}(H_1)_p}{(N-P)}}$$

where $\text{Var}(H_1)_p$ is the variance of H_1 for each prompt.

Mean Difference of Standardized Scores (MDSS)

For each prompt, the z-score for each test taker using the total group for the Human z_{jk_H} and Machine ratings z_{jk_M} :

$$z_{jk_H} = \frac{X_{jk_H} - \bar{X}_{total,k_H}}{SD_{total,k_H}} \; ; \; z_{jk_M} = \frac{X_{jk_M} - \bar{X}_{total,k_M}}{SD_{total,k_M}}$$

where *j* is for person *j*, *k* is for prompt, *total* is for all test takers who responded to the prompt.

The mean of the difference $(Z_{jk_H} - Z_{jk_M})$ by subgroup of interest is $SMD_{Males} = \overline{Z_{jk_H} - Z_{jk_M}}$.

Mean Squared Difference (MSD)

The mean squared difference between two human variables X_1 and X_2 is $E[(X_1 - X_2)^2]$ and is estimated with

$$MSD(X_1, X_2) = \frac{1}{N}\sum_{i=1}^{N}(X_{i1} - X_{i2})^2$$

Proportional Reduction in Mean Squared Errors (PRMSEs)

Mean Squared Error (MSE)

The mean squared error of a machine score *M* as a predictor of the variable *X* is equal to the Mean Squared Error $MSE(X|M) = \sum\left[(X-M)^2\right]$. If *X* is a single human score

$$\widehat{MSE}(H|M) = \frac{1}{N}\sum_{i=1}^{N}(H_i - M_i)^2$$

If X is an average human score.

$$\widehat{MSE}(\bar{H}|M) = \frac{1}{N}\sum_{i=1}^{N}(\bar{H}_i - M_i)^2$$

The Root Mean Squared Error is $RMSE(X|M) = \sqrt{MSE(X|M)}$.

Proportional Reduction in Mean Squared Error (PRMSE)

The $PRMSE$ due to predicting a score X with a predictor M is equal to

$$PRMSE(X|M) = 1 - \frac{MSE(X|M)}{Var(X)}.$$

When V is an observed score, the estimator for MSE described earlier is used

$$PRMSE(X|M) = 1 - \frac{\sum_{i=1}^{N}(X_i - M_i)^2}{\sum_{i=1}^{N}(X_i - \bar{X})^2} \times \frac{N-1}{N}$$

The factor $\frac{N-1}{N}$ can be ignored for large N.

PRMSE for True Scores

The $PRMSE$ for predicting a true score T with a machine score M is defined as

$$PRMSE(T|M) = 1 - \frac{E\left[(T_i - M_i)^2\right]}{Var(T_i)} = 1 - \frac{MSE(T_i|M_i)}{Var(T_i)}$$

Estimation of $MSE(T\backslash M)$ and $Var(T)$ require multiple human ratings for a random subset of the sample. With two human scores, define

$$V_E = \frac{1}{2N_2}\sum_{i=1}^{N_2}(H_{i1} - H_{i2})^2$$

The MSE of M as an estimate of T is estimated with

$$MSE(T_i|M_i) = \frac{\sum_{i=1}^{N}c_i(\bar{H}_i - M_i)^2 - NV_E}{N_1 + 2N_2}$$

where $C_i = 1$ or 2, is the number of human scores observed for individual i. $N = N_1 + N_2$ is the total sample size. The variance of the true score is approximated with

$$\mathrm{Var}\left(T_i\right) = \frac{\sum_{i=1}^{N} c_i \left(\bar{H}_i - \bar{H}\right)^2 - \left(N-1\right) V_E}{\left(N-1\right) + \dfrac{N_2\left(N_1 + 2N_2 - 2\right)}{N_1 + 2N_2}}$$

PRMSE When All Students Have Two Human Scores

If all students have two human scores, $N_1 = 0$ and $N_2 = N$.

$$MSE\left(T\,|\,M\right) = \frac{\sum_{i=1}^{N}\left(\bar{H}_i - M_i\right)^2}{N} - \frac{1}{2} V_E$$

Note that these criteria/thresholds are guidelines when determining if a scoring model is acceptable for operational use. If one or more criteria are flagged based on the preset thresholds, careful examinations on the scoring model need to be performed including, for example, comparison of human–human agreement, human–machine agreement, and machine–machine agreement.

Conclusions

Necessary steps for the validation of AS depend on the claims one wishes to make about the uses of the scores. If you want to claim that the automated score is useful in predicting performance on a real-world writing task, then evidence of the relationship between the automated score and the real-world task must be presented and critically evaluated. The initial claim made at the beginning of large-scale automated essay scoring was that the machine could replicate the scores assigned by a human rater (Page, 1966b). Although recently there has been some attention to other claims (e.g., Bridgeman & Ramenini, 2017; Weigel, 2011), the claim that AS models of both essays and speech could replicate human scores has continued to dominate operational research and procedures. Therefore, the primary focus of this chapter was on providing detailed descriptions of the many steps needed to evaluate this claim, including correlations with human raters, evaluating differences between human and machine scores in different population subgroups, and responses to gaming strategies. As models change and additional features are added or modified, all of these issues must be revisited and evaluated.

Scoring model deployment decisions are based on the analyses conducted in the evaluations. If the models built for a proposed engine performs better than a current model, the proposed engine is suggested for deployment. The evaluation must determine whether the proposed engine has similar or better performance when compared to a current operational engine as well as the effects of the changes on the association between human and machine scores.

In this chapter, we reviewed many of the analytical approaches and specific statistics that can be used in engine evaluations with the goal of establishing the relationship between human and machine scores: Pearson correlations, kappa and *QWK*, standardized differences between human and machine scores for different population subgroups, and proportional reduction in means squared error *(PRMSE)* for both observed and estimated true scores. Other evaluation strategies reviewed are relevant even when an automated system is designed to predict an external criterion rather than matching human ratings; identifying the appropriate sample for model building, the need to cross-validate regression weights, and evaluation of subgroup differences must all be addressed for models built to predict an external criterion. The key considerations in any validation effort of automated systems should start with a clear statement of the claims to be validated followed by the steps that would be needed to support these claims and an evaluation of threats to the validity of proposed analyses and procedures.

Author Note

We would like to acknowledge and thank our ETS colleagues who made significant contributions to the topics discussed in this chapter. The contributors for AS evaluation approaches and processes included: Jay F. Breyer, Jill Burstein, Aoife Cahill, Neil Dorans, Yoko Futagi, Lixiong Gu, Shelby Haberman, Matthew Johnson, Anastassia Loukina, Dan McCaffrey, Chunyi Ruan, André A. Rupp, Lili Yao, Klaus Zechner, Mo Zhang, and the ETS AS Technical Advisory Council (TAC) team. We thank Kim Fryer and Jim Carlson and our manuscript reviewers for their technical and editorial comments and suggestions.

17

Operational Considerations for Automated Scoring Systems: Commentary on Part II

David M. Williamson

No innovation is of value to society unless someone uses it – or a derivative of it – to create a better world. *Automated scoring* (AS) is no exception and the innovations in the development of AS systems only benefit the field of psychometrics once they are developed to the point that they – or the derivatives from them – can be used operationally. This part of the volume shares experiences of multiple organizations around putting innovations into practice for *constructed-response* scoring, with an emphasis on AS. Each chapter brings a perspective and experience in some aspects of what it takes to appropriately operationalize scoring, providing insight and value for the topics they offer. Taken together, the chapters paint a broader picture of the overall state-of-the-art of putting AS into operational practice and, with a little imagination, they create a vision for what the future of AS of constructed-response items might look like assuming we, as a profession, can continue to productively build in the direction implied by the work of these authors and their colleagues in the field.

The reason for AS, like any technical innovation, is to be able to do something better, cheaper, and/or faster than the current state-of-the-art. Since the current state-of the art for many forms of constructed-response scoring is with human raters then this is the methodology that must be bested by any AS system that would displace or supplement current practice. As such, a thorough understanding of *human scoring* is a prerequisite for any serious attempt at developing and deploying an AS system, at least for any use of the system for which there are *real consequences* for learners and other score users. Moreover, if there are real consequences because of the scores, then the criterion of 'better' (or 'at least as good') must take precedence over 'cheaper' and 'faster'. As such, human scoring represents both the methodology that AS is intended to improve upon and the most typical basis for both calibrating the AS system and for validating the results of the system once calibrated.

Chapter 10 by Ricker-Pedley, Hines, and Connell provides an overview of what it takes to operationalize human scoring of constructed-response tasks and is a valuable tool for anyone anticipating the development of a system for processing human scores for consequential tests. While they share many aspects of what is necessary to have a successful system, these can be roughly categorized into two areas:

1. **Managing responses.** Systems and procedures necessary to ensure that all responses are uploaded, are scored appropriately, have any scoring issues resolved, and are subsequently reported.
2. **Managing raters.** Systems and procedures are necessary to ensure that the recruitment, training, scheduling, monitoring, paying, and supporting the raters is complete and functional.

From this chapter, the reader gets an impression of the complexity, scope, and expense involved in creating and maintaining a system that can fulfill these requirements.

Specifically, *managing responses* requires a system capable of input cataloging, automatically routing individual responses to raters in ways that avoid any systematic effects of raters / sequences with other responses, tracking the number of raters, identifying the need for additional ratings for *score discrepancy resolutions* or problematic responses, feeding in *validity responses* to verify *rater accuracy*, selecting cases for *validity backreads* and managing those results, potentially identifying sets of responses that might need to be rescored based on some problematic scoring and reprocessing them through the scoring, and for resolving final scores for a reported result.

Layer upon this the need for a system that effectively *manages raters* by managing their qualifications, the administration and completion of training, the certification that they can score appropriately, the monitoring with validity responses and comparisons with other raters, the scheduling of shifts, the maintenance of communication and flagging systems between other raters and supervisors / trainers, and payment systems. All of this needs to be done within an *interface* that allows for effective engagement and one can start to appreciate the scope and complexity of what must be accomplished to conduct *high-quality scoring*.

These are significant challenges and the authors provide an appreciation for the complexities of both managing responses and managing raters effectively. However, there is so much involved in high-quality human scoring that even with all that the authors provide they had to omit or pay brief reference to some aspects of a successful system. For managing responses, this can include such aspects as when a response does not get only a single overall score but might get multiple component scores for different elements of the response that must also be managed appropriately or when an operational program uses a mixture of AS and human scoring such that the human scoring system needs to have input / output mechanism to route responses for AS and route results back into the workflow.

Similarly, in managing human raters there are additional considerations that include ensuring the qualifications of raters, recruiting and retaining a sufficient number of qualified raters to score large numbers of responses quickly, and addressing the fact that, in early stages of a program, the scoring rubrics may not be fully vetted. As a result of the latter situation, the system has to accommodate rapid scoring, rubric refinement, re-training, and re-scoring cycles until the alignment between final scoring rubrics, responses, and raters can be achieved.

Finally, from an even broader perspective, such a system of human scoring might need to be designed to accommodate more than one fixed model of scoring and a single system might need to be able to accommodate multiple testing programs with different response types, scoring models, and scoring policies, or a single program that might have multiple types of constructed-response items and formats such as essays, spoken responses, or even video. Designing a system for such flexibility in current demand and potential changes to testing programs in the future creates further complexity in producing a *robust* and *functional* scoring process.

This overview of human scoring sets the stage for what an operational AS system must also be able to accomplish if it is to match or improve upon human scoring. To be comparable to this human system it would have to have the same level of concern for ensuring the quality of scores in managing responses and potentially offer notable improvements in managing raters since the number of raters could be reduced, potentially dramatically, by a high-performing AS system. In fact, Chapter 11 by Lottridge and Hoefer provides an overview of just such a system, focused on the role of AS and including the mechanisms for building and validating AS models. Just as Ricker-Pedley et al. provide insight into the complexities of an operational human scoring system so, too, do Lottridge and Hoefer provide insight into the complexities of designing an AS system, building and calibrating AS models, validating them for operational use, and deploying them into practice.

In their presentation they emphasize aspects of AS that are not often represented in other descriptions of such systems, including the importance of the composition of team members responsible for building, evaluating, and deploying such systems. The performance of AS is moderated by the skills and priorities of those responsible for building and maintaining them and having a team that can incorporate the expertise and perspectives from multiple technical fields such as *psychometrics, natural language processing, machine learning,* and *information technology* to meet assessment needs is a key part of the success of the resultant system. Furthermore, they underscore that AS systems are not fixed and, like any other *software*, they may require regular updating that might range from relatively minor changes to incorporate new kinds of features from natural language processing or statistical models from machine learning or psychometrics to major re-coding efforts that are necessary to bring the entire *code base* up to modern coding practices and languages.

In an ideal world, the overview of processes from model creation to operational use provided in Figure 11.2 of Lottridge and Hoefer would be all that is needed for an operational scoring program, with all the complexities of human scoring from Ricker-Pedley et al. obviated and consolidated into the 'Score' phase of Figure 11.2, with the exception of some role for human raters in the early part of model building and evaluation. It is more common, however, for there to be some mixture of AS and human scoring when the results of scoring are consequential. These can range from a parity use of AS where the automated score is one of two scores – the other being human – for every response and discrepancies are adjudicated by humans. In that case, the automated score is more of a feed-in to the process outlined by Ricker-Pedley et al. to programs that rely heavily on AS where the operational use of human scoring is purely for validity *backreading* or resolution of *non-scorable responses*.

In any of these, there is still some elaboration, whether substantial or limited, of the 'Manage responses and score' part of Figure 11.2 to fully represent an operational scoring process resulting in reporting of scores. As such, the reader could envision that the right-hand part of Figure 11.2 feeds into some variation on the system defined by Ricker-Pedley et al., either similar to the description for the parity scoring model of large-scale assessments like GRE and TOEFL or a more limited version for narrow human scoring in *validity backreads* and issue resolution, creating some perspective on how a complete scoring system that includes automated and human scoring might be structured.

In a more complex version, the intersection of the Ricker-Pedley et al. system and Figure 11.2 from Lottridge and Hoefer would not be restricted to the right-hand part of the figure. In operational scoring in which there was no pre-testing and model building for AS prior to operational delivery, the results of early stages of operational human scoring might be routed from a Ricker-Pedley-like system to feed into the 'Responses and Human-applied Scores' portion of Figure 11.2 from Lottridge and Hoefer in a process where the AS models were being built and validated concurrently, potentially iteratively, with operational human scoring, creating a more integrated system between human scoring and AS.

This consideration leads the reader into considering a range of ways in which these separate considerations for human scoring and AS might be more *fully integrated* in an operational system; a speculation cut short by the fact that Habermehl, Nagarajan, and Dooley have already done the work for us and presented it as Chapter 14. In their chapter, we see the results of taking a *data perspective* on score processing instead of a *functional perspective*. This shifts the narrative from one around "scoring mechanism" to one centered on how "data flows" through the entirety of the operational scoring system, with all the elements of Chapters 10 and 11 still fully relevant and present but integrated into a single system designed to include these different elements, including iterative cycles through them.

This results in a coherent data management system that integrates human scoring, AS, and related components in much the same way that *educational modules* might be linked into a *learning management system* designed to be scalable to add or eliminate learning modules as appropriate for an *integrated curriculum*. In taking this perspective they expand the scope and conversation around scoring systems in several ways. One of the interesting aspects is to more fully integrate the role of human scoring and AS in *dynamic iterative model building* such that the AS process is not just passively receiving human scores as a batch result to build AS models, but actively directing the activities of human raters to score responses that would be most informative / valuable to building effective AS models in a continuous feedback cycle and therefore produce better-performing scoring models sooner than would otherwise be the case.

Another interesting addition is the expansion of scope of consideration for treating different consequences of scores (i.e., related to low, medium, and high-stakes use) not as completely independent use cases with different systems and processes, but, rather, as a *vertical continuum* within a single system. In this setup there is an encouragement to *borrow data* across the stakes of different programs to advance the interests of each in growing early-stage experimental AS systems from their initial applications in non-consequential use to more robust applications for score results that have impact on important decisions. The system described by Habermehl et al. is truly an ambitious undertaking. It expands the complexity and scope of an operational system to be one that combines the elements of Chapters 10 and 11 and extends them to explicitly span multiple testing programs of various stakes, multiple types of data (e.g., text, speech, video, and other forms of constructed-responses/performances), and a myriad of modeling techniques and validation criteria for AS systems.

With such an expansive perspective on what a scoring and reporting system might be, it is tempting to continue down the path of *data inclusiveness* suggested by these authors and imagine this system expanded even further to encompass characteristics of rubrics or the parameters for elements of items into the front-end of the system. It is interesting to consider how this might facilitate even earlier data to inform AS models or even *automatic item generation* such that the parameters describing aspects of items – from a design perspective rather than from an empirical calibration perspective – that are potentially predictive of item performance or rubric characteristics might be preliminary data for the development of initial versions of AS models before a single response has been provided for the items. Whether such a connection is real and/or feasible to implement is not immediately apparent, but the perspective they bring to integrated data management opens the imagination to connections not previously considered.

With a framework for data management that is potentially so all-encompassing then the next logical step would be to consider what elements are necessary within the system to make it fully functional. It is from this perspective of what it takes to make AS work well that Schneider and Boyer offer an overview of *quality assurance* for AS systems in Chapter 12. Their review provides further elaboration on how to build and evaluate AS systems to ensure that they are fulfilling their purpose in an assessment. One of the areas they emphasize that is not often part of the discussion of AS is the importance of the item and rubric design for AS to work well. It is commonplace for AS to be treated as an *afterthought* of test design, being examined only after items are written and often only after they are finalized and assembled into final test forms.

A far more productive approach is to design items with *foresight* that keeps AS in mind from the outset, so that the items might be better positioned to both address the construct as intended and to be readily amenable to AS. Of course, it is not just the item but the scoring rubric as well that demands this kind of forethought to improve the quality of AS models. Another area that is often under-represented in discussions of AS that they bring to the fore is the implication of *feature selection* and *relative weights* in scoring models for the scores of *special populations*. This is an area of special interest and importance for *fairness* to special populations that is not often directly examined or controlled in model building for AS.

In a similar vein, Chapter 13 by Shaw, Bolender, and Meisner provides an overview of quality assurance for AS systems. Where Schneider and Boyer provide an overview with some depth for typically under-represented issues in AS, Shaw et al. go even deeper into the feature selection aspects of model evaluation and provide several steps in designing features, evaluating them, selecting them, and being sensitive to the impact of each feature and the interplay among features on the subsequent scores. As such, it is a good companion piece to the same sensitivity that Schneider and Boyer have for the impact of these same kinds of issues for special populations. The two chapters also share a common interest in ensuring the quality of human scoring that AS models are constructed from and offer different sets of tools and policies for obtaining a high-quality score for model building. Finally, Shaw et al. go into greater depth in presenting sets of tools that can be used to evaluate the quality of the models build from perspectives that are somewhat different from the traditional evaluations that focus on overall agreement with human scores and would be beneficial in better understanding and designing AS models.

Some of these encouragements and approaches to model building and evaluation for AS are echoed by Yan and Bridgeman in Chapter 16. However, where previous chapters have been more in-depth in feature evaluations and tools for model checking, Yan and Bridgeman offer an emphasis on validation through association with external variables as validity criteria. They rightly point out that while there are some philosophical models of AS in which high correspondence with human scores is highly desirable (e.g., redundancy/prediction models), there are also philosophical approaches in which one might expect differences between automated and human scores. This is true whenever AS is better than humans at some aspects of scoring and humans are better than AS systems at other aspects. This is a consideration that is necessary before reflexively adopting the human scores as the single "gold standard" for evaluating AS performance.

They also broaden the conversation beyond the typical case of essay or other textual response scoring to include the scoring of spoken responses and reinforce the importance of examining the implications of AS for various *subgroups* of the test-taking population. Scoring model performance often varies by sub-population and subgroup evaluations are key characteristics to consider in ensuring that applications of AS are both valid and fair. Finally, they also offer some approaches for how to evaluate and approve an upgraded scoring engine when such engines are considered as replacements for currently deployed engines so that the continuing advances in NLP and psychometric modeling for AS can be deployed appropriately.

A final variation on design of automated systems for this part is offered in Chapter 15 by Khan and Huang, which shares an overview of advances in *deep neural networks* and their applications in education as scoring methodology as well as for generating dynamic interactive components of an educational or assessment interface. Where the other methodological chapters in this part focus on the approaches to scoring this chapter illustrates how deep learning networks are not just for scoring but might also be applicable for the dynamic generation of interactive avatars that can engage with human subjects in learning or assessment contexts. In so doing, the reader gains a broader perspective on both the possibilities and the complications of extending the current state-of-the-art beyond essays and simple spoken responses and into dynamic multimodal interactive communication that integrates audio, video, and content. The other chapters in this part help the reader anticipate the kinds of questions and issues that will need to be addressed from a design and a validation perspective over the continued evolution of such multimodal engagements, with some additional considerations related to the inherently black-box approach to the interpretation of the internal components of a deep learning network.

Collectively, the set of chapters comprising this part give the reader an appreciation of the range of tools and perspectives needed to appropriately design, build, evaluate, and deploy AS models and to conscientiously evaluate whether the use of such systems is truly better, cheaper, and/or faster than current practice. Further, the chapters allow the reader to have a broader perspective than just AS and to appreciate a complete system of data/human management that spans AS, human scoring, and integration of multiple assessment/learning programs of various designs, formats, and consequences in a single coherent system that incorporates both internal and external variables in building and evaluating approaches to scoring that blend the strengths of human scoring and AS. With such a perspective it is truly intriguing to imagine the potential for such an operational system that integrates all that these chapters offer.

It will be a topic of considerable curiosity to see whether the field continues to develop in this direction and whether such a system can be implemented in such a way that it can be both functional and maintainable over time as testing practice, programming techniques, and languages continue to evolve. In any case, these chapters are each useful in their own rights and the conscientious reader will find tools and techniques that are directly applicable to their immediate practice as well as inspirational directions for future research and development to help continue to expand the field of assessment through AS systems.

Part III: Practical Illustrations

18

Expanding Automated Writing Evaluation

Jill Burstein, Brian Riordan, and Daniel McCaffrey

CONTENTS

Automated writing evaluation (AWE) began in the 1960s with the *Project Essay Grade* (PEG) system (Page, 1966b), motivated by a perceived need to facilitate efficient grading of classroom writing assignments. PEG used a length measure as its primary *feature predictor* (i.e., the fourth root of the word count of the essay). Over the years, PEG has evolved into one of the leading commercial AWE engines (Shermis, Burstein, Elliot, Miel, & Foltz, 2015). Along with PEG (Shermis et al., 2015), two other systems developed in the mid-1990s – Pearson's *Intelligent Essay Assessor*[TM] (IEA) (Landauer, Foltz, & Laham, 1998; Landauer, Laham, & Foltz, 2003) and Educational Testing Service's (ETS) *e-rater*® (Attali & Burstein, 2006; Burstein et al., 1998; Burstein, Tetreault, & Madnani, 2013) – are now among the leading commercial AWE engines used for large-scale assessments. These systems have addressed the education community's drive to include performance-based writing assessments on standardized assessments that can be reliably and affordably scored.

Similarly, *Writer's Workbench* (WWB) (MacDonald, Frase, Gingrich, & Keenan, 1982), developed in the early 1980s, was the first system to address the need to help students edit their writing, and to provide *automated feedback* for writing mechanics and grammar. The application also provided information about topic sentences, which was the first attempt to address discourse. Picking up where WWB left off, Pearson's *Write-to-Learn*™ and ETS' *Criterion*® online essay evaluation service currently provide immediate automated feedback for student essay writing. These systems responded to a call from the education community to provide *real-time, diagnostic feedback* to student writers that provided more *explanatory feedback* above and beyond a score.

Of course, since the development of these earlier AWE feedback systems, a number of commercially-deployed systems are now available that support student writers, including *Grammarly*® and the Turnitin®'s *Revision Assistant* and *Feedback Studio*. Later in this chapter, we discuss more recent advances in AWE. Specifically, we discuss ETS' *Writing Mentor*® *Google Docs* add-on (Burstein et al., 2018), an innovative writing instruction app. The app is based on modern content-scoring research designed to address automated scoring of subject-matter material as well as exploratory AWE research that is responsive to a need to advance *personalized learning*. We also discuss domain-specific feedback within science curricula in the *Web-based Inquiry Science Environment* (WISE) (https://wise.berkeley.edu/; Gerard & Linn, 2017; Linn et al., 2014). Using the *c-rater*® content-scoring system, units in WISE combine automated scoring of the accuracy of short science explanations with expert-authored feedback designed to deepen and refine understanding of science concepts.

Broader Literacy Issues

In the United States (U.S.), literacy challenges exist in K–12 and postsecondary education. Specifically, the *National Center for Education Statistics* (NCES) reports that the average *National Assessment for Educational Progress* (NAEP) reading assessment scores characterize 12th-graders in the U.S. as "marginally proficient" (Musu-Gillette et al., 2017). Contributing to the U.S. literacy challenge is the large number of *English language learners* (ELL) enrolled in

US K–12 schools. In 2014–2015, it was reported that 4.8 million ELLs (or about 10% of the total student population) were enrolled in K–12 and were participating in ELL programs (United States Department of Education, n.d.).

In postsecondary contexts, millions of students reportedly lack the prerequisite skills to succeed (Chen, 2016). Consistent with this, more than 50 percent of students entering two-year colleges, and nearly 20 percent of students enrolled in four-year postsecondary institutions, are placed in math, reading, and writing *developmental courses* (Complete College America, 2012). Among the reasons noted for students not completing developmental classes and not graduating is the lack of preparation in reading and writing (Complete College America, 2012).

It is a challenge for instructors to provide writing instruction support at-scale. Specifically, it is not feasible for instructors to provide real-time, frequent, detailed feedback to their many students, and across multiple assignments. Paired with instructor guidance, AWE can provide real-time, frequent writing feedback and facilitates more writing opportunities for students in a self-regulated writing context. AWE can also be leveraged to develop personalized learning analytics to track students' needs and progress over time. Writing analytics could not only be helpful to students to understand their progress and support their individual learning needs, but also to their instructors to inform their curriculum.

In the remainder of the chapter, we first offer a general discussion of feedback and the importance of feedback before describing AWE systems that evaluate writing quality and content. We then offer a look at the future with regard to how we can expand AWE for personalized learning applications that may support students' immediate needs and extract meaningful information from their writing to help to gauge broader success outcomes.

Feedback Types

Human Feedback

Feedback, especially from experts such as instructors, is generally thought to be critical for learners to improve. Feedback can have positive effects on writing when provided by instructors, other adults, peers, and even computers (Graham, Hebert, & Harris, 2015). Research has begun to sketch the contours of effective general feedback practices by instructors (Shute, 2008). Reviewing the literature on how instructors promote learning in their classrooms, Gerard, Matuk, McElhaney, and Linn (2015) noted that effective instructors provided frequent, individualized guidance.

When providing feedback, they elicited students' ideas, helped students gauge their understanding of the material, and then decided on a learning path tailored to the student. In particular, effective feedback promoted reflection and self-monitoring by asking students to connect their knowledge to new material. These findings are echoed by the literature on feedback and memory. For example, Finn and Metcalfe (2010) found that "scaffolded" or elaborative feedback, wherein learners are given feedback hints but must generate the correct answer themselves, is more effective for later recall than standard feedback.

A similar theme runs through research on effective feedback on writing. Wang, Matsumura, and Correnti (2017) provide guidance to instructors on best practices for feedback on high-level writing tasks that require students to develop new ideas, such as analysis of literary elements, comparing and contrasting, and interpreting narrative themes. Their research suggests that feedback should be "specific, focused on content, and delivered as requests for more information" (p. 102). In other words, instructors need to identify specific difficulties and provide feedback as questions that help students reflect and generate improved ideas.

In addition to instructors, students' peers can provide effective feedback to improve writing. Similar to feedback from instructors, Patchan, Schunn, and Correnti (2016) show that peer-provided feedback is most effective when it focuses on higher level, substantive issues. In a large study of undergraduates, students who received peer comments on high-level writing issues (e.g., clarity, strength of arguments) and substance (e.g., missing or incorrect content) were more likely to improve their writing as they revised.

Writing Feedback

Writing is a complex process, typically involving multiple steps that include planning, drafting, and revising (Wilson & Andrada, 2016), and requiring significant practice to gain proficiency. *Intra- and inter-personal factors* may also affect writing achievement (MacArthur, Philippakos, & Graham, 2016). Feedback might take many different forms at each step of the process of writing. Moreover, there are many settings for writing instruction and practice across K–12, college, and beyond, each of which may call for tailored instructional and feedback practices (Graham et al., 2012, 2016). The complexity of the process and the multitude of instructional settings may provide many opportunities for automated guidance, but "getting it right" within each setting may prove challenging.

Automated Feedback

Holistic/Overall Score and Feedback

Virtually all AWE systems that support automated feedback predict a *holistic score* (i.e., overall score) derived from other predicted scores such as *analytic / trait scores*. High scores typically serve as an objective for students to achieve as they craft revisions. Given a set of scoring rubrics and a training set of student responses and human-assigned scores, predicting scores for new responses is "easy" for feedback systems. However, AWE systems do not usually provide users with the rationales for predictions and, in fact, such explanations are often difficult to produce from the highly complex *machine learning architectures* comprising most feedback systems.

Feedback in the form of a score usually attempts to provide general assessments of students' writing and broad suggestions for improvement. However, score-specific *qualitative feedback* is typically authored by experts. For instance, *content-scoring systems* can rarely provide rationales for their predictions and what constitutes effective automated feedback is an active area of research. As the accumulated work of Linn et al. (2014) on feedback in a *knowledge-integration approach* to science inquiry suggests, the form of qualitative feedback for a holistic or overall score can vary widely based on the feedback's purpose as well as content area.

Trait Scores, Features, and Feedback

Trait scores attempt to measure aspects of writing that are thought to be components of good writing on an *ordinal scale*. While what constitutes a trait for writing differs across systems, common content-related traits include 'organization,' 'ideas and elaboration,' 'style,' 'use of evidence,' and 'clarity' (Burstein, Chodorow, & Leacock, 2004; Foltz & Rosenstein, 2017; Wilson & Czik, 2016; Woods, Adamson, Miel, & Mayfield, 2017). Trait-scoring models are typically trained on *expert-assigned trait scores* to predict the human expert ratings on individual traits using a sample of written responses. These models are then used to provide *automated trait scores* for responses.

Like qualitative feedback for holistic scores, qualitative feedback on traits is typically authored by human experts. For instance, the PEG system uses a fixed set of feedback for each possible pairing of a trait and a score point, which describes features of a trait that good writers incorporate in their writing (Bunch, Vaughn, & Miel, 2016; Wilson & Czik, 2016). As shown in Figure 18.1, ETS' *Criterion* system provides feedback related to traits about English conventions, style, and organization and development (Burstein et al., 2004).

Pearson's *Write-to-Learn* system uses a similar set of content-related traits, which are labeled 'task and focus,' 'development of ideas,' and 'organization,' and feedback includes supplementary tutorials on how to improve each trait (Foltz & Rosenstein, 2017). The *Revision Assistant* system also uses trait-specific feedback, but feedback is more varied, targeting specific response characteristics (Woods et al., 2017).

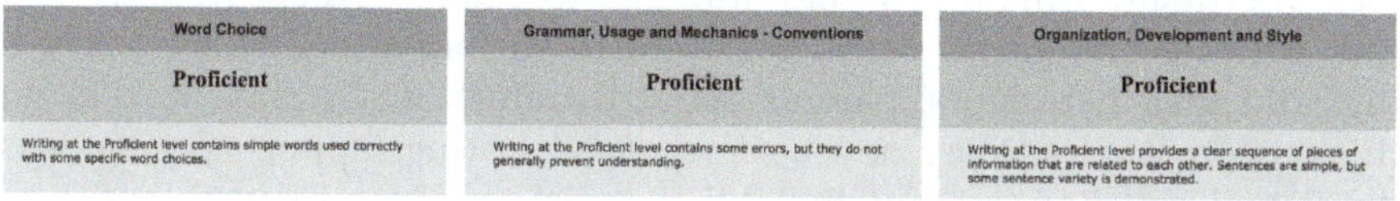

Word Choice	Grammar, Usage and Mechanics - Conventions	Organization, Development and Style
Proficient	Proficient	Proficient
Writing at the Proficient level contains simple words used correctly with some specific word choices.	Writing at the Proficient level contains some errors, but they do not generally prevent understanding.	Writing at the Proficient level provides a clear sequence of pieces of information that are related to each other. Sentences are simple, but some sentence variety is demonstrated.

FIGURE 18.1
Trait scores and expert-authored feedback from Criterion.

Localized Alerting and Feedback

The feedback within AWE systems has long incorporated localized feed-back on 'mechanics' – spelling, grammar, and sometimes usage – and these kinds of feedback continue to become more accurate and more widely offered. Yet recently, as technology in *natural language processing* (NLP) has developed, some systems have begun to offer content-related feedback at the sentence level and below; see also Cahill and Evanini (Chapter 5, this handbook).

For example, Woods et al. (2017) describe how the *Revision Assistant* sys-tem selects sentences from students' texts and tags them with trait-related feedback. The selected sentences are "influential" in the sense that their removal would significantly change one or more of the essay's trait scores. Selected sentences are classified by the traits that are relevant to the writing construct and are then tagged with feedback related to the classification of the sentence. Feedback is specific to the structural classification; it is not nec-essarily directly related to the sentence's content. The qualitative feedback that appears with these sentences is compiled from expert analysis of trait-related comments.

Feedback Systems

In this section, we provide illustrations of emerging AWE applications for *writing quality* and *domain knowledge* (content) feedback, which were designed to directly support student writers by generating real-time feedback. Systems may also be used by students to help improve their writing, by educators to better understand their students' learning needs, and by researchers to con-duct research that can inform the trajectory of next-generation feedback and curriculum.

Writing Mentor®: Writing Quality Feedback

Background

As we noted earlier, the education community has a strong interest in feedback for student learning, and analytics to support stakeholder under-standing on learner progress. AWE systems can generate feedback with meaningful information to help students to improve their writing, and can generate *educational analytics* to inform various stakeholders, includ-ing students, instructors, parents, administrators, and policy-makers. In this section, we describe a novel, *self-regulated* writing support app that was developed collaboratively with NLP scientists, writing experts, and *user interface / experience* experts.

Motivation

The *Writing Mentor* add-on to *Google Docs* (i.e., application / 'app') provides writers with 24/7 academic writing support "where they are" – referring to the fact that many students are already using *Google Docs* to complete writing assignments. Therefore, they have convenient access to the tool. The app was intended to support students in 2- and 4-year postsecondary settings, while also collecting user process and writing data to conduct writing achievement research*. It was designed to provide one-stop-shopping for writers who are looking for help with academic coursework assignments.

Other consumer-based apps such as *Grammarly* typically focus on English conventions. Applications such as ETS' *Criterion* (Burstein et al., 2004) and Turnitin's *Revision Assistant* (www.turnitin.com/products/revision assistant) provide feedback above and beyond English conventions, but require institutional subscriptions, which limits broader consumer access. The *Writing Mentor* design focuses on four key writing subconstructs expected in academic writing – specifically, 'credibility of claims', 'topic development', 'coherence', and 'editing'; see Table 18.1 for an overview of component features and how they map onto these subconstructs.

Writing Mentor embraces a *sociocognitive model* of writing achievement (Flower, 1994; Hayes, 2012b; Melzer, 2014), which emphasizes the complex contribution of skills to writing achievement, including writing domain knowledge, domain-general knowledge (e.g., critical thinking), as well as intra- (e.g., motivation) and inter-personal (e.g., collaboration) factors. Currently, the app addresses writing domain knowledge through feedback while also collecting user-reported *intrapersonal factor* information.

Specifically, the app collects information about *self-efficacy* (i.e., users' confidence about their writing skill) through a three-question optional entry survey. User responses examine the following factors that might be related to feedback preferences: (1) self-reported writing self-efficacy, (2) self-reported English language proficiency, and (3) reasons why learners utilized the app may be related to feedback preferences. The app also includes an optional *user exit perception survey* adapted from the *System Usability Survey* (Brooke, 1996), which gives users an opportunity to share what they think of the *Writing Mentor* app.

Application Design

As we noted above, *Writing Mentor* provides users with actionable feedback related to the writing being 'convincing' (e.g., claims and sources), 'well-developed' (e.g., topic development), 'coherent' (e.g., flow of ideas), and 'well-edited' (e.g., knowledge of English conventions). Feedback is presented

---

* No Personally-Identifying Information (PII) is deliberately collected by the Writing Mentor app.

TABLE 18.1

Writing Mentor Feature Description

Feature Name	Subconstruct	NLP-Based Feature/Resource Description
Claims	Convincing	Arguing expressions from a discourse cue and argument expression lexicon containing sets of discourse cue terms and relations (e.g., contrast, parallel, summary); arguing expressions, classified by *stance* (i.e., for/against), and *hedge* and *booster* status. Extension of Burstein et al. (1998)
Sources	Convincing	Rule-based scripts detecting in-text formal citations consistent with MLA, APA and Chicago style citation formats
Topic Development	Well-developed	Main topic and their related keyword detection (Beigman Klebanov & Flor, 2013b; Burstein, Klebanov, et al., 2016)
Flow of Ideas	Coherent	Topics are highlighted in the application to show how the topics are distributed (i.e., flow) across the text; see *Topic Development*.
Transition Terms	Coherent	Discourse cue and argument expression lexicon containing sets of discourse cue terms and relations (e.g., contrast, parallel, summary), and arguing expressions, classified by *stance* (i.e., for/against), and *hedge* and *booster* status. This is an extension of the *cluelex* from Burstein et al. (1998)
Long Sentences	Coherent	Sentences identified with a syntactic parser that contain one independent clause and >= 1 one dependent clause
Title & Section Headers	Coherent	Rule-based scripts detect titles and section headers
Pronoun Use	Coherent	Pronouns identified from a syntactic parser
Errors in Grammar, Usage, & Mechanics	Well-Edited	Nine automatically-detected *grammar* error feature types, 12 automatically-detected *mechanics* error feature types, and 10 automatically-detected *word usage* error feature types (Attali & Burstein, 2006)
Claim Verbs	Well-Edited	Verbs from a discourse cue and argument expression lexicon containing sets of discourse cue terms and relations (e.g., contrast, parallel, summary), and arguing expressions, classified by *stance* (i.e., for/against), and *hedge* and *booster* status. Extension of Burstein et al. (1998)
Word Choice	Well-Edited	Rule-based script detecting words and expressions related to a set of 13 "unnecessary" words and terms, such as *very, literally, a total of*
Contractions	Well-Edited	*Contractions* identified from a syntactic parser

by a friendly, *non-binary* persona* named "Sam" (a *gender-neutral* name in English). In addition to feedback, the app generates a report illustrating the different feedback types that the user viewed, which can be saved as a PDF file and shared with instructors.

Features in the app were informed by previous research with university faculty (Burstein, Beigman Klebanov, Elliot, & Molloy, 2016; Burstein, Elliot, & Molloy, 2016), the development of *Language Muse Activity Palette* app that automatically generates language activities targeting English learners (Madnani, Burstein, Sabatini, Biggers, & Andreyev, 2016), and collaborations with writing research experts and classroom practitioners.

Event Log Evaluations

Through the app we collect user *event logs* that can be used to better understand the types of feedback that users seek and how writing feedback promotes document revision. Event logs are used to conduct research that serves to inform personalized learning approaches with regard to writing achievement. Our event logs dynamically represent a rich source of information related to user survey responses, feature use, and document revisions, which may help to inform app development and, ultimately, personalized learning. Given the space constraints of the chapter, we provide a small analysis to illustrate what these log data can tell us about users and tool use.

Specifically, Figure 18.2 shows data on users' preferred feature use based on their self-reported self-efficacy. A *preferred feature* refers to the single feature on which the user spent the most time across one or more documents. The chart is based on approximately five months of data collected from late Fall 2017 through Spring 2018 ($N = 1,842$ documents) and includes users who responded to the entry survey.

The data presented in Figure 18.2 illustrate that the most preferred features across all users are grammar errors, claims, and topic development. In this sample, 'very confident' and 'pretty confident' writers' most preferred feature is grammar errors, followed by claims and topic development. The most preferred feature of 'Not very confident writers' is claims, followed by grammar errors and topic development. This example illustrates how the event log data can offer clues about student feature use choices relative to self-efficacy. Further evaluation of this kind of information can support the development of personalized learning in technology-based writing support environments.

---

* *Non-binary* in this context indicates that the gender of the character is not exclusively masculine or feminine.

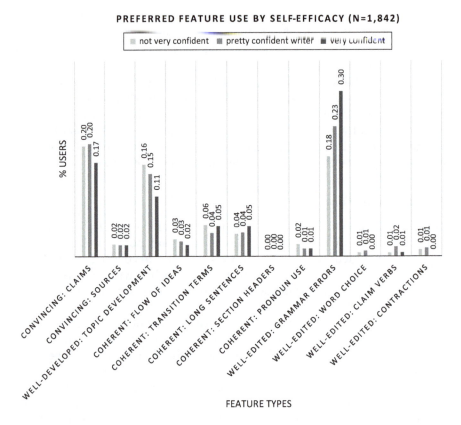

FIGURE 18.2
Users' preferred feature use by self-reported self-efficacy.

Development "Lift"

Technology innovation often requires a development "lift" (i.e., the additional work required to develop an idea or proof-of-concept into a real-life application). In this case, the NLP features already existed. Many of the features had been developed, for instance, as part of operational scoring engines or other deployed applications (such as Madnani et al., 2016). The heaviest lift came from three primary sources as follows. First, we needed to identify existing features that could be scaffolded in a meaningful way. Second, we needed to determine how to situate the features in an intuitive interface that users could easily access and understand. In the context of these first two points, a bigger and more critical issue is the general need for an operational backend for feature extraction / score computation that can ensure *stability* and *scalability* (Madnani et al., 2018).

Third, we needed to develop the app in the *Google* ecosystem. This required partnering with experts in writing studies, user interface /

experience design, and software development, respectively. Working with an experienced development team who embraced *agile and collaborative software development protocols* was essential to support an efficient and successful development process, resulting in the final app. The app was completed in eight months.

Briefly, informed by previous *prototyping research* (Burstein, Beigman Klebanov, et al., 2016), features were selected from the larger ETS portfolio, taking struggling writers' needs into consideration. Discussion with writing experts continued throughout app development during which time original scaffolding proposed in Burstein, Beigman Klebanov et al. (2016) was modified to align with the app interface design.

Discussion

The *Writing Mentor* app is a *Google Docs* application that uses AWE to provide self-regulated, academic writing support. The app is an example of a new, 24/7 AWE literacy solution that provides individualized learning support. Note that for users who are still developing writers, recommended best practice is to use the app in consultation with an instructor. While the initial version of *Writing Mentor* targeted a postsecondary population, there is an ongoing effort to collect event log data to inform app development for new populations and contribute to personalized learning research. Through the dynamic nature of the event log data collection, we can also observe changes in user writing over time and continue to revise the app based on usage trends and user feedback. In future work, as we continue to gather data in "known" instructional contexts (e.g., specific school sites and curricula), our goal is to have objective quality measures (e.g., quality scores) across revisions. This will support a comprehensive evaluation of relationships between app use and writing change.

C-rater®: Domain Knowledge Feedback

Background

Automated feedback for content is just starting to build on the best practices for content-based feedback from instructors and others. Of the main characteristics of effective human feedback discussed above, all automated feedback systems share the goals of enabling frequent feedback (Foltz & Rosenstein, 2017) and helping chart *individualized learning paths* for students. Current systems often differ in the extent to which they focus on content, elicit student ideas, and promote self-reflection. Systems also differ in the extent to which they are designed for individual study or to work in tandem with expert human feedback.

One successful approach to content-based feedback that attempts to tackle the goals of eliciting student ideas, promoting self-reflection, and providing feedback in concert with instructors is the system for automated diagnosis and feedback for science inquiry implemented in WISE. The WISE system leverages the *c-rater* automated scoring technology at ETS to "diagnose" levels of student achievement in writing (Liu, Rios, Heilman, Gerard, & Linn, 2016) and pairs it with theory- and pedagogically-driven guidance inspired by how expert instructors guide students. In this framework for writing practice and automated feedback, instructors design the science content, the scoring rubrics, and the feedback, while the actual scoring of responses and the assignment of feedback is automated (Linn et al., 2014).

C-rater is a general-purpose system for automated scoring of content in-text-based responses. The system analyzes responses by extracting a focused set of interpretable feature types and uses machine learning to create a *statistical model* of the relationship between the representation of each student's response and human-assigned scores. C-rater scores items that elicit and measure knowledge about specific content without explicitly considering aspects of writing quality such as the grammatical, stylistic, or organizational characteristics of responses. C-rater has been applied to content scoring in K–12 content areas including social studies, science, English language arts, and mathematics (Heilman & Madnani, 2015; Mao et al., 2018; Sakaguchi, Heilman, & Madnani, 2015; Zhu et al., 2017) as well as content scoring of longer responses in high-stakes settings such as the ETS *Praxis* tests for instructor licensure.

Working within the framework of science inquiry – where students engage in investigations, analyze data, and create explanations for complex phenomena – Linn and colleagues have explored how a pedagogical approach focused on *knowledge integration* can benefit student learning. A knowledge integration-based approach posits that students develop a set of ideas about a concept as they learn that must be sorted through and integrated to reach understanding. As they develop an explanation of the phenomenon, students need to weigh evidence, distinguish between ideas, and make links between concepts (Gerard et al., 2016; Linn & Eylon, 2011).

Specifically, students carry out science inquiry activities and take embedded formative assessments involving the construction of short scientific explanations via WISE. After submitting a draft explanation, students' responses are automatically scored against a knowledge integration rubric that assesses the extent to which responses use evidence and connect scientific ideas. Based on the scores, responses are assigned specific guidance related to knowledge integration. Students then revise their explanations to try to improve their scores before the responses are scored by the *c-rater* system. The knowledge integration rubrics are developed from a collaboration between researchers and science instructors (Gerard & Linn, 2017). The training data is annotated by researchers or expert science instructors experienced in the knowledge integration approach (Gerard & Linn, 2017; Liu et al., 2016).

Within this framework, across many studies, researchers have explored how instructor-authored targeted feedback on students' explanations –

automatically assigned – can improve their ideas and hence their science understanding. For example, Gerard et al. (2016) investigated whether pre-authored knowledge integration guidance can improve student learning over "generic" feedback (e.g., "Add evidence."). Knowledge integration guidance consisted of four parts: (1) acknowledge students' current beliefs, (2) focus students on a central concept, (3) direct students to relevant evidence, and (4) ask students to reformulate their responses. They found evidence that knowledge integration guidance was more helpful in improving students' short essays than other forms of feedback. Moreover, both students and instructors appreciated the knowledge integration guidance because it focused students on specific ways to improve their essays and allowed instructors to tailor their own feedback to students.

Gerard and Linn (2017) used a similar experimental design to explore the benefit of pairing automated guidance with instructor guidance. They found that combined automated guidance and instructor guidance – where the automated system provided an alert to instructors about struggling students – was effective for improving low-performing students' short essays. Tansomboon, Gerard, Vitale, and Linn (2017) investigated several forms of guidance to improve motivation. They found that *"transparent" guidance* that involved an explanation of how the automated scoring system scores student responses improved motivation to revise and improve scores. They also found that *"revisiting" guidance* – directing the student to revisit a part of the learning module – and *"planning" guidance* – prompting the student to plan specific types of revisions – can be effective.

While the approach of pairing human-designed curriculum, scoring rubrics, and feedback with automated scoring can yield tangible student learning results, students are limited to feedback tied to discrete knowledge integration scores. For instructors monitoring students' scores, more information about the gaps in students' thinking and how scores relate to discipline-specific standards would allow better decisions about how to help individual students and how to customize subsequent instruction.

System Development "Lift"

The full lifecycle of developing a system of formative feedback for classroom use encompasses design and development of a number of components, including curriculum, tasks (questions), automated scoring models, feedback, and the user interface. In the case of WISE and *c-rater* for formative feedback, curricula, tasks, and feedback are guided by knowledge integration theory. The user interface is similarly influenced by this theory and responsive to user (i.e., instructor and student) feedback (see Matuk et al., 2016, for an overview of the design of one component of the WISE platform). Scoring models are developed in a close collaboration between educational and NLP researchers, beginning at the task design stage and continuing through data annotation and error analysis. All components benefit from an iterative process of fielding curricula, tasks, and scoring models in actual classrooms and gathering feedback from instructors and students.

Beyond Feedback: Linking AWE and Broader Success Indicators

AWE feedback primarily provides writers with actionable guidance during the writing process by suggesting corrections and potential improvements. However, the field of educational technology is looking toward a future where AWE can expand. As part of this expansion, AWE could provide writing analytics to play a broader role in computer-supported personalized learning for writing in disciplinary studies where written communication is a critical competency and provide insights about connections between writing achievement and broader success outcomes (e.g., GPA and college retention).

Knowledge Gaps

In order to understand the potential for AWE in this broader context, deeper knowledge is required. We need a better understanding about the relationship between specific AWE features and someone's writing domain skills. Further, we need to know more about the contribution that the AWE features make to the fuller set of competencies necessary for successful writing as prescribed by the sociocognitive model of writing mentioned earlier, and how that might affect broader success outcomes.

The implication of the sociocognitive model in a postsecondary (college) setting is that some combination of writing domain, general knowledge domain, and intra- and inter-personal factors is essential for writing achievement at a level required for college success. However, there is little work that explores this area, and so these combinations are not well understood. What we do know is that students enrolled in postsecondary institutions who lack prerequisite writing skills may find it difficult to persist in the U.S. four-year postsecondary institutions (NCES, 2016).

There is research related to writing achievement measured by essay writing on standardized tests that examines the relationship between human rater essay scores and college success (i.e., broader outcomes; Bridgeman & Lewis, 1994); however, in this work, only the final overall essay score was evaluated. This score does not provide details about which aspect of a student's writing domain knowledge (e.g., knowledge of English conventions, discourse structure, or vocabulary choice) they need to improve. Such information in real-time could support writing instruction that leads to writing improvement.

Exploratory Research

Exploratory research designed to unlock relationships between writing achievement and broader success indicators in postsecondary education is currently underway. This research examines how AWE applied to postsecondary students' writing can improve our understanding on students' writing domain knowledge and how linguistic features in writing relate to broader measures of student success in postsecondary education. Findings from this exploration can help us to better understand how to use AWE to develop personalized learning systems to provide support to students during their college careers.

In one such study, Burstein, McCaffrey, Beigman Klebanov, and Ling (2017) explored how AWE features in students' responses to a standardized writing assessment (i.e., pilot test forms of ETS' *HEIghten® Written Communication Assessment*) related to their general domain knowledge and cumulative GPA – the success indicator – for postsecondary students from a sample of diverse institutions. The study revealed modest correlations between one or more AWE features measuring basic English conventions skills (e.g., errors in grammar and writing mechanics) and all of the external measures of general domain knowledge (*HEIghten® Critical Thinking Assessment* scores, *SAT Verbal* and *Math* scores, and *ACT Reading, Math, English, Science,* and *Composite* scores).

Features measuring these skills were also modestly correlated to writing domain knowledge as measured by scores on the multiple-choice section of the *HEIghten® Written Communication Assessment*, and success as measured by cumulative GPA. Similarly, one or more features related to vocabulary usage (e.g., sentiment term usage, metaphor usage, and preposition usage) were modestly correlated to the external criteria measures, as were features measuring the discourse structure and coherence and features assessing personal reflection in the writing. The personal reflection features were negatively correlated the external criteria suggesting that students whose writing included more indicators of personal reflection tended to score lower on criteria such as SAT or ACT scores. The strongest relationships were between features assessing vocabulary sophistication measures and *SAT Verbal* skills and *HEIghten® Critical Thinking Assessment* scores. Features from this standardized writing assessment were modestly correlated to students' self-reported cumulative GPA.

Building on Burstein et al. (2017), in another exploratory study, Burstein, McCaffrey, Beigman Klebanov, Ling, and Holtzman (2019) applied AWE to writing samples from student coursework in writing and subject-matter courses at six four-year postsecondary institutions. Burstein et al. (2019) explored how well AWE features identified in student coursework writing predicted their cumulative GPA. They found that various features measuring basic English conventions skills, discourse structure and coherence, and personal reflection were modestly correlated to GPA.

As with Burstein et al.'s (2017) study of *HEIghten* responses, indicators of better English conventions and more coherent writing were associated with a higher GPA, while increased usage of personal reflection was associated with a lower GPA. The strongest predictors of GPA were features assessing vocabulary richness (i.e., use of a variety of word types such as morphologically complex words) and vocabulary sophistication (i.e., use of lower frequency words; Burstein et al., 2019).

Using the Burstein et al. (2019) writing data, McCaffrey, Ling, Burstein, and Beigman Klebanov (2018) also found modest correlations between a subset of the AWE features and students' attitudes about writing based on responses to a student writing attitudes survey (MacArthur et al., 2016). This is consistent with previous work which has shown that both attitudes and writing domain skills play important roles in successful writing and possibly more general outcomes (MacArthur et al, 2016; Perin, Lauterbach, Raufman, & Kalamkarian, 2017).

Methodological Challenges

Burstein et al. (2017), McCaffrey et al. (2018), and Burstein et al. (2019) demonstrates promise while also highlighting challenges of using AWE-based analytics generated for authentic college student writing. Writing-domain-relevant linguistic features captured by AWE are related to general domain knowledge criteria that are predictors of success (e.g., *SAT Verbal* scores) and AWE feature scores themselves are predictive of GPA. What is important to consider as we move forward with this line of research is that AWE from multiple writing samples for each student is essential to make stable assessments of the student.

This relates to the issue of *feature stability* across different writing samples in the measures for any individual. If the scores for any given feature for the same student look notably different across writing samples, then any judgments made from a single writing sample could lead to inaccurate conclusions about the student. Using the Burstein et al. (2019) data set, McCaffrey et al. (2018) found only modest stability in the feature scores across writing samples for the same student. For example, variance among writing samples from the same students accounted for 53% of the overall variance in scores on the vocabulary sophistication feature, but up to 83% of the overall variance in a feature that measured usage of stative verbs (i.e., verbs that express a state, such as *feels*). These results suggest that data from multiple samples might be necessary to make accurate assessments of students' writing skills.

Moving forward, one idea is that AWE writing analytics can be used to efficiently contribute to technology that tracks success and obstacles throughout a student's college career. It remains an open question, however, how these AWE features from writing data can be used to generate meaningful and actionable analytics that, in turn, support student writing improvement. Burstein et al. (2017) and Burstein et al.'s (2019) foundational work is an exciting first step in what we predict will likely be an active AWE research direction in the coming years.

Conclusion

Expert instructors and peers can provide a variety of effective feedback on student writing. However, quickly supporting students with substantive, high-quality feedback for multiple assignments is not typically feasible (Gerard et al., 2015; Wang et al., 2017) and real-time feedback is even less feasible. Given the demands on instructors and classrooms and the benefits of personalized instruction, AWE systems could play an important role in evaluating student writing and domain knowledge needs, and providing targeted, well-crafted feedback, and personalized learning analytics on writing quality and domain knowledge in content-focused writing.

While advances in NLP have led to a larger portfolio of AWE features, research opportunities remain for these systems to provide more fine-grained analysis of student responses. For example, more complex feedback might include validating the presence of and connections between specific ideas, pointing out missing substantive information, identifying content inaccuracies, and flagging vague ideas. At a somewhat higher level, systems could assist instructors by identifying specific types of common problems in writing for higher-level thinking (Wang et al., 2017) where higher-level writing includes tasks such as summarizing, analyzing and critiquing texts, and synthesizing information from multiple sources (Shanahan, 2015).

In K–12, for example, the *Common Core State Standards* emphasize comparing and contrasting in many ELA standards (e.g., characters, events, properties of objects). One type of common misstep that students make in writing for comparing and contrasting is treating the objects of comparison separately and enumerating properties instead of directly comparing them. With more sophisticated NLP tools, a system should be able to recognize an example of this type of difficulty as a first step toward guidance to get students back on track. Still more ambitiously, since many writing opportunities at school are "writing-to-learn," systems could help diagnose weaknesses in reading or prior knowledge (Gerard et al., 2015; Wang et al., 2017). This is consistent with findings in Allen, Dascalu, McNamara, Crossley, and Trausan-Matu (2016) as well as Burstein et al. (2017), suggesting relationships between writing features and reading outcomes.

Even when students' weaknesses can be accurately diagnosed, what constitutes the best path for students and how to support them with guidance continues to be a challenge for instructors, especially inexperienced ones. To be effective, automated feedback on content must pair an accurate and deep diagnosis of student weaknesses with advice or requests for more information that build on what students have written so far. To the extent that these questions must be tailored to individual students' responses, rather than pre-authored for more general situations, systems will require fine-grained and robust language understanding and generation capabilities.

Finally, as students and instructors become more comfortable with AWE, they will expect increasingly natural and accurate suggestions for feedback. Emerging research that we described in this chapter discussed the need to provide feedback and analytics using the sociocognitive writing framework. We should have an eye on building AWE systems that offer (1) additional support to student writers to generate increasingly more meaningful and actionable personalized feedback across the writing construct related to content accuracy and (2) provide learning analytics to a variety of stakeholders to play a role in informing curriculum and illustrate student progress in writing proficiency and domain knowledge.

Author Note

Research presented in this paper was supported by ETS, and the Institute of Education Sciences (IES), U.S. Department of Education, Award Numbers R305A160115. Any opinions, findings, conclusions, or recommendations are those of the authors and do not necessarily reflect the views of the IES or ETS. We would like to acknowledge and thank our ETS colleagues who made significant contributions to research and applications discussed in this chapter. Core content of Writing Mentor and IES research reported here were presented at the 2018 Annual Meeting of the National Council for Educational Measurement (NCME) at the Coordinated Symposium organized by the first and third authors, entitled: *"What Writing Analytics Can Tell Us About Broader Success Outcomes."* Contributors included Norbert Elliot, Beata Beigman Klebanov, Guangming Ling, Nitin Madnani, Diane Napolitano, Maxwell Schwartz, Patrick Houghton, Hillary Molloy, and Zydrune Mladineo. We thank Hillary Molloy for her editorial contributions. We would also like to thank Mya Poe and Dolores Perin for their insights contributing to the Writing Mentor publication. We owe many thanks to our partner institutions who participated in the IES research: University of North Carolina, Wilmington; Slippery Rock University; Bloomsburg University; Jacksonville State University; Bowie State University; and, California State University, Fresno.

19

Automated Writing Process Analysis

Paul Deane and Mo Zhang

CONTENTS

Writing is a complex ability, drawing upon a mix of *social, conceptual, verbal*, and *orthographic skills* (Deane et al., 2018), particularly at higher levels of performance at which writers are expected to produce written texts within a *professional* or *disciplinary context* (Schriver, 2012). During the initial stages of writing development, educators focus on developing the ability to express ideas fluently and accurately in writing (Berninger & Chanquoy, 2012). As students move through the middle and upper grades into college and the professions, writers begin to write longer texts, over a more extended writing process, and need to develop flexibility and *strategic control* (Graham & Harris, 2000). More and more, their writing process focuses on *knowledge transformation* rather than the *direct expression* of preexisting knowledge

(Bereiter & Scardamalia, 1987) and reflects *disciplinary practices and values* (Prior & Bilbro, 2012).

Automated writing evaluation methods have historically focused upon on-demand writing and have examined features that can be extracted from the final, submitted text (Burstein et al., Chapter 18, this handbook; Crossley, Allen, Snow, & McNamara, 2016; Dong & Zhang, 2016; Shermis & Burstein, 2003). The features that can be extracted from the final submission provide direct information about some aspects of writing such as the writer's ability to generate relevant content, organize it in appropriate text structures, and produce orthographically and grammatically conventional text (Deane, 2013a). The features used in *automated scoring engines* capture important text properties, such as grammatical and mechanical accuracy, vocabulary diversity and sophistication, and syntactic variety. These features provide some level of predictive accuracy of the corresponding human evaluation of text quality (Cohen, Levi, & Ben-Simon, 2018; Yao, Haberman, & Zhang, 2019a) and broader academic outcomes such as scores on college admissions examinations or state accountability examinations (Burstein, McCaffrey, Klebanov, & Ling, 2017; Mayfield, Adamson, Woods, Miel, Butler, & Crivelli, 2018).

However, these features extracted from the written *product* typically are not capable of directly capturing critical aspects of the *writing construct*, including planning, review and evaluation, and revision skills (Deane, 2013a), which are most directly reflected in the choices that writers make during the writing *process*. The most common method of capturing the writing process is in a detailed *keystroke log*, which captures every change made to the *text buffer* – and often other writer actions – using a *digital interface*. It is worth noting, however, that it is also possible to capture intermediate steps taken by the writers during their writing process and the associated products such as initial plans as well as different versions or revisions, and to use these intermediate steps to provide *feedback* (Burstein et al., 2019; Foltz & Rosenstein, 2015).

Until recently, research into the writing process was largely disconnected from educational research on automated writing evaluation. This situation is changing (Chukharev-Hudilainen, 2019; Sinharay, Zhang, & Deane, 2019), but for many practitioners of automated writing evaluation, automated writing process analysis may still be an unfamiliar field. Our goal in this chapter is to provide a map of the domain, specifying the constructs that can be measured, identifying the types of evidence that can be adduced, and exploring major research directions and salient findings within this emerging field. We will discuss keystroke logging in depth and, where relevant, will discuss the evaluation of intermediate products as well.

Variations in Writing Tasks and Processes

The information presented in this section follows a line of analysis developed in previous publications (Deane, 2011, 2013a, b, 2018). Specifically, it is impossible to gain a clear understanding of what it means to measure the writing process without first sketching the ways in which writing tasks can vary and of the general range of activity types in which writers may engage. Many different forms and *genres* of writing exist; see Table 19.1 for an overview. Thus, it would be a mistake to speak generically about "the writing process" as if it were the same for all writers or across all forms of writing.

Consider, for example, the process someone might adopt in an online chat where participants expect a highly interactive mode of communication in which individual writing turns are short, participants may not know one another personally, interaction is (mostly) synchronous, and all writers can

TABLE 19.1

Dimensions of Variation among Writing Tasks

Dimension	Explanation
Social	• *Audience.* Whom is the writer addressing?
	• *Purpose.* Why is the writer writing addressing this audience? This includes broad categories (i.e., to inform or to persuade) but may be much more specific.
	• *Affiliation.* In what ways will the writer's social status affect how his or her writing will be received?
	• *Gatekeeping.* Who can publish? How is that decided?
	• *Collaborativity.* Is responsibility for text content or quality shared with other people?
	• *Interactivity and synchronicity.* Does the audience talk back? How often? What kind of time lag is there between contribution and response? How does interaction with the audience constrain what the writer chooses to write?
Conceptual	• *Analytical focus.* What kinds of reasoning are central to the writing task?
	• *Choice of topic(s).* What will the author be writing about?
	• *Sources.* Where will the author get the information included in the text?
	• *Presuppositions.* What information can the author take for granted?
Text Structure	• *Dominant organizing principle.* How should the author organize the text?
	• *Linguistic markers of coherence.* What kind of linguistic devices (such as transition words) should be provided as signals of text structure?
	• *Orthographic markers of coherence.* What kinds of formal devices (such as headings, tables of contents, footnotes, or hyperlinks) should be provided to signal connections among different parts of the text?
Linguistic Structure	• *Code base.* What language, dialect, or register is the writer expected to use?
	• *Complexity.* To what extent are the writer's choices constrained by the need to avoid difficult or unfamiliar language?
	• *Stylistic priorities.* What standard(s) of evaluation is the writer expected to prioritize (such as clarity, conciseness, or emotional expressivity)?
Orthographic Structure	• *Standardization.* Is the writer expected to follow specific orthographic or grammatical conventions (such as American or British spelling, or a particular stylesheet)?
	• *Modality.* How is the text transcribed – by hand, using a keyboard, on a mobile device, by dictation, or by some other means?
	• *Presentation.* How will the audience read the text? Does the presentation format differ in any way from the input modality?

contribute directly to the ongoing conversation. The writing process that is appropriate to this context is necessarily very different from the process a writer may adopt when submitting an article for publication in a journal, where the author is expected to produce a long text for a general audience that may read the text long after its initial publication. Writers usually learn how to produce this kind of extended writing through some form of *apprenticeship* (Prior & Bilbro, 2012). While writing in school is usually conceptualized as preparing students to write in formal, public contexts, schooling defines its own social space, which may impose different affordances than those that constrain writers in professional contexts (Deane, 2018).

Table 19.2 provides a quick overview of the range of activities that writers may carry out over the course of a writing task, which generally moves from planning to execution to evaluation, potentially resulting in revision. In the simplest forms of writing, the writer may proceed linearly from task analysis to discovery, from global planning to local planning, from local planning to translation, from translation to motor planning, and from motor planning to transcription. However, this is not the normal state of affairs (Chanquoy, 2009; Sommers, 1980). Most writers *shift fluidly* between activities as needed, especially on complex writing tasks that require the writer to solve difficult conceptual and rhetorical problems (Bereiter & Scardamalia, 1987).

As a result, writers often face a difficult *optimization problem*: they must decide which goals to prioritize because they cannot try to do everything at once (e.g., Anson, 2016) and simply do not have the required *working memory capacity* (Kellogg, Olive, & Piolat, 2007; McCutchen, 1996). *Novice writers* can be overwhelmed by this complexity. *Expert writers*, on the other hand, have effective strategies that help them manage and overcome their working memory limitations (Graham & Harris, 2012; Kellogg, 1987). However, even expert writers may solve the same writing problem differently, using different writing strategies and undertaking different writing activities in a different sequence, yet producing writing of comparable quality.

Sources of Evidence about the Writing Process

It is obviously a challenge to determine exactly what a writer is doing when the same individual may adopt different strategies from one task to the next or even within the same task on different occasions. Addressing this challenge may require consideration of a variety of sources of evidence, most notably:

- **Changes to the text**, including what changed, where the change happened, how quickly the change happened, how long and frequently the writer paused, where the pauses occur, what actions the

TABLE 19.2

An Inventory of General Writing Activity Types

	Planning	Text Production	Monitoring and Evaluating	Revising and Editing
Social Context	*Task Analysis* Set appropriate rhetorical goals and subgoals; identify barriers to carrying out the plan & devise solutions		*Rhetorical Review* Evaluate whether current text meets rhetorical goals and subgoals	*Rhetorical Revision* Set new rhetorical goals or subgoals and make changes to the text that will implement specific rhetorical goals
Conceptual Content	*Discovery (Idea Generation)* Identify what content needs to be produced to carry out the writer's rhetorical plan	Retrieve relevant knowledge; locate useful information; generate new ideas	*Content Review* Evaluate content for relevance and quality (such as the reliability of information and the strength of arguments)	*Content Revision* Insert, delete, and modify blocks of content
Text Structure	*Global Planning* Decide on the best organizational pattern to use given the writer's purpose and the content to be expressed	*Composition* Produce text segments in intended sequence with appropriate macrostructure markers	*Structural Review* Recognize structural problems such as information out of order, tangents, missing transitions	*Structural Revision* Move blocks of text; insert/modify linguistic markers of macrostructure
Linguistic Structure	*Sentence-level Planning* Decide which sentence structures and words will express the intended content	*Translation* Express ideas in words, phrases,2 and sentences	*Line Editing* Recognize problems in expression such as vagueness or ambiguity	Reword and rephrase existing content
Orthographic Structure	*Orthographic Planning* Determine the character sequence needed to transcribe a word, phrase, or clause	*Transcription* Write or type the correct character sequence	*Proofreading/Copyediting* Recognize typos, spelling errors, and other deviations from conventional standards	Correct typos, spelling errors, fix grammatical issues, and other deviations from conventional standards

writer took before or after the current event, and whether and how the change affected the overall quality of the text

- **Concurrent actions taken by the writer**, including accessing other texts (for instance, source articles, references on the internet), using literacy tools (for instance, a spell-checker, a planning tool or a graphic organizer), and marking up a text or taking notes
- **Participation in collaborative writing activities and response to feedback**, either human or automated
- **Other observable behaviors** such as direction of eye gaze and facial expressions
- **Self-reports** such as interviews and think-aloud protocols

We discuss each of these sources in the following subsections.

Changes to the Text

Keystroke log analysis is the major method applied when writing process analysis focuses on the nature and timing of changes to the text (Leijten & Van Waes, 2013). A *keystroke record* in JSON format may look like below:

$$\{"p": 10, "o": "", "n": ".", "t": "6.14"\}$$

where "p" refers to the position in the response box, "o" refers to the current text at that position, "n" refers to change made to that position, and "t" is the time elapsed since the student began the writing task. In this example, the student inserted a period at position 10 at a time stamp of 6.14, computed relative to when the writing starts (i.e., at 0 elapsed seconds). Keystroke logs may also provide additional details about *keystroke mechanics* such as the time a key was pressed and the time that key was released.

The nature and location of the changes that writers make to their texts directly support inferences about where the writer is in the writing process. If a writer types long sequences of characters and thereby adds new content to the text, it is reasonably safe to assume that they are primarily engaged in translation and transcription. On the other hand, if a writer deletes and inserts a few letters at a time, and, by so doing, eliminates orthographic errors, it is reasonably safe to assume that they are primarily engaged in proofreading. Similarly, if most of the changes that a writer makes happen at the end of the text, it is reasonable to assume that they are primarily engaged in composition, translation, and transcription. On the other hand, if the writer jumps from the end of the text to make changes somewhere in the middle, they are more likely to be engaged in revision and editing (Galbraith & Baaijen, 2019).

The *timing of changes* also matters. When a sequence of actions is performed in rapid succession, it is reasonable to group them together. For example, transcription is associated with bursts of fast typing or fluent handwriting with pauses in between, reflecting an alternation between transcription and planning (Hayes, 2012b; Limpo & Alves, 2014b). The timing of keystroke events also reflects properties of the word being typed (Ostry, 1983).

Longer pauses are particularly significant since they represent the portion of the writing process where planning and evaluation / reviewing activities are most likely to occur. For example, global planning is most probable early in the writing process or before major episodes of text production (Flower & Hayes, 1981) whereas sentence-level planning is most likely at major text junctures such as paragraph, sentence, and clause boundaries (Medimorec & Risko, 2017). On the other hand, long pauses associated with monitoring and evaluation may be more frequent later on, when the writer has more text to review (Galbraith & Baaijen, 2019).

Preceding and following actions may also affect how a specific action is interpreted. For example, sometimes people *backspace over text* in order to reach and correct an error, in which case they may end up retyping the text they just erased while retaining the same content. In other cases, people may backspace over text because they had a change of mind and rejected the content they just produced. The former behavior can be classified as a form of proofreading and editing; the latter, as a type of content revision. Backspacing for the purpose of error correction may appear even in *copy-typing tasks*, whereas backspacing for content revision purposes is specifically associated with drafting new content (Deane et al., 2018).

Other Actions

A keystroke log indicates how the text changed moment by moment. It does not indicate what was going on while the text remained unchanged. However, in a digital environment, many of the writer's actions can be recorded and made available for analysis. For example, if an individual navigates to (and, presumably, reads) specific texts or scrolls to particular locations in a longer text, this information can be captured and linked to other aspects of task performance (Knospe, Sullivan, Malmqvist, & Valfridsson, 2019), along with student use of other digital literacy tools such as *highlighters* or *note-taking software*. Similarly, if writers make use of specific digital tools such as *graphic organizers* to support planning or *spell-checkers* or *thesauri* to support editing, the writers' actions with those tools can be recorded and integrated with other evidence about their behavior (White, Kim, Chen, & Liu, 2015).

Collaborative Writing Activities and Feedback

Many digital writing tools provide support for *collaborative writing*. As word processing technologies have matured, typical offerings have come to include the ability to track text changes, approve or reject proposed changes, and put comments on the text. Some technologies such as *Google Docs* directly support simultaneous editing by a group of collaborators, which has made it attractive as a classroom tool for *project-based learning* (Chu, Kennedy, &

Mak, 2009) and collaborative writing (Blau & Caspi, 2009). In addition, many online learning-support systems include tools that enable *peer review* as well as teacher review. Peer review plays a critical role in improving student writing performance (Graham & Perin, 2007). Moreover, it can readily be measured in an online system where hundreds or thousands student reviews and peer evaluations can be compared and analyzed (Cho & MacArthur, 2010; Moxley & Eubanks, 2015).

In fact, *responsiveness to feedback* may be one of the key differentiators between students who are likely to improve their writing and those who will not (Patchan & Schunn, 2015). Automated forms of feedback also have a measurable impact on the quality of student revisions (Wilson, 2017). To gain a full understanding of student writing processes, it is therefore critical to measure their participation in collaborative writing and peer review activities and to analyze their responsiveness to feedback during revision and editing.

Other Observable Behaviors

Thus far, we have considered evidence that can be registered by logging writer actions within a digital writing system. Such evidence is, by its very nature, partially ambiguous. For example, writers might pause for a variety of reasons, including the following:

- because they are searching for the correct key to press;
- because they are uncertain of how to spell a word;
- because they have not yet decided which word to type;
- because they need to plan the next clause or paragraph;
- because they need to reread the text to scan for errors;
- because they are reading a source text;
- because they are frustrated or confused; and
- because they have completely disengaged from the task.

Information about observable behaviors can help address some of these questions. For instance, *eye-tracking software* can provide information about where the writer's gaze is directed during the writing process (Wengelin et al., 2009) and can reveal whether the writer is attending to the keyboard, the text buffer, or some other location on- or off-screen. Similarly, *facial recognition software* or other methods of monitoring affect can also be used to track changes in the emotional state of the writer during the writing process (Kolakowska, 2013; see D'Mello, Chapter 6, this handbook).

Self-Report and Annotation

Much of the early research on the writing process exploited *think-aloud protocols* in which subjects were asked to verbalize what they were doing as they wrote, either *concurrently* with the writing task (Hayes & Flower, 1981; Perl, 1979) or *retrospectively* (Ransdell, 1995; Rose, 1980). It continues to play an important role in studies of writing processes, both for native (Beauvais, Olive, & Passerault, 2011; Breetvelt, Van den Bergh, & Rijlaarsdam, 1994) and second-language writers (Stapleton, 2010).

Self-report data plays a critical role in *validating inferences* drawn from other types of writing process data since it can help to confirm whether specific behavioral patterns are associated with a specific type of writing activity. In addition to self-reports, when a detailed record of an individual's writing process exists, some studies have used a *trained annotator* to make inferences about what the writer is doing at various stages of the writing process (Lindgren & Sullivan, 2002). When human annotation of this type is reliable, it provides an additional source of information about how to interpret writer actions. This kind of information may make it possible to build models that *automatically tag* sequences of keyboard activities to identify specific writing process states (e.g., text production, planning, editing).

Major Analytical Approaches

The literature includes several major approaches to analysis of writing processes, which vary along a few key dimensions that we discuss in the following subsections.

Data Sampling

Analysis of the writing process can be conducted at different *grain sizes*. At the highest level, writers often produce intermediate products such as outlines and drafts, which may be circulated and reviewed. When presented as an external revision history or as text markup, the changes between drafts may themselves be reviewed by the author or discussed among collaborators as part of an extended writing process (Purdy, 2009). The specific modifications that writers and their collaborators have made to a text between drafts can be classified and used to describe writers' revision and editing processes (Jones, 2008). Alternatively, two drafts of an essay may be compared using features from an automated scoring engine to identify how revisions have affected essay quality (Burstein et al., 2019; Wilson et al., 2014).

At a finer grain size, a writing process log can capture the timing and sequence of edits and other actions, including the use of digital tools such as a *spell-checker* or a *thesaurus*. At the finest grain size, a keystroke log can capture the content, location, and timing of individual keystrokes (Deane et al., 2015; Leijten & Van Waes, 2013).

Summary Features

Writing process analysis frequently defines *summary features* intended to capture a behavior of interest. For example, typing behavior can be characterized using the concept of *burst*, which is a sequence of characters rapidly typed without a long pause (Hayes, 2012a). Once bursts are identified (Rosenqvist, 2015), summary features for a log can be calculated such as the frequency and length of bursts, the speed of typing within bursts, the amount of time spent inside bursts, or the amount of time spent pausing between bursts (Almond, Deane, Quinlan, Wagner, & Sydorenko, 2012; Feng, 2018). Various other constructs are commonly used to define summary features, including *inter-key intervals* (i.e., pause time between key presses) and *inter-word intervals* (i.e., the time between typing the last character in one word and the first character of the succeeding word) (Chenu, Pellegrino, Jisa, & Fayol, 2014; Zhang & Deane, 2015).

Once a process feature inventory has been defined, it is also common practice to apply *dimensionality-reduction techniques* such as *principal component analysis* or *factor analysis* to identify groups of features that appear to measure the same construct (Baaijen et al., 2012; Grabowski, 2008). These can be used to create indicators for *latent traits* (Zhang, Bennett, Deane, & van Rijn, 2019). Alternatively, individual features can be selected so as to minimize intrinsic *redundancy* in the feature space (i.e., collinearity) (Allen, Jacovina, et al., 2016).

Finally, once features have been defined and selected, the analyst can examine their relation to *extrinsic variables*. Commonly examined constructs include writing quality (Allen, Jacovina, et al., 2016; Sinharay, Zhang, & Deane, 2019; Zhang & Deane, 2015), engagement (Bixler & D'Mello, 2013), and emotional states (Kolakowska, 2013).

Time-Series Analysis

Since writing process data often consists of many similar events occurring at semi-regular intervals, another process analysis approach treats a sequence of such events as a *time-series* and to apply analytic methods that have been developed for that purpose such as *spectral analysis* (Alshehri, Coenen, & Bollegala, 2018; Wallot, O'Brien, & Van Orden, 2012). Less complex forms of temporal analysis examine differences between writing processes at different stages of composition. For example, some methods contrast the beginning, middle, and end of the composition process to see whether particular processes or features are more or less frequent at different stages of writing (Breetvelt et al., 1994; Zhang, Hao, Li, & Deane, 2016).

Probabilistic sequence models are potentially one of the most powerful methods for building temporal models of writing process data. In this type of model, the types of actions that writers undertake can be modeled as reflecting underlying user states and the pattern of behavior displayed by the writer can be modeled by assigning probabilities to *state transitions* (Reimann & Yacef, 2013).

Some writing process work has been done using this kind of model. For instance, Zhu, Zhang, and Deane (2019a) examined patterns of keystroke actions (i.e., insertions, deletions, cuts, pastes, and jumps) and showed that sequence patterns associated with editing behaviors were most likely to discriminate between high- and low-scoring writers. Benetos and Betrancourt (2015) combined writing action data with human annotation to define a *transition probability matrix* for the process of creating a written argument. This is potentially one of the most powerful approaches since it combines process data with labels from self-report or annotation. More sophisticated approaches such as *semi-Markov models* and *hidden Markov models* have also been used to model the writing process (Chen & Chang, 2004; Guo, Zhang, Deane, & Bennett, 2019). However, the application of sequential probabilistic modeling techniques to writing process data is still in its infancy.

Salient Findings

To the extent that writing process data can be extracted automatically, it can support automated assessment of student writing. In this section, we briefly summarize some of the most salient findings in the field.

Variation across Writing Tasks

Relatively little work has examined how writing process data varies across different types of writing tasks. However, the available literature indicates that both the distribution and meaning of writing process features can be affected by changes in the writing task.

Grabowski (2008) examined keystroke patterns for three different writing tasks: copying text from memory, copying from text, and generating text from memory (by writing a description of the route they take from their apartment to the university they attended). Grabowsky had a group of 32 students perform all three tasks and studied the following metrics for each task: the number of deletions, the number of cursor movements, transition time (inter-key interval) within words, other transition times (roughly, inter-word intervals), keyboard efficiency (the ratio between the number of keystrokes in the log and the number of characters in the final text), the number of orthographic errors in the final text, and a measure of final document length. For the copy tasks, document length was the difference between the length of the copied text and the text produced; for the generation task, it was the length of the generated text in characters.

Exploratory factor analysis demonstrated significantly different factor structures across the three tasks, relatively low correlations across task scores for most of the features, and several significant differences between means for the same factor across factor structures. In particular, Grabowski found that the transition time within words was a highly reliable measure, demonstrating *cross-task correlations* between 0.91 and 0.96. The only dimensions that were significantly correlated across tasks were the ones that transition time between words loaded on, which Grabowski interpreted as measures of typing speed. In addition, Grabowski found that that keyboarding efficiency was highest for copying from memory and lowest for text generation; that the proportion time spent pausing was smallest for copying from memory and greatest for text generation; and that individual pauses were slightly but significantly longer for copying from sources than for either of the other two conditions.

In a follow-up study with a slightly larger sample, Wallot and Grabowski (2013) analyzed the complexity of *keystroke latency patterns* across the same three tasks while applying spectral time-series analysis methods to generate features that characterized the complexity of the time-series patterns in each task. Their results indicated that text generation displayed significantly more complex typing patterns than was observed in the two copy tasks.

Deane et al. (2018) collected data from 463 middle-school students, contrasting copying text from sources, drafting an essay from a plan, and editing the same essay before submission. This study confirmed many of Grabowski's findings. Typing was significantly slower in copy-typing than in text generation, even though text generation involved proportionately more time spent pausing as a function of total writing time (see Figures 3-5 and Table 3 in the original publication). Once again, within-word transition times provided a reliable measure of typing speed with a 0.81 correlation between the speed measures across the copy-typing and drafting conditions. The correlation coefficient between these measures further went up to 0.84 when the in-word typing speed was calculated using only most common English words that were typed without modification (Zhang, Deane, Feng, & Guo, 2019), suggesting that within-word typing pattern is a reliable measure of individual's keyboard typing skill. Otherwise, most of the individual features displayed very low correlations across tasks.

Deane et al. (2018) also observed significant differences in the distribution of specific keystroke events across tasks. In brief, backspacing was much more frequent in drafting than in copy-typing or editing, presumably because writers were rejecting and backspacing over significant amounts of the text they generated; pause time spent before jumping to a different location in the text to make an edit occupied a much larger proportion of total time in drafting than in copy-typing; and, in editing tasks, pausing before a jump was even more salient, accounting for 70% of total time. These results imply that there is considerable variation in the patterns of keystroke features across tasks. These results also suggest that, while certain features such as in-word typing speed tend to be *robust* (i.e., consistent across writing tasks), some process features are *sensitive* to different task requirements.

Recently, writing processes were compared between when an essay task is presented at the end of a scenario-based assessment and when an essay task was given alone. Zhang, Zou, Wu, and Deane (2017) found that, by giving extensive support on task and topic preparation, students appeared to be able to spend more time and cognitive resources on editing and revising related activities. In contrast, when students who did not receive as much support on the writing task, essay performance appeared to be driven more by general writing fluency.

Even more recently, Gong et al. (2019) examined the relation between measures of typing speed and essay performance on a set of NAEP writing prompts and found that there is a threshold below which decreases in typing speed would have a considerable negative impact on essay quality. However, this typing speed threshold varied by genre with a higher threshold for persuasive and informational writing tasks than for narrative. This observational result can be explained by the cognitive model of the writing process in which subprocesses of writing including idea generation, task planning, translating ideas into language, and putting language on paper / on screen, all compete for the limited cognitive resources during writing. Severe deficiencies on one dimension such as keyboard typing would take up much of the short-term memory that could have otherwise been allocated for other tasks, thereby harming the overall writing performance. Much more research needs to be done to understand how writing processes vary, or are consistent, across tasks. In the meantime, caution should be exercised in generalizing results from writing process studies across different kinds of writing tasks.

Variation across Persons

Most of the data available from writing process analyses are drawn from single writing tasks or sets of writing tasks that share broad similarities (e.g., writing items from the same assessment). These datasets can be used to study how individuals vary in their behavior on specific writing constructs and a number of studies have tried to build an array of features that jointly characterize several aspects of the writing process (Baaijen et al., 2012; Van Waes & Leijten, 2015; Zhang & Deane, 2015). These studies suggest that it is possible to identify feature sets that provide information on specific writing activities such as transcription, revision, and planning (Galbraith & Baaijen, 2019). They also indicate that the distribution of pauses may reflect the relative degree of attention writers are devoting to revision or to global, sentence-level, or orthographic planning processes (Medimorec & Risko, 2017). Many of these features predict differences in writing quality (Allen, Jacovina, et al., 2016; Sinharay et al., 2019).

Orthographic Planning and Transcription

In addition to the keystroke logging literature, there is an extensive literature on typing behavior whose findings are relevant to orthographic planning and transcription (e.g., Cooper, 1993; Feit, Weir, & Oulasvirta, 2016). There are also process studies that focus on the production of *handwritten text* (e.g., Alamargot, Chesnet, Dansac, & Ros, 2006; Johansson, Johansson, & Wengelin, 2012; Limpo, Alves, & Connelly, 2017). Orthographic planning requires the writer to construct a letter sequence and hold it in working memory while transcription processes guide the execution of the hand and finger movements needed to get that letter sequence onto the page or screen.

The literature indicates that these processes are sensitive to linguistic and orthographic structure, regardless of modality. Typically, orthographic planning happens at major lexical and phonological junctures such as word, morpheme, and syllable boundaries (Pinet, Ziegler, & Alario, 2016; Will, Nottbusch, & Weingarten, 2006; Zesiger, Orliaguet, Boe, & Maounoud, 1994). It is also heavily influenced by orthographic knowledge, such as control of spelling patterns (Lambert, Alamargot, Larocque, & Caparossi, 2011). As transcription skills become more automatized, it becomes less necessary for writers to monitor hand movements during handwriting or typing (Alamargot, Plane, Lamber, & Chesnet, 2010). However, even among fluent typists, there can be a significant subgroup who watch their fingers during typing, which may prevent them from monitoring their output while typing (Johansson, Wengelin, Johansson, & Holmqvist, 2010).

Several measures appear to be particularly sensitive and stable measures of transcription skills, most notably, *within-word transition times* (Gong et al., 2019; Zhang & Deane, 2015; Zhang, Deane, et al., 2019), and the speed and length of bursts of text production (Limpo & Alves, 2014b; Zhang et al., 2018) where a burst is defined in this work as long stretches of text production without interruptions. If typing speed is measured primarily using common words that all writers are likely to know how to spell, the resulting metrics may provide a relatively pure measure of keyboarding skill (Guo, Deane, van Rijn, Zhang, & Bennett, 2018).

Sentence-Level Planning and Translation

The natural unit for sentence-level planning is the *clause* or sometimes, the *phrase* (Ford, 1979; Ford & Holmes, 1978; Holmes, 1988; Wagner, Jescheniak, & Schriefers, 2010), though pauses may also appear after a formulaic phrase or a sequence of function words, reflecting an interaction between grammatical and lexical choices (Schneider, 2014). Like fluent transcription, fluent translation processes lead naturally to the appearance of bursts of text production, with pauses between the bursts at natural locations for sentence-level planning (Alves & Limpo, 2015; Hayes, 2012a). Thus, the variability of between word pauses, and the amount of time spent pausing at natural locations for sentence-level planning provide natural measures of sentence-level planning (Zhang & Deane, 2015).

Global Planning and Composition

Global planning is a critical part of the writing process (Flower & Hayes, 1980). It normally happens early in the writing process, though it may happen later if the author decides on a major revision that entails significant changes to the way the text is organized. Global planning is often associated with idea generation, though idea generation can be more broadly distributed throughout the writing process (Van den Bergh & Rijlaarsdam, 2007). For example, some writers may develop a global plan up front, but alternate between composing and idea generation, pausing between paragraphs and other major text junctures to generate the ideas for the next section of the text (Spelman Miller, 2002). However, the literature suggests that novice writers do very little global planning before they start to write, and that as writers gain expertise, they are more likely to revise (Berninger et al., 1996).

In a writing process log, time spent on global planning is most likely to be reflected in a long pause before the writer starts writing fluently, long pause associated with large chunks of text production such as before sentences and between paragraphs, or by the writer's use of a graphic organizer or outline to produce initial notes (Deane, 2014; Zhang & Deane, 2015).

Evaluation, Revision, and Editing

In cognitive writing theories, revision and editing behaviors are triggered by monitoring actions in which the writer rereads and evaluates their own text (Hayes, 2004; Quinlan, Loncke, & Leijten, 2012). The same skills are critical in peer review and other collaborative processes in which people provide feedback about other people's writing (Stellmack, Keenan, Sandidge, Sippl, & Konheim-Kalkstein, 2012). However, novice writers tend to focus on surface errors and surface revisions, without doing significant amounts of substantive or content revision (Limpo & Alves, 2014a), though they are more likely to do substantive revision if the task is structured to focus their attention on rhetorical goals and quality of content (Midgette, Haria, & MacArthur, 2008).

Coarse measures of reviewing and monitoring behaviors include the ratio of deletion actions to insertion actions, or percent of words that were edited as a function of total number of words produced during the text production process. Research has found that even these high-level coarse measures of editing and reviewing are indicative of the final writing quality (Zhang, Zhu, Deane, & Guo, 2019). Higher-scoring essays exhibited greater extent of deletion relative to insertion actions and showed more editing and reviewing activities than lower-scoring essays.

One of the most distinctive and representative indicators of monitoring behavior occurs when the writer changes the cursor position, especially if the shift in position is accompanied by edits. Such "jump edits" may result in different kinds of edits – content corrections, changes to wording, or spelling corrections. However, to the extent that such edits occur, they suggest that the writer has reread the text, found a specific problem at a specific location in the text, and shifted gears to correct it (Deane et al., 2018; Grabowski, 2008; Zhang & Deane, 2015).

A recent study analyzed jump-edit behaviors in great depth where the jumps must happen across a sentence boundary (by eliminating local jumps)

(Zhu, Zhang, & Deane, 2019b). The authors found that higher-scoring students showed significantly more of such jump-edit behavior than lower-scoring students. These results aligned with the previous findings and existing literature in that much of the shift in behavior between novices and expert writers is linked to increases in the metacognitive knowledge that supports text evaluation, and in the willingness to revise texts in response to such evaluations (Butterfield, Hacker, & Albertson, 1996; Chanquoy, 2009).

However, it may be harder to determine why a writer made specific changes to the text without a deeper linguistic analysis and tracking of text/content changes. In some cases, such as typo-correction, the reason for the change may be self-evident. This kind of proofreading behavior is relatively easy to track in a writing process log and can easily be linked to writing scores (Deane et al., 2015). Deeper revisions may be harder to classify and may require various forms of automated analysis of content and linguistic structure (Jones, 2008).

Task Analysis and Discovery

Studying the preparatory phrases of writing, in which writers develop and understanding on the writing task and gather information they will need, is inseparable from the larger problem of understand the processes by which students develop research and inquiry skills. This process typically takes place over an extended period and cannot easily be observed unless the process takes place in a digital environment in which student actions and decisions can be recorded. A variety of digital systems have been developed for gathering such evidence, including *Digital IdeaKeeper* (Quintana, Zhang, & Krajcik, 2005), *Knowledge Integration Environment* (Linn, 2000), *WISE* (Linn, Clark, & Slotta, 2003), *Progress Portfolio* (Loh et al., 1997), *Sourcer's Apprentice* (Britt & Aglinskas, 2002), *Multiple Source Comprehension Assessment* developed for *Project READi* (Goldman & Scardamalia, 2013), *ORCA* (Coiro, Sparks, & Kulikowich, 2018a), and *SAIL* (Coiro, Sparks, & Kulikowich, 2018b).

These systems capture evidence about brainstorming processes, source finding and evaluation, note preparation, and other aspects of the discovery process. Sources of evidence about student processes may include such information as searches performed, links followed, texts read or selected for further use, notes taken during reading, among others. These studies suggest that there is considerable variation among students in their ability to deploy digital literacy skills. However, there has been relatively little work that explicitly links research and inquiry processes to other aspects of the writing process.

Variation across Groups

There are persistent differences in achievement among social and economic groups in the U.S. *Traditionally-underserved groups* (low SES and minorities) demonstrate significantly lower scores on standardized assessments and are less likely to remain in school, graduate from high school, attend college, or obtain a college degree (Bennett et al., 2004; Lacour & Tissington, 2011). This *achievement gap* is particularly large when it comes to writing (NCES, 2012). There is a similar achievement gap based on gender; in particular, female students outperform male students on standardized writing assessments (e.g., NCES, 2012). Writing process data can provide important information about these kinds of group differences, since it may, for example, provide evidence about mastery of component skills such keyboarding fluency, or reveal the extent to which different groups engaged in specific writing practices.

Thus far, relatively few studies have examined group differences in writing processes. However, the results to date suggest that these differences may be significant. Liu and Zhang (2017) found that the socioeconomic status of schools could moderate the effect of teacher professional development on student writing performance. As part of an effort to evaluate the causal effects of a teacher professional development program that intends to improve middle-school student writing, they examined the impacts of the program on not only final essay scores, but also writing process indicators extracted from keystroke logs. The authors reported significant positive effect on one writing process indicator, the amount of local and word-level editing, but the effect was significant only for high SES schools.

Guo et al. (2019) found different writing behaviors for different socioeconomic groups, and some of the differences were statistically significant even after matching for ability between subgroups. For example, both low-SES students and Black students showed significantly lower text production efficiency; that is, they produced significantly more keystrokes than the number of characters in their final texts when compared to high-SES and White students, respectively. At the same essay quality level, the low-SES and Black student groups also spent more time writing but were more likely to take significant amounts of time pausing between bursts.

Several studies have systematically analyzed gender differences, with generally consistent results. Overall, female students are more fluent than male students. Zhang, Bennett, et al. (2019) found that female students not only obtained higher average essay scores than male students, but also differed in their text production processes (see Tables 6–7 in the original article). Female students entered text more fluently, engaged in more macro and local editing, and showed less need to pause at locations associated with planning (e.g., between bursts of text, at sentence boundaries).

Zhu, Zhang, and Deane (2019a, b) further analyzed gender differences in writing process on several specific writing behaviors and found that female students spent significantly longer times on writing, used their time more efficiently, showed greater writing fluency, and conducted more jump-edit activities than the male students (see Table 2 in Zhu et al., 2019a). In Guo et al. (2019), gender differences persisted even after students were matched by overall writing score. On average, female students produced relatively longer essays (10 words longer) in less time (40 seconds shorter), typed faster, spent more time in continuous text production without long pauses, and were more likely to make short, quick edits, when they were compared to males who achieved the same essay scores.

Since process features can be used to predict final performance, and process differences can be observed across subgroups, *differential predictions across subgroups* will be an area of interest, as *fairness* is a central issue in educational assessment. This future direction is mentioned in Sinharay, Zhang, and Deane (2019). Another possible issue worth investigating is *differential feature functioning* as studied in Zhang, Dorans, Li, and Rupp (2017), since it is important to determine whether the interaction of the process features with human scores demonstrates a decent level of consistency among population subgroups.

Conclusion

As we have demonstrated with this review, writing process data makes it possible to develop enriched forms of automated writing evaluation because features of the student writing process enable us to draw inferences about critical supporting skills such as keyboarding fluency and editing as well as to evaluate how writers allocate effort between different aspects of the writing process. It may help us to achieve a better understanding on the relationship between the cognitive demands of writing tasks, the differences between novice and skilled writers, and the ways in which writers from different backgrounds or with different formative experiences may make different decisions about how to manage the writing process.

20

Automated Scoring of Extended Spontaneous Speech

Klaus Zechner and Anastassia Loukina

CONTENTS

While *automated scoring* (AS) of written constructed responses – essays in particular – has been explored for more than 50 years (e.g., Page, 1966b), AS of speech has a much shorter history. It was not until around 1990, when *automated speech recognition* (ASR) technology had reached a certain level of maturity, that first investigations into speech scoring and evaluation were published (e.g., Bernstein, Cohen, Murveit, Rtischev, & Weintraub, 1990). Early research into speech scoring focused on *predictable speech* such as reading a passage aloud or repeating a sentence (Bernstein et al., 1990). Research and development into automated evaluation of *spontaneous (i.e., less predictable) speech* was not undertaken until about a decade later (Cucchiarini, Strik, & Boves, 1997). As spontaneous speech provides a better and more authentic representation of language proficiency than predictable speech, it has more dimensions that can (and should) be evaluated by both human raters and automated systems.

Specifically, for predictable speech, the evaluation is mostly concerned with aspects of fluency, pronunciation, and prosody, in addition to *accuracy measures* such as words omitted, inserted, or substituted. For spontaneous speech, in contrast, one can also look at other *dimensions of speaking proficiency* such as range, diversity, and precision of vocabulary used; range, sophistication, and accuracy of grammatical constructions; accuracy of spoken content; quality of aspects of discourse (e.g., cohesion, coherence, organization); and quality of pragmatics (i.e., appropriateness in terms of social context and role expectations) (Zechner, Higgins, Xi, & Williamson, 2009; L. Chen et al., 2018).

Automated speech scoring systems typically have four main components: (1) an ASR system, (2) speech feature computation modules, (3) a statistical scoring model, and (4) a filtering model. Specifically, the ASR system produces a so-called "hypothesis," a string of words that the machine "thinks" was most likely spoken by the test taker based on both the properties of individual speech sounds and the likelihood of word sequences. This *ASR hypothesis* – in combination with other information extracted from the speech signal, such as pitch and energy – is then used by *feature computation modules* to compute speech features related to fluency, pronunciation, vocabulary, grammar, and content. The *statistical scoring model* then computes a score for a spoken response given a set of feature values. Finally, throughout the process, *nonscorable responses* are *filtered* using different *rule-based* or *statistical models* as it is essential to avoid scoring of responses that clearly should not obtain a regular score (e.g., an empty response).

Similar to *automated essay scoring* systems, automated speech scoring systems can be deployed operationally in a variety of scenarios. For example, in an all-human or so-called "confirmatory" scoring approach, automated speech scores can be used as one quality-control mechanism for human raters. In this approach, a second human rater could be called in for final score adjudication whenever the system identifies a human–machine score discrepancy that exceeds a pre-set threshold.

In contrast, in a so-called "contributory" scoring approach, the machine score and the human rater score could be combined by using a pre-set or empirically derived weighting function to result in a final item score. Finally, machine scores can also be used as sole scores with or without human rater review of a subset of item scores. This scenario would require that very high psychometric standards are met for deployment in high-stakes assessment and is therefore more likely to occur in low-stakes or no-stakes practice tests.

Some of the major considerations around deploying automated speech scoring technology are related to the assessment criteria of *validity, reliability,* and *fairness* that are the foundation of many *industry standards* (e.g., AERA, APA, & NCME, 2014). Put briefly, it is typically argued that an automated speech scoring system has a sufficient degree of *construct validity* if it can be empirically demonstrated that the features it uses to predict an item score measure aspects of the speaking construct that are considered relevant, are part of the human scoring rubrics, and are related to key *cognitive processes* engaged in by human raters.

In terms of reliability, the main criterion is typically the *agreement* between automated scores and both observed human scores as well as *estimated true scores* – if such estimation is possible – as well as the *precision* of any reported scores that utilize or rely on the automated scores. Finally, automated speech scores are typically considered fair if they do not exhibit any *bias* when item-level scores or higher-level aggregate scores are compared with a *"gold standard" benchmark* within different *population subgroups* defined by demographic variables such as gender, country of residence, native language, and the like.

Table 20.1 provides a few dimensions that affect the difficulty of automated speech scoring. From this table, we can see that quite a few considerations have to be taken into account at the outset when designing a speaking assessment that will be scored automatically in order to maximize its effectiveness and the likelihood that it can conform to validity, reliability, and fairness standards. These include the nature of task, the characteristics of test-taker population, and the technical aspects of test administration.

We have organized this chapter into three main content sections. In the first section, we discuss the nature of speaking tasks in a practice test that is currently scored automatically by *SpeechRater*, which we use as an example

TABLE 20.1

Some Dimensions Affecting the Scorability of Spoken Responses with Automated Speech Scoring

Dimension	Easier to score	Harder to score
Speech predictability	Highly predictable (e.g., read aloud)	Highly unpredictable (e.g., spontaneous speech)
Response based on stimulus/source	Tightly connected to stimulus materials (e.g., prompts with lectures, images, and readings)	Not connected to particular stimulus materials (e.g., prompts ask for personal knowledge, experience, or opinions)
Task interactivity	Monologic	Dialogic
Test-taker's native languages	Only one or very few native languages spoken	Many different native languages spoken
Test-taker age	Adults	Children
Test-taker proficiency levels	Mix of proficiencies	Fairly similar proficiency
Speech capture	Professional recording equipment such as close-talk noise-cancelling microphone	Consumer devices such as mobile device with built-in microphone
Overall recording environment	Low noise (e.g., quiet room, soundproof recording booth)	High and variable noise (e.g., café, street, classroom)
Voice contamination during recording	Only single speaker recorded	Potentially multiple speakers around or in background

application in this chapter. In the second section, we describe the overall system architecture and its main components in more detail. In the third section, we provide recommendations for best practices when designing automated speech scoring systems. We then close this chapter with a discussion and outlook on future research.

Task and Response Characteristics

While this chapter is about automated speech scoring in general, our specific use context for this chapter is the *SpeechRater* system developed at Educational Testing Service (ETS), which is used to provide sole scores for spoken responses in the TOEFL® Practice Online (TPO) Speaking test. The TPO test provides students who are planning to take the TOEFL iBT English language assessment with the opportunity to practice *retired* (i.e., non-operational) TOEFL iBT items in an environment very similar to the operational TOEFL iBT. While the reading and listening sections consist of multiple-choice items and, hence, can easily be scored automatically, the writing and speaking sections elicit constructed responses that are much less straightforward to score.

The speaking section of the TOEFL iBT consists of two *independent* and four *integrated* items eliciting spontaneous speech situated in an academic and college campus context. The independent items elicit responses that can be answered from a test-taker's general knowledge and experience (e.g., "Describe a person who you admire") and allow for 45 seconds of speech in the recorded response. In contrast, the integrated items consist of various reading and/or listening stimuli that the test taker needs to refer to in the spoken response and allow for 60 seconds of speech.

The six speaking items in the TOEFL iBT assessment are scored operationally by trained *human raters* on a scale from '1' (lowest proficiency) to '4' (highest proficiency) with special codes for responses that are not scorable due to technical difficulties with the recording equipment or other logistical issues. Human raters are trained to follow rubrics for both item types, which describe typical speech characteristics for each score band in three main dimensions: 'delivery' (fluency, pronunciation, prosody); 'language use' (vocabulary and grammar); and 'topic development' (content and discourse; ETS, 2018). Scores are assigned *holistically* and a section score based on all speaking item scores is computed as the sum of the six-item scores and is then finally scaled to the '0' to '30' scale score range for the TOEFL iBT Speaking section.

For TPO, the basic setup is almost identical to TOEFL iBT except that the test taker can choose between a 'timed' and an 'untimed' mode. As the names suggest, the timed mode simulates the timing of the TOEFL iBT Speaking section precisely while the untimed mode allows for extra time in preparation and/or breaks between items. Moreover, if *SpeechRater* scores a response as being non-scorable, the test taker has a total of three trials to re-submit his/her spoken response to the system.

System Architecture Design

System Overview

In this *system architecture*, a digitally recorded test-taker response ('audio file') is first sent to an ASR system and, simultaneously, to various *signal processing modules*. The ASR system generates a word hypothesis – with start and end times for each word – for the test-taker's response, and the signal processing modules generate statistics on *energy and pitch distributions* over the entire recording (Figure 20.1).

The second stage comprises a large set of feature computation modules, which take as input the outputs of the first stage (i.e., the word hypothesis as well as pitch and energy statistics) to generate features related to the various sub-dimensions of the speaking proficiency construct: features measuring aspects of fluency, pronunciation, prosody, vocabulary, grammar, content

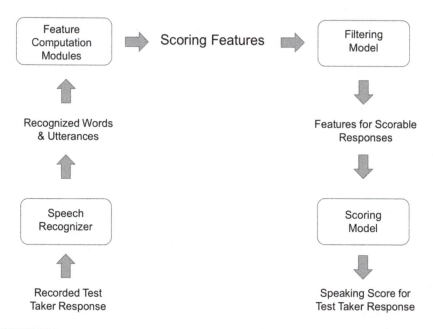

FIGURE 20.1
High-level system architecture of *SpeechRater*.

and discourse. Currently, *SpeechRater* generates more than 100 features related to the TOEFL iBT Speaking construct in all of these construct sub-dimensions as well as features that are appropriate for use in other speech assessments, including those with more restricted speaking tasks such as the *Test of English-for-Teaching* (TEFT) (Zechner et al., 2014).

The third stage is a *filtering model*, which determines, based on speech features and other information related to the test-taker's response, whether it is likely that this response can be scored by *SpeechRater*'s applicable scoring model or whether this response is likely to be non-scorable due to technical issues (e.g., high noise levels, no speech present) or other problematic issues (e.g., off-topic response); in the latter case, the response is labeled as "TD" (for "technical difficulty") and set aside.

If the response can be scored, then the *statistical scoring model* is applied in the fourth stage of processing. *SpeechRater* uses a *multiple linear regression model* to predict a score for a test-taker's response, which is trained on a *human-scored corpus* and typically undergoes *automatic feature selection* (Loukina, Zechner, Chen, & Heilman, 2015) with occasional *manual fine-tuning*. The final output of *SpeechRater* is a numeric real-valued score between 0.5 and 4.5, indicating *SpeechRater*'s prediction of the item score based on its features and scoring model. We describe all of these components in more detail later in this section.

On the one hand, the motivation for the architecture of *SpeechRater* is derived from a similar architecture of previously developed systems for AS of essays (see Lottridge & Hoefer, Chapter 11, this handbook). On the other hand, its design is driven by the necessity to first compute evidence for certain aspects of speaking proficiency by means of speech features and then to *aggregate evidence* to predict a response score. That is, unlike written texts, digitized speech recordings require an additional *pre-processing step* where an ASR system is used to generate the most likely sequence of words spoken by the test taker before feature values for the response in question can be computed.

While it is conceivable in principle to combine filtering and scoring models into a single component, keeping filtering and scoring models separate has turned out to be advantageous from a practical point of view since they can be independently trained and maintained as well as flexibly combined (e.g., some assessments may require more diverse filtering models than others).

In summary, this architecture is essentially a *linear pipeline* of four main components whose order is fixed by necessity: the ASR engine needs to run before speech features are computed; features need to be computed before filtering or scoring models can be run; and filtering models need to be run before responses can be sent (or not sent) to the scoring model.

Automatic Speech Recognition

Most speech features in *SpeechRater* are dependent on which words were spoken by the test taker, so it is essential that the words of a test-taker's utterance are first obtained via an ASR system. Such systems use two statistical models to achieve this task, the *acoustic model* and the *language model*. The acoustic model describes the statistical properties of speech sounds ("phones") in higher-dimensional acoustic space and is typically trained on a corpus in which the recorded speech was manually transcribed verbatim. The language model captures the likelihood of word sequences in English and is typically trained on a large corpus of transcribed speech and additionally, in some cases, written language.

The actual act of producing a word hypothesis using the ASR models is an *optimization process* which aims to find a word sequence that sounds as similar as possible to what it finds in the digitized recording (using the acoustic model) and is most plausible in terms of what word sequences in English look like (using the language model). The ASR system employs a search through various possibilities of word sequences ('lattice') and finally determines one hypothesis with the overall *highest probability / likelihood* of being produced by the test taker.

The acoustic and language models of an ASR system are ideally trained on a corpus with very similar characteristics to the data that is expected to be processed during actual system operation. The larger the amount of training data, the higher the performance of the ASR system will be. For many decades, the acoustic models of ASR systems were based on *Gaussian mixture models*; in recent years, these models have been, for the most part, replaced with more powerful *deep neural networks* that led to substantial improvements in performance (Cheng, Chen, & Metallinou, 2015; Khan & Huang, Chapter 15, this handbook).

In automated speech processing research, ASR performance is usually measured using *word error rate* (WER), which is computed based on *automated string alignment* of the ASR's word hypothesis and the reference transcription produced by a *human transcriber* (i.e., the words that the system produces compared to the words that a human transcriber produces). Specifically, WER is defined as the sum of all *alignment errors* (i.e., word substitutions, insertions, and deletions) *normalized* by the length of the reference (human) transcription. For spontaneous non-native speech, WERs above 50% in the past had been typically in the 2000s (Zechner et al., 2009); more recently, when using deep neural network-based ASR systems, researchers were able to obtain WERs in the range of 20% to 30% (Tao, Ghaffarzadegan, Chen, & Zechner, 2016; Cheng, Chen, & Metallinou, 2015), but this is dependent on the particular use context.

For non-native spontaneous speech, WERs have been found to be around 15% when comparing pairs of human transcriptions (Evanini, Higgins, & Zechner, 2010; Zechner, 2009); these results suggest that it may not be possible to improve ASR WER much below 15% on average for this particular type of spoken input. Still, compared to native speech ASR where WERs lower than 5% have been observed, this ASR performance is substantially worse, which is mainly related to the diverse population of test takers in terms of native languages and proficiency levels.

Finally, there are three other factors that impact the processing of spoken input by an ASR system and make it different from the processing of written text for other automated systems. First, ASR word hypotheses do not contain punctuation; therefore, *clause boundaries* have to be inserted post-hoc for features requiring clauses as input. Second, the vocabulary of an ASR system is always limited, so some words spoken by a test taker may not be in the ASR dictionary and will be misrecognized by the system, which can be particularly an issue for proper names. Third, because of the influence of the language model within the ASR system, some errors by the test taker can be hyper-corrected by the ASR system; for instance, an incorrect preposition could be replaced by a correct one – in particular if the two prepositions sound alike – or a missing determiner might be inserted by the ASR if the language model predicts a high likelihood for this short word in a given context.

Computation of Speech Features

The architecture of the *SpeechRater* system was conceived to align with different dimensions of language proficiency in general and the human scoring rubrics for speaking assessments in particular while aiming to reduce the system's susceptibility to factors that produce *construct-irrelevant variance* (Zechner & Bejar, 2006). For example, each feature is expected to have a clear interpretation and to have a solid grounding in research in second language acquisition. As we mentioned above, *SpeechRater* currently extracts more than 100 features developed in close collaboration with experts in English language teaching and learning, which provide extensive coverage of the three dimensions of language proficiency considered critical by the human raters: 'delivery', 'language use', and 'topic development'.

Features related to 'delivery' cover general fluency, pronunciation, and prosody. The aspect of fluency is measured through the general speaking rate as well as features that capture pausing patterns in the response such as mean duration of pauses, mean number of words between two pauses, and the ratio of pauses to speech production. Pronunciation quality is measured using the average confidence scores and acoustic model scores computed by the ASR system. The final group of delivery features evaluates a speaker' use of prosody by measuring patterns of variation in time intervals between stressed syllables as well as the number of syllables between adjacent stressed syllables (L. Chen & Zechner, 2011a; Lai, Evanini, & Zechner, 2013) and variation in the durations of vowels and consonants (Sun & Evanini, 2011).

To improve the validity of these features, *SpeechRater* uses a *two-pass system* (L. Chen, Zechner, & Xi, 2009; Herron et al., 1999); under this approach, the spoken responses are first transcribed using an ASR system trained on speakers with different proficiency levels. Next, a separate ASR system trained on proficient English speakers is used to compute *acoustic similarity scores* between native pronunciation and a speaker's pronunciation. While such an approach increases the time necessary to compute the score, it increases the validity and accuracy of the final scores. Put differently, the ASR system trained on non-native speakers ensures higher transcription accuracy while the ASR system trained on native speakers provides a more valid model for evaluating a test-taker's pronunciation.

Features related to aspects of 'language use' cover vocabulary and grammar. Vocabulary-related features include the average log of the frequency of all content words and a comparison between the response vocabulary and several *reference corpora* (Yoon, Bhat, & Zechner, 2012). Grammar is evaluated using a *content vector analysis* comparison computed based on part-of-speech tags, a range of features which measure occurrences of various syntactic structures and the language model score of the response (M. Chen & Zechner, 2011a, b; Yoon & Bhat, 2018). Finally, the most recent version of *SpeechRater* also includes features related to 'discourse structure' and 'topic development' with further work under way (Wang, Bruno, Molloy, Evanini, & Zechner, 2017).

One of the greatest challenges in developing features for measuring proficiency in spontaneous speech is the fact that the content of test-taker responses is not known in advance. For example, many approaches to automatic evaluation of pronunciation require prior knowledge of the words and phones uttered by the speaker. For spontaneous speech, such information has to be obtained automatically using ASR. As we discussed earlier in this chapter, automated – and even human – transcriptions of non-native speech are likely to contain at least 10% to 15% of incorrect words. As a result, test-taker pronunciation may be flagged as incorrect simply because it is compared to a wrong reference model due to an error in automatic transcription.

While substantial work exists on measuring grammar and vocabulary in essays, this research is not always directly applicable to spoken responses (Loukina & Cahill, 2016). First of all, not all such features are robust to ASR errors. Some of the grammatical errors or incorrect word choices could be an artifact of inaccurate ASR transcriptions. In other cases, the ASR system may erroneously "correct" an existing error by a test taker. Second, many grammatical features rely on knowledge of sentence boundaries in order to parse the response into syntactic constituents. In written responses, the sentence boundaries can be established based on punctuation. In spoken responses, however, these have to be estimated using *machine learning algorithms* such as the ones described in L. Chen and Yoon (2011). Furthermore, sentence boundaries in speech are often ambiguous. All of these factors may lead to a decrease in feature performance.

Third, spoken and written discourse differ in what is considered appropriate in terms of language use (Biber & Gray, 2014; Chafe & Tannen, 1987). For example, sentence fragments are typically considered inappropriate for written language but are generally quite common in unscripted spoken responses. Finally, spoken responses such as those submitted to TPO or TOEFL iBT tend to be much shorter than an essay – usually around 100 words – which can be insufficient for obtaining reliable measurements using standard approaches developed for essay scoring (Crossley & McNamara, 2013). For these reasons, the development of *SpeechRater* features for the dimensions of 'grammar', 'vocabulary', and 'topic development' requires substantial additional research to adapt the existing essay scoring algorithms to speech contexts or to develop new algorithms especially for the spoken modality.

Scoring Model

A scoring model defines what features are used to compute the final score, what the relative contribution of each feature is, and what procedures are used to derive the score. In all operational versions of *SpeechRater* deployed so far (L. Chen et al., 2018; Zechner et al., 2009), the features are combined into a *linear equation* where each feature is assigned a certain *weight* and the weighted values are *summed* to obtain the final score. With this scoring setup, the final score can be traced back to individual feature values and should only be affected by considerations immediately related to language proficiency.

The scoring models for recent versions of *SpeechRater* were built using the *hybrid method* of feature selection described in Loukina et al. (2015). In this approach, an expert first identifies a subset of *SpeechRater* features that are applicable to a given task type. A combination of *Lasso regression* (i.e., a type of regression that favors simple / sparse models with small numbers of parameters) and *expert judgment* is then used to identify the optimal set of features using the set of responses designated as the model training set. The coefficients / model parameters for the selected features are then estimated using *non-negative least-squares linear regression*. Finally, the model performance is evaluated on a separate set of responses (i.e., held-out or evaluation set) that were not used for model training or fine-tuning (i.e., there is no overlap of test takers between the training and evaluation sets).

As of 2019, the current operational scoring model deployed for TPO uses 13 features. The model includes nine features that cover various aspects of delivery and account for 64% of the final score as well as four features related to language use and topic development. Specifically, the *Pearson product-moment correlation* between section-level score computed based on human scores and those computed based on automated scores is $r = 0.79$ ($R^2 = 0.62$) compared to $r = 0.88$ ($R^2 = 0.77$) for agreement between two human raters.

Agreement with human scores is by far not the only criterion considered when evaluating *SpeechRater* scoring models. As laid out in Williamson, Xi, and Breyer (2012) for example, AS models are evaluated on several dimensions including construct validity as well as subgroup differences. In addition to subgroups considered for all AS engines such as groups defined by test takers' native language or gender, an additional group considered in the context of speech scoring is speakers who exhibit signs of speech impairments such as stutter to ensure that the automated scores provide a valid estimate of their proficiency as well (Loukina & Buzick, 2017). Yan and Bridgeman (Chapter 16, this handbook) discuss details of AS validation.

As can be seen from this section, building and evaluating *SpeechRater* scoring models is a complex process that requires experts from many fields including English language learning, AS, and psychometrics. This work requires special support systems to manage the computational aspects for evaluation analyses. At ETS we developed a customized program (*RSMTool*) to streamline the many steps involved in making sure all procedures are followed correctly for each new application (for more details, see Madnani, Loukina, Davier, Burstein, & Cahill, 2017).

Filtering Model

The scoring model of *SpeechRater* makes a few assumptions about the properties of the speech input. For example, it assumes that the response contained speech, that it was produced in a good faith effort by the test taker, and that the features generated as input for the scoring models are all available and computed correctly. Yet, in operational applications, some responses submitted to actual spoken language proficiency assessments have atypical characteristics that violate the aforementioned assumptions and, hence, may lead to inaccurate scores if scored with the existing scoring model.

Yoon et al. (2018) distinguishes two types of such non-scorable responses: system-initiated and user-initiated. For *system-initiated non-scorable responses*, critical system errors during test administration or AS make it impossible to assign fair scores. As we alluded to earlier, these can occur at the response recording stage where microphone failure or other equipment malfunction meant that no sound was captured (i.e., the recording is silent) or the recording has serious audio quality problems such as distorted speech or constant noise that obscures a candidate response.

However, technical difficulties may also occur at other stages of the AS pipeline beyond the recording. As the discussion so far has suggested, automated speech scoring systems use a wide range of natural language processing technologies such as automated parsers for syntactic structure analysis used for evaluating syntax and grammar. Each of these components may fail to produce a valid output during feature generation resulting in feature values that are not representative of test-taker proficiency (see also Yoon & Zechner, 2017).

In contrast to such system-initiated non-scorable responses, *user-initiated non-scorable responses* contain sub-optimal characteristics due to test-taker behavior. These include responses that are entirely not relevant to the task, responses that were memorized from external sources, or responses in a language other than the target language. An automated speech scoring system that is not configured to detect such responses may assign an inappropriately high score to test takers who used one of these gaming strategies since their speech may be otherwise fluent and similar in its feature values to that of genuinely proficient test takers.

To address the issues above, *SpeechRater* has always included a separate filtering model that is used to flag non-scorable responses (Higgins, Zechner, & Williamson, 2011). In low-stakes contexts such as TPO, the majority of non-scorable responses are ones with low audio quality or for which the test-taker produces no speech. In contrast, gaming behaviors are extremely rare (Yoon & Zechner, 2017). Therefore, the filtering models used for TPO are configured to detect these two types of non-scorable responses.

Specifically, the baseline filter uses various audio quality features and a statistical *decision tree classification model* to identify responses that would be scored as 'technical difficulty' or '0' by human raters. In addition, several rule-based filters flag responses with characteristics that are likely to lead to invalid scores such as a low confidence of the ASR hypothesis, a very short ASR hypothesis, or extremely loud or quiet recordings. These filters correctly identify 90% of non-scorable responses annotated by human raters at the time of this writing.

While only these two types of filters are used in operational *SpeechRater* as of 2019, research has also shown that it is possible to use natural language and speech processing technologies to accurately identify other types of user-initiated non-scorable responses such as responses with off-topic content (Yoon & Xie, 2014; Yoon et al., 2017), canned responses (Evanini & Wang, 2014; Wang, Evanini, Bruno, & Mulholland, 2017), as well as non-English responses (Yoon & Higgins, 2011). Although rare in the TPO context, these types of responses may be more likely to occur in high-stakes contexts such as TOEFL iBT and thus would need to be identified by the filtering model, particularly if *SpeechRater* were ever to be intended for use as a sole score.

An important consideration when designing all such systems is the trade-off between the accuracy of the automated scores and the percentage of *scoring failures*. Yoon et al. (2018) consider an example from a recent version of *SpeechRater* where the ASR system showed a very high WER for a small number of responses which would have resulted in highly inaccurate scores. To prevent this, a new filter was added to the filtering module to flag such responses as non-scorable. This filter increased the performance of the overall system as measured by agreement between automated and human scores. However, it also increased the percentage of responses flagged as non-scorable, which meant that more test takers were asked to re-submit their responses before they could receive the final score.

Finally, once non-scorable responses have been identified by the filtering model, a decision needs to be made how these responses should be scored given that they will not be sent to the standard AS model pipeline. Yoon and Zechner (2017) describe a hybrid scoring scenario that combines human and AS with filtering models in the context of a low-stakes English language proficiency assessment. A set of filtering systems were used to automatically identify various classes of non-scorable responses, and these responses flagged by the filtering models were then routed to and scored by human raters.

Yoon et al. (2018) provide an overview of other possible scenarios such as when no score is reported to a test taker or if a score may be reported with a warning in a low-stakes context. Specifically, in TPO, a score from a single non-scorable response is imputed as a median of the other five scores; if two or more responses are deemed non-scorable, no final score is reported to the test taker.

Feedback

Until recently, TPO test takers only received a single score for all six responses they submitted to *SpeechRater* plus a *prediction interval* that indicated the expected range of human scores for their submissions. A small-scale user study by English language learners and their teachers showed that there was interest in obtaining additional information about a test-taker's spoken proficiency beyond a single numeric score (Gu et al., submitted). Such diagnostic information could provide test takers with a more detailed perspective on their differential performance across different areas of the speaking construct and could also further motivate, focus, and inform their English language learning activities.

Therefore, since 2018 *SpeechRater* produces a feedback report in addition to the score report for TPO, which contains performance information derived from a few *SpeechRater* features across several sub-dimensions of speaking proficiency. Selection criteria for the features included their empirical performance (e.g., correlations with human holistic scores), their interpretability by non-experts, their discriminatory properties across the range of speaking proficiency, and the extent to which they may inform subsequent language learning activities. For this reason, the features selected for feedback may be different from those used in *SpeechRater*'s scoring model where different feature selection criteria apply.

Table 20.2 presents the seven features which were eventually chosen for this feedback report, along with their descriptions for the test takers that are shown on the feedback report.

The operational feedback report also provides a graph for each feature indicating the *percentile standing* of the test taker, which represents the average value across all six TPO responses of a particular feature relative to a reference corpus of TOEFL iBT test takers. In addition to the feature percentile, four colored bars are displayed in each feature graph representing the 25th-to-75th percentile ranges (i.e., the *interquartile ranges*) for four scaled score ranges that represent learners with low to high proficiency, again computed based on the TOEFL iBT reference corpus.

The idea is that test takers should more easily be able to interpret the meaning of their specific feature percentile since they can place it in the context of the performance of other test takers with different ability levels. Finally, the language in the feedback report encourages test takers to work with their tutors and teachers to decide on their language learning activities based on

TABLE 20.2

Features Displayed in the TPO Speaking Feedback Report

Feature/Sub-construct	Explanation for Test Takers
Speaking Rate (Fluency)	Speaking Rate is a measure of how many words you speak per minute. Stronger speakers tend to speak faster. But be careful: if you speak too fast, it may be difficult for others to understand you.
Sustained Speech (Fluency)	Sustained Speech is a measure of the average number of words you say without: (1) pausing or (2) using a filler word such as "um." Stronger speakers tend to say more words without pausing or using a filler word.
Pause Frequency (Fluency)	Pause Frequency is a measure of how often you pause when speaking. Stronger speakers tend to pause less frequently. But, keep in mind that other aspects of pausing are also important. For example, pausing at the end of a sentence is better than pausing in the middle of an idea.
Repetitions (Fluency)	Repetitions is a measure of how often you repeat a word or phrase, as in "I need to go to, to go to the library." Stronger speakers tend to have fewer repetitions.
Rhythm (Pronunciation)	Rhythm is a measure of whether syllables are stressed appropriately. Stronger speakers tend to put clear stress on appropriate syllables.
Vowels (Pronunciation)	Vowels is a measure of how you pronounce vowels compared with the pronunciation of a native speaker. Specifically, it is a measure of vowel length. Stronger speakers tend to pronounce English vowels more like a native speaker.
Vocabulary Depth (Vocabulary)	Vocabulary Depth is a measure of your vocabulary range. Stronger speakers tend to use a variety of words. A higher score indicates that you probably used words that are less common and/or are more precise.

the obtained feedback report. The report also contains links to useful language learning resources that test takers may review to help improve their English speaking proficiency.

Recommendations for Best Practices

In this section, we discuss a few best practices to guide colleagues who may wish to develop automated speech scoring systems.

Creation of a Shared Understanding of Speech Scoring Capability

Whenever a potential new deployment of speech scoring technology is considered, it is very important to establish an understanding of the currently available speech scoring capabilities in the field. This includes feasibility of

AS of various items or item types in the assessment, the extent to which sub-construct areas are represented by automatically extracted features, or the capability to automatically identify pertinent classes of non-scorable responses. This discussion should involve interdisciplinary teams consisting of representatives from the testing program, assessment development, English language learning, psychometrics and statistics, and AS. These colleagues work under different constraints and bring complementary as well as conflicting perspectives to the table that need to be resolved in order to make the AS deployment successful at scale. Put differently, rather than discussing AS in the abstract, the AS capabilities have to be placed in the context of a likely use case scenario that includes the contribution of the automated score to the final score reported to the test taker and other score users. These considerations affect design decisions for the systems as well as associated evaluation procedures with their embedded standards for acceptability in light of reliability, validity, and fairness of assessments.

Integration of Automated Speech Scoring into the Assessment Development Cycle

There is a substantial advantage for using automated speech scoring in language assessment if considerations of automated speech scoring are brought to the table as early as possible in the assessment design process. If automated speech scoring systems are designed for assessments that were developed without any consideration about automated speech scoring, it is likely that some aspects of the targeted construct – as represented in the scoring rubrics – and some core aspects of certain items or item types will pose substantial challenges for automated speech scoring systems since they exceed the current state-of-the-art.

However, if considerations about realistic capabilities of automated speech scoring systems are brought to the table early in the assessment design process, it is more likely that items and task types that lend themselves more naturally to AS will be designed properly from the outset. In cases where certain rubric dimensions are indispensable for the purpose of the speaking assessment, but automated speech scoring technology is likely unable to provide sufficiently meaningful evidence for them, a hybrid scoring system should be considered with human and machine scores both contributing to a final item score in complementary ways.

As an example of a more superficial design choice to support the use of automated speech scoring, we found that automated speech scoring systems typically perform better if test takers are provided with a brief period of preparation time before their response begins. For items without such preparation time, the initial part of a test-taker's response in particular may include substantially more hesitations and pauses since there is more planning going on during a spoken response rather than before it. This may result in a reduction in feature reliability by the automated speech scoring system.

Design of a Modular and Flexible System Architecture

An automated speech scoring system contains multiple complex components and, if it is to support a variety of different speech assessments with varying item types and target populations, it is imperative that it be designed in a highly *modular* and *flexible* way. For instance, based on the target population, different acoustic models may be used by the ASR system; different scoring models and filtering models may be needed for different assessments; and different sets of features may need to be included in the scoring system dependent on the sub-constructs targeted by the various item types of the speech assessments in question. Over time, software systems tend to become more complex, so the more flexible and modular the system architecture of a speech scoring system is, the less likely it will be that the entire system or large parts thereof will need to be re-designed based on the needs of a new application.

Evaluating Operational Scalability

When deploying automated speech scoring systems in the real world, many issues need to be considered ahead of time. One aspect concerns the run-time behavior of the scoring system: ideally, a response should be scored at least in *real time* (i.e., at most one minute of processing time for a one-minute response) so that AS does not take much longer than a human rater would. In addition, costs for server hardware need to be limited and memory consumption should be bound to manageable magnitudes.

Other considerations concern provisions for any *item-dependent processes* (e.g., reference texts for read-aloud passages or item-type-specific language models). Some of these parameters and input data may be known ahead of operational launch time but others may be known only a few days before an operational administration when a decision has been made about which items to use for this particular operational administration. In the latter case, efficient computational architectures with appropriate checking and error handling need to be in place to update the AS system in time before the administration so that AS can be done with all needed resources in place.

Planning for Regular Updates and Upgrades

Once an automated speech scoring system is operational, a plan should be in place to monitor the quality of automated scores continuously and automatically, for example by using human rater check scores. It is also a good idea to plan regular updates and upgrades of the AS system that allow for the integration of newly developed features, updates of the scoring and/or filtering models based on changes in population and test-taker behavior, and the like. Ideally, a standardized data set should be used for such regular system updates, which can be done, for example, by using a sample of test takers from a select number of previous years on a rolling basis. This ensures that system evaluations, as well as models trained for system upgrades, are based on recent test-taker population characteristics.

Conclusion

In this chapter, we discussed general principles for the design and implementation of automated speech scoring systems via a case study of the *SpeechRater* system, which is in operational use for the TPO Speaking section. We described the overall system architecture and its main components and discussed a number of issues when automated speech scoring is used operationally for speaking assessments from a practitioner's point of view.

Many aspects of the discussion are similar in nature to those of other constructed response scoring systems – most closely essay scoring systems. However, one of the main differences between scoring speech and scoring written text is the fact that the words spoken by the test taker are not available directly from the response unlike the words typed by the test taker in case of automated essay scoring. Thus, the essential and most distinguishing component of *SpeechRater* is its ASR system that converts the speech signal recorded in a test-taker's spoken response into a sequence of words, which can then be further processed to determine a wide array of linguistic speech features that measure various aspects of spoken proficiency.

Current state-of-the-art speech scoring systems are able to generate reliable scores for speaking assessments; however, their *construct coverage* is typically much more comprehensive for items with predictable responses (such as read-aloud items) than items with spontaneous speech responses (such as TOEFL iBT Speaking items). Thus, improved construct coverage in areas such as content and discourse is essential for increased validity of automated speech scoring for spontaneous speech. While progress has been made in recent years, measuring aspects of topic development concisely and accurately is likely going to remain a challenge for AS of constructed responses for some time to come since it ultimately would involve a deep understanding of the propositional content rendered in an essay or a spoken response.

Furthermore, improvement in scoring and filtering models may lead to additional improvements in the reliability, validity, and fairness of the information produced by such automated speech scoring systems. This could be accomplished, for instance, by reducing score differences for key population subgroups or by being able to automatically flag responses that are off-topic or plagiarized more accurately so as to route them to human raters for a secondary review.

Similarly, even though the WER of *SpeechRater* on TOEFL responses was reduced from around 70% in 2006 to around 20% in 2017 (when recognizing responses to prompts that were seen in the ASR training data), further ASR improvement is both possible and necessary for improved feature accuracy, particularly when considering sub constructs such as vocabulary, grammar, content, or discourse where correct word identities are essential for high-quality features. Compared to the near-perfect ASR performance that has been obtained recently on some data sets consisting of native speech, it is likely to remain a challenge for some time to achieve a similar level of performance on non-native speech, in particular when the system is required to process test takers from many different language backgrounds and proficiency levels.

While we have only dealt with the scoring of open-ended spontaneous speech in this chapter, speaking assessments may include more restricted speaking items such as reading a passage aloud, repeating a sentence presented acoustically, or building a sentence based on a list of words. *SpeechRater* was once operationally used for the TEFT speaking assessment, which included predominantly highly- and moderately-predictable item types (Zechner et al., 2014) and, as a result, is able to also generate a number of features that are geared toward these more predictable speaking items (e.g., determining how close the test-taker's response is to the reference text in a read-aloud item).

We also want to point out that recent advances in machine learning related to deep neural networks make it theoretically feasible to build end-to-end speech scoring systems where the digitized spoken response is used as input and a speaking proficiency score is directly generated at the output without using any intermediate representations such as ASR outputs, signal processing outputs, or linguistic speech features. Research in this area is still preliminary but promising (see L. Chen et al., 2018); it is conceivable that the performance of such black-box end-to-end systems could reach or even surpass that of traditional speech scoring systems described in this chapter.

Certainly, questions of score validity and the explanatory power of such a model would loom large. Whether and to what extent it would be possible to link internal representations of such deep neural networks to linguistically meaningful speech features to convey to test takers and other stakeholders how key aspects of targeted competencies are captured is currently also an open research question. Nevertheless, there may be some low-stakes applications (e.g., in the formative language learning area) where such opaque but highly effective scoring systems could be deployed.

Furthermore, it is also conceivable that a hybrid approach, whereby a traditional speech scoring system such as *SpeechRater* is combined with a deep neural network scoring system, would yield even superior performance without losing its overall validity and explanatory power. Unlike for summative scoring, however, it is hard to imagine how such deep neural networks would be able to generate useful feedback outputs for language learners; the use of more traditional approaches in feature development may need to continue here for quite some time to come.

21

Conversation-Based Learning and Assessment Environments

Arthur C. Graesser, Xiangen Hu, Vasile Rus, and Zhiqiang Cai

CONTENTS

Starting in 1997, the *Institute for Intelligent Systems* (IIS) at the University of Memphis developed a number of *learning environments with conversational agents* that attempted to facilitate learning and assessment by holding a conversation in natural language. Most assessment systems give the student or test taker one chance to provide an answer to a question or a response to a prompt, as in the case of automated assessments of essays or answers to open-ended questions. In contrast, *conversation-based learning and assessment systems* have multiple conversational turns between the human and system on each task (e.g., problem, main question, item) so more detailed answers evolve iteratively and more information is available for assessment over the course of the activity.

The first system of this kind was *AutoTutor* (Graesser, 2016; Nye, Graesser, & Hu, 2014), which helped college students learn *science, technology, engineering, and mathematics* (STEM) subject matters; Figure 21.1 presents an example screenshot of *AutoTutor* on the topic of computer literacy. In this figure,

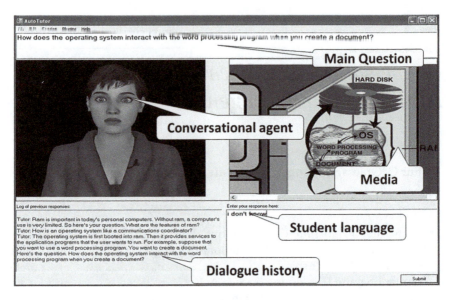

FIGURE 21.1
Interface of *AutoTutor* with a task targeting skills in computer literacy.

different parts of the screen show windows with information that refer to the main question being asked, the conversational agent, the media information available for review (i.e., information on computer components for this task), the dialogue history, and input capabilities.

AutoTutor presents difficult questions or problems that require the learner to produce one to seven sentences for an ideal, expected answer and engages them in a multi-turn tutorial dialogue until a correct response is covered. Sometimes learners express errors and misconceptions in natural language, which are corrected whenever *AutoTutor* detects them. *AutoTutor* has a *talking head* with speech generation, facial expressions, and some gestures. Some versions of *AutoTutor* have used *automated speech recognition* (D'Mello, Dowell, & Graesser, 2011; Zechner & Loukina, Chapter 20, this handbook) while others have been sensitive to the learners' emotional states through *multimodal analysis* focusing on multiple communication channels such as natural language, facial expressions, and body posture (D'Mello & Graesser, 2012).

Systems with conversational agents like *AutoTutor* have evolved over the past two decades in the IIS on a broad set of topics: computer literacy, physics, biology, research ethics, scientific reasoning, comprehension strategies, electronic circuits, medical reasoning, and group problem-solving (Graesser et al., 2018; Nye et al., 2014; Rus, D'Mello, Hu, & Graesser, 2013). These conversation-based *intelligent tutoring systems* (ITSs) with agents have been shown repeatedly to help students learn. Specifically, *AutoTutor* and other systems developed in the IIS showed learning gains of approximately 0.80 standard deviations (between 0.40 and 2.00) compared with reading a textbook for an equivalent amount of time (Graesser, 2016).

The number of agents interacting with the student has been an important consideration in the design of agent-based systems. In a *dialogue*, a single agent interacts with a single human, as in the case of the original *AutoTutor*. In a *trialogue*, there is a three-party conversation among two computer agents and a human learner. The two agents take on different roles, but often serve as tutors and peers of the learner. There are alternative trialogue designs that address different pedagogical goals for particular classes of learners (Graesser, Li, & Forsyth, 2014; Graesser, Forsyth & Lehman, 2017).

For example, learners can observe two agents interacting so that the learner can model the actions and thoughts of an ideal agent. Agents can argue with each other over issues and ask what the learner about the argument (D'Mello, Lehman, Pekrun, & Graesser, 2016; Lehman & Graesser, 2016). A tutor agent can pit a learner against a peer agent in a game scenario in order to enhance motivation (Millis, Forsyth, Wallace, Graesser, & Timmins, 2017). It is also possible for a single agent to communicate with a small group of learners as in the case of *AutoMentor*, which was designed to facilitate group learning and problem-solving (Morgan, Keshtkar, Graesser, & Shaffer, 2013).

Agent-based systems have also been used in assessment in addition to learning. In particular, components of *AutoTutor* trialogues have been developed at Educational Testing Service (ETS) in assessments of reading, writing, science, and mathematics (Zapata-Rivera, Jackson, & Katz, 2015). Collaborative problem-solving has been assessed by *tetralogues* that have two agents interacting with two humans (Liu, Von Davier, Hao, Kyllonen, & Zapata-Rivera, 2015). Agents are also used in the *Programme for International Student Assessment* (PISA), an international assessment of collaborative problem-solving in 2015 (Fiore, Graesser, & Greiff, 2018; Graesser et al., 2018; OECD, 2013).

Although agent-based systems have been used in assessment contexts, in this chapter we focus on the use of systems like *AutoTutor* in the context of tutoring. We have organized the chapter into two sections. In the first section, we discuss the structure of *AutoTutor* conversations whereas in the second section we discuss methodological challenges in developing *AutoTutor* lessons.

The Structure of *AutoTutor* Conversations

The structure of most *AutoTutor* systems is quite similar. The easy part involves choreographing the agents and associated media (e.g., diagrams, tables, images, animations) in a rigidly scripted delivery of information. This is needed to present didactic information and to introduce topics or instructions to the student. The more difficult part, essentially the essence of all agent-based systems like *AutoTutor*, is to design a flexible conversation that tracks what the student knows as the conversation proceeds and adaptively generates *discourse moves* to advance the learning process. In the following, we address the adaptive components of *AutoTutor* and similar agent-based systems such as *DeepTutor* for physics (Rus, D'Mello, Hu, & Graesser, 2013), *ElectronixTutor* for basic electronic circuits (Graesser et al., 2018), and *Operation ARA* for scientific reasoning (Millis, Forsyth, Wallace, Graesser, & Timmins, 2017).

Expectation and Misconception-Tailored Dialogue

In flexible conversations, *AutoTutor* presents to the student a series of main questions to answer (or problems to solve) that require reasoning and that cover a set of sentence-like *conceptual expressions* (e.g., semantic propositions, claims, main clauses) in an ideal response. *AutoTutor* attempts to get the student to articulate the answer (or solution to the problem) through *collaborative constructive conversation* over several turns.

There is a systematic conversational mechanism called *expectation and misconception-tailored (EMT) dialogue* (or alternatively a trialogue) that guides the process. Specifically, a list of *expectations* (e.g., anticipated good answers, steps in a procedure) and a list of anticipated *misconceptions* (e.g., bad answers, incorrect beliefs, errors, bugs) is associated with each main question. As the students articulate their answers over multiple conversational turns, the contents of their contributions are compared with the expectations and misconceptions.

As one might expect, students rarely articulate a complete answer in the first conversational turn. Instead, their answers are spread out over many turns as the tutor generates hints and other conversational moves to enable the students to express their knowledge. The students' answers within each turn are typically short, vague, ungrammatical, and not semantically well formed. *AutoTutor* compares the students' verbal content to expectations and misconceptions through *pattern-matching* processes with semantic evaluation mechanisms that are motivated by research in *computational linguistics* (Cai et al., 2011, 2016; Rus et al., 2013; see Cahill & Evanini, Chapter 5, this handbook). Listed below are tutor dialogue moves that frequently occur in the EMT dialogues in *AutoTutor*:

Main Question or Problem This is a challenging question or problem that the tutor is trying to help the student answer.

Short Feedback The tutor provides feedback to the student's previous contribution that is either *positive* ("yes," "correct," head nod), *negative* ("no," "almost," head shake, long pause, frown), or *neutral* ("uh huh," "okay").

Pumps The tutor gives nondirective pumps ("Anything else?" "Tell me more.") to get the student to say more or to take some action.

Hints The tutor gives hints to get the students to talk or take action by indirectly nudging the students along some conceptual path. Hints vary from generic statements or questions ("What about X?", "Why did that occur?", "How is X related to Y?") to statements that more directly steer the student toward a particular answer.

Prompts The tutor asks a leading question to get the student to articulate a particular word or phrase. Sometimes students say very little so these prompts are needed to get the student to say something specific.

Prompt Completions The tutor expresses the correct completion of a prompt.

Assertions The tutor expresses a fact or state of affairs.

Summaries The tutor gives a recap of the answer to the main question or solution to the problem.

Mini-Lectures The tutor expresses didactic content on a particular topic.

Corrections The tutor corrects an error or misconception of the student.

Answers The tutor answers a question asked by the student.

Rather than simply lecturing to the student, however, the tutor provides *conversational scaffolding* for the student to articulate the expectations through a well-designed sequence of dialogue moves. Specifically, *AutoTutor* provides a cycle of *pump -> hint -> prompt -> assertion* for each expectation until the expectation is covered; the cycle is discontinued as soon as the expectation is articulated by the student or the cycle is exhausted. Consequently, an excellent assessment of what the student knows is reflected in the distribution of the dialogue moves that *AutoTutor* generates (Jackson & Graesser, 2006).

More knowledgeable students tend to receive *AutoTutor* pumps and hints whereas low-knowledge students tend to receive *AutoTutor* prompts and assertions. As the student and tutor express information over many turns, the list of expectations is eventually covered and the main task is completed. Moreover, the student may articulate misconceptions during the multi-turn tutorial dialogue. When the student content has a high match to a misconception, *AutoTutor* acknowledges the error and provides correct information. Low-knowledge students have a comparatively high incidence of *AutoTutor* corrections (Jackson & Graesser, 2006).

Structure of Tutor Turns

Most tutor turns have three informational components. The first component is *short feedback* (positive, neutral, negative) on the quality of the contribution in the student's last turn. The second component is a *dialogue advancer* that moves the tutoring agenda forward with either pumps, hints, prompts, assertions with correct information, corrections of misconceptions, or answers to student questions. The third component, *floor shifts*, manages the conversational turn-taking by ending the turn with a question, gesture, or other cue for the student to "take the floor" and provide information.

Segmenting and Classifying Speech Acts

AutoTutor needs to interpret the natural language of the student in order to assess what the student knows, give feedback, and intelligently advance the conversation. As in all dialogues, the tutor and student take turns taking the conversational floor. *AutoTutor* segments the information within each student turn into so-called *speech acts* (i.e., utterances that serve particular functions in communication) and classifies the speech acts into different categories. Early versions of *AutoTutor* had only six categories of student contributions: questions, short responses (e.g., "yes", "okay"), statements, metacognitive expressions (e.g., "I'm lost", "Now I understand"), meta-communicative expressions (e.g., "What did you say?"), and expressive evaluations (e.g., "This is frustrating", "I hate this material").

More recent versions such as *DeepTutor*, whose advances have been fueled by automated analyses developed by Rus and his colleagues, have had a more fine-grained classification of student contributions (Rus, Moldovan, Graesser, & Niraula, 2012). The accuracy of these systems is moderately impressive, but not perfect. Fortunately, a high percentage of student turns have only one or two speech acts so the segmentation problem is manageable. Regarding the performance on speech classification, the automated classifier is approximately 50% to 90% accurate (above chance) when compared to the classification of human raters (Olney et al., 2003; Samei, Li, Keshtkar, Rus, & Graesser, 2014). *AutoTutor* uses these classifications to respond adaptively to the content of the student's previous turn. For example, when the student asks a question, *AutoTutor*'s response is very different than when the student expresses statements and short answers.

Semantic Analysis of Student Statements

The student speech acts that are classified as statements are analyzed semantically by *AutoTutor* when assessing their knowledge of the subject matter in the EMT dialogue. Their knowledge is classified as 'high' whenever their statements have a close match to the expectations but not the misconceptions. Their knowledge is classified as 'low' whenever their statements are vague, have little overlap with expectations, and/or have higher overlap with misconceptions.

For illustration, consider the physics problem below:

PHYSICS QUESTION: If a lightweight car and a massive truck have a head-on collision, upon which vehicle is the impact force greater? Which vehicle undergoes the greater change in its motion, and why?

There are three expectations and two misconceptions associated with this physics problem; E1 is one of the three expectations and M1 is one of the two misconceptions:

E1: The magnitudes of the forces exerted by A and B on each other are equal.
M1: A lighter object exerts no force on a heavier object.

The statements expressed by the students in the tutoring exchange are semantically matched to expectations and misconceptions like these. The resulting *semantic match scores* are computed on a scale from 0 to 1; the score after each student turn is based on the cumulative set of statements the student expresses up to that point in the conversation for the main question. Match scores that meet or exceed some threshold for a particular expectation indicate that the expectation is successfully covered by the student. When a match score is below the threshold, an expectation is not covered, so *AutoTutor* adapts and presents scaffolding dialogue moves as a result (i.e., pumps, hints, prompts) in an attempt to achieve successful coverage of the expectation. After all of the expectations for the task are covered, the task is finished and *AutoTutor* gives a summary.

The dialogue moves generated by *AutoTutor* depend on what the student has expressed at a particular point in the conversation. Suppose an expectation has content words A, B, C, and D and the student has expressed A and B but not C and D. In this case, *AutoTutor* generates hints and prompts to attempt to get the student to express C and D so that the threshold for the expectation is reached. *AutoTutor* may end up giving up trying to get the student to express C and D after a few hints and prompts and may generate an assertion to cover the expectation in the conversation.

The assessment of the student's performance on a task can be computed in a number of ways. One way is to compute *initial match scores* for each expectation immediately after the student gives their first response to a question or problem. That is, there is an initial match score for each of the set of n expectations ($E_1, E_2, \ldots E_n$) and an overall mean initial match score over the set of expectations. A second way to assess performance is to compute *cumulative match scores* after the student has received all of the scaffolding moves of *AutoTutor* during the course of the conversation. A third way is to systematically generate the scaffolding dialogue moves for each expectation and then compute how much scaffolding was attempted. As we discussed earlier, in an EMT dialogue there are cycles of *pump -> hint -> prompt -> assertion* for each expectation after the student's initial response to the main task. The score between 0 and 1 decreases to the extent that the student needs more scaffolding with these tutor dialogue moves to get the student to articulate an expectation.

AutoTutor's semantic match scores depend on how accurately the verbal responses can be analyzed through ongoing advances in computational linguistics and we have evaluated many semantic match algorithms over the years (Cai et al., 2011; Graesser, Penumatsa, Ventura, Cai, & Hu, 2007; Hu et al., 2014; Rus, Lintean et al., 2013). These algorithms have included keyword overlap scores, word overlap scores that place higher weight on lower frequency words in the English language, scores that consider the order of words, *cosine values* from *latent semantic analysis* (LSA) (Hu et al., 2007, 2014; Landauer, McNamara, Dennis, & Kintsch, 2007), *regular expressions* (Jursfsky & Martin, 2008), and procedures that compute *logical entailment* (Rus, Lintean, et al., 2013). It is beyond the scope of this chapter to describe the semantic match algorithms in greater depth; however, we do offer three conclusions.

First, the SEMILAR system (Rus et al., 2013) provides the most accurate tool for analyzing the semantic overlap between the students' contributions and expectations. Second, a combination of LSA and regular expressions go a long way in providing semantic matches in *AutoTutor*, with accuracy scores that are almost as high as trained experts (Cai et al., 2011). Third, a high percentage of the students' contributions are telegraphic, elliptical, under-specified, and ungrammatical so there are limited payoffs in implementing deep syntactic parsers and analytical reasoning components on the students' verbal contributions. Instead, *AutoTutor* tracks the content through *pattern-matching* processes and delivers dialogue moves in attempts to achieve *pattern completion* of the expectations.

Challenges in Developing *AutoTutor*

It takes considerable time and expertise to design, develop, test, and revise agent-based learning environments as well as other classes of ITSs, which has motivated efforts to develop authoring tools, utilities, and standards to streamline the process (Sottilare, Graesser, Hu, & Brawner, 2015). In this section, we discuss some of the major challenges in the development cycles of *AutoTutor* and other agent-based learning environments. For each challenge, we describe how we have attempted to solve or minimize the major hurdles.

Integrating Multiple Forms of Expertise

AutoTutor development requires a research team with multiple areas of expertise. In addition to the content experts, needed areas of expertise include curriculum design, knowledge engineering, computational linguistics, discourse analysis, computer science, artificial intelligence, data mining, learning science, educational measurement, multimedia, and human-computer interaction. It is often difficult to fill the different forms of expertise so there is a risk that the quality of the learning environment suffers as a result. Even when these different types of expertise are available, there are challenges of coordinating the multifaceted contributions, communicating with individuals who have radically different backgrounds, and adhering to packed production schedules.

There are two important approaches to minimizing this first challenge. The first approach is to organize research and development teams with *distributed expertise* that includes high-caliber experts that focus on their particular piece of the system. A particular area of expertise such as measurement or artificial intelligence may require only 10 hours of effort per month. It would be better to have an accomplished expert than to assign the work to a person with modest training or with the expectation that the person will learn it on the job. Of course, an accomplished expert will need to have a sufficient understanding of the project as a whole but accomplished experts are accustomed to doing that.

A second approach is to use digital technologies that facilitate communication, coordination, production schedules, and products. The research teams in the IIS have used a variety of technologies to achieve these goals such as *Google Drive, Assembla, Trello, Blue Jeans*, and so on. The members of the team of course need to use the technologies reliably, which sometimes requires nudging, pestering, and other forms of leverage. Moreover, technologies with metadata are needed to store the history of research and development efforts and to identify those who are slackers or produce low-quality work.

Creating Usable Authoring Tools

We have created *AutoTutor Script Authoring Tools* (ASATs) to populate the content for the subject-matter knowledge for both learning environments (Cai, Graesser, & Hu, 2015) and assessments (Zapata-Rivera, Jackson, & Katz, 2015). That is, subject-matter experts and pedagogical experts create curriculum scripts with expectations and misconceptions, the tutor's dialogue moves (e.g., main questions, pumps, hints, prompts, prompt completions, assertions, summaries, mini-lectures, corrections), frequently asked questions and their answers, and media (e.g., diagrams, tables, pictures, videos).

This content needs to be put in a specific format in ASAT that can be accessed and used by the *AutoTutor* processing mechanisms. In practice, however, it is typically difficult for subject-matter experts to be precise in providing this analytical detail and to follow the formatting constraints because they lack expertise in data base organization, linguistics, and/or the learning sciences. Consequently, they often make a large number of errors in ASAT and there is a risk that the *AutoTutor* system does not function properly as a result. Script writers often do not know how to write good hints and to provide the correct number of prompts to cover the important content words (i.e., nouns, adjectives, verbs) in the expectations.

Figure 21.2 shows an example screenshot of *AutoTutor*'s ASAT. Along the left window there are rigid conversations with opening and closing exchanges for dialogue episodes. In between there are tutoring packs that specify the

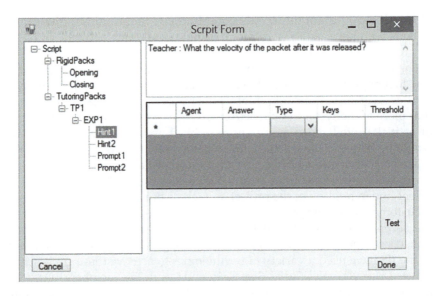

FIGURE 21.2
Screenshot of *AutoTutor* script authoring tool (ASAT).

dialogue moves that try to get the learner to articulate content as a result of the tutor giving hints such as "What is the velocity of the packet after it is released?" Notice that there is an error in the articulation of that hint in the top widow, so the authoring tool needs to allow the author to add the word "is" after "What." There are multiple hints and multiple prompt questions that can be added by the author in this screenshot of *AutoTutor*.

Next, the middle right window has labels to access information about agents, the correct answer, the type of speech act, keywords, and the thresholds on a good semantic match for the expected answer to the question. In the bottom right window, the author can type in a description and test how well the description matches the expected answer. This on-line testing by the author is extremely important so they do not have to wait weeks or months later to find out whether the script is performing as specified in the authoring tool.

Our ideal vision was to have the subject-matter experts simply type in the main questions, expectations, and other ASAT content in natural language. Then there would be a small number of automated utilities or human experts in computation who can concentrate on a limited number of technical components to create a symbolically enhanced representation of the script. Thereafter, *AutoTutor*'s bundle of processing components, called the *AutoTutor Conversation Engine* (ACE), would input the enhanced script and execute the *AutoTutor* processing during the interaction with the student or test taker. That vision ended up being too idealistic, however, because the amount of effort in creating the symbolically enhanced script ended up being substantial.

Nevertheless, we have taken a number of steps to improve the completeness and fidelity of the subject-matter content in ASAT. First, we have written ASAT with *constrained formatting structures* to minimize the likelihood of careless errors. Second, we have simplified the *interface* of ASAT to incorporate the insertion of external media and conversational agents. Third, we have created simple lists of *functionally equivalent expressions* for particular discourse moves. For example, there may be many ways to express a pump (e.g., "What else?", "Tell me more", "Anything else?") or negative feedback (e.g., "no", "not quite", "not really") that are randomly selected in the ACE component during tutoring. These "bags" of functionally equivalent expressions are stored in *discourse libraries* so they can be migrated to many different *AutoTutor* systems with the option of adding or deleting expressions for specific applications.

Fourth, attempts have been made to automate specific discourse components using advances in *natural language generation*. As an example, prompts can be automatically generated from the sentence-like expectations by having a question generated to elicit each of the content words in the expectation (Cai et al., 2006). These generation systems are imperfect but they give the content experts candidate prompts to select and refine. Fifth, there are various testing utilities in ASAT that allow the tester to see how the agent pronounces the various tutor discourse expressions and how parts of the conversation would end up occurring.

Sixth, there have been proposals to include an *AuthorTutor* facility to train the script writers how to author good content (Rus, D'Mello, Hu, & Graesser, 2013). These various attempts have been made to assist the curriculum script developers in content creation, but it is important to acknowledge that the authoring tools of ASAT as well as other ITSs have considerable room to improve and constitutes one of the major bottlenecks in the development of such systems (Sottilare et al., 2015).

Overall, the most difficult feature of authoring tools is the *production rules* component. A single production rule is an *IF <state S>, THEN <tutor/system action A>* expression that specifies that action A occurs whenever state S occurs. For example, when a student asks a question, then *AutoTutor* answers the question or makes some other suggestion. Sets of production rules in a production system map alternative actions onto a finite set of alternative states. To illustrate, *AutoTutor* is expected to give a short feedback (i.e., positive, neutral, negative) to the statements in a student's previous turn that vary in quality (high, medium, low).

A set of production rules for short feedback might map transitions as follows:

[high quality → positive feedback; medium quality → neutral feedback; low quality → negative feedback].

This setup might make sense for some populations of students but not others. For example, students with low self-esteem such as struggling adult readers might get dispirited and give up from too much negative feedback. For them, the mapping might be as follows:

[high quality → positive feedback; medium quality → neutral feedback; low quality → neutral feedback]

Curriculum developers thus need to have different mapping rules at their disposal and declare them in the authoring tools. As a result, these production systems can end up extremely complicated and run the risk of outstripping the analytical capacity of the curriculum developers.

One approach to handling the production rule complexity problem is to automate the production systems in a library used by the ACE. This is a reasonable solution to the extent that the production systems are generic and reusable. However, complications arise when the curriculum developers want to have atypical production systems. As an example, there was a playfully rude version of *AutoTutor* that responded to high-quality student contributions with negative feedback from a special bag of negative feedback expressions (e.g., "Aren't you the little genius", "You finally have managed to say something smart"). The team needed to figure out how these atypical production rules can be declared in the authoring process.

Changing ACE is mighty complex and requires serious computer programming experience, including expertise in artificial intelligence. The curriculum developer needs to either dive into this dense code or have some facility to change the default rules with minimal complications. For example, the default rules in a production system could be presented to the developers in a simple format that they can select or modify. There are tradeoffs in these two approaches that need to be explored further in the future.

Making Discourse Visible

Conversation is experienced daily by humans but the conversation patterns and mechanisms are invisible to most people and sometimes even researchers who investigate discourse processing. Thoughtful discourse analyses are needed to identify the alternative speech acts that a person would likely produce at different points in conversation (Schober, Rapp, & Britt, 2018). For example, person B is expected to generate an answer in a dialogue when person A asks a question; it is impolite to be silent or to generate a statement that changes the topic. In contrast, silence and topic change are appropriate after person A yells at person B with curse words. Some points in a conversation are highly constrained whereas others allow many options. *State-transition networks* can be constructed that specify the alternative discourse moves that are reasonable at particular points in conversation (Morgan et al., 2013). Unfortunately, state-transition networks can be quite complex and beyond the capabilities of most curriculum developers.

One approach to reduce this complexity is to make the state-transition networks visible through *chat maps* with alternate discourse contributions at each point in the conversation between the human and computer agents. The developer can inspect the chat map and make modifications with a digital facility. Another approach is to automate common discourse exchanges that the developer can select as needed. For example, two common conversation patterns are exchanging greetings at the beginning of a meeting and the *initiate-response-evaluate* (IRE) sequence in classroom discourse (Sinclair & Coulthard, 1975). In a typical IRE conversation pattern in tutoring, the tutor asks a question, the student generates an answer, and the tutor gives feedback on the quality of the answer. The ACE component has a library of the common discourse patterns that the developer can access, reuse, and modify.

Linking External Media

Agent-based tutoring environments periodically interact with different types of media that can range from pictures, tables, and diagrams to interactive simulations and virtual reality (Johnson & Lester, 2016). The authoring tools can accommodate these external media by simply linking them into its content and specifying when ACE launches them at particular points in time. Complications arise when particular aspects of the media need to be highlighted on the media display or pointed to by the agent at particular points in time.

This becomes particularly difficult when the x-y coordinates shrink, expand, or fluctuate in other ways that depend on the screen resolution of the digital device. The x-y spatial composition might be fine for one computer platform but fail entirely for another. Programmers can normalize and adjust to the spatial resolution, but that is beyond the scope of the vast majority of designers who lack those computational skills. Similarly, the temporal resolution can create problems when there is animation, simulations, and virtual reality. Sometimes the speed of the computer's processing components influences the speed of the media display. If there is no universal atomic clock as a reference, the different media and ACE can lack coordination. These challenges in integrating and coordinating with the external media require that software engineers with particular expertise be part of the development team.

One of the important forms of media interaction in the *AutoTutor* system is the *Point & Query* (P&Q) facility (Graesser et al., 2018; Graesser, McNamara, & VanLehn, 2005), which allows the student to point to a particular element on a display by clicking on or mousing over the element, at which time a set of questions are listed; the student then selects a question and then a good answer to the question appears. The P&Q facility supports self-regulated exploration and learning that can be tracked by the computer and provide relevant assessments. Students with high self-regulation skills are prone to ask questions with the P&Q facility and the selection of deep questions (e.g., why, how, what-if) is a good measure of deeper levels of comprehension (Graesser & Olde, 2003).

Representing World Knowledge

AutoTutor has represented world knowledge associated with the subject matter with two approaches. The first approach is to represent the content as a set of *knowledge components*, which are "an acquired unit of cognitive function or structure that can be inferred from performance on a set of related tasks" (Koedinger, Corbett, & Perfetti, 2012). For example, in physics, one of Newton's laws of motion – "force equals mass times acceleration" – would serve as a single knowledge component.

There were approximately 40 knowledge components in the *AutoTutor* system for Newtonian physics (VanLehn et al., 2007) but there were approximately 100 knowledge components in the *ElectonixTutor* system that integrated multiple ITSs and more static resources into one integrated learning environment to cover electronic circuits (Graesser et al., 2018). However, the same set of knowledge components are referenced for the problems and learning resources in the *AutoTutor* system.

Specifically, a *Q-matrix* is constructed that specifies which of the K knowledge components are relevant to each of the P problems and R learning resources. Performance on each knowledge component is tracked throughout the learning experience and serves as a foundation for assessment. More specifically, whenever a learning resource is used by the student, a value between 0 and 1 is returned after the problem is completed and designates how well the student performed on the relevant knowledge component for a particular problem and resource. Performance on a particular knowledge component integrates these previous 0-to-1 values across the problems and resources.

One of the primary challenges around knowledge components has been to come up with a principled method of identifying the knowledge components for a particular subject matter. *Subject-matter experts* can be asked to generate these but there is no assurance they will agree on the list of knowledge components and they often disagree on what grain size to formulate them. Some experts list words, others list sentences, and others list richer conceptualizations.

The disagreement between knowledge components arises primarily because subject-matter experts are not trained on representational systems in cognitive science. One approach to solving this challenge is to declare one expert as ground truth after training the expert on a suitable representational format for articulating knowledge components. Another approach is to train three to five subject-matter experts, to obtain *inter-judge reliability* metrics, and to require them to come to a consensus on the knowledge components. At this point, there is no automated method of extracting knowledge components from a corpus of documents.

The second approach to representing subject-matter knowledge in *AutoTutor* is LSA, which "learns" from a very large corpus of documents the meaning of words associated with the subject matter and represents words by *vectors* (Foltz, Kintsch, & Landauer, 1998; Landauer et al., 2007). The vectors of words are more similar to the extent that they appear in the same documents in naturalistic corpora. Weighted sums of word vectors are used to represent ideal answers and student utterances. The *cosine value* between an expectation (or misconception) vector and a student utterance vector is used as a match score when assessing whether the student contribution matches an expectation. The cosine value ranges theoretically from –1 to 1, but in practice negative cosine values are often treated as 0.

LSA vectors represent the meaning of words in the sense that words that are similar in meaning have similar vector values and, thus, have larger cosine values between vectors. It is important to acknowledge that the meaning of words often depends on the specific subject matter. For example, the meaning of the word "force" in the domain of Newtonian physics could be very different from its meaning in the domain of the military.

Since the vectors in an LSA space are learned from a given *corpus* of text, forming a good corpus that can best represent the targeted domain is a very important step in the process of representing world knowledge challenging. This requires expertise in both *corpus analysis* and the mathematical foundations of LSA but finding experts with these skills can be challenging. One way out of this would be to have a universal corpus. For example, *Wikipedia* provides articles covering almost any domain one can think of and it is still growing from which one can sample articles to form a satisfactory corpus to generate LSA vectors. However, there are times when the subject matter is very esoteric and requires the design of a corpus with greater targeted effort.

Tuning Semantic Matching

Despite the computational advantages of LSA, it does not go the distance in providing accurate match scores between student verbal input and expectations. Currently, LSA-based similarity scores correlate only between 0.45 and 0.65 with human judgments whereas human judgments have correlations around 0.7 to 0.8. LSA has trouble with capturing word order, negation, quantifiers, deictic expressions (e.g., here, there, you, this), and logical deductive reasoning. LSA is equipped to handle similarity of verbal expressions with distinctive content words but not these other more subtle language features. Once again, student language is often vague, telegraphic, ungrammatical, and underspecified so a perfect natural language analyzer would have limited value. As a result, it is important to take steps to improve the semantic matching component beyond LSA.

Importantly, beyond scientific undesirability, there are practical costs to having imperfect semantic matches that can affect the quality of student responses and behavior with a system like *AutoTutor*. Students are prone to become irritated when an imperfect semantic matcher does not give the student credit for what the student believes is a correct answer. For example, when an expectation has four distinct words (A, B, C, D) and the student expresses A and C (but not B and D), the student may want full credit for covering the answer because they believe that they have adequately covered the answer if they can generate a couple of key words.

However, *AutoTutor* expects a more complete answer, typically setting the threshold between 0.6 and 0.7 of the important content words. One could lower the coverage threshold from 0.7 to 0.5 and give the student full credit in such a case but the fallout of that relaxation of the threshold is that the student will sometimes get credit for very poor answers. The problem is exacerbated by a frequent semantic blur between an expectation and a semantically similar misconception. For example, the phrase "A lighter object exerts some force on a heavier object" is very similar to "A lighter object exerts no force on a heavier object" yet the second statement reflects a misconception.

Examples like these encourage improvements in the semantic matching facility of *AutoTutor* but the semantic matching component of *AutoTutor* will never be perfect. However, humans are not perfect in making these judgments either so it is not surprising that the automated components are not perfect. One method of improving system performance in this regard is to add additional computational linguistics facilities to improve the semantic match score accuracy. For example, regular expressions have greatly improved the match scores. In one study, a combination of LSA and regular expressions had scores that were nearly the same as pairs of human judges (Cai et al., 2011).

A regular expression is a type of symbolic code that can accommodate alternative spellings of distinctive words (e.g., *expl** can successfully match 'explanation', 'explains', 'explaining', 'explained', and other words that start with 'expl'). Regular expressions can also specify ordered combinations of words and embedded structures. Unfortunately, it takes a large amount of human training and experience to create regular expressions for each expectation and misconception. We have made attempts to automate the generation of regular expressions from natural language sentences, but this is still work in progress.

There is a second method of handling potentially imperfect semantic matches, which is to engineer conversation to attempt to improve the matches. For example, when the match score is near the threshold value, *AutoTutor* would have low confidence that an expectation is covered. When that happens, *AutoTutor* could generate a pump to get the student to elaborate. Similarly, when there is a semantic blur between an expectation and a semantically similar misconception, *AutoTutor* could ask a question to see which alternative the student believes is correct.

Classifying Speech Acts

An important task in dialogue-based educational systems is the detection of student intentions from their typed or spoken utterances (Rus, Graesser,

Moldovan, & Niraula, 2012). As we discussed earlier, an important computational step in *AutoTutor* is to segment the verbal contributions within each turn into speech acts and to classify speech acts into basic categories. The speech act categories do not go the full distance in capturing bona fide intentions, but they are basic level approximations of these intentions. When questions are asked, the speaker has the intention of obtaining information from the addressee whereas requests have the intention of getting the address to do something. Sometimes errors occur, however, which can compromise conversation coherence; for example, a turn segment may not be assigned the correct speech act category and/or the correct intention.

We have pursued two computational approaches to improve the accuracy of speech act analyses, both of which consider the context of the discourse when classifying a particular speech act. The first approach has been to track the previous sequence of speech acts that lead up to the speech act being classified (Samei, Li, Keshtkar, Rus, & Graesser, 2014). There is a sequence of 5 to 10 previous speaker-speech-act categories that lead up to speech act N in a conversation. Presumably, this prior context should improve classification over and above the linguistic features of speech act N.

The second approach is to analyze the higher-level rhetorical structures, which are called *dialogue modes* (Rus, Niraula, Maharjan, & Banjade, 2015), that surround the speech act category in question. For example, *AutorTutor* may be (a) introducing the main question to which the student responds, (b) scaffolding the student to generate a better or more complete, (c) challenging the student in a debate mode, or (d) requesting the student to summarize an answer. A student's speech act N may then be classified very differently in a way that depends on the dialogue mode. This potential solution would of course require *AutoTutor* to perform dialogue mode recognition, yet another computational component to implement.

Closing Comments

In this chapter, we have described the computational mechanisms and methodological challenges of developing *AutoTutor* for learning and assessment environments. We have identified many of the obstacles that we have faced and some of the strategies we have used to minimize or circumvent them. Accordingly, we have summarized major challenges, solutions, and considerations in Table 21.1. The good news is that it is possible to create conversational tutors that can spin conversations and have some modicum of coherence and value to the students – this is no small accomplishment! The disappointing news is that it typically takes a very long time to develop these conversation-based systems because of the multifaceted expertise that needs to contribute to building these artifacts. Perhaps the future will have breakthroughs that cut down the time and costs for development.

TABLE 21.1

Challenges and Solutions in Building AutoTutor

Challenge	Solution	Considerations
Developing a digital tutor that helps people learn through conversation	*AutoTutor* is a conversation-based learning environment with one or more agents that interact with the human in multiple turns for each task	Advances in computation linguistics, AI, discourse processing, and technology have solved many but not all problems
Identifying a conversational mechanism that helps students learn	In *expectation-plus-misconception-tailored tutoring*, the tutor attempts to get the student to articulate good answers in a task through hints and questions, along with generating conversation moves to correct student errors	Automating components of the tutoring mechanism has some technical difficulties
Identifying conversational components of tutoring contributed by students	*AutoTutor* segments the student turns into speech acts and classifies each speech act into categories, such as questions, statements, short responses, metacognitive expressions, meta-communicative expressions, and expressive evaluations	Automating speech act categories depend on advances in computational linguistics and discourse
Computing semantic matches between student contributions and expectations or misconceptions	*AutoTutor* computes the semantic similarity between student statements and the expectations and misconceptions through algorithms in computational linguistics and high dimensional semantic spaces	Automated semantic matches may be imperfect and there can be a semantic blur between expectations and misconceptions
The development of new materials with *AutoTutor* requires multiple forms of expertise	Digital technologies are needed to develop content (through authoring tools) and coordinate the development process and communication.	Many subject-matter experts have modest knowledge in advanced digital resources, pedagogy, language, and discourse
Subject-matter experts need to develop the learning materials	Authoring tools need to be developed and improved to minimize the difficulty of usage by experts	Subject-matter experts often have modest expertise in computer science so many of the components of *AutoTutor* need to be automated
The creation of *AutoTutor* lessons requires making discourse visible and specifying links to external media	Authoring tools allow content developers to create conversations and link speech acts to external media	Substantial training is needed to specify conversation components and links to external media

Acknowledgments

The research was supported by the National Science Foundation (SBR 9720314, REC 0106965, REC 0126265, ITR 0325428, REESE 0633918, ALT-0834847, DRK-12-0918409, 1108845; ACI-1443068), the Institute of Education Sciences (R305H050169, R305B070349, R305A080589, R305A080594, R305G020018, R305C120001), the Army Research Lab (W911INF-12-2-0030), and the Office of Naval Research (N00014-00-1-0600, N00014-12-C-0643; N00014-16-C-3027). Any opinions, findings, and conclusions or recommendations expressed in this material are those of the authors and do not necessarily reflect the views of NSF, IES, or DoD. Examples of *AutoTutor* can be found at http://www.*AutoTutor*.org).

22

Automated Scoring in Intelligent Tutoring Systems

Robert J. Mislevy, Duanli Yan, Janice Gobert, and Michael Sao Pedro

CONTENTS

Digital simulation environments open the door to a radically new paradigm of learning, characterized by the interaction and *personalized adaptation* seen in *intelligent tutoring systems* (ITSs). Making sense of the learners' interactions with a simulation environment to manage the challenge and to provide feedback and support is inherently a form of assessment. Concepts from *principled assessment design* offer support to ITS designers and users. In conjunction with principled design, *automated scoring* (AS) processes can provide for less constrained interactions, more authentic situations, and better targeted support than encapsulated assessment tasks can provide.

In this chapter, we discuss the role of AS in *simulation-based ITSs*. In the first section, we provide background on simulation-based ITSs. The section includes brief introductions to the two examples we look at in more detail, the *Hydrive* ITS to help US Airforce Force trainees learn to troubleshoot aircraft hydraulics systems (Steinberg & Gitomer, 1996) and the *Inq-ITS* system to foster middle school students' skills in science inquiry (Gobert, Sao Pedro, Raziuddin, & Baker, 2013b; Gobert, Moussavi, Li, Sao Pedro, & Dickler, 2018) In the second section, we provide concepts and terminology from *evidence-centered (assessment) design* (ECD; Mislevy, Steinberg, & Almond, 2003) that facilitate the discussion of system design and the coordinated role of AS in simulation-based ITSs. We follow this up with closer looks at how the ideas play out in *Hydrive* and *Inq-ITS* and then provide some concluding comments.

Background

Novices in a wide variety of domains face similar difficulties. They do not know what information is relevant, how to integrate information, or what to do next. Nevertheless, novices routinely display some amazing capabilities by working with patterns: patterns of perceiving, patterns of thinking, and patterns of acting, which we assemble flexibly in real-time once sufficiently practiced (Greeno & van de Sande, 2007). Expert performance is made possible through the continual interaction between the external patterns that structure peoples' interactions in situations, such as language and professional practices, and a person's internal cognitive patterns for recognizing, making meaning of, and acting through these patterns (Ericsson, Charness, Feltovich, & Hoffman, 2006; Greeno & van de Sande, 2007).

We develop our internal cognitive patterns through experience; that is, by participating in activities that are structured around these patterns, discerning regularities, and seeing what happens as we and others act. Through *reflective practice* – best with feedback, often starting from simplified situations, usually with feedback and support – persons build their capabilities and overcome the pervasive limitations that plague novices (Ericsson et al., 2006).

Digital simulations enable learners to engage in these kinds of rich experiences, create artifacts, explore complicated environments, solve problems, and interact with others. A simulation emulates some features of actual situations but not others. It may speed up, slow down, change size, or simplify aspects of real-world situations. A designer of a simulation environment for learning must determine what aspects of real situations to include and what aspects to modify or omit, for the purposes and the populations of interest. The features of the simulation environment must present challenges that involve important features of the domain, engage concepts and representations a learner must learn to work with, and call up the actions, tactics, and strategies one needs to become proficient in the domain. Opportunities to engage with problems in such an environment helps learners develop these capabilities, as the reactions of the simulation itself provide feedback on their actions, thus on the knowledge and skills they have employed.

An ITS built around a digital simulation environment offers further learning benefits by providing one or more of the following types of information: feedback on actions, selection of challenges, adaptation of scaffolding, explication of situations and actions in terms of domain concepts, or some combination of these. To do so effectively requires more than a simulation of the domain; it requires theories of proficiency in the domain, of how a learner's actions evidence the nature of the thinking that led to given actions in a given simulation state, and of what, among its own capabilities, the ITS should do to optimize the chances of improved learning by providing immediate evaluation feedback and opportunities for learners to try additional tasks until mastery is achieved. AS plays a role in the determination of how actions evidence learners' thinking. The trick is to be able to make sense of learners' actions in terms of all these theories and all these needs, in real time, without human intervention.

Although AS methodology has advanced rapidly during the 25 years that separate our two examples. The advances in the particulars of the methods they employ are striking. Yet, more striking are the similarities in how both are grounded in the constructs of interest, the purpose of the system, and the coordinated design of all its elements through which AS succeeds. The two examples we present are:

- **Hydrive** (Steinberg & Gitomer, 1996; Mislevy & Gitomer, 1996). *Hydrive* was an ITS designed to help Air Force trainees learn to troubleshoot the hydraulics systems of the F-15 aircraft. The simulator enabled trainees to carry out test procedures and effect repairs; the system state and its responses would be adjusted accordingly. The problems, the interfaces, and the feedback were built around Newell and Simon's (1972) theory of problem-solving: defining an active path in a problem space, performing tests, and applying strategies such as space-splitting and serial elimination. Assessment occured continuously as trainees carried out information-gathering and repair actions in the simulation space. AS was implemented as rules that evaluated sequences of actions according to the information they provided about the system, given the trainee's previous test results and repair actions. These evaluations were used in two ways. First, descriptive feedback and comments were provided at troubleshooting proceeded, and evaluations across sequences were synthesized in a *Bayes net* (e.g., Almond et al., 2015) to initiate recommendations for review and practice on topics in relevant strategies and relevant and subsystems.

- **Inq-ITS** (*Inquiry Intelligent Tutoring System*; Gobert et al., 2013b, 2018). *Inq-ITS* enables middle and early high school students to carry out scaffolded investigations to learn and to demonstrate competency at science practices identified in the *Next Generation Science Standards* (NGSS) (NGSS Lead States, 2013). As students conduct their investigations, the work products they create by conducting their investigations and processes they use are automatically assessed using integrated AS algorithms. The cognitive model for sub-skills underlying the inquiry practices was derived from the research literature and focused *think-aloud* case studies. Task activities are conducted in a simulated *microworld*, and AS methods from *knowledge engineering* and *data mining* are used to evaluate students' inquiry practices elicited from their work products (i.e., end-state products) and processes (i.e., actions/behaviors as indicated in their *log files*). The results are aggregated into performance indicators that provide evidence of students' proficiencies for each inquiry practice and their respective sub-skills. *Bayesian knowledge tracing* (BKT) (Corbett & Anderson, 1995; Sao Pedro et al., 2013a) is used to study progress across problems.

ITSs using Evidence-Centered Design

ECD is a principled assessment design framework that supports the development of not only familiar forms of assessment, but also game- and simulation-based assessments and ITSs (e.g., Clarke-Midura, Code, Dede, Mayrath, & Zap, 2012; Gobert et al., 2013b, 2018; Mislevy et al., 2014; Shute, 2011). The assessment design patterns guide not only the documentation of the *evidentiary argument* explicitly, but also the decisions related to *evidence identification, aggregation,* and *interpretation* (Mislevy, et al., 2009, 2003b).

Three chapters in the present volume examine aspects of the role of AS in complex digital simulations: In Chapter 3, DiCerbo, Lai, and Ventura address the larger perspective of assessment design for AS. In Chapter 9, Mislevy locates AS and its relationship to other assessment elements in arguments for assessment design, score interpretations, and score uses. Finally, in Chapter 23, Almond examines delivery and processing frameworks to implement designs for such assessments in streaming environments. While our chapter is self-contained, it draws on ideas discussed more fully in these chapters. For present purposes, we briefly review terms and concepts that are central to our discussion of AS in ITSs.

ECD is structured in terms of *layers* of distinct kinds of work in the design and operation of an assessment (including those embedded in ITSs, games, or other products requiring user modeling in the service of learning). The three that are relevant in this chapter are *Domain Modeling,* the *Conceptual Assessment Framework* (CAF), and the *Four-Process Delivery Architecture.*

We need to say little about Domain Modeling other than to note that it encompasses the grounding in literature, experience, necessary research, product requirement, and constraints needed in the design process. For ITSs, this includes the theories of the domain, cognition, learning, and evidentiary relationships mentioned above. The examples show how a designer draws on this foundation to design the elements of the system. Similarly, we need say little about the details of Delivery Architecture as Almond, in Chapter 23, provides the connection to the design of delivery systems in interactive systems including ITSs.

We will, however, give descriptions and key terms from the CAF that we use in the following discussion. They are where the coordination of AS with interface, affordances, task situations, and cognitive models is articulated in the design. The principal design objects of the CAF in ECD provide a framework for specifying the knowledge and skills to be assessed, the conditions under which observations will be made, and the nature of the evidence that will be gathered to support the intended inference. They detail the materials, the requirements, the computational objects, and the processes that are carried out in the assessment delivery system. The principal models of the CAF that figure into the design of simulation-based ITSs that use AS appear in Table 22.1.

Example 1: *Hydrive*

Hydrive is a simulation-based tutoring / assessment system for developing troubleshooting skills for the F-15 aircraft's hydraulics systems (Gitomer, Steinberg, & Mislevy, 1995; Steinberg & Gitomer, 1996). In the following, we address the cognitive foundations of the system as well as considerations in designing the aircraft simulator and the ITS interface, defining student-model variables, crafting the AS process, and synthesizing evidence about student-model variables with a Bayes net. We describe how AS and probability-based inference support claims about aspects of student proficiency through the combination of epistemic analyses of particular actions. We see how the psychology of learning in the domain and the instructional approach play crucial roles in *Hydrive*'s design and operation.

Activity Design

The hydraulics systems of the F-15 aircraft are involved in the operation of flight controls, landing gear, the canopy, the jet fuel starter, and aerial refueling. The simulation environment in *Hydrive* is designed to simulate many of the important cognitive and contextual features of troubleshooting during flight. A problem starts with a video sequence in which a pilot describes some aircraft malfunction to the hydraulics technician; for example, the rudders do not move during pre-flight checks. *Hydrive*'s interface then offers the student several options such as performing troubleshooting procedures by

TABLE 22.1

Principal Design-Specification Objects in Evidence-Centered Design

Object	Definition
Student model	A collection of variables representing the knowledge, skills, and abilities of an examinee about which inferences will be made. A Student Model includes (1) Student Model Variables that correspond to aspects of proficiency the assessment is meant to measure and (2) Model Type that describes the mathematical form of the Student Model (e.g., classical test theory, item response theory, Bayes net). *In an ITS, student-model variables are keyed to the nature and grainsize of feedback and support the ITS is intended to provide.*
Evidence model	A set of instructions for interpreting performance. It is the bridge between the Task Model, which describes the task, and the Student Model, which describes the framework for characterizing what is known about a student's state of knowledge. An Evidence Model generally has two components:
	• *Evaluation processes* (e.g., answer keys, human raters and rubrics, automated scoring processes) that describe how to identify and characterize essential features of performances and/or products. They produce values of "observed variables."
	These processes operate on realized performances, to provide evidence about those performances. These are the primary uses of AS in simulation-based ITSs. The results can be used for feedback about the realized performances.
	• Statistical/psychometric *measurement model(s)* that synthesizes evaluated of features of task performances across tasks, expressed in terms of student-model variables. These models synthesize evidence about realized performance, which are expressed as values of observed variables, in terms of values (or posterior distributions) of student-model variables, as evidence about students' capabilities as inferred across multiple (realized or hypothetical) performances.
	In an ITS they can be used to determine scaffolding, feedback, and instructional activities.
Task model	A generic description of a family of tasks that contains (1) task-model variables which are used to describe key features of the tasks, (2) presentation material specifications that describe material that will be presented to the examinee as part of a stimulus, prompt, or instructional program, and (3) work product specifications that describe the material that will be evaluated. *The conception of task-model variables is generalized for assessments (including ITSs) in which action is continuous and the context evolves as the student(s) and simulation interact (Almond, this volume; Mislevy, 2018, Chapter 16). These so-called run-time task-model variables may be needed to evaluate students' actions, as they are in the AS processes of both Hydrive and Inq-ITS.*
Assembly model	Provides the information required to control the selection of tasks for the creation of fixed form or adaptive systems consisting of discrete tasks, or, in interactive assessments, it includes how many of what kinds of tasks, task features to balance or select on, and adaption or routing schemes. *In ITSs, the conception of the assembly model is generalized to Context Determination, which provides information both to AS routines and to selection of instructional modules or other learning activities (Almond, this volume).*

accessing video images of aircraft components and acting on those components; reviewing on-line technical support materials; and accessing instruction at any time, in addition to (or instead of) the system's recommendations.

The state of the aircraft system, including the fault to be isolated and any changes brought about by user actions, is modeled by *Hydrive's system model*. The student's performance is monitored by evaluating how that student uses available information about the system to direct troubleshooting actions. *Hydrive's AS processes* are used to evaluate specific troubleshooting actions, and its *student model* characterizes student understanding in terms of more general constructs such as knowledge of systems, strategies, and procedures that are associated with troubleshooting proficiency.

Cognitive Model

The cognitive theory that underlies *Hydrive* traces back to Newell and Simon's (1972) *information-processing* research on problem-solving, as applied to troubleshooting in *finite systems* (Jonassen & Hung, 2006) and as detailed in the specific domain of troubleshooting F-15 hydraulics (Steinberg & Gitomer, 1996). The systems consist of many interacting components, each with inputs and outputs that vary with conditions and settings, when operating correctly or under various fault conditions. The presence of a fault is signaled by unexpected output; for example, the pilot wiggles the stick but the rudder does not move signaling that something is wrong along the so-called *critical path* of components between the stick and rudder actuator.

The primary learning goal is for technicians to develop proficiency in carrying out troubleshooting actions and repairs as needed along critical paths that arise in the F-15 hydraulics system. A secondary goal is becoming proficient with troubleshooting strategies more generally (i.e., a capability that can transfer to other systems); we say more about this later. Effective strategies include 'space-splitting', which is most efficient when it is applicable; 'serial elimination' along the path; and 'remove-and-replace' along the path. In-expert actions provide information that is irrelevant or that is redundant to information that is already knowable, although the student may not know it. Applying effective troubleshooting strategies requires not only understanding the strategies conceptually, but it is also necessary to have sufficient knowledge of the system and of the procedures necessary to carry them out in the problem at hand; that is, a *conjunction* of these capabilities is required.

Hydrive presents problems and evaluates action sequences in terms of inferred strategy use in the context. It provides local feedback and suggestions for reflection based on AS processes described below. Further, the measurement model synthesizes information across interpreted actions across multiple sequences, upon which to suggest instructional modules. There are 15 instructional modules, organized in terms of strategies, test procedures, and subsystem knowledge, encompassing their components, operations, and interactions. We will see that the variables in the student model correspond to these modules to trigger higher-level instructional suggestions at the level at which options are available.

Automated Scoring

Hydrive's approach to AS is knowledge engineering, which means that the designers used analyses of F-15 hydraulics, cognitive theory about troubleshooting, think-aloud studies of expert and novice solutions, and the intended instructional approach to craft rules that would identify and interpret cognitively relevant sequences of actions. The resulting rules run in real time on the accumulating log of actions a trainee takes in a problem, starting from the initial symptom of the fault. As actions within the simulation environment are virtually unlimited, it is necessary to be able to make sense of any sequence of actions a trainee carries out. *Hydrive*'s produces values of *interpreted action* variables using its *system model, action evaluator,* and *strategy interpreter.*

The *system model* appears to the trainee as an explorable, testable aircraft system in which a failure has occurred. It is built around sets of components connected by inputs and outputs. Connections are expressed as pairs of components, the first producing an output which the second receives as an input, qualified by the type of power characterizing the connection. The system model processes the actions of the trainee and propagates sets of inputs and outputs throughout the system. A trainee activates the system model by providing input to the appropriate components and can then examine the results for any other component of the system. A trainee can move the landing gear handle down and then observe the operation of the landing gear. If the landing gear does not descend, the trainee may decide to observe the operation of other components to begin to isolate the failure.

The *action evaluator* considers every troubleshooting action from the point of view of the information that has been conveyed to the trainee about the problem area, regardless of the trainee's actual understanding. When a trainee acts to supply power and input to the aircraft system, the effects of this input spread throughout the system model, creating explicit states in a subset of components. These form the *active path*, which is comprised of the points from which input is required to initiate system function to its functionally terminal outputs and all the connections in between. Amid this sequence of unconstrained potential actions the set of actions of powering up the system to see the result defines an *evidence-bearing opportunity*. This is akin to an emergent task in a traditional assessment, although these opportunities differ in number and particulars from one performance to the next in simulation-based activities like this one.

At this point the *action evaluator* updates its problem area as if the trainee correctly judged whether observations reveal normal or abnormal component states. If, having supplied a set of inputs, a trainee observes the output of a certain component that the system model "knows" is normal, then it is possible for the trainee to infer that all edges on the active path, up to and including the output edge, are functioning correctly and remove them from the problem area. If the trainee, in fact, makes the correct interpretation and draws the appropriate inferences, then the problem areas that the student model and the trainee hold will, in fact, correspond and troubleshooting continues smoothly. But if the trainee decides that the observed component output was unexpected or abnormal, then, at least in the trainee's mind, all the edges in the active path would remain in the problem area, any others would be eliminated, and the problem area maintained by the student model would begin to diverge significantly from the one present in the trainee's mind. Irrelevant and redundant actions become more likely.

The *strategy interpreter* evaluates changes to the problem area (denoted as **k**), or the entire series of edges belonging to the system in which the problem occurs. As a trainee acts on the system model, **k** is reduced because the results of action sequences, if correctly interpreted, eliminate elements as potential causes of the failure. If the trainee inspects any particular component, the system model will reveal a state which may or may not be expected from the trainee's perspective. *Hydrive* employs a relatively small number of strategy interpretation rules (~25) to characterize each troubleshooting action in terms of both the trainee and the best strategy.

An example is the rule for detecting power-path space-splitting:

> IF *the active path which includes the failure has not been created and the student creates an active path that does not include the failure and the edges removed from the problem area are of one power type, THEN the student strategy is power path splitting.*

Note that applying such rules requires several kinds of information:

1. the demarcation of the action sequence to evaluate from the previous evaluated action sequence to the present one; this is what Almond in Chapter 23 calls 'watermarks' that signal the beginning and end of a scorable segment of context and activity;
2. the marked sequence of trainee actions, which together may change the state of the system model;
3. the change in the system model that result from the actions; and
4. the interpretation of this specific change in light of its comparison with a valued strategy (in this rule, space splitting, here with respect to the power path) or effect (in other cases, no change in the problem area due to irrelevance or redundancy).

Local feedback on actions, both effective ones, as when this sample rule is satisfied, or ineffective ones, as when a redundant-move action sequence is satisfied, can be provided immediately or at the end of a problem session.

Psychometric Model

The AS processes described above identify and evaluate sequences of log file actions in terms of cognitively relevant descriptors of performance. In addition to the local feedback based on specific observed action sequences, a psychometric model was piloted that synthesized evidence about interpreted actions across distinct but cognitively equivalent situations. The psychometric model provided two benefits. First, it reflected tendencies across similar situations for interpretations and recommendations at the level of constructs since even experts sometimes miss opportunities to space-space and novices sometimes act in ways that are consistent with space-splitting. Second, additional cognitive theory could be brought to bear to help sort out the possible reasons that a trainee has difficulties with various aspects of knowledge and skill. Recall that effective troubleshooting requires the conjunction of strategies, system knowledge, and troubleshooting tools and tactics.

In the resulting Bayes net, observable variables are defined in terms of the "identified actions" produced by the AS rules. For example, when an emergent task in which space splitting is possible is identified, the value of the corresponding observable variable is the result of mutually exclusive rules (i.e., 'space-splitting', 'serial elimination', 'remove-and-replace', 'redundant', or 'irrelevant'). Observable variables like these are modeled in the Bayes net as dependent on student-model variables for 'space-splitting', 'serial elimination', 'subsystem knowledge' (whichever applies in the problem at hand), and 'familiarity with test procedures'; Figure 22.1 provides a graphical example of such a network.

This network is described more fully in Mislevy and Gitomer (1996) and Mislevy (2018) but note that the observable variables are on the far right, the student-model variables are on the left, and the middle column of student-model variables corresponds to available instructional modules. The observable variables depend on the conjunction of relevant student-model variables so that the contributions of knowledge that are intertwined in performance may be sorted out for further practice. For example, suppose a trainee carries out space-splitting consistently in canopy problems but not in landing gear problems. The updated network will then show a high probability of strong canopy knowledge and space-splitting knowledge, but a low probability for landing gear knowledge. Such a pattern across student-model variables triggers the recommendation to spend time with the landing gear refresher module. Again, the AS needs to be considered from the system design phase.

Example 2: *Inq-ITS*

Inq-ITS (*Inquiry Intelligent Tutoring System;* see Gobert, Sao Pedro, Baker, Toto, & Montalvo, 2012; Gobert et al., 2013b; Gobert & Sao Pedro, 2017) is an interactive environment that assesses middle and early high school students'

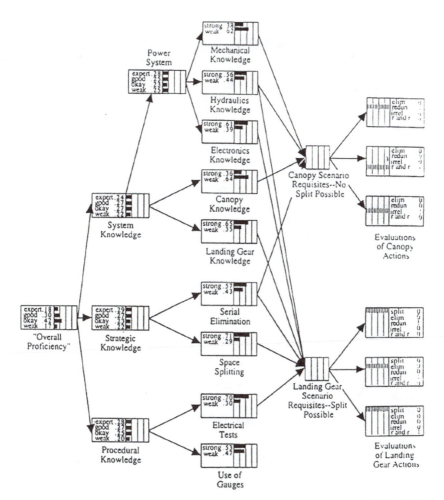

Note: Bars represent probabilities, summing to one for all the possible values of a variable. A shaded bar extending the full width of a node represents certainty, due to having observed the value of that variable; i.e., a student's actual responses to tasks.

FIGURE 22.1
Status of a student after observation of three in-expert actions in canopy situations and three more expert actions in landing-gear situations.

inquiry aligned to the *Next Generation Science Standards* (NGSS) (NGSS Lead States, 2013); see Figure 22.2 for an overview.

While students conduct their investigations, the work products they create by conducting their investigations and processes they use are automatically assessed in a fine-grained manner using integrated AS algorithms. Automated assessment provides three key benefits for students and teachers. First, for students, *Inq-ITS* has an integrated *pedagogical agent* that provides immediate, personalized feedback based on the specific ways students

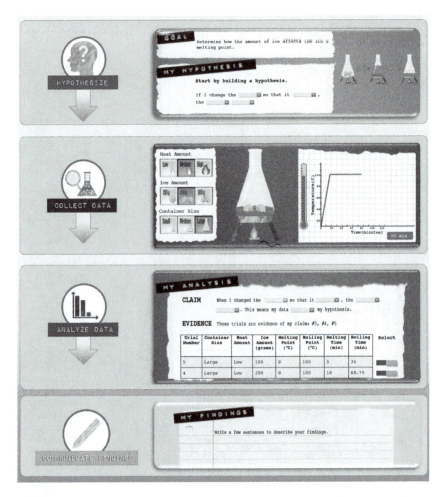

FIGURE 22.2
With *Inq-ITS*, students investigate a scientific phenomenon using a simulation and representational tools. Students form questions, collect data, analyze data, and communicate their findings. As they work, all of their work products and inquiry processes are assessed as they work using a combination of knowledge-engineering-derived and data-mining-derived autoscoring algorithms.

struggle to complete their investigations as they are working (see Sao Pedro et al., 2013a; Gobert, Moussavi, et al., 2018). Second, for educators, *Inq-ITS* provides integrated reports that summarize class-level and student-level performance at the practices over time. Finally, also for educators, *Inq-ITS* also has an integrated real-time alerting dashboard, *Inq-Blotter* (Sao Pedro et al., 2018; Dickler et al., 2019) that alerts educators about exactly how students are struggling. As described in the cited articles, *Inq-ITS* was designed through the ECD framework, the terms of which we use in the following discussion.

Activity Design

A key first step to developing *Inq-ITS* and AS algorithms was to break down the NGSS science practices into more concrete, appropriately grain-sized sub-skills/sub-practices. These were created by reviewing the NRC and NGSS frameworks, interviewing teachers, conducting cognitive think-aloud studies with students using hands-on and virtual materials, and reviewing relevant research on students' difficulties with inquiry at the middle school and early high school levels (Gobert & Sao Pedro, 2017). Example sub-skills for key practices are as follows:

- *Asking Questions* (NGSS Practice 1) entails formulating a testable question, which requires students to be able to distinguish independent (manipulable) variables from dependent (outcome) variables and posit a relationship between them (Gober, Sao Pedro, Baker, Toto, & Montalvo, 2012).

- *Planning and Carrying out Investigations* (NGSS Practice 3) entails collecting controlled data as well as data to enable the supporting or refuting of testable questions specified (Sao Pedro, Baker, Gobert, Montalvo, & Nakama, 2013).

- *Analyzing and Interpreting Data* (NGSS Practice 4) entails interpreting the relationship between the independent and the dependent variable as well as linking the interpretation back to the original testable question (Gobert et al., 2018).

- *Constructing Explanations* (NGSS Practice 6) entails forming a claim based on the data collected and warranting the claim by selecting appropriate evidence (data trials collected) that back up the claim (Gobert et al., 2018).

- *Communicating Findings* (NGSS Practice 8) entails describing the results of the investigation in a claim-evidence-reasoning (MacNeill & Krajick, 2011) open response format against which several sub-skills are checked (Li et al., 2018).

Based on the sub-skill breakdown, activities were developed with a similar look-and-feel that combine simulations that represent science phenomena with *representational tools* that elicit and structure students' thinking. The simulations and representational tools provide the *data streams* over which AS algorithms can score students' inquiry practices. Each activity provides students with a goal aligned to a specific NGSS 'Disciplinary Core Idea' and content misconception (Gobert & Sao Pedro, 2017), marrying the content and practices as envisioned by the NGSS.

To elaborate, the example *Inq-ITS* activity for 'Phase Change' shown in Figure 22.2 begins with an explicit goal to determine if a particular independent variable (e.g., container size, heat level, substance amount, and cover status) affects various outcomes (e.g., melting point, boiling point, time to melt, and time to boil). Students then use representational tools and the simulation to create work products and execute processes that can be assessed. The tasks within the activity are semi-structured into tasks that provide students with a fair amount of flexibility and choice commensurate with conducting an investigation (Quintana et al., 2004), but still structured such that competencies at the practices can be inferred from students' work products and processes (Gobert & Sao Pedro, 2017).

Students first attempt to generate a testable question that will address the goal using a forming questions tool. Next, students use a simulation to collect as much data as needed to support or refute the question they posed. A data table is provided to summarize their data. Once completed, students analyze their data by constructing claims, warranting their claims and linking their findings back to their original question using an analysis tool. Finally, students use open response fields to summarize their findings in a claim-evidence-reasoning format.

Automated Scoring

AS algorithms were developed to score each sub-practice, some of which are garnered from students' work products, and others from their experimentation processes. *Product data* are akin to students' stated hypotheses and results of the analyses of their data while *process data* arise from students' entire stream of data collection activities. The product data are evaluated using knowledge-engineered rules whereas the process data are evaluated using more complex models derived using techniques from data mining (e.g., Baker & Yacef, 2009). AS was considered from the beginning of the system design along with ECD design patterns.

Evaluating Work Products

We focus specifically on sub-skills associated with the 'Asking Questions' and 'Analyzing and Interpreting Data' NGSS practices. We obtain evidence for these practices from data that students enter specifically as responses. These work products are the hypotheses students generate using the hypothesis widget and the analyses they construct about their data using the analysis widget. These products lend themselves to evidence identification processes using knowledge-engineered rules because the sub-skills are relatively well defined and the relationship of elements of the work products to the sub-skills is relatively straightforward.

For example, when a student generates an analysis of the data that they collect, the system uses knowledge-engineered rules to check whether the analysis has the following components: Did they identify the correct independent variable and dependent variable? Did they specify the correct relationship between them? Lastly, did the appropriate trials support their analysis? Binary evaluations (0-1) for each sub-skill are scored each time a student creates work products.

Evaluating Processes

In this sub-section, we focus specifically on sub-skills associated with the 'Planning and Carrying Out Investigations' NGSS practice. A data-mining approach is better suited to process-oriented sub-skills that can be demonstrated in several different ways (Gobert et al., 2012; Sao Pedro et al., 2012; Sao Pedro, Baker, Gobert, et al., 2013; Gobert et al., 2013b). Such constructs have variable log signatures across students and would thus be subject to expert blind spots (and subsequent mis-scoring) under a knowledge-engineering approach (e.g., Baker et al., 2006; Gobert et al., 2012; Sao Pedro, Baker, Gobert, et al., 2013).

For example, one of the sub-skills we assess is whether students design controlled experiments (Sao Pedro, Baker, & Gobert, 2013a, b; Sao Pedro, Baker, Gobert et al., 2013), which students can demonstrate in a variety of manners. Some students collect only controlled data. Some play with the simulation, then collect controlled data. Others seek for covariation effects in addition to running controlled trials (Sao Pedro, 2013a). In all these cases, students understand how to design controlled experiments but are also interweaving other kinds of valid exploration procedures. Similarly, there are many ways in which students may struggle, which makes authoring knowledge-engineered rules very difficult. Data-mining, on the other hand, allows for taking multiple features of log file data and the interactions between these features into account.

To build data-mining-based AS algorithms, we combined *text replay tagging* with *supervised data mining* (Baker et al., 2006). Text replay tagging distills student log data into easy-to-read "playbacks" that enable human scorers to quickly score students' processes. For designing controlled experiments, human scorers labeled a replay as 'designed controlled experiments' or not by looking at what hypotheses the student made, which simulation variables the students change and which trials they ran. Next, in line with the *text replay methodology* (Baker et al., 2006), a supervised data-mining algorithm is used to train a *computational model* (i.e., the AS algorithm) to score as a human expert would base on student interaction data and the labels derived from the text replay tagging.

To train such a computational model, the interaction data are first distilled into a *feature set*. Features can be thought of as potential indicators of whether the student has demonstrated at the sub-skill level. For designing controlled experiments, features included how many trials students ran, counts of the number of pairwise controlled experiments students had, and counters for which variables students changed (more information on specific features can be found in Sao Pedro et al., 2012; Sao Pedro, Baker, Gobert, et al., 2013). Then, the feature sets and human-scored labels are combined and a data-mining algorithm from which the AS algorithm is constructed "learns" optimal features and feature cutoff values to autoscore students' sub-skills. In our prior work, we found that choosing features that have high theoretical alignment with the sub-skills of interest make for autoscoring algorithms that are more human-interpretable and better agree with human scoring (Sao Pedro et al., 2012).

Psychometric Model

To amalgamate students' performances, *Inq-ITS* has used BKT, a technique known to successfully measure proficiency in other environments that provide explicit learning support (e.g., Corbett & Anderson, 1995; Koedinger & Corbett, 2006). Technically speaking, a BKT model is a *two-state hidden Markov model* that estimates the probability a student possesses latent skill (L_n) after n practice opportunities to conduct inquiry within *Inq-ITS*.

In this case, observable student performance is demonstration of data collection skill within a sequence of log file entries ('clip') as assessed by the detectors. BKT models are characterized by four parameters, G, S, L_0, and T, which are used in part to compute latent skill L_n. The *guess parameter* (G) is the probability the student will demonstrate the skill despite not knowing it. Conversely, the *slip parameter* (S) is the probability the student will not demonstrate the skill even though he/she knows it. Finally, L_0 is the initial probability of knowing the skill before any practice and T is the probability of learning the skill between practice attempts. Within the BKT framework, these four parameters are assumed to be the same for all students and knowledge of the skill is assumed to be binary (i.e., either the student knows the skill or he/she does not).

Using these four parameters, the probability that a student knows the skill, $P(L_n)$, and the probability that a student will demonstrate that skill in his/her next practice opportunity, $P(Demonstrate\_Skill_n)$, can be computed as follows:

$$P\left(L_{n-1}|\,\mathrm{Obs}_n\right) = \begin{cases} \dfrac{P\left(L_{n-1}\right)*\left(1-S\right)}{P\left(L_{n-1}\right)*\left(1-S\right)+\left(1-P\left(L_{n-1}\right)\right)*G}, & \text{Demonstrated\_Skill}_n \\[4mm] \dfrac{P\left(L_{n-1}\right)*S}{P\left(L_{n-1}\right)*S+\left(1-P\left(L_{n-1}\right)\right)*\left(1-G\right)}, & \sim\text{Demonstrated\_Skill}_n \end{cases}$$

$$P\left(L_n\right) = P\left(L_{n-1}|\,\mathrm{Obs}_n\right) + \left(\left(1-P\left(L_{n-1}|\,\mathrm{Clip}_n\right)\right)*T\right)$$

$$P\left(\text{Demonstrate\_Skill}_n\right) = P\left(L_{n-1}\right)*\left(1-S\right)+\left(1-P\left(L_{n-1}\right)\right)*T$$

Note that $P(Demonstrate\_Skill_n)$ is an *a priori estimate* of demonstrating a skill since it depends on the prior estimate of knowing the skill, $P(L_{n-1})$. An example application of these equations for a students' performance profile associated with the designing controlled experiments sub-skill is shown in Table 22.2. In this example, the student engaged in nine data collections across activities. Each engagement was autoscored using the data-mining-based AS algorithm as demonstrating skill or not (i.e., was scored 0-1). From there, estimates for $P(L_n)$ and $P(Demonstrate\_Skill_n)$ can be found by applying the equations above.

In this example, a *brute force search* (Baker et al., 2010b) was used to train a BKT model with data from a set of *Inq-ITS* activities on phase change micro-world. In other words, the model learns the best fitting parameter estimates (values for G, S, L_0, and T) given the data.

Validity and Reliability

For validity of the AS algorithms, whether knowledge-engineering-based or data-mining-based, we focus specifically on *face validity* and *construct validity*. This enabled us to ensure that the sub-skills we score have meaningful inter-pretation for application for *formative assessment* and scaffolding. That is, the AS must seem interpretable but actually first "earn" its interpretation and then prove useful in practice. Face and construct validity were established by ensuring the sub-skills aligned with NGSS and current research and that they also had meaning and usefulness to teachers. We also rigorously inspected the algorithms developed, particularly those generated from data mining, to ensure their components matched teachers' expectations and had plausible features and relationships to autoscore students' work products and processes.

For reliability of the AS algorithms, we focused specifically on *inter-rater reliability* (human-computer) to ascertain how well the algorithms agree with human scoring. To illustrate, for the designing controlled experiments data-mining-based AS algorithm, we collected and had human experts' hand-label additional student data to form a *held-out dataset*. We then compared the algorithms' scores with the human scores on new data that were not used

TABLE 22.2

Example Student Practice Profile with BKT Estimates

Designing Ctrl'd Exp's Practice Opportunities	1	2	3	4	5	6	7	8	9	Final
$P(L_{n-1})$	0.077	0.387	0.819	0.97	0.795	0.965	0.769	0.959	0.994	0.999
P(Demonstrate_Skill)	0.191	0.429	0.761	0.877	0.742	0.873	0.723	0.869	0.895	
Observable: Demonstrated Skill?	Yes	Yes	Yes	No	Yes	No	Yes	Yes	Yes	

BKT Model: $\{L_0 = 0.077, G = 0.132, S = 0.100, T = 0.038\}$

Note: This student engaged in nine data collection activities and their final estimate of knowing this skill is $P(L_n) = 0.999$.

to build the models. Several goodness metrics were used: *precision, recall, A',* and *Cohen's kappa* (Hanley & McNeil, 1982; Yan & Bridgeman, Chapter 16, this handbook).

These were computed both at the *student level* (to determine that they are not biased against particular groups of students) and the *topic level* (to determine if/how the algorithms generalize across different *Inq-ITS* science topics; Sao Pedro, Jiang, Paquette, Baker, & Gobert, 2014). The AS algorithms proved to be *robust* across inquiry activities for different science topics and diverse students, strongly matching human scoring (Sao Pedro, Baker, & Gobert, 2013b; Sao Pedro et al., 2014; Sao Pedro, Baker, Gobert, et al., 2013; Gobert et al., 2012, 2013b; Gobert et al., 2015).

Regarding the BKT psychometric model, reliability and validity were determined in two ways. First, we predicted performance within the environment, $P(Demonstrate\_Skill_n)$, providing a measure of the *internal reliability*. Using this metric, we found that BKT could estimate students' proficiency at the sub-skills at each opportunity to demonstrate them within *Inq-ITS* (Sao Pedro, Baker, Gobert, et al., 2013). Second, we used the BKT probability estimates of knowing the sub-skill, (L_n), to predict performance on transfer tasks requiring their usage, providing a measure of *external validity*. Overall, the BKT estimates of each skill obtained from students' performance within *Inq-ITS* were significantly, albeit modestly, correlated to skills estimates from a corresponding transfer test (Sao Pedro, Baker, Gobert, et al., 2013).

Real-Time Feedback

As mentioned earlier, *Inq-ITS* also has an integrated pedagogical agent ("Rex") shown in Figure 22.3 that synthesizes the AS assessment data into immediate feedback for students. This agent provides help proactively (as opposed to students asking the agent for help) because students may lack the appropriate metacognitive help-seeking skills to recognize when they should ask for help (Aleven et al., 2004) and to prevent unproductive, haphazard inquiry (Gobert et al., 2006).

To determine what feedback should be given, a hint message is generated by determining which sub-skills were assessed as incorrect by the AS algorithms and generating an appropriate help message. Messages are layered into multiple levels, with each level providing a more specific hint, similar to *cognitive tutors* (e.g., Anderson & Reiser, 1985; Corbett & Anderson, 1995; Koedinger & Corbett, 2006). Thus, if students continue to struggle, more directed help is provided to the student.

As a concrete example, consider the case when students are collecting data with the simulation, scored using data-mining-based AS algorithms. Two sub-skills are scored: whether students collect data to test their questions and whether they design controlled experiments. Students could be assessed as demonstrating either one, both, or neither. To account for these

FIGURE 22.3

Example feedback given by pedagogical agent Rex when students are not designing controlled experiments nor collecting data to test their question. This feedback is driven in part by assessment data generated from data-mining-based AS algorithms.

possibilities, *Inq-ITS* has different scaffolding levels for each of these cases with each level providing more specific help.

For example, while collecting data, if the system detects that a student is designing controlled experiments but not collecting data to test their hypothesis, the pedagogical agent will tell them "It looks like you did great a designing a controlled experiment but let me remind you to collect data to help you test your hypotheses." If students continue to struggle in this fashion, a bottom-out hint is given: "Let me help some more. Just change the [Variable] and run another trial. Don't change the other variables. Doing this lets you tell for sure if changing the [Independent Variable] causes changes to the [Dependent Variable]", where contents between brackets are specific to the students' choices.

Inq-ITS also decides when the proactive feedback should be given. For example, when students are collecting data, scaffolding may be triggered when students signal that they have completed the data collection task, or (more proactively) as they are collecting data with the simulation. As students are running the simulation, the autoscoring algorithms score the students' processes and if they are deemed to be off-track, the system will jump in and provide support. This scaffolding approach has been shown to support students' acquisition and transfer of inquiry practices within and across multiple science topics, as measured by performance within *Inq-ITS* (e.g., Sao Pedro, Baker, Gobert, et al., 2013; Gobert, Moussavi, et al., 2018).

Conclusion

In the course of discussing AS in simulation-based ITSs, we have had to consider a larger picture of designing simulations, simulation-based learning, and simulation-based assessment. When the purpose of a digital simulation environment is learning, attention is on the cognitive patterns and activity patterns that people develop to be able to perform effectively in the relevant situations. Thus, we have discussed in some detail not only AS processes but the *evidentiary-argument considerations* that go into designing the environments in which the interfaces, problems, affordances, and situations are realized, producing the activities that AS will address in the service of a system's purposes. Yan and Bridgeman (Chapter 16, this handbook) discuss the evaluations of AS systems including ITSs.

In the special case of the embedded assessments that are central to simulation-based ITSs, a long-standing lesson applies: to design a rich simulation environment to collect data without consideration of how the data will be evaluated, and hoping analysts will somehow "figure out how to score it," is a bad way to build assessments. Data-analytic techniques in general, and those for AS in particular, are limited by the strength of the relationship between what is in the data and its connection to the targeted inferences (Mislevy, Behrens, Dicerbo, & Levy, 2012). Close collaboration and interaction from the very start of the design process is preferable – among users, who understand the purposes for which the assessment is intended; domain experts, who know about the nature of the knowledge and skills, the situations in which they are used, and what test takers do that provides evidence; data scientists and psychometricians, who know about the range of situations in which they can model data and examine its evidentiary value; and software designers, who build the infrastructure and implement the AS systems to bring the assessment to life.

23

Scoring of Streaming Data in Game-Based Assessments

Russell G. Almond

CONTENTS

Consider the game *Physics Playground* (Shute, Ventura, & Kim, 2013; Kim, Almond, & Shute, 2016). In this game, players manipulate a scene by drawing new objects or manipulating physical parameters to guide a ball to a balloon. The educators' goal is to use this game to draw inferences about the players' understanding of physics. More important than the *product* of the game level (essentially, whether or not the player solved the level) is the *process* by which the player solved or attempted to solve the level. As is done in many game- and simulation-based assessments, the process is captured by *logging* a series of *events* to a database.

Figure 23.1 shows a screenshot from the *Physics Playground* tutorial. The goal of each level is to get the ball to the balloon by using the laws of Physics. In this case, drawing a lever following the dotted lines solves the level. Figure 23.2 shows a manipulation level, in which the player manipulates sliders and clicks on the puffer to solve the level. The game generates an event for each player action ('object drawn' or 'control manipulated'), as well

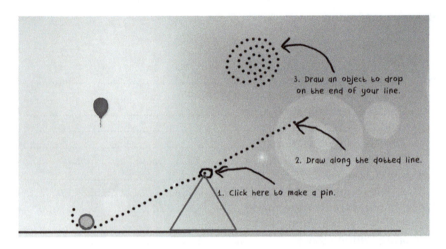

FIGURE 23.1
Screenshot of tutorial segment from *Physics Playground*.

FIGURE 23.2
Screenshot of manipulation level from *Physics Playground*.

as some higher level events, such as the system recognizing a drawn object as a lever, or the player solving the game level.

The *delivery environment* for *Physics Playground* also differs from traditional test administration. The game is written in Unity® and played in the browser of a remote computer. When a player starts, the *game software* is downloaded onto a local computer which then sends log messages back to the *server* over a school network (which sometimes has *capacity limits* and *firewall issues*). The game is also played over several days, so that the stream of data from any one player is split into several *sessions*. Consequently, the scoring engine may not know if more data will come later. Completing the game may not even be important; if the goal is to guide instruction, or to use adaptive presentation of tasks, it may be necessary to make periodic inferences about the players' abilities based on the data received so far.

The data whose analyses are addressed in this chapter have two key differences that distinguish them from traditional psychometric data. First, the data are *streaming data* (Akidau, Chernyak, & Lax, 2018) in that they are potentially infinite and arrive in increments at the processing server. Because of networking issues between the client and server, the data may arrive out of order or some portions may be lost. Furthermore, approximately real-time inferences are required based on the available data. Second, the event data are *voluminous*; for example, nearly two hundred students testing *Physics Playground* for five class periods generated almost four million events. These data streams must be summarized before they can be sent on to the psychometric models that will be used to draw inferences about student abilities.

In this chapter we walk through the processing stream from events generated by the game to final reported scores. The first section presents the *four-process architecture* for assessment delivery modified for streaming systems. The second section defines *scoring windows* that act like *tasks* in a streaming system. The next two sections describe the *evidence identification* (EI) and *evidence accumulation* (EA) processes, respectively, which form the heart of the scoring. The final section summarizes some lessons learned from the *Physics Playground* field test.

Four Process Architecture

This chapter adapts the four-process architecture (Almond, Steinberg, & Mislevy, 2002) to work with streaming data. The four-process architecture was originally designed to communicate some of the processing requirements that would come from new types of learning and assessment created through *evidence-centered assessment design* (Mislevy, Steinberg, & Almond, 2003) to the IMS Consortium Working Group on Question and Test Interoperability. In particular, Almond and Steinberg (as working group contributors) tried to encourage the IMS working group to leave room for constructed response tasks that might have complex work products, including event logs, and complex *Bayes nets* (Almond et al., 2015) or *diagnostic classification models* (e.g., Rupp, Templin, & Henson, 2010) that might have multivariate scores. The original four processes (Almond et al., 2002) were arranged as a cycle, with the fourth EA process feeding back into the first (activity selection) process to choose the next item or task. For streaming processing the four processes are best viewed as a pipeline however in which data are successively summarized.

The Four Processes of the Evidence Pipeline

Figure 23.3 shows the four-processes rearranged as a pipeline. In this view, the processing stream starts with the *context determination* (CD) process (called 'activity selection' in Almond et al., 2002), which determines the task or item presented to the examinee. This signals the *evidence capture* (EC) process (called 'presentation' in Almond et al., 2002) process to start collecting evidence. For the purposes of this chapter, the raw evidence is in the form of *log events*. These events flow into the EI process, which summarizes the event stream as a stream of *observables:* variables that summarize what was done in a particular *context* or *scoring window*. These observables flow into the EA process, which synthesizes evidence across contexts, often with the help of a psychometric model. The EA process produces *statistics* – scores – summarizing what is known about the examinee. In an *adaptive system*, these statistics feed back into context determination to select the next activity.

The following is a detailed description of the four processes.

Context Determination (CD)

The CD process is responsible for sequencing the activities or tasks that the examinee sees. It is responsible for starting and stopping the assessment. Automatically stopping an assessment implies that enough evidence has

FIGURE 23.3
Pipeline view of the four-process architecture.

been gathered to make sufficiently reliable inference about the domain of interest. In an adaptive assessment, the CD process may use statistics emitted by the evidence accumulation process to determine the next activity. The key output of this process is *watermarks* that identify the beginning and end of a scorable context (see Section 22.3). As noted above, Almond et al. (2002) called this process *activity selection*. The name is changed here because in the context of a game, simulation, or observation of a natural activity, the context determination process may be defining the boundaries of the scoring windows rather than scheduling the tasks.

Evidence Capture (EC)

The EC process presents some kind of stimulus material to the examinee, and then captures the response or *work product*. In game-based assessment, the EC process is usually (part of) the game engine. This chapter focuses on the case where the key work product is an event log of interactions with the game or simulation. Thus, the output of this process is a stream of *events*. Note that in cases in which the task is to produce a constructed response of some kind, the submission of that response can be logged as an event, so the architecture is general enough to encompass both process and product scoring. Almond et al. (2002) called this the *presentation process*; the name evidence capture emphasizes its role in capturing the work product rather than its role in presenting the stimulus.

Evidence Identification (EI)

The EI process takes care of the summarization of data within a single scoring window. The event data can contain arbitrarily complex data objects, but typical psychometric models assume that the input variables are collections of nominal, ordinal, interval, and scale variables. The EI process can be as simple as determining whether the offered solution is a "success" or a much more complex process that determines a large number of scorable features of the work product, called *observable variables*. The messages from the EI process to the EA process contain vectors of observables associated with a particular scoring window.

Evidence Accumulation (EA)

The EA process is responsible for summarizing the evidence across the scorable windows. Typically, it maintains a *student model* that tracks what is known about a specific examinee. The student model is updated when the observables arrive from the EI process. In turn, it outputs a message containing a vector of *statistics* – summaries of the updated student model. These are usually the outputs of interest of the system; however, the CD may also use them for determining the next task in an adaptive system.

Proc4 Messages

The package `Proc4` (https://pluto.coe.fsu.edu/Proc4/) provides a reference implementation of the four-process architecture. In particular, it provides a data structure to use as a template for messages sent between processes. The messages sent in `Proc4` are split into a *header* and a *data* portion. The header has a series of standard fields that are used to route and prioritize the message. The data portion is a series of tagged values; the allowed tags and the ranges for the values are defined by the application and the message. Note that the values themselves can be complex objects. Listing 23.1 shows a prototypical message in JSON (JavaScript object notation; Bassett, 2015) format.

The header fields have the following definitions:

app

> A globally unique identifier (guid) for the assessment application. The URL-like structure is intended to allow each organization to issue its own app IDs.

uid

> An identifier for the student or player.

context

> An identifier for the scoring context (scoring window) in which the message was generated. In *Physics Playground* this corresponds to game levels, but it might have other meanings in other applications. Section 22.4 explores this in more detail.

sender

> An identifier for the process generated the message.

```
1   {
2     app: "ecd://epls.coe.fsu.edu/PP",
3     uid: "Student 1",
4     context: "SpiderWeb",
5     sender: "Evidence Identification",
6     mess: "Task Observables",
7     timestamp: "2018-10-22 18:30:43 EDT",
8     processed: false,
9     watermarks: {"TaskComplete":"2018-10-22 18:30:43 EDT",
10                  "EICompleted":"2018-10-22 18:33:43 EDT"},
11    data:{
12      trophy: "gold",
13      solved: true,
14      objects: 10,
15      agents: ["ramp","ramp","springboard"],
16      solutionTime: {time:62.25, units:"secs"}
17    }
18  }
```

LISTING 23.1
A typical message in JSON format in *Proc4*.

mess

A subject line for the message indicating its content.

timestamp

A timestamp for the message. Generally, messages for the same uid need to be processed in chronological order. The timestamps are usually related to event time in the EC process.

processed

A flag that can be set after the message has been processed. This allows databases to be used as a queue.

watermark

This is a newly added field (not yet in the reference implementation). Watermarks allow tracking the progress of the scoring, as well as helping to ensure completeness of the data. This is a list of fields with timestamps.

The data field is an unrestricted container; the subfields of the data field and their possible values are defined by the app and contextualized by the header fields. As the contents of the data field are not determined by the Proc4 message format, the data field is an extension field; and the only one in Proc4. This differs from the way the xAPI format (Betts & Smith, 2018) supports extensions in nearly every field (making each field a compound object).

The xAPI format also makes other header fields into guids, making the messages longer and harder to read; while Proc4 uses only one guid field (app). This reflects a difference in approach: the xAPI approach tries to make the event data log a complete record including all necessary metadata while the Proc4 approach assumes that there is an external code book for the data which can be found using the app id. The Proc4 approach results in more compact storage and display of events.

The xAPI format does introduce two additional header fields, verb and object, particularly useful for messages coming from the EC process. In xAPI these are complex elements consisting of both data and metadata defining the vocabulary. In Proc4 these are simple identifiers, with the assumption that the application id will point to a code book defining the vocabulary. Listing 23.2 shows an example.

Databases, Queuing, and the Dongle

Each of the four processes has a queue of messages that come from the preceding processes. One simple way to implement that queue is to use a database. Note that Proc4 messages have a processed field that allows the process to tell that the message has been processed. Thus, the database forms a natural queue. The next message to process is the unprocessed message with the oldest time stamp. This even provides some amount of buffering against late arrive messages.

```
1   {
2     app:"ecd://epls.coe.fsu.edu/AssessmentName/StudyCondition",
3     uid:"Student/User ID",
4     timestamp:"Time of event",
5     verb:"Action Keyword",
6     context: "Context ID",
7     object:"Object Keyword",
8     processed: false,
9     watermarks:{...},
10    data:{
11      field1:"Value",
12      field2:["list", "of", "values"],
13      field3:{part1:"complex", part2:"object"}
14    }
15  }
```

LISTING 23.2
An event record in *Proc4*.

Physics Playground used the Mongo® no-SQL database (MongoDB, Inc., 2018). Mongo is a document database, which stores JSON documents like the ones in Listings 23.1 and 23.2. This kind of database allows storing semi-structured data. In the `Proc4` implementation, the header fields have a well-defined schema and specified types but the data field can contain arbitrary objects. In particular, `Proc4` builds indexes on the header fields making queries faster.

Each of the four processes also sends its output to the database. When a process encounters a *trigger* in its processing instructions, it sends an output, often in the form of a message. There are two roles for these trigger messages. One is to join the input queue of another process and the other is to update a collection in the database that collects information about what is known about an examinee.

Physics Playground uses these collections of information to buffer communications going from the scoring processes back to the game engine. The database collections are attached to the web server through a series of simple scripts called a *dongle*. When the game engine needs information from the scoring processes, it posts a message to a designated URL and the web server script responds with the most recent record for that player in the corresponding database collection. *Physics Playground* uses this to restore the status of the players when they restart after a pause, to update the scores in the players' *backpacks* (i.e., a display of their progress on the physics skills, see Figure 23.4) and to get information about the players' skills for use in adaptive selection of game levels.

The dongle queries always return quickly because they do minimal processing, and consequently are only subject to network delays. The information may not be up to date, in the sense that there may still be unprocessed events that will update the player status, but the information will be the best available at the time it is requested. This means that the player should not experience delays waiting for the scoring processes.

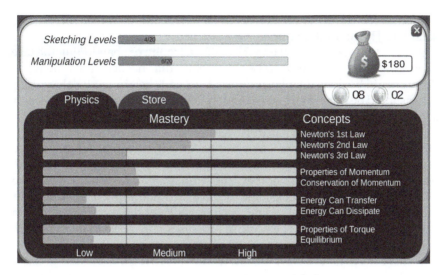

FIGURE 23.4
Screenshot of backpack/progress indicator from *Physics Playground*.

Finally, caching the message queue in the database provides some measure of security against system failures. Most databases have built-in redundancy and support data mirroring and logging. Consequently, when the messages are stored in a database it is easier to recover more from system crashes and rerun scoring if the scoring models are updated.

Tasks, Windows, and Watermarks

Evidence-centered assessment design used the word *'task'* to encourage test developers to think beyond simple items in developing assessments (Mislevy et al., 2003). In the original four process paper (Almond et al., 2002), a task was defined as the unit of data for messages propagating around the system. In the pipeline view of the four-processes, this definition no longer seems appropriate. The EC, EI, and EA processes all input data at a finer grain size than they output. In this section we take a closer look at the definition of task, which may be slightly different for each assessment. Two terms from the theory of streaming data, *windows* and *watermarks* will be useful in expanding the concept of task for scoring data.

Predefined vs. Emerging Tasks

Physics Playground consists of a number of game levels. Each level consists of a single puzzle with a single goal: getting a ball to a balloon. Successful players earn a trophy (a gold or silver coin, depending on the efficiency of a solution). The natural definition for a *task* is the period of time between the player starting the level and either earning a trophy or abandoning the level. Even this simple definition raised questions that needed to be answered: What should be done if a player quits a level and then immediately restarts? What should be done if a player abandons a level and then later returns to try again, or to earn a gold coin after previously earning a silver coin?

Physics Playground has a natural level structure corresponding to tasks, but other games and simulations may not have such a clear structure (Mislevy et al., 2015). Consider a flight simulator task in which the goal is to fly the plane from one airport to another. The simulation will pass through many phases (e.g., pre-flight checks, take off, cruising, handling a weather event, more cruising, approach, landing) each of which will have different scoring rules (evidence models). The scoring processes may need to determine the task boundaries from the stream of events.

In short, 'task' is no longer a good word to describe this bundle of events. In the flight simulator examples, each of the scoring contexts share the same overall goal (reaching the destination airport) and therefore from the point of view of the activity selection process all the contexts are part of the same task. Almond, Shute, Tinger, and Rahimi (in press) call these *(scoring) contexts* (this is how they are notated in the Proc4 message, see Listing 23.1). The word 'contexts' is less than ideal as it has too many other uses; however, the theory of streaming data processing has a useful term: *window*.

Scoring Windows

In stream processing, *windows* are a way of dividing a time stream into non-overlapping intervals in which processing takes place (Akidau et al., 2018). In particular, a process gathers all of the events in a window and then outputs a summary. Event windows can be defined in either processing time or event time, with the latter usually having better validity for scoring. The most critical window is the one used for the EI process (the EC process outputs events as they are generated and the EA process accumulates across scoring windows).

We define a *scoring window* for the EI process as a region in event time (for a certain user and a certain assessment application) which has the following properties:

1. It does not overlap with any other scoring window (with the same user and application).
2. The observables to be output from the EI process can be calculated without information from any other scoring window.
3. The EA process can regard the observables from different scoring windows as conditionally independent given the variables in its proficiency model.

The first property is simply part of the definition of a window. The third property is required for many psychometric models. Particularly, this corresponds to the expanded definition of the task independence which is necessary for correct inferences from *Bayesian scoring models* (Almond et al., 2015) and other types of *latent-variable models* commonly used in educational measurement. Note that this property is often assumed when it does not hold in practice. Consider a game-based assessment given to young children. Their eye-mouse coordination skills may affect the performance in many scoring windows, but motor skills may be omitted from the proficiency variables in the EA process as they are not a focus of the assessment. This could cause the model to underestimate the skills being measured when the real problem was one of unmeasured motor skills.

The second property is critical for establishing the robustness of the computations; in particular, robustness to events arriving out of sequence. In *Physics Playground* the player's bank balance is tracked through a series of rules in the EI process that adjust the balance whenever the player earned or spent game money. According to the scoring window definition, the bank balance should be the responsibility of the EA process as it crosses scoring windows. In practice, it was easiest to simply update an observable variable in the player's state in the EI process every time the bank balance changed. This required the bank balance from one scoring window to be carried over into the next, a violation of the second property. The implemented solution is not robust to late-arriving events (which fortunately was not an issue in the pilot testing). Moving the bank balance functions to the EA process would improve robustness.

Watermarks as a Tool for Establishing Data Completeness

A *watermark* is "an assertion that no more data with event times less than E will ever be seen again" (Akidau et al., 2018, p 56). These can be perfect (known true) or heuristic (probably true). When a trigger posts a message updating the status of the player, the message is early, on time or late with respect to the watermark. Late messages occur when late arriving data cause a need for a correction. They are a way of letting the downstream consumers of process output know about the completeness of the inferences.

First, let us take a look at the role of watermarks in the EI process. The EI process must be able to make inferences about when all events from a particular scoring window have been observed. In *Physics Playground*, the watermark was established by looking at the 'initialized game level' event. If this event indicated the player had moved to a different level, a watermark was issued and an 'on time' message containing the observables was sent to EA.

This presents a possible solution to what to do if a player goes back for a second attempt at a game level. In this case, the watermark is a heuristic one (as the computer cannot predict the players' future choices). The updated observables would be sent to the EA process, but be marked as 'late.' The

EA process would then be responsible for properly replacing the out of date information.

The EA process, in contrast, requires a new kind of watermark that comes from the assembly model (the part of the evidence-centered assessment design responsible for the completeness of an assessment form). The joint standards (AERA, APA, & NCME, 2014) state that the reliability and validity of an assessment must be sufficient for the purposes. To help ensure sufficient reliability, a sufficient quantity of evidence must be gathered about the reported proficiency variables to have an adequate standard error of measurement. To help ensure validity, the tasks or contexts of the assessment must adequately cover the domain of interest. A *psychometric watermark* indicates that these requirements are met.

The EA process can output statistics at any point in the process. Before the psychometric watermark, results are 'early.' These may be good enough for formative purposes (e.g., planning instruction or sequencing tasks in an adaptive assessment), but not for more formal inferences. The statistics issued with the psychometric watermark are the 'final' scores of the assessment. These are the ones that the test users need for drawing inferences. If data arrive late, then 'late' scores are also a possibility, this is an indication that records for the student containing those scores should be updated.

The *EIEvent* Language for Streaming Processing

In this section, we take a deeper dive into the EI process used for *Physics Playground*; in this case a new software system called EIEvent. Following HyDRIVE (Mislevy & Gitomer, 1996; Mislevy et al., Chapter 22, this handbook) it is a *rule-based system*. Each player is given a state that tracks that player through the scoring window. This makes the system a finite state machine, similar to the system used by Almond et al. (2012) to label *keystroke events* in writing process analysis (see Deane & Zhang, Chapter 19, this handbook).

EIEvent is a streaming system. Its main processing loop works as follows:

1. Fetch the next event—oldest unprocessed event—from the queue. (Note that separate queues/loops can be used for each application and user.)
2. Fetch the cached state for the corresponding user.
3. Find the rules that are applicable to the current event and state.
4. Run the rules to update the state.
5. If there are applicable trigger rules, these generate output messages which are stored in the database (including the queue for the EA process).

6. Update timestamps and watermarks.

7. Repeat.

Prototype code and more complete documentation are available at https://pluto.coe.fsu.edu/Proc4.

States and Events

Events are just messages from the EC process (see Listing 23.2). Following the xAPI protocol (Betts & Smith, 2018), they have the additional *verb* and *object* headers. These headers are used to quickly filter rules (e.g., there are rules that are applicable only with certain verbs and objects). Note that the author of the rules knows what data come with a particular verb and object combination, and thus can write rules that exploit those data.

The *status* or *state* object tracks a player through a scoring window. It contains three collections of variables:

Observables

These are values that will be output in the future.

Flags

These are values that are intended for internal state tracking, but will not be output.

Timers

These are like flags but are specifically used to determine the amount of time a player spends in a certain state.

Listing 23.3 shows a generic status object.

Both flags and observables are variables of arbitrary type (categorical, numeric, vector valued, or a more complex data object). The difference between an observable and a flag is merely one of usage. A variable is an observable for one of three reasons: (a) it will be used as an input to the EA process, (b) it is

```
1   {
2       app:"Global identifier for the application"
3       uid: "Identifier for player",
4       context: "Task or context identifier",
5       oldContext: "Task or context identifier.",
6       timestamp: "Timestamp of last event processed.",
7       watermarks: "List of watermarks.",
8       timers:{...},
9       flags:{...},
10      observables:{...}
11  }
```

LISTING 23.3
A generic status (state) object in *EIEvent*.

monitored by a system which tracks the status of a student, or (c) it is logged to a database for future research, either into the construct being studied or performance of the software. At the end of a scoring window, a trigger rule causes the observables to be output from the EI process. Additional trigger rules can output a subset of the observables for special purposes.

The context field of the status is also used to filter rules, but it is slightly more complex as there is a hierarchy of contexts. In *Physics Playground*, game levels are divided into sketching levels and manipulation levels, each of which forms a context set. The sketching levels set contains subsets for ramp, lever, springboard, and pendulum levels depending on which agents of force and motion are useful for solving the problem. These sets are not distinct. A rule may be applicable for a specific context or for a set of contexts (including the set of ALL contexts).

The inclusion of both context and oldContext (see Listing 23.3) is used for detecting the end of a scoring window. Scoring windows are marked by a change in scoring context. When this is detected, context is updated, but oldContext is left unchanged until the scoring window EI processing is complete. Testing oldContext and context for equality is often used for marking the completion of a scoring window. The oldContext field is set to the value of context at the penultimate step of the event processing loop.

Rules

EIEvent is actually a programming language with the program written using a series of rules. The rules are expressed using JSON notation as in Listing 23.4. The condition field contains a number of tests written a language based on the Mongo query language (MongoDB, Inc., 2018). The predicate field gives instructions to be executed—usually updates to the status—if the condition is satisfied. The language for predicates is again based on the Mongo query language.

A rule is *applicable* if its verb and object fields match those of the event and if its context field matches the current status for the player (i.e., the level which is currently being scored). The rule can reference a context set instead of a context. For example, if the rule's context was 'Sketching Levels' any sketching level would match. The verb, object, and context fields can also contain the wildcard 'ALL' in which case any event or context would match.

If the rule is applicable, its condition is checked. The condition is a series of logical tests, all of which must be true for the condition to be satisfied. If the rule is applicable and the condition is satisfied then the predicate is run.

Rule Types

In a classic rule-based system, the order in which the rules are run is arbitrary, but, in practice, the programmer needs some control over order. The ruleType field divides the rules into five types:

```
1   {
2     app:"Global identifier for the application",
3     ruleName:"Human readable identifier",
4     doc:"Human language description",
5     context:"Context or Group Keyword",
6     verb:"Action Keyword or ALL",
7     object:"Object Keyword or ALL",
8     ruleType:"Type Keyword",
9     priority:"Numeric Value",
10    condition:{
11       <field1>:<value1>,
12       <field2>:[<value-list>],
13       <field3>:{"?op":<value3>},
14       ...
15    },
16    predicate:{
17       <update operator1>:{ <field1>: <value1>, ... },
18       "!set":{ state.flags.<logical>: true},
19       "!set":{ state.timers.<name>.running: true},
20       "!incr":{ state.flags.<count>: 1},
21       ...}
22  }
```

LISTING 23.4
A basic rule schema in *ElEvent*.

1. *Status Rules*—Update flags and timers
2. *Observable Rules*—Update observables
3. *Context Rules*—Check for changes in context (end of scoring window)
4. *Trigger Rules*—Send messages to other processes
5. *Reset Rules*—Clean up state at end of scoring window

Status rules are always run before observable rules which are always run before context rules and so forth. The `priority` field provides a mechanism for sequencing rules within a type: rules with lower priority numbers are always run first. The state and observable rules provide a mechanism for updating the state. Both types of rules can update fields in the flag, timer, or observable categories in the status; however, in most processing the flags and timers need to be updated before the observables. Therefore, status rules always run before observable rules.

The context, trigger, and reset rules deal with end of scoring window processing. The context rules check to see if the scoring window has closed. If so, they change the `context` field of the status (but not `oldContext`). Trigger rules can run at any time, but they usually run at the end of a scoring window. They output a message from the EI system, which other systems listen for. In particular, they provide the observables to the EA process. The reset rules are run after the trigger rules, but only when the scoring window changes. They reset the values in the status to the initial values for the new scoring window. Almond et al. (in press) provide some examples of the EIEvent rules in action. Additional examples are available in the EIEvent documentation.

Performance Issues

The streaming four-process architecture assigns a different grain size to the EC, EI, and EA processes. The EC process inputs user actions and outputs messages at the level of raw events. The EI process input raw events and outputs scoring-window specific observables. The EA process inputs scoring-window specific observables and outputs assessment level scores. Thus, a theoretically clean implementation of the processes would put scoring rules that happen within a window in EI and across windows in EA.

The clean theoretical implementation is not necessarily the best one in practice. Often scoring steps that theoretically should be placed in EI can be more efficiently implemented in EC or EA. Two examples from *Physics Playground*:

- One set of evidence rules determines whether an object drawn by a player is a lever, pendulum, ramp, springboard, or something else. These rules require detailed knowledge of the movement of the object; thus, it was easier to implement them in the EC process which could use its physics engine to calculate object characteristics. The EC process then generated an event that it had 'identified' a 'game object.' The EI process simply looked for that event rather than trying to reproduce detailed game mechanics on its own.

- The EA process evidence models include a node for time spent on the level, but this was expressed as an ordered categorical variable. The final step is to transform the continuous measure of time into the category. In this case the EI process outputs the duration as a continuous observable. The EA process is responsible for the categorization. This means that the EI code does not need to be updated if the scoring model using by EA is changed to use a different cut point.

In a system that uses adaptive sequencing of tasks such as *Physics Playground*, the processing speed of the EI process must be high. *Physics Playground* used a prototype implementation of the process which was not optimized. The first day of testing in our large field trial was a Friday and included only a half day of game play because of the need for pretesting. Even so, monitoring the progress of the event processing queues, it was unclear if the processes would finish over the weekend. Faster processing was necessary.

One technique that resulted in a large increase in processing speed was filtering the events. There was a large number of events (e.g., "ball position") which were simply not used in any of the evidence rules. Filtering the event stream to remove these events yielded a dramatic speed up. Note that another way of speeding processing is to split evidence rules, using just those rules necessary for the observables that are required in real-time in the first set, and the observables that are used for diagnostic and research purposes in the second set. The second set of observations can be then calculated in a second, lower priority process.

Clearly *parallel processing* is necessary for a system at scale. In *Physics Playground* each of the three study branches was given a separate process, providing a threefold increase in speed. Ideally, the processing could be split using (user, application) as the key. This requires the event processing loop to be robust to the possibility of multiple worker threads working on data for the same user. Also, there is a bigger chance that data will arrive out of sequence when the queues are shorter. Because the queues are processed according to event time rather than arrival time, if the queue is long enough it provides a buffer for late arriving data.

One natural way to get more processing power is to put more of the EI processing onto the client machine which runs EC. There are two difficulties here. First, this strategy moves the load from the server to the client, so it is important to make sure that the client is still responsive to the user input. Second, when events are sent from the client to the server, they are naturally logged. If this logging is done later at a higher grain size, then the system is less robust to failures of the client and the network, and problems in the EI processing that take place on the client using unlogged inputs are harder to debug.

Evidence Accumulation

As the EI process does scoring within a scoring window, the EA process, following its name, does accumulation across scoring windows. It produces a final score, or collection of scores, for the entire assessment. Chapter 13 of Almond et al. (2015) describes a general scoring mechanism for Bayes nets. The basic algorithm described in that chapter can be adapted to a number of other scoring models.

An important constraint on the data coming into the EA process is that observables from different scoring windows are independent. In this case, the EA algorithm only needs to consult its stored state and observables from the scoring window currently being processed. It also implies that the EA process is robust to the order in which the observables arrive. In the case of Bayesian models, this constraint amounts to the familiar local independence assumption of items (or tasks, or contexts).

Basic Scoring Algorithm

A *Bayesian scoring model* consists of the following parts: a core *proficiency model*, a collection of *evidence models*, and a set of *statistics*. The proficiency model defines a number of variables that describe the proficiencies to be measured and the population distribution of the variables. The evidence model provides the likelihood of the observables from a particular scoring window context or task, given the proficiency variables. The statistics are functionals of the proficiency model distribution that produce the scores of interest.

Each (user, application) pair is assigned a *student model*. As new evidence arrives, *Bayes theorem* is applied with the current student model distribution as the *prior* and the appropriate evidence model as the *likelihood*. The student model is updated to contain the resulting *posterior*. Thus, the student model summarizes the evidence considered so far. For streaming data, this algorithm can be expressed as follows:

- Process next (lowest timestamp) message which contains a collection of observables from a given scoring window.
- Fetch the appropriate student model for the uid and app in the message header. If no student model is available, create a new one from the proficiency model.
- Fetch the appropriate evidence model given the context field of the message.
- Incorporate the evidence into the student model by an application of Bayes rule using the student model as prior and evidence model as likelihood.
- Save the student model.
- Calculate the statistics using the student model, and send a message indicating the new statistic values.
- Repeat.

Many existing psychometric models fit into this Bayesian framework. For example, item response theory models fit the framework if a prior distribution is placed over θ. Many diagnostic classification models also fit if a prior distribution is placed over the attributes (proficiency variables).

While the Bayesian theory provides justification for the algorithm, a Bayesian scoring model is not necessary. All that is necessary is for the scoring model to have a mechanism for performing Steps 2, 4, and 6. Consider weighted number-right scoring. For Step 2, the proficiency model defines a number of counters and their initial values. For Step 4, the evidence models define which counters get updated and by how much based on the observable values. For Step 6, the statistics define summary functions of the accumulator variables.

Retraction: Late and Duplicated Data

Part of the correctness of the scores coming out of the EA process relate to the independence of observables from different scoring windows. This assumes that the context associated with two different scoring windows is always different. In a game, scoring contexts are often associated with repeatable parts of the code. For example, in the user-controlled mode for *Physics Playground*, the player could go back and revisit a previously attempted level. The player may or may not have earned a coin (solved the puzzle) in the previous attempt.

There are several ways of handling the duplicated data:

Score First Attempt

Only count the first attempt; discard later attempts.

Score Last Attempt

Count the most recent attempt. When a duplicate context is detected retract evidence from the previous attempt and add evidence for the new attempt.

Summarize Across Attempts

Combine the observables from the previous attempt with the ones from the current attempt to make a new set of observables. As with the previous case, evidence from the previous observables must be subtracted and replaced with the new evidence.

Treat Attempts as Independent

Ignore the dependence between multiple attempts at the same problem and treat them as if they are independent. Here all scoring windows are accumulated.

Down-Weight Past Evidence

Down-weight past attempts by multiplying the likelihood by some number λ which is less than one. The immediately previous attempt is given the weight λ and the one prior to that λ^2 and so forth. This produces an exponential filter (Brockwell & Davis, 2002). Implementing this strategy requires retracting the previous evidence and replacing it with the new filter weights.

None of these strategies are ideal. There is more information in two attempts than in one, so the first two strategies are too conservative. On the other hand, the fourth strategy, treating duplicated scoring windows as independent, is too liberal as the attempts are not truly independent. The last strategy is a compromise that should produce an amount of evidence somewhere between the first two strategies and the fourth, but it is the most difficult to implement and lacks a strong theoretical foundation.

The third strategy, combining observables, requires opening up the EI process and shifts the responsibility of summarizing across attempts to that process. Another way to think about this strategy is that the two attempts represent the same scoring window and so the scoring windows should be merged to produce a new window. This is the hardest strategy to implement because it requires defining the rules for merging the observables.

Note that three of the strategies require the ability to retract evidence as well as add new evidence. While it is certainly possible to replay the scoring using cached data, that method is not the most efficient.

Early, On-Time and Late Scores

The theory of streaming and watermarks (Akidau et al., 2018) provides three more terms for talking about scores: *early*, *on-time*, and *late*. In a streaming system, these are used to mark outputs from the stream. The watermark is set to a time when the stream thinks it has seen all the data for the window. Outputs issued before the watermark are early, and those issued after the watermark (because data arrived unexpectedly after the watermark) are late. The output issued with the watermark is on-time.

From a psychometric standpoint, scores are suitable for a particular purpose only when enough evidence has been gathered to make them sufficiently reliable for that purpose. This often means that the contexts in which evidence is gathered must span some construct of interest. This is often expressed as a test blueprint. In adaptive systems, this coverage requirement is often expressed as a series of constraints.

When these constraints have been satisfied, the EA process (or maybe the CD process) can issue a watermark stating that the scores are valid for a particular purpose. Scores computed before that watermark are early. They may be useful for formative purposes, particularly selecting tasks in an adaptive assessment, but they should not be reported to end users. Scores that are on-time with respect to the watermark *can* be used for the formal purpose of the test. Scores that are marked as late indicate revisions to the scores because of new evidence. Many high-stakes testing systems try very hard to avoid such revisions, but this often means that the scores are delayed to the point where they cannot be used formatively.

The instructionally embedded model used by the *Dynamic Learning Maps®* (DLM) *Consortium* (DLM Consortium, 2019) is an interesting application of this idea. The DLM assessments consist of a number of testlets, each keyed to a topic (called an essential element) and a difficulty band (called a linkage level). In the year end model, the student is assigned a testlet from each topic, with the difficulty band selected by an adaptive algorithm. This procedure always results in a complete form of the assessment. In the instructionally embedded model, the teacher is free to present the testlets at a time and place where they fit within the instruction. The teacher can give more than one testlet per topic (often in different difficulty bands) if this seems justified for the particular student. The teacher can use early scores to make instructional decisions. Then there is an end of the year window in which the student is assigned testlets to fill out the blueprint. This provides the on-time score at the end of the school year for accountability purposes.

Psychometric watermarks provide another way to think about accountability testing. Schools and teachers could give testlets that provide evidence about a subset of standards at various points in the school year. These could be positioned closer to the point of instruction, avoiding long reviews before the end-of-year test. The testlets would be shorter, and hence easier to schedule. In many cases, they could be replaced by portfolios of student work (e.g., writing samples produced in class rather than timed essays written to a prompt). Meanwhile, the system would track the quantity and kind of evidence that was gathered. When sufficient evidence was gathered, on-time scores could be released for accountability purposes. Before then, early scores could be used for formative purposes by teachers, schools, students, and parents.

Discussion

Physics Playground was built using the four-process architecture. Employing a database to cache the messages passed through the system made it robust to the inevitable bugs that needed to be fixed in a prototype system. In particular, if a bug was discovered in the EI process, it could be fixed and then the input stream replayed to correct the scores. The use of the dongle meant that even though there were performance delays in the scoring, the game play continued.

Scoring performance was an issue with *Physics Playground* field trial. Some of the issues were due to the nature of prototype code, which was optimized for safety and debugging rather than speed. Addressing some issues will require rethinking how parallel and streaming processes can be used to speed the process. In particular, the prototype evidence identification code* needs to be rewritten to support more parallel processing.

Rethinking the four-process architecture as a streaming system is an important development in its history. While in some ways it is useful to think about it as a cycle, it other ways it is useful to think about it as a funnel. The EC process generates lots of low level events; the EI process summarizes the events within a scoring window as a collection of observables and the EA process summarizes the observables across scoring windows as a score.

Although the four-process architecture is a useful model for assessment design, it is not always optimal for assessment construction. In particular, it may be more efficient to do the work of the EI process in the EC process (e.g., game engine) or EA process. Also, the EI process itself may be a complex pipeline with several stages. Consider ETS's *e-rater*® system for scoring essays (Attali & Burstien, 2006; Cahill & Evanini, Chapter 5, this handbook). First the essay goes through several processing stages designed to look at grammar, usage, mechanics, style, vocabulary, and other aspects of document and

---

* Available at https://pluto.coe.fsu.edu/Proc4/

discourse structure. These could be – and probably are – implemented as parallel processes all producing intermediate observables. These are then combined into a final score for the essay. In many assessments, that is combined with scores from other essays and multiple-choice portions of the test.

The theory of streaming processes brings with it a number of useful concepts—*triggers, windows*, and *watermark*—as well as computer architectures designed to handle the large volume of events. This chapter is only a start of how to apply these ideas in automated scoring, because both the theory of streaming processes is still young and because I am just learning its subtleties. However, while it opens the door for more robust scoring in a distributed environment, it also raises critical questions about how reliability and validity are defined in a streaming world.

Acknowledgments

Work on this chapter was supported by the National Science Foundation grant DIP 037988, Val Shute, Principle Investigator. Version 1 of *Physics Playground* was supported by the Bill & Melinda Gates Foundation U.S. Programs Grant Number #0PP1035331, and concurrent work on *Physics Playground* is supported by the Institute of Educational Science, Goal 1 039019, Val Shute, PI. Additional support was provided by National Science Foundation Grant #1720533, Fengfeng Ke, PI.

In addition to Val and Fengfeng, many people worked on the *Physics Playground* scoring including Weinan Zhao, Seyedahmad Rahimi, Chen Sun, Xi Lu, Ginny Smith, Zhichuan Lukas Liu, Buckley Guo, Seyfullah Tingir and Jiawei Li.

We would also like to thank the teachers, administrators and especially the IT staff of the schools at which we did our playtesting. Their help has made this game a valuable educational experience for their students.

24

Automated Scoring in Medical Licensing

Melissa J. Margolis and Brian E. Clauser

CONTENTS

The *National Board of Medical Examiners* (NBME) was founded in the early 20th century with the goal of developing "a comprehensive examination of such high standards that [all] state licensing authorities could justifiably accept its results without the need for further examination" (Hubbard & Levit, 1985; p. 4). The NBME develops, administers, and scores the *United States Medical Licensing Examination* (USMLE®), whose successful completion is required of physicians with an M.D. degree who want to become licensed to practice medicine in the United States.

Medical licensing in the United States has a long and storied past (Mohr, 2013) during which several different assessment formats have been used with varying degrees of success. Early licensing examinations were developed and administered by individual states and included written, oral, and practical components. Increases in the number of physicians seeking licensure and advances in the science of testing created both practical and technical pressures that led to an evolution of approaches to assessment for medical licensing.

Until the 1950s, the first two parts of the then three-part examination were *multiple-day written tests* and the third part was a one-day *bedside oral examination*. After several years of study, in 1954 a decision was made to discontinue the use of essays and replace them with multiple-choice questions because research had demonstrated that the scores based on multiple-choice items were both more reliable and showed a stronger relationship with external criteria than those based on essays (Hubbard & Levit, 1985). At that time, the third part of the examination remained unchanged because there was a belief that the proficiencies assessed represented the true essence of medical practice: the only appropriate way to evaluate whether a candidate was able to appropriately perform these key practice activities was to have a physician observe and evaluate the candidate engaging in these activities (Hubbard & Levit, 1985).

In spite of the belief that the bedside oral examination provided an important and unique assessment of the skills required for the practice of medicine, by the 1960s it became obvious that its continued use was unsustainable. *Psychometric evaluations* raised serious questions about the standardization of the examination, and practical issues related to recruitment of sufficient numbers of examiners and patients led to the ultimate decision to discontinue use of this examination format. At that point, a new format was introduced that was intended to provide a more standardized assessment of some aspects of patient management that had been evaluated as part of the bedside oral. These new items were referred to as *patient management problems* (PMPs).

The design of PMPs was based on the belief that the management of a patient requires dynamic integration of new information, including changes in the patient's condition, the results of laboratory tests, and so on. The format used complex paper-and-pencil tasks in which examinees selected management options from lists and then used a special pen to reveal a latent image or uncovered a certain area within the test book to gain access to the results of tests and treatments. Based on these results, new options were provided as the examinee moved through the case.

Although test developers initially viewed PMPs as an important addition to the licensing examination, careful evaluation of the format led to the conclusion that it was seriously flawed. There were two main problems: First, because the format was presented in a *printed test booklet*, it was possible for examinees to page ahead and examine the available options as the case unfolded. This provided an unfair advantage for *test-savvy examinees* who understood that the options included in later items provided clear indications about the correct answers to earlier items. Second, there were serious problems with the *scoring procedures* that undermined the *validity* of the resulting scores (Webster, Shea, Norcini, Grosso, & Swanson, 1988).

The bedside oral examination and PMP test formats reflected the importance of assessing the physician's ability to formulate questions in an undefined problem space and to integrate information over changing conditions. This perceived need was the driving force behind the development of a more modern and technologically-advanced assessment of physicians' patient management abilities: *computer-based case simulations* (CBCSs). This assessment allows physicians to demonstrate their ability to manage patients in a simulated patient-care environment. It is both delivered and scored by computer and, at the time it was introduced, represented a revolutionary development in the field of assessment.

In this chapter, we describe these CBCSs and focus specifically on the *automated scoring* (AS) approaches that were developed to support them. We do this in four main sections. In the first section, we describe the development of the simulations, including how the construct was defined and operationalized and how the interface was designed. In the second section, we describe the program of research that was used to inform the development and refinement of regression-based and rule-based AS procedures. In the third section, we discuss four key issues that motivated research as we refined the scoring approach. In the fourth section, we discuss key lessons learned in the process of developing and implementing an AS process for a high-stakes licensure examination. We conclude the chapter with a few brief reflections on the tension between technology-enhanced assessment, automated scoring, and validity.

Computer-Based Case Simulations

Construct Definition

Initial plans for the simulations were ambitious, and, in retrospect, they also were unrealistic. It was hoped that the range of the proficiencies that had been assessed in the bedside oral – from the ability to collect patient history information to the ability to integrate new information to update diagnostic and treatment decisions and to make management decisions – could be assessed similarly using this new format. Unfortunately, though current technology may have allowed for simulating an interactive encounter in which a physician asks questions to elicit a patient history and uses a computer interface to collect physical examination information, much of the basic development work for these simulations occurred in the 1980s and early 1990s, when technology was a significant limiting factor. At that time, realistic simulation of such an encounter was not feasible.

This reality led to an appropriate *narrowing of the construct*. The new construct was limited to patient management as assessed by evaluation of the tests, treatments, consultations, and monitoring that an examinee ordered in the simulated environment. In some respects, the new model was similar to that of the PMPs. There were, however, important distinctions. As we mentioned previously, the *paper-and-pencil administration* of the PMPs allowed examinees to look at options that would become available later in the case and evaluate present options with this perspective. The *computer administra tion* eliminated that limitation.

More importantly, the new simulation format avoided one particularly troubling flaw of the PMPs: prompting the examinee about the correct path based on the options that were available. At each decision point in a PMP, examinees were given a list of options from which they selected the orders they wanted to place. Both the options and the decision points created an unrealistic prompting for the examinees. With the simulation format, the examinees were required to use free-text entry to place orders; the examinees also had to make decisions about the point in simulated time at which they would re-evaluate the patient or place new orders. These changes eliminated the majority of concerns related to prompting.

Another important component of this new assessment system was that it allowed for collecting data about an aspect of the construct that until this point had been unexplored: the extent to which the examinee ordered *potentially dangerous actions*. The main focus of the assessment was on the examinee's ability to take the correct management steps in a timely manner, but inherent in the design is that examinees can order tests and treatments that are *not* indicated for the patient's specific condition and that may carry some significant level of risk. This information provides important insight into an examinee's overall performance.

Simulation Interface

The structure of the simulation interface works in conjunction with the structure of the individual tasks to elicit scorable evidence about the construct. The tasks are designed to create a dynamic simulation of the patient-care environment in which the patient's condition changes based on both the underlying condition and the examinee's actions. Each case begins with a brief opening scenario as shown in Figure 24.1.

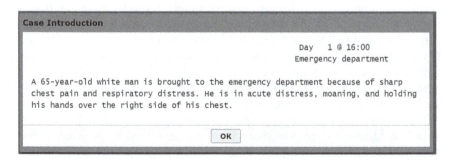

FIGURE 24.1
Sample opening scenario. It provides the patient's location, the simulated time at which the case begins, and the chief complaint.

This initial screen provides information about the patient's chief complaint and location. After the examinee reviews the opening scenario, s/he must decide on the next steps in managing the patient; the available options are grouped into four main categories of actions that are presented as four icons at the top of the screen: (1) requesting information from a physical examination, (2) entering orders into the chart, (3) advancing the case in simulated time, and (4) changing the patient's location; Figure 24.2 provides an annotated screenshot of this main screen.

Figure 24.3 provides an example of the screen for requesting physical examination information. Examinees can select as many components of the physical examination as they deem necessary; after the components are selected and the request is submitted, the examinee receives a message indicating the amount of simulated time that each component will require. If s/he chooses to proceed, the clock advances to the indicated time and a text-based report on the results of the physical examination is provided. A common reason for initially choosing not to proceed with the physical examination components and associated clock advance is that in more acute situations, for example, it may be necessary to order other diagnostic or laboratory studies or treatments before too much time passes.

Figure 24.4 shows the interface for entering orders. Examinees type the desired orders using free text; the system recognizes nearly 9,000 terms (e.g., abbreviations, acronyms, brand, and generic names) representing approximately 2,500 unique actions. If the entered term does not have an exact match

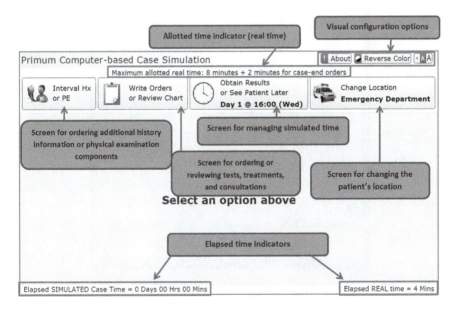

FIGURE 24.2
Computer interface for CBCSs. Visual configuration options are a recent addition to the interface and were not available in the early versions of the case simulations.

History	
	☐ Interval/follow up

Physical Examination

(Order VS from order sheet)

☐ General Appearance	☐ Heart/Cardiovascular
☐ Skin	☐ Abdomen
☐ Breasts	☐ Genitalia
☐ Lymph Nodes	☐ Rectal
☐ HEENT/Neck	☐ Extremities/Spine
☐ Chest/Lungs	☐ Neuro/Psych

FIGURE 24.3
Screen for ordering physical examination components. After the request is submitted, the system generates a message indicating how long each order will take. The examinee then may choose to order other tests, treatments, or consultations or advance the clock to the simulated time at which the desired result is available.

FIGURE 24.4
Order sheet. Examinees use free-text entry to order tests, treatments, and consultations. After the orders are confirmed, the system generates a message indicating how long each order will take. The examinee then may choose to order other tests, treatments, or consultations or advance the clock to the simulated time at which the desired result is available.

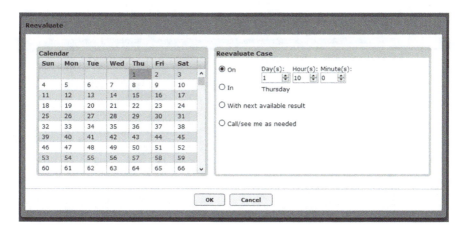

FIGURE 24.5

Screen for managing case time. Examinees may advance the clock between one minute and one year, advance to the point at which the next result becomes available, or select "Call/see me as needed" if there is no further need to actively manage the patient.

in the system, the examinee is shown a list of terms that are similar to the typed entry. If there is a match, the system may request that the examinee select from options that provide additional detail about the specific order and then confirm the order. For example, a medication order may require identification of the route of administration (e.g., oral, intravenous).

While the examinee is entering orders, the clock displaying simulated time does not move forward. To advance *simulated time* (and have orders take effect), the examinee must access the screen shown in Figure 24.5. The examinee can advance the clock between one minute and one year or may choose to advance to the point at which the next result becomes available. The movement of simulated time also can be interrupted by messages from a nurse or a member of the patient's family reporting a change in the patient's condition (e.g., the patient is having more trouble breathing).

The final screen shown in Figure 24.6 allows examinees to change the patient's location. Depending on the severity of the problem, examinees can be sent from the office to the emergency department, admitted to the hospital, sent to the intensive care unit, or discharged. This screen also allows them to schedule a follow-up office visit.

Developing the Scoring System

During the nearly two decades that CBCSs have been a part of the USMLE, the intended construct and the simulation system have changed relatively little. Although the CBCSs have remained relatively unchanged, the USMLE

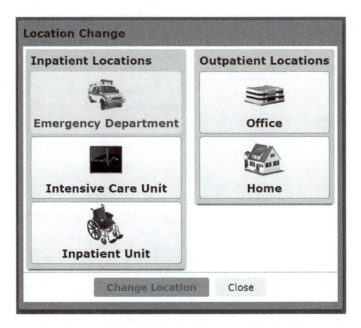

FIGURE 24.6
Screen for changing the patient's location.

overall has been substantially impacted by technology and an interest in broadening the scope of the constructs that are assessed for licensure. For example, computerization has allowed for increased use of pictures as well as the addition of audio and video within the multiple-choice components.

As an additional example of broadening the scope, an examination was introduced in 2004 to assess clinical proficiencies that could not be assessed effectively using computer-administered formats. In this examination, examinees interact with standardized patients and then document the relevant results in a structured patient note using free-text entry. These notes are currently individually scored by content experts, but plans are underway to use AS procedures based on natural language processing beginning.

The scoring system for the CBCSs has undergone a similar evolution. Throughout this evolution, two things have remained unchanged: The *input to the scoring engine* (i.e., the information captured by the simulation and used as the basis for scoring) and the *intention to model expert judgment* as the basis for scoring.

Initial Scoring Considerations

The *output from the simulation* (i.e., the *input for scoring*) is referred to as a *transaction list*. This list includes all actions ordered by the examinee along with the simulated time at which the actions were ordered, the simulated time at which the results of tests became available along with the specifics of those results, and the time and specifics of other events from the simulation (e.g., a nurse's note indicating that "the patient is having more trouble breathing"). This latter information is important because these prompts to the examinee typically mark time points representing more or less timely care; as such, they help to define the scoring algorithms. The computer administration also produces *log files* that include process data such as individual key strokes and time stamps, which are used for understanding how examinees navigate the system but do not impact scoring.

To the extent that evidence or other standards of practice are available to guide management choices in the context of a simulated case presentation, content experts will be able to identify an optimal approach to management for each case. After examining a transaction list, content experts similarly are able to confirm the extent to which the course of management reflected in the transaction list corresponds to that optimal approach.

The nature of the simulations focuses the expert judgments on *process* rather than *outcome*. The simulated patient's condition changes as a function of both the underlying problem and the examinee's actions; this means that a patient might improve over time even though the examinee took no action or the patient's condition might deteriorate even with optimal care. A good outcome for the patient is not necessarily linked to excellent performance by the examinee; similarly, an unfortunate outcome for the patient is not necessarily linked to poor management. As such, patient outcome will not provide an appropriate measure of performance.

Instead, scoring focuses on the process of patient management by evaluating the appropriateness and timeliness of examinee actions. At a very basic level, the examinee is expected to order important actions that are indicated for treatment given the patient presentation and to avoid potentially dangerous non-indicted actions. In some instances it is also required that actions are done in a specific order. For example, to be a useful diagnostic tool, a bacteria culture must be ordered *before* an antibiotic is started. Similarly, an examination demonstrating the presence of a collapsed lung must precede the insertion of a chest tube. Even if the chest tube turned out to be the needed treatment, it would be viewed as inappropriate if it were done without prior evidence that the associated risks were justified. Finally, in many instances it is better to take critical steps sooner rather than later.

Because the transaction list can be evaluated based on expert review, an AS system is not an absolute requirement for use of the simulation as an assessment. For example, the transaction lists could have been reviewed and rated by experts following the model for scoring essays that was in large-scale use at the time. In fact, during the development period the simulations were used at a small number of medical schools and faculty at the schools scored the resulting transaction lists. Although in principle an AS system was not required, it was assumed from early in the development process that, practically speaking, it would be impossible to operationalize the system if it depended entirely on human scoring. Hundreds of thousands of transaction lists would need to be scored each year if the simulations were included as part of the licensing examination and the expense of having physicians rate each transaction list and the associated logistical complexity of scoring that number of transaction lists in a timely fashion seemed prohibitive. This led to a decision from the outset to develop an automated approach to scoring.

Development of the Scoring Key

Early efforts in development of the scoring system were reasonably simplistic, but they laid a foundation for later work. One important step in this process was the development of a *scoring key* that specified the actions that content experts identified as important indicators of performance. The idea was that each of the possible actions could be classified in terms of the importance of that action for management. For example, actions could be categorized as:

- essential for adequate treatment,
- important for optimal treatment,
- desirable for optimal treatment,
- non-indicated and carrying minimal risk,
- non-indicated and carrying moderate risk, and
- non-indicated and highly intrusive or carrying substantial risk.

At different times, minor variations on this structure were used; differences generally were driven by preferences on the part of the content experts and are not relevant in understanding the AS approaches.

As we noted previously, the importance of an action in the context of the simulations cannot be evaluated independent of the sequence and timeliness of the action. The sequence issue can be handled in a reasonably straightforward manner by including critical sequence issues in the definition of the action. For example, rather than defining 'chest tube' as an action, the action can be defined as 'chest tube after supporting diagnostic examination.' By contrast, the timeliness issue has the potential to change each of the actions from a dichotomous to a polytomous variable: Instead of simply asking whether or not the examinee ordered a chest tube, we now need to consider whether it was ordered in timeframe A, B, C, or not at all. Moreover, these timeframes are case-dependent: in some cases, they may represent ordering an action in the first 30 minutes, first two hours, or first eight hours, while, in other cases, they may reflect an action ordered in the first day, first week, or first month.

Even with the complexity introduced by considering issues of sequence and timing, the scoring model was reasonably straightforward. Methodological challenges came not from the complexity of the scoring model, but from the scale of what could be included: there are thousands of unique actions that can be ordered one or more times in managing a case. Content experts developing a key of this sort would need to consider each possible action in the context of each case-specific key.

Most of the thousands of possible actions will be seen as non-indicated for any individual case. In fact, it may be very unlikely that an examinee will order many – if not most – of these actions in the context of any specific case (e.g., it is unlikely that an examinee will order a heart monitor for an otherwise healthy patient with a fractured wrist). Regardless of the fact that examinees may be unlikely to order a given test or treatment, the scoring system must account for the possibility of its occurrence. To respond to this need, content experts created a default classification for the *level of intrusiveness or risk* associated with each of the actions available in the system.

For example, they would begin by considering the case of a healthy woman between the ages of 40 and 60 and evaluating the level of risk associated with a specific diagnostic test or treatment. They then would repeat the process for younger and older women and for men. Although this was a laborious task – requiring a judgment for each of the several thousand actions that an examinee can order for multiple patient groups defined by sex, age, and in the case of women, pregnancy status – it needed to be completed only once. Having created this default *database*, the key development for a specific case then can be limited to consideration of those actions that are indicated for treatment and a (hopefully small) subset of those non-indicated actions for which the associated risk changes significantly because of the patient's specific condition.

The scoring key described in the previous paragraphs was driven largely by the structure of the simulations and the value judgments of content experts, which followed from the construct and the design of the simulation to provide evidence about that construct. We have described this first-level scoring key in some detail because previous descriptions of the AS approaches for this simulation – and of AS more generally – often have taken these details for granted, focusing on the next *level of aggregation*. This approach may mislead practitioners about the level of *pre-processing* that is required to provide the input for AS.

What we have described was, in effect, the first "automated" scoring system for the CBCSs. Two scoring approaches were considered as part of this system. The first simply credited the performance with a point for each of the beneficial actions ordered and deducted a point for each of the non-indicated potentially risky actions ordered. A second approach, more in line with the structure of the scoring key, credited the performance with three points for each action classified as 'essential for adequate treatment,' two points for each action considered 'important for optimal treatment,' and one point for each action considered 'desirable for optimal treatment.' One, two, or three points similarly were deducted for each non-indicated action ordered based on the three levels of increasing risk or intrusiveness.

The mechanical implementation of the scoring key was an effort to capture the *policies* expressed by content experts using a *rule-based algorithm*. From the outset it was clear that this algorithm represented a gross simplification of the complex judgments made by experts when they rate transaction lists; the challenge at that point was to find better ways to *approximate* these expert judgments. The various efforts to improve on this approximation led to an evolution in AS for the CBCSs. The first step in that process focused on developing and evaluating *regression-based procedures*.

Regression-Based Scoring

The history of the use of *mathematical algorithms* – typically *regression-based procedures* – to approximate human judgments is long, rich, and largely beyond the scope of this chapter. Nonetheless, some context is appropriate for understanding the direction taken in this project. An early contribution in this area came from Karl Pearson in an obscure paper with the title *"On the Problem of Sexing Osteometric Material"* (Pearson, 1915). In that paper, Pearson suggested that a mathematical algorithm, using numerous measurements as inputs, might outperform human judgments in identifying whether a given bone intended for anthropologic study was from a man or a woman. A few years later, Thorndike (1918) made a similar suggestion about assessing the fitness of men for a specified task.

By the second half of the last century, the usefulness of this approach was demonstrated using a number of applications in education and psychology. In his 1954 book *"Clinical vs. Statistical Prediction,"* Paul Meehl demonstrated that linear models could outperform trained psychiatrists in combining information to make psychiatric diagnoses. In the 1960s, Ellis Page began work on computer scoring of essays based on linear models (Page, 1966a); this work showed that automated essay scoring algorithms could produce scores that agreed with human raters more closely than humans agreed with each other. In 1974, Dawes and Corrigan published their now classic paper *"Linear Models in Decision Making"* that provided additional examples of the potential superiority of linear models over expert judgment.

This brief historical review makes it clear that the use of regression-based algorithms to replace human raters was not a new idea at the time the simulation system was being developed. That said, there were few, if any, examples in the literature to guide thinking about the use of these approaches. Even fewer of these examples presented alternative models, and there were no examples describing applications of these approaches to scoring dynamic simulations. In this context, we began to experiment with potential models.

Clauser et al. (1993) presented a simple regression model based on a variation on the scoring key described in the previous section. Stated briefly, the model included two levels of actions that were appropriate for treatment and three levels that were inappropriate for treatment. The most important of the appropriate items then were recoded to reflect the examinee's omission of these critical treatment steps and the recoded variable was combined with the variable representing the most dangerous inappropriate actions.

The overall intention was to replace simple counts of actions with *estimated weights* that maximized the relationship between the scores and expert ratings of the same performance. The approach assumed a single model for all cases. The estimates were produced by fitting a *linear model* with the mean rating across a sample of cases acting as the dependent measure and the mean count of actions in each of the previously described categories acting as the independent measures. *Cross-validated results* showed a substantial (approximately 0.10) increase in the correlation between the scores and the ratings when the regression weights replaced counts of actions as the basis for the score.

The initial findings provided evidence for the usefulness of a regression-based approach in this context, and thus began efforts to refine the methodology. The next step was to examine the potential advantage of constructing separate estimates for each case. Clauser, Subhiyah, Nungester, Ripkey, Clyman, and McKinley (1995) used data from seven cases that had been administered at a medical school and reported results. Counts of actions in the six categories described previously – three levels of beneficial actions and three levels of non-indicated actions – served as the independent measures in the model and the expert ratings served as the dependent measure. A separate regression was estimated for each of the cases included in the study. The results, which showed higher correlations between the ratings and the computer-based scores, supported this case-specific approach to developing automated algorithms.

The next step in the development process was to examine the characteristics of transaction lists for which there were large *residuals*. This process often led to identification of *errors* by the raters, but it also highlighted the fact that aggregating the most essential beneficial actions into a single variable was not optimal. In many cases, the model improved significantly when the essential actions were treated as *separate variables*. That is, rather than using the sum of the scores for the essential actions as a variable in predicting ratings, the scores for each of the essential items were used as separate variables. Subsequent studies examined alternative approaches to further improving the models, such as the inclusion of *interaction terms* in *individual case models* (Ross, Clauser, Fan, & Clyman, 1998); none of these subsequent studies led to improved correspondence between the ratings and the scores.

By this point in the development process, we had succeeded in demonstrating that it was feasible to score the simulations with a computer. Content experts were needed only to develop the scoring keys and provide the original sample of ratings that were used to fit the regression model; we typically had approximately 200 to 300 transaction lists rated by at least three experts. The approach was efficient, but it still had limitations. Although the key for each case (i.e., the categorization of actions by level of benefit or risk) reflected expert opinion and the weights were modeled on expert judgment, the overall scoring process had a "black box" quality. That is, content experts could not transparently review and evaluate the process. This motivated research to examine whether algorithms could be developed that not only approximated expert judgment but actually implemented the rules articulated by experts. An optimal procedure would not simply produce the same scores as humans; it would produce them using the same logic. This perspective led to renewed research on rule-based scoring procedures that, in turn, spurred a series of related methodological inquiries that we discuss in the next section.

Evaluating Score Properties

In this section, we discuss four key issues that drove research as we refined our scoring procedures: (1) *transparency*, (2) *generalizability*, (3) *efficiency*, and (4) *quality control*.

Transparency

The process used by the content experts to create a structure for rating performance on a case generally begins with having the experts familiarize themselves with how the case plays out with different approaches to treatment. Once they are familiar with the case, they work as a group to define the actions that are required for 'optimal management,' including the necessary sequence and timing for each action; this pattern receives the highest rating. They then define the actions required for 'just acceptable management'; these patterns receive a moderate rating. Scores at the bottom end of the scale represent a significant 'failure in management' with additional exposure of the patient to substantial risk and/or discomfort. Next, they discuss the types of actions that might be present or absent to justify the intervening scores. This approach, which begins with a description of optimal performance, is workable because the cases are relatively straightforward; the same approach might not be applicable for a simulation of a highly complex or unusual patient presentation where best practice is ill-defined.

The resulting scoring rules are related to the scoring rules for the transaction lists described earlier. Both types of expert-based scoring rules are based on identification of the appropriate and inappropriate actions that can be ordered in the context of a case, including issues of timing and sequence. The *rubrics* for the ratings discussed here extend the keys by specifically mapping combinations of actions onto a specified numeric scale. Obviously these rules do not account for every pattern that might appear in a transaction list; although the management displayed by competent examinees may share much in common, there are endless ways to be less than competent.

Nonetheless, because the rating structure used by the content experts is defined entirely in terms of actions and related timing and sequencing, it was able to be directly and accurately translated into rule-based scoring algorithms. These algorithms then went through a review and refinement process during which (1) the transaction lists for which there were substantial discrepancies between the ratings and the computer-based scores were individually reviewed, and (2) the results were used to refine the rule-based scoring algorithms. This refinement step uniformly improved the correlation between the ratings and scores for an independent sample of performances (i.e., ones that were not used in the refinement process). For some cases, these improvements were modest; in others, the increase exceeded 0.20 (Clauser et al., 1997).

This research demonstrated the potential utility of rule-based procedures, but a direct comparison of the rule-based and regression-based approaches showed that the regression-based procedure yielded scores that had higher correlations to ratings (Clauser, Margolis, Clyman, & Ross, 1997). Although transparency remained a goal as we prepared for the incorporation of the CBCSs into the licensing examination, this finding returned the focus of research to regression-based scoring.

As work continued on efforts to improve the regression algorithms, there was particular concern for the practical limitations of collecting the large numbers of expert ratings that were needed to build regression models before the first scores could be reported. Efforts to optimize the parameter estimates included both statistical (Clauser, Ross, Fan, & Clyman, 1998) and content-driven alternatives. For example, models were developed that aggregated subscores representing 'diagnosis', 'treatment', 'monitoring', and 'timeliness' to evaluate potential advantages over the previously described regression models that included categories based on importance of the actions. In addition to this practical issue associated with implementing the regression procedure, effort also went into a more careful examination of the characteristics of resulting scores.

Generalizability

An obvious metric for evaluating the usefulness of an AS algorithm is the correlation between scores produced by the algorithm and expert judgments of the same performances. Correlations are the measure of choice for these analyses because mean differences between these scales can be adjusted statistically whereas differences in rank order cannot. In most circumstances, a reasonably high correlation would seem to be a necessary – but not sufficient – condition for adopting the AS algorithm as a replacement for human scoring. Throughout the process of developing and evaluating the regression-based and rule-based procedures, we paid considerable attention to other characteristics of the score scales as well. Much of this work was done within a *generalizability theory* framework (Brennan, 2001).

Generally speaking, the correlational analysis examining the correspondence between automated scores and expert judgments takes place at the level of the individual case. The next level of analysis takes place at the level of the score scale that results from aggregating scores across cases. At this level, there are three questions of particular interest:

1. How generalizable (reliable) are the resulting aggregate scores?
2. To what extent does the aggregate score measure the same construct as the aggregate of expert judgments (e.g., what is the true-score correlation between the two scales)?
3. To what extent is there an effect associated with the specific group of experts used to model the AS algorithm?

Efforts to answer these questions were summarized in two papers. Clauser, Swanson, and Clyman (1999) compared the generalizability of scores based on expert judgment to that of scores produced by a regression-based algorithm and a rule-based algorithm. The comparison showed that the generalizability of the regression-based scores was similar to that for expert judgments; the generalizability coefficient for the automated algorithm was higher than that for a score based on a single human judgment per case and lower than that produced when four experts independently judged each performance. In addition, both the expert judgments and the regression-based scores were more generalizable than the rule-based scores.

That same paper reported on the relationship between the scores from the three scales. The observed correlations between the three scale scores were in the 0.95 to 0.96 range. *Corrected for unreliability,* these correlations ranged from 0.98 to 1.00. These results provided support for the idea that the regression-based algorithm could be used to replace expert judgments: the *reliability* of the scores was similar and the constructs being assessed essentially were identical.

The second paper (Clauser, Harik, & Clyman, 2000) reported on a study designed to answer the third question about expert differences. In this study, three independent groups of experts judged performances for the same set of cases. The results showed that scores produced using a regression-based algorithm displayed minimal variability across groups of experts except for a main effect associated with group stringency. This effect can be reduced (or eliminated) by using different groups to develop the algorithms for different cases or through scaling procedures.

The same paper also reported on a *multivariate generalizability analysis* designed to assess whether the scores produced by different groups of experts measured the same construct (i.e., the extent to which the true or universe scores resulting from the group-specific algorithms were perfectly correlated). The results showed that the *universe scores* – which are analogous to *true scores* in *classical test theory* – were perfectly correlated, thereby indicating that algorithms modeled on different groups of experts did in fact measure the same construct. In other words, the results supported the use of a single regression-based algorithm.

Efficiency

Even though the initial research findings clearly were supportive of the decision to implement the regression-based scoring procedures, research evaluating alternative approaches continued after the CBCSs were introduced into the live examination. The primary motivation for this work was interest in maximizing the efficiency of the scoring algorithm development process; with the initial process, the time it took content experts to produce the judgments that were used to build the regression models created a bottleneck that limited the ability to rapidly introduce new cases into the pool.

To develop each model, content experts recruited from around the country had to meet as a group to develop rating criteria and then rate a sample of 200 to 250 transaction lists after returning home. Not surprisingly, scheduling the meetings, running the meetings, and completing the associated activities (i.e., collecting the ratings from content experts who were completing the work at home) was extremely resource intensive; simplifying the process would have obvious advantages.

The first step in the process-improvement research was to evaluate the viability of simpler and less resource-intensive procedures. Harik, Clauser, and Baldwin (2013) began this work by comparing the case-specific regression models that were being used to score the operational examination to a number of alternative models:

- case-specific models based on small targeted samples of examinee performances,
- regression models based on average weights produced across numerous cases, and
- two different non-regression approaches using fixed weights.

Results of this first set of analyses indicated that the then-current operational approach continued to show a stronger correspondence between the computer-based scores and expert judgments.

At the same time that this work on regression-based scoring was completed, rule-based scores were developed for a sample of new cases. Direct comparisons between rule-based and regression-based scores similar to those described in Clauser, Margolis, Clyman, and Ross (1997) were made based on the new cases. This time, the results were somewhat different. There was little difference between the magnitude of the correlations between the expert judgments and regression-based scores and the correlations between the expert judgments and the rule-based scores (Harik, Clauser, & Baldwin, 2010).

Although there was no empirical evidence to explain the relative improvement in performance for the rule-based procedure, it seems likely that much of the change was associated with the increased structure in the training of the content experts who both provided the judgments and were the basis for the rule-based models. Details about the performance of expert raters in this context, which affects the reliability and validity of the resulting ratings, is beyond the scope of this chapter, but the interested reader is referred to Clauser et al. (1999) and Clauser et al. (2000).

For a variety of reasons related to efficiency – largely stemming from a reduction in burden placed on the content experts – and transparency, the rule-based approach was adopted as the operational approach for scoring in 2012 and is still in use today. Between the regression-based and rule-based procedures, well over half a million examinee performances have been scored since the examination was launched in 1999.

Quality Control

One final area that warrants comment in understanding the development and implementation of AS for CBCSs is the challenge of quality control. The use of computers eliminates a range of potential human errors that may impact scores based on expert judgment: once the algorithm is developed, the computer will implement it flawlessly and without the effects of fatigue. If there is an error in the algorithm, however, the computer will also flawlessly repeat the error.

As one might imagine, implementing either the regression-based or the rule-based algorithms required extensive and complex computer code. The algorithms also needed to be case specific, which required the development process to be ongoing – a new algorithm needed to be developed each time a new case was introduced. An error in programming could be catastrophic for both the examinees and the assessment program, which meant that careful checking and re-checking of the code was a must. Similarly, evaluating the code by scoring performances designed to have predetermined performance levels was essential. Unfortunately, neither of these approaches could guarantee that important errors had not been missed. In the end, it was decided that it was necessary to independently replicate each case-specific scoring algorithm and much of the system-wide code that supported it.

To do so, the notes that were produced when content experts met to develop the basic scoring keys were given to two key developers who independently coded the scoring keys. In addition, two independent systems were developed to read the output from the simulation, extract the scoring information, and apply the scoring rules. This approach allowed simultaneous, independent parallel (re)scoring of every examinee-case performance. Comparison of the results provided quality control of the scoring process. In other contexts, where the scores are used for lower-stakes decisions (e.g., formative training), such extensive and resource-intensive approaches to quality control may not be necessary or justified. We highlight this aspect of the scoring system to alert test developers to one of the possible *hidden costs* of taking advantage of the efficiency of automated scoring.

Lessons Learned

Over the many years in which we developed and implemented the AS system(s) for the CBCSs, we had the opportunity to learn (and re-learn) numerous lessons. In the following pages, we highlight three of these lessons that we think can benefit practitioners who may be considering or using AS approaches in their testing contexts. The first lesson relates to the question of *fidelity* and how a focus on scoring has helped to shape our thinking about the construct we intend to assess. The second area in which we think we have learned a valuable lesson relates to the specifics of the simulation interface and the importance of carefully evaluating *assumptions* about the meaning of the evidence we collect as the basis for scoring. Finally, we discuss some of what we have learned about the strengths and limitations of *expert judgment*.

Fidelity

As technology has advanced, the potential to increase the fidelity of the simulation has similarly increased. Early in the development of the case simulations, *video disc technology* made it possible to show a short video of the patient. At the time, a decision was made that this stimulus material would not be introduced into the live examination because the *specialized hardware* required would limit the use of the simulation. With current technology, introducing video that would allow examinees to evaluate aspects of the patient's condition such as gait or posture would be trivial. Images could allow examinees to see a rash or an X-ray rather than reading a text-based description of such physical or diagnostic findings. Heart and lung sounds similarly could be introduced.

It may be attractive to think that any addition to the simulation that enhances the fidelity of the stimulus will enhance the interpretation of the scores. There are a number of reasons why this sort of reasoning can take us down the wrong path. The design of the medical licensing examination does not simply reflect how physicians spend time in practice. If it did, a very large part of the assessment would focus on common complaints such as minor viral illnesses, headaches, and hypertension. Instead, the structure of assessment for physician licensure breaks the requirements for safe practice down into a number of separate constructs and sub-constructs. Some of these sub-constructs are combined into a single score and some produce distinct scores that are reported separately.

Given this reality, enhanced fidelity either could contribute or detract from the resulting scores. In evaluating this sort of stimulus material, there are at least two important questions that should be carefully considered. First, does including this material improve assessment of the construct we intend to measure? Second, does incorporating this material *in this way* represent an efficient use of testing time?

To explore these questions, we can use a single example such as the addition of X ray images as opposed to text descriptions of X-ray findings. It seems almost too obvious that interpreting the X-ray must be part of the construct until we consider the implications of this decision. When the simulations were first included in the examination, examinees completed nine unique patient cases; the examination currently includes 13 cases. Having multiple cases allows for a more precise measure of the construct, and each case assesses the examinee's proficiency in making diagnostic and treatment decisions based on accumulating evidence and evaluation of changes in the patient's condition. Because X-rays are not relevant to many (perhaps most) of the simulated cases, the one or two cases on any particular examination that include X-rays would be measuring a secondary proficiency. Because failing to correctly interpret the X-ray(s) could result in a meaningful impact on the score for the primary construct, increasing the fidelity of the simulation could undermine score interpretation.

In addition to the potentially unintended consequences related to score interpretation, measuring the ability to interpret X-rays in this context is likely to represent an inefficient use of testing time. Images such as X-rays can be (and are) presented as part of the multiple-choice component of USMLE. When they appear on the multiple-choice examinations, the expectation is that these items can be completed relatively quickly (i.e., in less than two minutes); in contrast, the case simulations require 10 to 20 minutes to complete.

The example in the previous paragraphs aside, there may be good reason to enhance fidelity in the simulation. For example, it may be important in some contexts to enhance fidelity to avoid unnecessary prompting or to avoid uncertainty that would not exist in the actual practice setting. Consider a case in which a child has a significantly elevated temperature. The examinee must decide on the next steps in diagnosis, which may include potentially invasive testing (e.g., a lumbar puncture). A short video clip of the child might provide the examinee with a general sense of how ill the child is; a text-based description likely would provide unintended prompting.

Artifacts of the Interface

Scoring rules can interact with examinee behavior in unexpected ways, and it is important to evaluate and monitor patterns of examinee behavior to ensure that the rules have the intended outcome. When content experts develop scoring rules, they are likely to focus on the actions that would be important in a real-world setting. The actual question of interest is somewhat different: what would a competent examinee do in this necessarily artificial setting?

Even the most sophisticated and fully-immersive simulations create a somewhat *artificial assessment context*. This is true because in an assessment setting the physicians are trying to maximize their scores. In the real world, physicians are trying to care for patients. Though it is hoped that strategies to maximize one's score will overlap substantially with those for taking care of the patient, it is not likely that the two sets of behaviors are identical.

This point brings to mind pertinent feedback that we received from medical students who were participating in a *think-aloud protocol* to evaluate a new item format. The format was designed to assess the examinee's ability to write hospital admission orders. The participants repeatedly stated that they had considerable experience with the task and that they felt competent to complete it. They also clearly indicated that how they responded as part of a test would be driven by how the item was going to be scored. These differences in behavior between the assessment and the practice setting may be exacerbated by characteristics of the interface and more generally by aspects of the real world that are not reflected in the simulation.

The physical examination interface shown in Figure 24.2 at the beginning of this chapter is a case in point. An earlier version of this interface provided the components of the physical examination as well as a check box allowing the examinee to request the patient history. The content experts developing the scoring rules typically would insist that a proficient examinee would request the history information and at least some components of the physical examination. This is, no doubt, sensible practice. Nonetheless, we observed that a significant number of examinees who otherwise managed the case impeccably failed to request the patient history or any physical examination information. Our conclusion was that the artificial nature of the interface led to *game-based behavior* rather than *practice-based behavior*.

In practice, the physician typically is in the same physical space as the patient. The physician will greet the patient and likely will ask questions that elicit the relevant components of the history. Unless s/he averts her/his eyes, it is unavoidable that the component of the physical examination associated with general appearance also will be completed. In the context of the simulation, the same physician may take a very different approach, failing to access the physical examination interface. Such examinees may believe that they can narrow their list of likely diagnoses based on the case presentation and that diagnostic tests will lead directly to the final diagnosis and treatment. With this mindset, patient management within the simulation becomes more like a computer game and less like an imitation of real-world practice. In the end, a decision was made to provide the initial history to all examinees at the beginning of the case and to make much more judicious use of the physical examination in scoring.

Another example of such an artifact also comes from the physical examination interface. In a previous version of the interface an examinee could select the individual components of the examination or, for convenience, they could select a 'complete physical' which included all components. This led to scoring problems, because the content experts considered it inappropriate to perform, for example, a rectal and genital examination on a young child brought to the office because she injured her arm at soccer practice. It seems unlikely that examinees ordering a complete physical are considering the full implications of the order and even less likely that they would actually complete all components of the examinee in practice. To resolve this problem, the 'complete physical' option was removed from the interface.

These examples of the mismatch between expectations for real-world behavior and the way examinees interact with a simulated environment may seem trivial. Nonetheless, they highlight the point that creating a functionality in the simulation that allows the examinee to achieve outcomes that would be important in the real world does not mean that the examinee's behavior will mirror that examinee's performance in the real world. Again, this can be exacerbated by the fact that motivation is different in the real world and in the context of an assessment.

Expert Judgment

As we mentioned previously, all scoring algorithms that we have described are intended to model the judgments that experts would make in reviewing the same examinee performances. Early work suggested that these judgments are relatively stable across independent groups of experts and that the regression-based algorithms produced by modeling different groups of experts were similar in the sense that they were reasonably highly correlated. For this early work, the experts developed general rules for mapping performance onto the score scale. These general rules allowed for substantial use of professional judgment in evaluating specific performances.

Over time, there was an interest in limiting that professional judgment to increase standardization. The level of specificity in the scoring rules that the content experts used was increased. This had two attractive results. First, it increased the correlation between the scores from different judges. Second, it provided more detailed information to use in building rule-based scoring algorithms. Both of these results led to higher correlations between the expert judgments and the computer-based scores. At the same time, we also observed an apparent reduction in the level of correspondence between both judgments and scores produced by AS algorithms developed by independent groups of experts.

These results led to a kind of paradox in which the correlation between the individual case scores and associated expert judgments increased – supporting the use of the AS algorithm as a replacement for expert judgment – while at the same time the reliability of the composite score based on aggregating multiple case scores decreased. Although we do not have empirical evidence to demonstrate causality, it appears that variables reflected in the more general rules used earlier in the process generalize better across independent groups of content experts than do the more detailed rules that are currently in use and, by extension, that these variables produce scores that generalize better across different cases. Put another way, the *score variance* that is specific to the group of experts that developed the rules appears to be relatively uncorrelated with the general construct assessed across cases. This seemingly paradoxical result has implications both for developing AS algorithms and more generally for the training of experts for scoring complex tasks.

Conclusions

The research and development activities described in the previous pages unfolded over an extended period of time. The early stages of this development took place when computer simulations and computer-based testing were relatively new technologies. Both of these technologies are now commonplace and there is considerable enthusiasm for the potential of gaming technology to impact – if not revolutionize – learning and assessment. Though the potential for technology to enhance assessment can hardly be overstated, if there is one take-home message from our experience with the AS of CBCSs it is that the technology is the easy part.

Making valid interpretations based on technology-enabled assessment is not about the technology – it is about the skillful use of technology in the service of assessment. Getting the use of technology right requires understanding the impact and implications of a seemingly limitless number of small decisions that are made along the way. This requires constant attention to detail and equally constant skepticism. Cronbach's comment that a "proposition deserves some degree of trust only when it has survived serious attempts to falsify it" (Cronbach, 1980) has rarely ever been more apt than when applied to the proposition that technology will enhance assessment.

In some instances, it appears that test developers begin the development process by identifying the technology that will be used to deliver the assessment. Development of an assessment must begin by identifying the inferences we want to make about the examinees. Technology may help to better assess the construct of interest and may make the assessment process more efficient. In either case, technology should be a tool for achieving an end – not an end in and of itself. In the context of CBCSs, the efficiency of AS made it practical to deliver an assessment that likely would not have been implemented as part of large-scale high-stakes testing if it had been necessary to have content experts score each examinee.

The computer technology for delivering the CBCSs also allowed for assessing an aspect of patient care that would have been difficult if not impossible to assess with multiple-choice questions. In the process of developing and implementing these technologies, we learned a great deal about the validity of the resulting scores – whether those scores were based on expert judgment or produced by computer algorithms. We also learned that although technology can provide useful and valuable answers to difficult assessment challenges, it may not provide easy answers.

25

At the Birth of the Future: Commentary on Part III

John T. Behrens

CONTENTS

The chapters in the third part of the *Handbook of Automated Scoring* represent a sample of snapshots reporting the birth and early toddler years of computational and *artificial intelligence* (AI)-enhanced assessment empowered by *automated scoring* (AS). These endeavors consist of complex social processes that require the melding of perspectives from multiple disciplines along with the alignment of complex social organizations that move from the world of ideas to working tools embedded in the fabric of end-user social contexts, distant from the world of the developers. It is, in fact, quite "magical" that the ideas from the learning-, data-, and computational sciences can be integrated and operationalized into software that promotes learning in the lives of millions of unseen users potentially around the world.

To accomplish this magic, the authors of the different chapters report undertaking a wide range of seemingly disparate activities and considerations to reach the desired goals. Across these activities we distill seven dimensions of focus in this commentary that are considered to bring AS from theory to practice:

1. the theory of action in the socio-cognitive domain of interest,
2. the intended purpose of the application,
3. the ecosystem of use,
4. the limits (and affordances) of computation and mathematics,
5. the limits (and affordances) of the software development processes employed,
6. the availability (or lack of availability) of data, and
7. the economic and organizational pragmatics that support or constrain the endeavor.

Our approach in this commentary is social, pragmatic, and sequential. A *social perspective* understands methodology and development as social behaviors subject to social and organizational interpretations (Behrens & Smith, 1997). By a *pragmatic perspective* we mean to respect the practical actions individuals take in new uncharted contexts where theory may be limited or nonexistent. As Rosenberg and Scott (1991) quipped: "The difference between theory and practice is that, in theory, there is no difference between theory and practice" (p. xvi). By a *sequential perspective* we mean to call out that each of the seven concerns will be treated in turn because time and space are sequenced, but the areas are highly interdependent. A good illustration of this concept is Chapter 20 by Zechner and Loukina in which the authors note that speech scoring reliability is higher (a computational concern) when users are given some time in advance (a user ecosystem concern) to allow planning (a cognitive domain concern).

Seven Elements

Theory of Action in the Socio-Cognitive Domain of Interest

The logic of a *software application* is driven by an explicit or implicit framework for understanding some aspect of human activity. In the contexts of AS, the relevant domains are typically those used to understanding an area of proficiency (e.g., literacy, language, speech, writing, logical reasoning, mathematics, or troubleshooting) as well as more general domains that study the nature of proficiency and its social function (e.g., the learning sciences and educational or policy studies). In Chapter 19 by Deane and Zhang, for example, the authors discuss the different types of domain related issues they are concerned with. In Chapter 24 by Margolis and Clauser, the authors provide examples of how the specific dynamics of medical thought and behavior lead to numerous design and implementation decisions. In Chapter 21 by Graesser et al., the authors attack a range of proficiency domains while creating a new methodological domain of the *AutoTutor* conversational management approach.

Importantly, these software systems can create a feedback cycle with activity in the proficiency domain. The process of developing software requires the formalization and specification of logic, rules, and representations relevant to the domain. In undertaking that formalization, many assumptions are often raised and require additional thought and research. In addition, continuous evolution of the systems can serve as a series of instructional experiments informing the methodological domains of interest as happens in application of *intelligent tutoring systems* (Anderson, 1987) and is evidenced in Graesser's long-term research and development program; see also Chapter 22 by Mislevy et al..

Intended Purpose of the Application

While the proficiency and methodological domains give a starting point for the problems to be addressed by the AS software, a key decision point is the intended use, or problem to be solved, by the system. As reflected in the set of chapters in this part, the range of potential applications continues to grow from the more traditional summative and formative assessments to instructional systems that are driven by assessment-informed inference; see Chapter 23 by Almond and the chapters by Mislevy et al. and Graesser et al. again. Specifically, in Chapter 18 by Burstein et al., the authors note that developments in the field have evolved over time from the highly constrained contexts of summative assessment to the more fluid contexts of instructional systems and that the ability to re-use component pieces across those purposes was an important design requirement.

Ecosystem of Use

A common theme across many of the chapters is the importance of *user experience design* (Levy, 2015) as a critical lens to bring to the application development process. Strong theory and good will are simply insufficient to design and validate appropriate representations and user flows. This stems from the fact that software users vary dramatically in their needs, assumptions, expectations, interpretations of the interface appearance, semantics and flow from screen to screen. Mismatches between how a user expects to use a system and how they interpret their encounter of the system can lead to unwanted *cognitive load* and corresponding *construct-irrelevant variance* unaccounted for from the domain perspective.

Margolis and Clauser provide an excellent narrative regarding the nature of the many *user-experience trade-offs* that they were required to make to obtain the balance between *surface-level fidelity* (i.e., it looks like the real-life task) and *inferential fidelity* (i.e., it properly reflects the construct). The value of deeply understanding the user's perspective in approaching the software is well represented in their insight that "Our conclusion was that the artificial nature of the interface led to game-based behavior rather than practice-based behavior." (p. xx). Burstein et al. explicitly recognize that their embedding of their capabilities in the *Google* writing suite opens both new possibilities and challenges and emphasizes the interdisciplinary approach required.

Limits (and Affordances) of Computation and Mathematics

For many, AS is the application of the technology to the theory of psychometrics or assessment in general. Accordingly, the psychometric world view and sense-making empower and constrain what is possible in the system writ large.

Margolis and Clauser, for instance, discuss the interplay between highly specific custom-designed scoring rules that bring about specific observations of interest while using *generalizability theory* and various aggregation approaches to evaluate the impact of different rules. Mislevy et al. apply complex modeling approaches like *Bayes nets* or *Bayesian knowledge tracing* to synthesize automatically scored observations while Zechner and Loukina describe an entire pipeline of computational activities to make use of speech data.

These chapters provide excellent examples of how the best computational methods are highly informed by methodological and proficiency domain information. For the most part, we see that these applications lean toward simpler and well-tested statistical or computational approaches as is befitting operational applications where the behavior of the computational system must be well understood. There is only passing reference to *deep neural networks* in the chapter by Zechner and Loukina and reference to *reinforcement learning* which constitute two of the latest darlings of computational research. This is not a failure of innovation, but, rather, a guarding of the end user until all the relevant information is in.

Limits (and Affordances) of the Software Development Processes

Modern software development is a complex endeavor whose processes constitute a discipline unto itself (Abbott & Fisher, 2015). Software teams aiming for large-scale deployment often employ specialists in user experience, software programming, system architecture, program management, and computing infrastructure. Insofar as the applications of AS presented in this part are software applications, the constructs and lessons of this industry are ever more intertwined with the needs of educational endeavors employing AS. We briefly touch on a number of these issues.

Computational and Operational Scale

The more computationally ambitious an endeavor is, the more computational scale and associated stability must be considered. In this context, 'scale' refers to the ability of the system to function with appropriate speed under high or distributed use while 'stability' refers to the degree to which the system maintains overall availability across different use patterns. For example, Almond discusses how his applications were revised to be more efficient and scalable by ensuring that the algorithms considered only the most relevant paths and that the computers and algorithms allowed parallelized execution instead of older sequential approaches. While computing power is continuing to improve for fixed costs, the ambitions of its application often require attention to the software design to conserve processing power.

Modularity and Interoperability

Zechner and Loukina suggest that if systems are to be used for multiple purposes, they should be built in highly modular ways that allow for the swapping and combining of various components. The *four-process delivery architecture* that is mentioned by Mislevy et al. as well as discussed and illustrated by Almond throughout his chapter provides a starting place for understanding how the logical decomposition of assessment delivery can be translated into requirements for system design. Almond walks us through a wide number of design decisions that were made in the context of developing a process analysis and scoring system in a game playing context. This level of detail gives a window into the evolution of that language as it addresses new forms of user experience.

Missing from the chapters in this part are references to the emerging shift for many activities from *web-based computing* using a laptop or personal computer to *mobile computing* using a phone-like device. While web-based computing remains the predominant mode of computing in educational institutions, mobile devices are the predominant mode of computing for individuals.

Availability (or Lack of Availability) of Data

As computing systems become increasingly complex and automated, the role of data becomes more central. The chapters by Deane and Zhang as well as Almond argue for tight integration between domain goals and data needs. A *process-focused approach* to understanding writing requires a process-enabling user ecosystem and concomitant data collection and processing. These chapters show the tight interplay between the possible and the required: as we build new systems that provide additional process data, it gives us new experience that suggests new opportunities and reveals current limitations following the adage that "the more I know, the more I know I don't know."

While much of the assessment community has evolved from *fixed-response scoring* to *essay-based scoring* (i.e., still fixed in time and scope), Zechner and Loukina give us an enticing peek into the relatively uncommon world of speech data. Here we see the opening of a door to entirely different range of behaviors and constructs that require end-to-end data management and analytic systems.

It should be remembered though that there are often many forms of data other than user performances in large-scale systems. Margolis and Clauser, for example, discuss the numerous pieces of operational data that must be managed including design specifications, potential task elements, and scoring rules. For many of the systems described in this part, the statistical models are likewise treated as data that are versioned, stored, and called up as necessary. Graesser et al. also discuss the importance of curating an appropriate corpus to train *latent semantic analysis* models tuned to particular instructional purposes. Because the meaning inferred by the analysis is functionally derived from the text used, the input textual data constrains the behavior of the analytic system and the appropriateness of subsequent inferences.

Economic and Organizational Pragmatics

The development of applications and approaches described in this part have not occurred in an economic or organizational vacuum. They are sponsored by organizations that perceive importance to the work and managed by individuals who must cultivate that sponsorship potentially for decades. Few classes in graduate schools prepare the academic leaders for the management and financial challenges hidden behind these successful applications. Likewise, these programs also require sophisticated strategies to manage individuals at lower levels of the organizational hierarchy. In this vein, Zechner and Loukina as well as Graesser et al. discuss the importance of heterogeneous teams with high expertise and well fit tools and processes. These are large human endeavors that do not happen by chance.

Conclusion

Efforts around AS in assessment represent the primary application of advanced computing and the application of AI. The application of AS is a complex and multidimensional problem for which a growing body of theory and praxis is emerging, which we expect to be used to accelerate the quality, availability, and scope of new technologies.

26

Theory into Practice: Reflections on the Handbook

André A. Rupp, Peter W. Foltz, and Duanli Yan

CONTENTS

Putting together a book on a highly multidisciplinary field that is at a stage of transformation is always a daunting task and, as editors, we cannot help but feel that we accomplished some things we set out to do well but that there are others for which we wished we had had more time and space. In this chapter, we provide a synthesis, a reflection, and an outlook in order to take stock of the key lessons that the handbook has to offer; we discuss key implications of these lessons for scientific and practical work around *automated scoring* (AS); we note selected supplementary topics that a second edition might cover in the future; and we offer some suggestions for how readers might stay abreast of developments in this field. To keep the narrative from being too convoluted, we do not point to individual chapters since many chapters draw multiple connections between the following themes.

Common Themes

Despite the variety of issues that were discussed in the three parts in this handbook, a few common themes that we outlined in the introductory chapter have, indeed, been consistently reiterated by the contributing authors. In this section, we reflect on these themes and offer a few suggestions for what they might mean for the future.

Multidisciplinary Engagement

Throughout the book, authors have underscored the *multidisciplinary* nature of work around AS. Even in fields that feed practical advances such as *natural language processing* (NLP), *cognitive sciences*, and *machine learning*, there is typically a strong multidisciplinary engagement for how work gets done. Many authors have cited this as a challenge that practitioners of this work should be ready to deal with since it can be both enriching and frustrating to deal with these multitudes of perspectives, experiences, and styles that colleagues bring to projects around AS.

Specifically, it seems apt to draw some parallels here with the processes of learning foreign languages and acculturating into new cultures in which these languages are spoken. Researchers in second language acquisition, various subfields of linguistics, ethnography, and related fields have long studied how people from different cultural and personal backgrounds wrestle with the tensions between the home and target cultures, the differences between the home and target languages, and the pressures between social and individual drivers of the acculturation processes in order to be successful, active, and empowered participants in the new culture.

For example, just like learning a new language, it is important to learn the core vocabulary, grammatical structures, speech acts, discourse moves, and metacognitive schemata that govern effective communication when engaging in multidisciplinary work. To some degree, this requires treating these aspects much like studying core content in any other discipline. However, to achieve competency in areas such as pragmatic, rhetorical, and situational awareness, which is needed to effectively communicate and collaborate with others, it takes more targeted practice of these studied facts. How difficult these tasks are for individual learners and how effective they end up becoming critically depends on factors such as intrinsic motivation, general cognitive ability, the development of noncognitive skills, and so on.

Connecting this to the area of AS, this means that organizations need to afford individuals with resources that allow them to engage in similar kinds of learning pathways. For example, in multidisciplinary teams it is generally important to develop *common ground* around the use of key terms and core methodological approaches. In order to appreciate the perspective of others and understand the constraints that others are under when engaging in a common workflow with independent yet complementary pieces of information, it is important to create *artifacts* that allow others to share, discuss, and modify their views.

However, there may be simple steps that can be done in order to enhance the use of artifacts for multidisciplinary projects. To use a simple example, it may be helpful to create annotated timelines in which each step of the process is expressed in transparent language that makes it clear to all members of the multidisciplinary team who is responsible for what parts, what each party is doing as these steps are being completed, what the interdependence of different pieces of information is, and what kinds of risk exist for potential breakdowns in the process. This can be amended by a glossary, multiple visual representations that represent the work flow with its intricacies in ways that align with the past experiences, work culture practices in different areas, and managerial goals within the subunits. For example, some colleagues in areas like computer science and NLP may prefer to use portals like *GitHub* to communicate and manage workflows while other colleagues in areas such as assessment development may use customized databases and relatively simple tools like *Microsoft Office Teams*; many other tools exist.

Successful multicultural engagement also involves the acceptance of errors and miscommunication in the mutual engagement with one another, which is more of a mindset issue than anything else. Professional organizations are typically primed for achieving abstract goals such as "100% accuracy" or "timely delivery," which are of course objectively desirable. However, the way of designing AS systems to achieve these goals is typically littered with design and implementation challenges that make the process messier than the metrics may seem to suggest. Just as in learning a new language or culture, colleagues may use terms inappropriately, may misrepresent ideas, may misunderstand processes outside of their main area of responsibility, may misjudge their role and responsibilities, and so on, even despite the creation of initial artifacts like project management charts or digital collaboration tools. They typically bring their past experiences in their professional and personal cultures to the multidisciplinary teams and will try to adapt them to the process at hand.

Consequently, effective multidisciplinary engagement also involves time and space for frequent exchanges of ideas across team members, especially when key design decisions are being made. This includes particularly the early phases of projects during which the vision for the AS project is articulated, refined, and encoded. We are reminded here of a story from a professional development workshop in which the instructor noted how there is often a belief that a "quick start" to a project signals efficiency when, in fact, the opposite can be true. That is, when insufficient time is being given early on to bring perspectives together and agree on common goals, workflow principles, and methodological approaches, many substantive questions subsequently arise and are being managed "on the side," often leading to longer execution and project completion times.

To the contrary, when more time is given early on to settle on these issues, less subsequent confusion may arise and the project may actually be completed faster than under the quick-start approach. Efficiency needs to be viewed systemically, not piece-meal. Generally speaking, process management matters and it is advisable that an AS organization is led by someone who has deep experience with multidisciplinary processes, is competent in project management, has a vision that is aligned with the organization, and has excellent communication and interpersonal skills to oversee the process.

Learning and Assessment Systems

Applications of AS for educational purposes have traditionally been associated with large-scale assessments, which are designed to be relatively broad snapshots of a learner's competencies through one-time administrations of test forms. These test forms may contain some *technology-enhanced item types* (e.g., graphing items, microsimulations) or some *extended constructed-response* tasks (e.g., essay or speech prompts, constrained content questions). However, they are neither as complex, immersive, nor comprehensive in their skill coverage as more mature *game- and simulation-based assessments* nor as geared towards *diagnostic feedback* in the support and management of *personalized learning pathways* as more comprehensive learning and assessment systems.

In this handbook, there are a few examples of more mature systems that provide a snapshot of how far the field has come in the last 10 to 15 years, sometimes longer. Yet, these kinds of activities and systems have been typically developed and used in specific research projects and are often deployed as *prototypes*. Uses of a more formal large-scale game-based learning and assessment suite that uses AS for *construct-relevant feedback* are still the exception, rather than the rule. Perhaps the best opportunities for the creation of these is in well-funded professional domains such as medical licensure and the military although the general *educational technology sector* is certainly booming. Similarly, while there are certainly various *commercial computer games* with sophisticated automated routines that may be usable in smartly designed curricula for learning, they are typically not designed to formally and comprehensively evaluate learners on relevant educational constructs.

In learning and assessment systems, the collection, analysis, and interpretation of *process data* captured in *log files* plays a critical role. This has been perhaps most apparent when *keystroke log analyses* for extended written responses are used to provide diagnostic information, when *clickstream analyses* are used to identify aberrant responses for quality-control in large-scale assessment, or when various *activity traces* are used to detect socio-emotional states such as boredom, engagement, or flow. However, analyzing these data in an efficient way that yields information that can be used reliably and meaningfully for instructional guidance still remains a challenge. While some conceptual and computational frameworks for data management of log files have been proposed, end-to-end demonstrations of the use of process data at scale that are generative for other contexts where they can be easily adapted or adopted are still relatively rare.

Computational Models

Related to the previous points about new languages and cultures, the maturation of the field of AS has underscored the critical role of *computational thinking*, one of the oft-touted complex competencies in modern learning and assessment spaces. The design and implementation of AS systems is, by definition, heavily computational in nature since it relies on the creation, evaluation, and integration of computer code at all stages. Moreover, the management and integration of AS systems into larger organizational systems are similarly computationally intense and require constant management, analysis, and interpretation of data from different sources.

Another flavor of computational operations relevant to AS systems is apparent in the changing nature of the statistical modeling techniques that are being used to process and integrate various sources of information to produce holistic scores, trait scores, diagnostic feedback, routing decisions, or other forms of informational management in the service of learning and assessment needs. On the one hand, tried-and-true techniques from statistics are still in operational use today and sometimes are viewed as having benefits over more modern techniques. For example, variants of *multiple linear and logistic regression* modeling as well as some basic variants such as *classification and regression trees* continue to be used today for certain kinds of prediction and classification problems. Sometimes, even simpler *rule-based approaches* are utilized as well. The amounts of data that are available for model training and evaluation has grown rapidly, however, and so certain aspects around model estimation and model-data fit evaluation are being treated differently nowadays. For example, small-sample inference through *confidence intervals* and *hypothesis tests* for regression parameters is often less relevant than *multi-level cross-validation analyses* coupled with the computation of *uncertainty bands* around predicted values.

In some application areas, traditional models may be superseded by more advanced computational modeling techniques from the *data science* and *computer science* fields for instance. This may take the form of algorithms like *support vector regression* or *gradient boosting* being used as well as various *neural network algorithms*. When there is a genuine choice by an analyst with regards to which algorithms are suitable candidates for analysis, there is typically a tension between measures of *predictive accuracy* and the "transparency" of the algorithm that helps to convey to stakeholders how evidence from responses and processes is being identified and integrated for reporting. Perhaps no other use case is more eye-opening in this regard than the analysis of *multimodal data* for which various types of signals such as audio, video, facial expressions, gestures, and clickstream behavior may be analyzed jointly. For these areas of work, there is often relatively little choice about which classes of statistical method can be used, since implementation is simply not feasible with some older tried-and-true methods. For example, the use of various forms of *deep learning* architectures is critical for processing multimodal data and the "black-box" nature of these component systems is less of an issue than the overall transparency in how their summary output is used to undergird various types of decisions around learning and assessment.

Educational Transformation

Along with the field of AS, the field of education is changing as well and the role of "ed tech" is becoming increasingly prominent. Larger organizations are at the risk of being replaced – sometimes in certain business spaces, sometimes wholesale – by smaller but more *agile start-ups* that are able to create targeted learning and assessment solutions in more comprehensive and cost-effective ways. For these companies, re-envisioning a certain educational implementation model, re-designing incentive structures within that space, and re-designing assessment and learning systems to support this vision is simpler since they are starting from scratch to implement this vision. More established companies can be at risk of not being able to shift their business model sufficiently quickly or comprehensively in order to accommodate newer visions because established systems, processes, and mindsets are too deeply established. This is a fascinating, complex, and ever-evolving race in a highly competitive space to provide solutions that are not only mission-oriented but also "keep the lights on" internally in the long run.

This leads to a larger degree of multi-organizational collaborations that attempt to leverage the complementary expertise of the partner institutions. With the rapid development of educational learning and assessment solutions, there is an abundance of inspiring ideas in the world that await multi-faceted vetting by scientists, practitioners, and users. One important aspect of this change is the transformation of the space of educational pathways throughout K–12 and post-graduate education. More traditional higher education institutions are feeling the push to offer digital courses with a variety of organization now offering multitudes of digital certificates such as *badges, microcertificates,* and *degrees* that help individuals tell their story very differently than in decades past. This has created a need for new types of learning and assessment systems that increasingly blurs the boundary between what used to simply be called *'summative assessment'* and *'formative assessment.'*

Similarly, the increasing permeation of technology into classrooms has opened up opportunities for all learners to access learning and assessment tools that incorporate AS in some form "under the hood." While certain organizations still care very much about the comprehensive evaluation of information for their assessments since high stakes are still associated with them, many organizations are starting to free themselves from this potentially limiting rigor. Instead, they are focusing on producing more coherent digital learning academies with continual, embedded *stealth assessment* coupled with selected supplementary stand-alone moments of evaluation.

From the perspective of AS, this means that the kinds of scoring solutions that will be created in the future will pose *evidence identification* and *evidence integration* challenges that are completely novel at a finer level of detail even if some of the broader desiderata surrounding them will remain familiar. For example, it will still be meaningful to consider *informational quality* since statistics that are almost all noise are probably not very helpful (see the concept of *reliability*), it will still be meaningful to consider how claims about learners, teams, and systems can be supported in a defensible manner (see the concept of *validity*), and it will still be meaningful to evaluate whether learners with different profiles that affect performance and learning pathways are treated equitably at different levels (see the concept of *fairness*).

However, the kinds of research studies, metrics, and value judgments will likely continue to change to accommodate more *transformational views* of the educational systems in which these questions are being asked. Consequently, multidisciplinary teams will need to be both creative and principled and will have to reinvent, transform, adopt, and otherwise engage with AS systems. This, in turn, has implications for how organizations need to prepare themselves in order to be able to handle this transformative workflow. There needs to be an increasing room for *agile development cycles* with frequent iterative tryouts, novel evaluation and implementation schemes, and multiple layers of automation within individual components of an AS system, as well as its integration with other technological systems and workflows.

Human Score Quality

At a more technical level, it may seem that AS is the sole future for scoring. However, in the current state-of-the-art, many authors repeatedly highlight the need for adequate training of AS algorithms. Even though we are starting to see a world in which *artificial intelligence* writ large can help us recognize and create artifacts such as images, text, and sound bites without labeled data, for the moment there are still quite a few instances in which human input is still required. Importantly, this does not only apply to the prototypical case of *automated essay scoring* but to other instances as well. For example, human raters may be needed to code certain kinds of *linguistic errors* that can help with the development of a linguistic feature, certain *misconceptions* in a log file from a simulation-based environment to develop an automated routing algorithm, certain *critical discourse moves* or *critical incidents* to develop a feedback mechanism, or certain *metacognitive, emotional, or engagement states* to develop an automated detector.

Even if their coding is not necessarily always required for the development of an AS algorithm, it may be required for external *validation studies* in which it is important to show that the algorithm captures mostly *con-struct-relevant* – rather than *construct irrelevant* – sources of variation in the responses. Generally speaking, the old adage of "garbage-in-garbage-out" still very much applies whenever human ratings are required for data and despite all kinds of technological and computational advances, no sophisticated AS algorithm can overcome the noise inherent in poor human rating or even unlabeled data.

The issue of human rating quality also connects with the issue of how the information from human raters and AS systems is best used. The current state-of-the-art still supports the notion that machines are best used in complementary fashions to humans, taking over rote diagnostic or evaluative duties that they can perform at least as well – sometimes even better – than the human counterpart. However, in the immediate development horizon, it seems unrealistic to expect AS systems to be able to fully replace human raters when it comes to comprehensively evaluating those qualities of performance that get to the *humanistic core* – creativity, subtlety, and nuance, to name just a few – or relatively *complex skill applications* such as those that support effective *collaborative problem-solving* for instance. Over the years we will surely see a *stronger approximation* of complex characterizations of learners along these kinds of dimensions but there are always going to be aspects that will be impossible to assess with similar levels of *fidelity*.

This is particularly relevant when we think about systemic implications around the use of technology for learning and assessment more generally and AS in particular. In current applications, AS systems are mostly used in STEM disciplines for learning and assessment even though analytics have infused and transformed other fields fundamentally as well. For example, archaeologists now have access to more sophisticated scanning systems and digital libraries, sports analytics is changing the way teams are being managed, and multimodal technologies are allowing musicians to create sounds through movement. In other words, there will clearly be a future in which AS systems are used for learning and assessment in the arts as well in a broader STEAM implementation space.

For example, while the above applications are themselves not formal learning and assessment opportunities per se, it is clear that there is a strong implied need to develop authentic, targeted, and construct-relevant learning and assessment opportunities that allow professionals-in-training to upskill and demonstrate their competency. To some degree, this will take on the form of richer and more diversified *digital portfolios* with associated *digital credentials* in which simply "doing" certain kinds of well-designed tasks will become a possible way to demonstrate competency similar to what is done already in many creative and performance-oriented industries. However, if additional diagnostic profiles are to be desired, then it will be important to develop AS solutions for certain environments and activities that allow the evidentiary tracing of learner log files in these new spaces in a principled manner.

Changing Evaluation Narratives

One result of the emergence of new learning, assessment, and AS systems within the context of our changing educational and professional landscape is the change in procedures for approaching the evaluation of the information that AS provide. As we noted above, traditionally, assessments are evaluated along three complementary dimensions: reliability, validity, and fairness. Again, assessment developers want to convince the users of the information provided by these systems that they can trust that the scores or diagnostic information that is produced contains more signal than noise (reliability), that it helps them to make meaningful decisions about learners based on how it characterizes them through scores and other information (validity), and that it does not unduly disadvantage learners from a particular background or with particular experiences (fairness).

The evidence that needs to be amassed for stakeholders has consisted traditionally of a wide variety of studies that were often centered around *quantitative score pattern analyses, correlational analyses* of scores from different instruments or different conditions, and supplementary *verbal protocol studies*, perhaps along with the occasional *user testimonial*. In the modern world, the narrative around how we evaluate our learning and assessment systems is changing. It seems more and more counterproductive to focus exclusively on key indicators of statistical quality at the expense of more *systemic considerations* around design, implementation, and decision-making surrounding the implementation.

Nevertheless, it remains true that noisy information contained in certain scores is undesirable, especially if few such pieces of information are available and the stakes associated with the ensuing decisions are high. For example, it is not surprising that simple rule-based scoring algorithms can sometimes be the method of choice for high-stakes assessments even in a world where deep neural nets start to prevail. However, having a more diversified and richer portfolio of information from each learner across a longer available time span may critically "offset" some of the information contained in each individual score. Moreover, it also alleviates the information burden placed on individual scores and acknowledges that the surrounding evaluations and decisions made by individual learners, instructors, and other systems can mitigate such limitations.

Consequently, the way we collect evidence about various aspects of validity is changing as well. In the modern digital age, validity evidence may go beyond the types listed above and may also take the form of testimonials from learners and teachers, for example, who can attest, in a more informed manner, to the fact that a particular learning environment has helped them with information-management tasks, supported effective learning, and resulted in higher engagement. Learners may be happy to know that they have reasonably targeted resources at their disposal that allow them to attend to some weaknesses that they might have, even if they may not fully agree with the automated evaluation from a system in all cases. Similarly, teachers may be happy to know that they can be informed about which learners to pay more attention to due to real-time alerting mechanisms even if not all of the learners need the same kind of attention. This, of course, requires more comprehensive efforts to help all teachers and learners become digitally and quantitatively literate so as to be the most informed users they can be in these information-rich settings.

Opportunities for Future Editions

As the introduction to this handbook and the previous section in this concluding contribution have made clear, it is not really possible to cover a space as complex as the one around AS comprehensively within a single volume. In this section, we want to mention a few areas that are somewhat underrepresented in our book.

Fairness and Bias

This handbook did not include a targeted chapter on how to frame, evaluate, and act upon empirical investigations specifically around fairness and *biases* even though the concept came up repeatedly in several chapters. Certain kinds of statistical analyses that can provide quantitative evidence about equitable treatment of learners are natural components of the state-of-the-art of AS system evaluations nowadays but broader inquiries into the kinds of instructional experiences, technology access considerations, and system-level uses of evaluation information that can affect how equitable we treat diverse populations of learners are just beginning to mature. In a future edition we thus expect a stronger emphasis on fairness and bias and more complete case studies that demonstrate how to amass relevant evidence.

Customer Perspectives

There is typically a notable disconnect between *customer expectations* – learners, teachers, state representatives – about what AS should be able to do and what kind of evidence is actually available to demonstrate that this is true. For example, with artificial intelligence being in the media on a daily basis and many technological innovations being disseminated through various social media platforms, it is no wonder that the general public expectations around AS are high. This is actually a very healthy phenomenon since it pushes the associated industries to "live up to the hype" and demonstrate the power of novel methodologies and systems. It is also a great opportunity for engaging with different customers in meaningful conversations about their expectations and the kinds of evidence they want to see for their AS solutions. Many established practices for development, evaluation, and implementation of AS systems can be justly challenged. We expect that quite a few instances generally emerge within any given organization in which things are done out of a principled habit that could use some empirical and rational revisiting. We thus expect customer perspectives to get more treatment in a future edition, especially as the space of learning and assessment systems with AS solutions continues to grow and gains a larger end-user base.

User-Interface/-Experience

Especially in the area of immersive or highly interactive digital learning systems, design decisions around the digital delivery platforms and interfaces that learners and instructors engage with is critical for the acceptance of these systems in practice. Ironically, without such acceptance it will be impossible to gather new data to improve these systems in the long run, which is critical for the AS components under the hood. Different learner populations have different types of exposure to and familiarity with technological interfaces, the appropriate interpretation of novel types of information they present, and the information-management demands of engaging with these systems. It is critical that sufficient time be spent on efforts such as playtesting, focus group evaluations, and iterative data-driven tryouts, guided by multidisciplinary collaborations with experts in areas such as computational cognitive modeling, user-interface/user-experience design, and the learning sciences – to name but a few. In future editions of the handbook, we certainly expect to see a stronger emphasis on this topic as the maturity of the digital systems increases further.

Further Learning

Finally, with an area as vibrant as that of AS, it is perhaps more critical to stay abreast of ongoing developments than it is to have a single reference book on the shelf. Consequently, in Table 26.1 we have assembled a collection of relevant networks/organizations that readers might want to explore. In addition, new websites that offer comprehensive professional development via training modules, courses / MOOCs, seminars, and other events are popping up frequently (see, e.g., https://www.theschool.ai or https://ncme.elevate.commpartners.com as two examples). This content is

TABLE 26.1

Selected Organizations in Areas Related to Automated Scoring in Education

Organization	Website	Conference/ Workshop	Journal
Artificial Intelligence in Education (AIED)	https://caed-lab.com/aied2019/index.html	X	
Association for Computational Linguistics (ACL)	https://www.aclweb.org/portal/	X	X
Association for the Advancement of Artificial Intelligence (AAAI)	https://www.aaai.org/	X	X
Division D on Measurement and Research Methodologies within the American Educational Research Association (AERA)	https://www.aera.net/	X	X
Educational Data Mining	http://educationaldatamining.org/	X	X
European Association of Artificial Intelligence (EurAi)	https://www.eurai.org/	X	X
International Artificial Intelligence in Education Society (IAIED)	https://iaied.org/	X	X
International Conference on Computer and Information Technology (ICCIT)	http://iccit.org.bd/2019/		
International Conference on Intelligent Tutoring Systems	https://its2019.iis-international.org/	X	
International Society of the Learning Sciences	https://www.isls.org/	X	X
Learning at Scale	https://learningatscale.acm.org/las2019/	X	
National Council on Measurement in Education (NCME)	http://www.ncme.org	X	X
NLP4Call	https://spraakbanken.gu.se/eng/research-icall/8th-nlp4call	X	
Special Interest Group (SIG) for Building Educational Applications (SIGEDU) within the Association for Computational Linguistics (ACL)	https://sig-edu.org/	X	
Special Interest Group (SIG) for Speech and Language Technologies in Education (SLaTE) within the International Speech Communication Association (ISCA)	https://www.isca-speech.org/iscaweb/index.php	X	
User Modeling (UM)	https://www.um.org/	**X**	

often made available for free or, at most, for a small fee so that upskilling becomes a much more accessible endeavor for a wide variety of interested professionals.

That being said, keeping up with the fields of AS and artificial intelligence overall at a finer level of detail will essentially be challenging for any single person as the proliferation of knowledge and experiences continues to grow. Therefore, relying on collated resources and on a smartly assembled and managed multidisciplinary team is key. We sincerely hope that the chapters and the glossary in this handbook as well as the myriad sources referenced throughout prove to be a useful starting point for multidisciplinary teams to engage in the process of designing, evaluating, implementing, and monitoring AS systems within the context of broader learning and assessment systems in this new transformative world of educational technology.

Glossary

21-st century skill: A complex constellation of competencies (e.g., critical reasoning, collaborative problem-solving) that is brought to bear in complex performance tasks

Accuracy/inaccuracy: A rater effect that results in an increase/decrease of the agreement between assigned scores and target scores

Acoustic model: A component model within an automatic speech recognition system, which represents the relationship between the audio signal and speech sounds (i.e., phones)

Active learning: A special approach to machine learning in which a learning algorithm is able to guide the sampling and labeling processes for supervised learning

Adjudication: The process of resolving a large difference between two or more scores from human raters or automated scoring models, typically through the use of expert human raters

Application program interface (API): A set of routines, protocols, or tools to specify how software components interact with one another

Architecture (for software): The design of software that considers internal and external entities and the relationship among those entities

Artificial intelligence (AI): An area of computer science that emphasizes the creation of intelligent algorithms that work and react like humans

Audio file: A digitized speech recording from a learner/test taker, which is sent to an automatic speech recognition system for processing

Authoring tool: A tool for developers to create content that can be incorporated into a computer-based learning environment

Automated rater: An alternative term for the automated scoring model produced during training and validation that is used as the production model for operational use due to its optimal performance

Automated scoring: The process of submitting constructed responses or performances to computer algorithms in order to produce reportable information for stakeholders for the purposes of learning and assessment

Automated scoring model: A set of mathematical and algorithmic functions that receive a response and produce a score or set of scores

Automated scoring pipeline: The sequence of procedural steps and associated technical and human systems that comprise the development and implementation of an automated scoring system

Automated scoring system(s): The totality of computational architectures, human team members, work flows, artifacts, and documentation to accomplish defined objectives in support of the automated scoring pipeline

Automated writing evaluation (AWE): Computational methods developed for purposes of evaluation (e.g., scoring, feedback), in which linguistic features are identified that represent key aspects of the writing construct

Automatic speech recognition (ASR): A system that takes digitized speech recording as input and generates a digital representation (i.e., word hypothesis) as its output

AutoTutor: A learning environment with computer agents that help students learn by holding a conversation in natural language

Back-reading: A rater monitoring/feedback process through which human scoring supervisors review and potentially replace the scores assigned by human raters

Basic Automated BS Essay Language (BABEL) Generator: On-line software that can produce meaningless, gibberish, counterfactual essays given three content word inputs

Calibration: The process of evaluating the competency of human raters in order to allow them to score operationally

Calibration pool: The sample of responses that are scored and reviewed by human raters for the purposes of training and evaluating the automated scoring models

Categorization: The process of creating groupings of things to help make sense of the world

Category learning: The process of developing the ability to classify things into a new or predefined set of categories, often with feedback

Centrality/extremity: A rater effect that results in a decrease/increase in the dispersion of assigned scores relative to some benchmark

Classical test theory: A common measurement framework that focuses on the analysis of test scores and decomposes them into observed scores, true scores, and error scores

Cognitive load: The overall demand on cognitive processing that is imposed through stimuli such as learning or assessment activities

Cognitive science: The study of the mind and brain using a multidisciplinary approach that includes cognitive psychology, philosophy, computer science, neuroscience, linguistics, anthropology, and other fields

Computer agents: A computer-generated talking head or chat participant that interacts with a human in natural language

Conceptual assessment framework: The combination of the student model, evidence model, and task model

Conditional agreement by scorepoint: The rate of agreement for the responses from a single scorepoint on the rubric scale, computed between two human raters or between a human rater and a machine

Conditional generative adversarial networks: A special case of generative adversarial networks that include explanatory/predictor variables that are used for conditioning

Conference on College Composition and Communication (CCCC): An annual meeting established by the National Council of Teachers of English (NTCE) that has twice released criticial statements on automated essay scoring

Consensus scores: Numerical ratings assigned to responses by expert human scorers, which are intended to represent a gold standard against which scores assigned by operational human raters can be compared

Construct/subject-matter expert: An expert in the area of knowledge, skills, or abilities that a test is designed to assess (e.g., a researcher in reading comprehension who helps design reading assessments)

Construct-irrelevant variance: Sources of variation in scores that are not of primary interest for learning and assessment and thus unduly affect the performance of different learner groups

Data science: A multidisciplinary field of study that considers all aspects around data analysis, including defining research questions, extracting and processing data, modelling data, and communicating results

Data-level fusion: The combination/integration of signals at the level of the raw data with minimal processing

Decision-level fusion: The combination/integration of outputs rather than features or data

Deep generative networks/models: A family of statistical models that can learn complex patterns from data without requiring labeled training data (i.e., they are an unsupervised modeling technique)

Deep learning: A family of machine learning methods based on neural networks, which are concerned with emulating human information-processing approaches for recognizing complex patterns in data

Deep sequential networks: A family of statistical models that can accommodate the sequential dependencies in data

Defensibility: A characteristic of an argument that articulates whether effective procedures and standards have been followed in the development and execution of an AS pipeline

Discourse parsing: The process of identifying conversational elements in a larger text along with the relationship between them

Distributed scoring: A setup for scoring in which raters can evaluate responses across various locations and time windows, which is typically implemented via digital score management systems

Ensembling: An algorithmic or statistical method that transforms outputs from two or more statistical models to predict score or sets of scores

Equally weighted Conditional Agreement: A single-value metric that conveys the marginal accuracy of machine scores, which is computed as the average conditional agreement across scorepoints

Evaluation criteria: A set of statistics for the evaluation of automated scoring systems

Evidence model: The means of identifying, scoring, and aggregating evidence from learner response in order to make inferences about the targeted constructs captured in the student model

Evidence-centered design: A common framework for principled assessment design that helps with the specification and implementation of key characteristics for tasks, evidence, and reporting

Exemplar responses: A set of constructed responses that have been selected by experts during rangefinding work as typical examples of each score point in the rubric for training raters and monitoring scoring accuracy

Expected response/expectation: In the context of automated dialogue management, a theoretically grounded appropriate answer to a problem or main question in an assessment task

Explainable artificial intelligence: Approaches to artificial intelligence in which the basis for the results can be understood by human experts

Extended constructed response: An alternative name for constructed responses, which is used to underscore that a prompt allows a free response that is typically one to several paragraphs in length

Fairness: An umbrella term for a wide variety of considerations around the equitable treatment of learners/examinees, which can be supported with various types of rational and empirical evidence

Feature engineering: The process of researching and selecting computational features that can capture construct-relevant evidence from response text

Feature importance table: A table of values that reflects the degree to which each feature in an automated scoring model contributes to the reduction of model prediction error

Feature-level fusion: The combination/integration of signals at the level of featurism which are higher order abstractions of raw data

Feedback: For evaluation of writing or content, information identified in text that is used to praise or critique a test-taker or student response, which can be generated by humans or machines

Filtering model: A rule-based or statistical model that predicts whether a learner/examinee response is scorable or not (i.e., whether it is "regular" or "aberrant" in some sense)

Forced alignment: The process of producing sequential time stamps for all of the phonemes and words included in a transcription of a digital speech signal

Free text entry: Words, phrases, or longer written responses produced using a keyboard

Game-based assessment: The process of making inferences about learners' knowledge, skills, and attributes based on performance in a game designed to elicit evidence about these characteristics

Gaming/gamed responses: Explicit response actions that are unrelated to the intended demands of an item/task/activity, but are intended to result in higher scores

Generalizability: The extent to which the observed score on an examination reflects the score that would be observed if it were measured without error

Generalizability theory: An extension of true score/classical test theory that takes into account multiple sources of systematic measurement error

Generative adversarial networks: A family of deep generative networks that consist of two component networks (i.e., Generator, Discriminator), which are pitted against each other using principles from game theory

Gestalt psychology: A school of psychology interested in developing simple principles to explain how people perceive meaning from a complex percepts

Halo effect: The existence of spuriously high correlations between multiple scores assigned to by a single human rater

Hidden layer: A series of unobserved (i.e., latent) variables that are embedded within a multi-level modeling architecture

In-person scoring: A setup for scoring in which raters meet in a common physical location to perform their scoring duties in a common timeframe

Instructors: Educators who teach in K-12, post-secondary, adult education, or workplace settings

Intelligent tutoring system: A computer-guided system that provides diagnostic feedback to learners and automatically selects items for presentation that are best matched to learner needs

International context (for scoring): Development and research on machine scoring taking place in languages besides English and countries outside of the US

Inter-rater agreement/reliability: A statistical measure capturing the consistency between two sets of human scores

Item response theory: A common measurement framework that focuses on the analysis of item-level scores and subsumes a large number of complex models used in operational large-scale testing

kaggle.com: A data science competition web site, which hosted two competitions related to automated scoring in 2012

Knowledge component: A piece of information on a subject matter that is tracked during learning sessions and that is assessed on the extent to which the student has mastered the piece of information

Language model: A component model within an automatic language processing system (e.g., speech recognition), which contains the likelihoods of word sequences in a language

Latent semantic analysis: A content-based approach using natural language processing for analyzing semantic relationships between a set of texts via the terms they contain using dimension reduction techniques

Latent trait models: An umbrella term for a wide variety of statistical models with unobserved (i.e., latent) predictor variables that are used to predict observable score patterns

Learning analytics: A field of study concerned with modeling construct-relevant information from student data that can be used to evaluate student progress

Learning progression: A description of the stages or steps that learners go through as they move from novices to experts in a particular domain

Machine learning: A field of study that uses statistical methods to identify patterns in and predict values from data

Massive open on-line course (MOOC): A relatively comprehensive course delivered via a mature learning management system sometimes with the possibility of earning badges/microcertificates/credits

Misconception: An incorrect understanding of key concepts during evolving stages of subject-matter understanding, which leads learners to produce erroneous responses

Model building: The process of training an automated scoring algorithm using reliable human ratings and textual features to provide accurate score predictions

Model evaluation: The process of evaluating the quality of scores produced by an automated scoring model for accuracy and construct validity using a variety of statistical measures

Morphology: A subfield of linguistics concerned with analyzing the internal structure of words

Multimedia principle: An articulation of the idea that people learn better form words and graphics than either alone

Multimodal analytics: A field concerned with the application of statistical and analytical methods aimed at making sense of data captured from multiple modalities and/or multiple sensors

Multimodal data: A stream of data from multiple sensory inputs that may include audio, video, text, log files, and so on

Multisensor: A term referring to multiple sensors in humans (e.g., eyes, ears) and machines (e.g., cameras, microphones)

Multisensory/Multimodal integration: The process of integrating information from multiple senses by either humans or machines

Natural Language Processing: An inter-disciplinary field of research that is concerned with computational approaches to the extraction analysis, understanding, and generation of human language

Neural network: A statistical computational architecture inspired by human brains

Node: A component of a network that is either observed or unobserved (i.e., latent) as part of an input, output, or intermittent layer, which is connected to other nodes via paths/edges

Non-productive scoring time: The totality of time spent on tasks that do not lead to operational scores such as training, breaks, and review

Open-source software: Software that is freely available for inspection, modification, and use as determined by its license

Partnership for Assessment of Readiness for College and Careers (PARCC): Testing entity formed in 2010 as a state collaborative in response to US Federal Race-to-the-Top grants

Part-of-speech tagging: The process of assigning a label (e.g., noun, verb, adjective) to a sequence of tokens (e.g., words)

Personalized learning: An umbrella term for various methods that evaluate, for a given domain, the individual needs of students and tune feedback to that student's needs

Phonetics: A subfield of linguistics concerned with the empirical analysis of speech production and perception

Phonology: A subfield of linguistics concerned with how the sounds of a language are represented as abstract discrete units in a speaker's mind

Pitch Extraction: The process of using speech processing technology to compute a sequence of measurements for pitch (i.e., the relative highness or lowness of a tone) at regular intervals through a digital speech signal

Pragmatics: A subfield of linguistics concerned with how language is used in different contexts

Pre-processing (of inputs): The preparation of digital text for feature computation, which may include parsing, spelling correction, removal of unrecognized characters, and tokenization

Principled assessment design: A broad family of frameworks or approaches to designing assessments with the goal of linking constructs of interest to tasks and to evidence

Probabilistic graphical models: A framework for modeling dependencies across multiple variables via graphical representations and associated statistical relationships

Production rule: An "IF <state> Then <action>" expression that specifies the specific conditions under which a particular action is generated by the computer

Productive scoring time: The totality of time spent on producing scores

Prosody: Non-local aspects of pronunciation (e.g., pitch contour, stress patterns)

Psychometrics: A field of study that is concerned with the measurement of skills, knowledge, abilities, noncognitive factors, and educational achievement

Python: An open-source software programming language often used in automated scoring because of its readability and easy integration into various natural language processing and machine learning toolkits

Quality control: The process of ensuring that various performance characteristics of an automated scoring system are functioning as intended during operational deployment

Random sample: An ideally large and representative subset of population data that is used to build and evaluate automated scoring models

Rangefinding: The process wherein human scoring experts read and discuss a range of responses to a task, provide consensus scores, and identify typical examples of each score point in the rubric

Rater attrition: The expected or observed numbers of raters that drop out of an operational scoring process due to problems in their performance or other factors

Rater bias: A systematic type of rater error that manifests itself as response scores that are either overly severe (i.e., generally too low) or overly lenient (i.e., generally too high)

Rater characteristics: Features of human raters such as education and experience that may influence the quality of the scores that they assign

Rater effects: Various patterns of error in human-assigned scores

Rater inconsistency or unreliability: Random rater error that may have many contributing sources arising from the interaction of rater characteristics, response characteristics, and rating method/context characteristics

Rater pool: The total number of raters that are available at any given time for operational scoring from which systems can sample and assign raters to different shifts

Rating: An evaluation of a constructed response made by a human rater or automated scoring system, usually resulting in a numeric value or categorical code

Rating context: The environment and process through which human ratings are collected, especially as it relates to the quality of the resulting scores

Rating quality: An umbrella term for the accuracy, reliability, and precision of scores

Redundancy (of signals): The case when two or more signals convey the same information

Regression-based scoring: Procedures that weight quantifiable aspects of examinee performance to maximize the relationship between the aggregate of the scorable components and some dependent measure

Reliability: An umbrella term for the consistency of scores or classifications across multiple conditions (e.g., timepoints, forms, raters), which can be quantified with a variety of statistical indices

Remediation: The process of identifying raters that are not performing according to expectation, retraining them, and then re-evaluating them for possible continuation of operational scoring work

Response content: Characteristics of the responses being scored that may influence the quality of the scores assigned by human scorers

Rubric: A scoring guide used by human raters to provide ratings along a numbered scale (e.g. 1-4, 1-6) in a consistent fashion

Rule-based scoring: Procedures that implement logical rules such as "IF/THEN" statements to aggregate quantifiable examinee actions to produce a score

Scorability: A qualitative or statistical index that can be used to determine whether or not a particular automated rater is capable of accurately scoring a particular test item, task, or collection of items or tasks

Score reproducibility: The ability of an automated scoring system to predict the same score or set of scores for a given response across defined conditions

Scoring model: A statistical model that predicts a score for a learner/examinee response based on a set of linguistic (or other) features

Scoring rules: A set of rules that is used to convert all available rating(s) from humans and machines into a final item score

Scoring system evaluation: A process to evaluate the performance of an automated scoring system

Semantic evaluation: The process of determining the extent to which one verbal expression has similar meaning to another verbal expression

Semantics: a subfield of linguistics concerned with analyzing the meaning of words, sentences and larger texts

Severity/leniency: A rater effect that results in consistent decreases/increases of assigned scores relative to a benchmark

Short constructed-response: An alternative name for constructed responses, which is used to underscore that a prompt allows a free response that is typically no more than one or several sentences in length

Signal processing: The process of extracting various types of basic information from an audio signal (e.g., computing the energy and pitch for a short speech segment)

Smarter Balanced Assessment Consortium (SBAC): A testing entity formed in 2010 as a state collaborative in response to US Federal Race-to-the-Top grants

Sociocognitive perspective (on assessment): A theoretical framework that helps researchers and developers to be aware of various situational, cultural, and linguistic factors that affect student performance to create fairer assessments

Sociocognitive writing framework: A theoretical model of writing achievement that emphasizes the complex contribution of skills to writing achievement, which include writing domain knowledge, general knowledge, and various personal factors

Speaking construct: The articulation of the knowledge, skills and other abilities that need to be measured by a speaking assessment (e.g., the dimensions of speech to be evaluated by human raters or automated scoring systems)

Speech act: A category of language contribution that reflects its function such as questions, statements, expressive evaluations, and metacognitive expressions

Speech feature: A measure of speaking proficiency such as the average duration of inter-word silences/pauses or vocabulary complexity

Stakeholder: A person who is using information from learning and assessment systems to make decisions about themselves, others, or systems

Student model: An articulation of the knowledge, skills, and attributes analysts are interested in assessing

Superadditivity: The creation of a new type of signal from two or more input signals, which carries novel information that the component signals alone do not carry

Syntactic parsing: The process of assigning a linguistic structure to a sentence

Syntax: A subfield of linguistics concerned with analyzing the internal structure of clauses and sentences

Task model: The activity that learners engage with that produces evidence related to the constructs of interest in an assessment

Test operations: The processes surrounding the preparation, administration, scoring and reporting of assessements/tests

Tokenization: The process of separating a large text into smaller units, which include paragraphs, sentences, words, phrases, symbols, or other meaningful elements (i.e., tokens)

Transaction list: A list of all the scorable actions taken by an examinee in managing a patient as part of the computer based case simulations used as part of the United States Medical Licensing Examination (USMLE)

Transcription: The process of creating a word-based representation of a speech recording, which can be accomplihsed by human transcribers or by an automatic speech recognition system

Trialogue: A three party conversation that includes a human interacting with two computer agents

True score/classical test theory: A measurement framework that focuses on observed composite scores and the error associated with them

Turn: A contribution in a conversation by a single person or computer in between contributions of other participants in the conversation

United States Medical Licensing Examination (USMLE): The examination sequence required for physicians with an MD degree to practice medicine in the United States

User experience design: A field of study concerned with understanding, investigating, and improving the way learners and other stakeholders interact with technology in order to maximize the goals of the system

Validity: An umbrella term for a wide variety of considerations and needed evidence to support desired interpretations from scores or feedback produced by a learning or assessment system

Validity agreement: A measure of the consistency between human scores and consensus scores

Validity responses: A set of responses that have been assigned consensus scores by experts, which are embedded into operational responses so that scoring supervisors can evaluate the performance of human raters

Vector: A statistical object that contains numerical or categorical information from statistical variables, which is a main component of almost all statistical modeling techniques

Waterfall approach: An approach for assessment, software, or machine learning development characterized by a linear and sequential set of phases development

Word error rate: A performance measure for automatic speech recognition systems, which is the sum of individual word errors divided by the length of the transcribed speech in words

Word hypothesis: The output of an automatic speech recognition system, which is the word sequence that is most likely to have been spoken by the test taker based on individual sounds and word sequence likelihoods

Writing analytics: The process of automatically deriving construct-relevant information from student writing samples or process data that can used to evaluate student progress

References

Abadi, M., Agarwal, A., Barham, P., Brevdo, E., Chen, Z., Citro, C., … Zheng, X. (2015). TensorFlow: Large-scale machine learning on heterogeneous distributed systems. *Google Research White Paper.* Retrieved from http://download.tensorflow.org/paper/whitepaper2015.pdf

Abbott, M., & Fisher, M. (2015). *The art of scalability: Scalable web architectures, processes, and organizations for the modern enterprise.* New York, NY: Addison-Wesley.

Agirre, E., Cer, D., Diab, M., Gonzalez-Agirre, A., & Guo, W. (2013). *SEM 2013 shared task: Semantic textual similarity. In M. Diab, T. Baldwin, & M. Baroni (Eds.), *Second joint conference on lexical and computational semantics (*SEM). Proceedings of the main conference and the shared task: Semantic textual similarity* (Vol. 1, pp. 32–43). Retrieved from https://www.aclweb.org/anthology/S13-1

Ajay, H. B., Tillet, P. I., & Page, E. B. (1973). *Analysis of essays by computer (AEC-II).* Washington, DC: U.S. Department of Health, Education, and Welfare, Office of Education, National Center for Educational Research and Development.

Akidau, T., Chernyak, S., & Lax, R. (2018). *Streaming systems.* Sebastopol, CA: O'Rielly Media, Inc.

Alamargot, D., Chesnet, D., Dansac, C., & Ros, C. (2006). Eye and pen: A new device to study reading during writing. *Behavioral Research Methods, Instruments, and Computers, 38,* 287–299. doi:10.3758/BF03192780

Alamargot, D., Plane, S., Lamber, E., & Chesnet, D. (2010). Using eye and pen movements to trace the development of writing expertise: Case studies of a 7th, 9th, and 12th grader, graduate student, and professional writer. *Reading and Writing: An Interdisciplinary Journal, 23,* 853–888. doi:10.1007/s11145-009-9191-9

Aleven, V., McLaren, B., Roll, I., & Koedinger, K. (2004). Toward tutoring help seeking: Applying cognitive modeling to meta-cognitive skills. In J. C. Lester, R. M. Vicario, & F. Paraguacu (Eds.), *Proceedings of the seventh international conference on intelligent tutoring systems* (pp. 227–239). Berlin, Germany: Springer-Verlag.

Allalouf, A. (2007). Quality control procedures in the scoring, equating, and reporting of test scores. *Educational Measurement: Issues and Practice, 26*(1), 36–46. doi:10.1111/j.1745-3992.2007.00087.x

Allalouf, A., Gutentag, T., & Baumer, M. (2017). Quality control for scoring tests administered in continuous mode: An NCME instructional module. *Educational Measurement: Issues and Practice, 36*(1), 58–68. doi:10.1111/emip.12140

Allen, A., D'Astolfo, E., Eatchel, N., Sarathy, A., & Schoenig, R. (2016, March). *Reclaiming the conversation: Influencing how the world talks and thinks about testing.* Orlando, FL: Closing general session at the annual meeting of the Association of Test Publishers.

Allen, L., Dascalu, M., McNamara, D. S., Crossley, S., & Trausan-Matu, S. (2016, July). Modeling individual differences among writers using ReaderBench. In *EDULEARN16 Proceedings, 8th international conference on education and new learning technologies* (pp. 5269–5279). Valencia, Spain: International Academy of Technology, Education, and Development.

Allen, L. K., Jacovina, M. E., Dascalu, M., Roscoe, R. D., Kent, K. M., Likens, A. D., & McNamara, D. S. (2016). {ENTERING}ing the Time Series {SPACE}: Uncovering the writing process through keystroke analysis. In T. Barnes, M. Chi, & M. Feng (Eds.), *Proceedings of the 9th international conference on educational data mining* (pp. 22–29). Retrieved from http://www.educationaldatamining.org/EDM2016/proceedings/paper_126.pdf

Allen, M. J., & Yen, W. M. (2002). *Introduction to measurement theory.* Prospect Heights, IL: Waveland Press.

Almond, R. G., Deane, P., Quinlan, T., Wagner, M., & Sydorenko, T. (2012). *A preliminary analysis of keystroke log data from a timed writing task* (Research Report No. RR-12-23). Princeton, NJ: Educational Testing Service. doi:10.1002/j.2333-8504.2012.tb02305.x

Almond, R. G., Mislevy, R. J., Steinberg, L. S., Williamson, D. M., & Yan, D. (2015). *Bayesian networks in educational assessment.* New York, NY: Springer-Verlag. doi:10.1007/978-1-4939-2125-6

Almond, R. G., Shute, V. J., Tingir, S., & Rahimi, S. (in press). Identifying observable outcomes in game-based assessments. In H. Jiao, & R. W. Lissitz (Eds.), *Applications of artificial intelligence to assessment.* Charlotte, NC: Information Age Publishing.

Almond, R. G., Steinberg, L. S., & Mislevy, R. J. (2002). Enhancing the design and delivery of assessment systems: A four-process architecture. *Journal of Technology, Learning, and Assessment, 1*(5). Retrieved from https://ejournals.bc.edu/index.php/jtla/article/download/1671/1509/

Alpaydin, E. (2010). *Introduction to machine learning* (2nd ed.). Cambridge, MA: MIT Press.

Alshehri, A., Coenen, F., & Bollegala, D. (2018). Spectral analysis of keystroke streams: Towards effective real-time continuous user authentication. In P. Mori, S. Furnell, & O. Camp (Eds.), *Proceedings of the 4th international conference on information systems security and privacy* (Vol. 1, pp. 62–73). doi:10.5220/0006606100620073

Alves, R. A., & Limpo, T. (2015). Progress in written language bursts, pauses, transcription, and written composition across schooling. *Scientific Studies of Reading, 19,* 374–391. doi:10.1080/10888438.2015.1059838

American Educational Research Association, American Psychological Association, & National Council on Measurement in Education. (2014). *Standards for educational and psychological testing.* Washington, DC: American Educational Research Association.

American Institutes for Research. (2017). *South Dakota Smarter Balanced summative assessments, 2016–2017 technical report: Addendum to the Smarter Balanced technical report.* Retrieved from the South Dakota Department of Education website: http://doe.sd.gov/Assessment/documents/17-TechReport.pdf

Amorim, E., Cançado, M., & Veloso, A. (2018). Automated essay scoring in the presence of biased ratings. In M. Walker, H. Ji, & A. Stent (Eds.), *Proceedings of the 2018 conference of the North American chapter of the Association for Computational Linguistics: Human language technologies (long papers)* (Vol. 1., pp. 229–237). Retrieved from https://aclweb.org/anthology/volumes/N18-1/

Anderson, J. R. (1987). Methodologies for studying human knowledge. *Behavioral and Brain Sciences, 10*(3), 467–477.

Anderson, J. R., & Reiser, B. J. (1985). The LISP tutor. *Byte, 10,* 159–175.

Anson, C. M. (2016). The Pop Warner chronicles: A case study in contextual adaptation and the transfer of writing ability. *College Composition and Communication, 67,* 518–549.

Armer, P. (1963). *Computers and thought*. New York, NY: McGraw-Hill.

Aryadoust, V. (2010). Investigating writing sub-skills in testing English as a foreign language: A structural equation modeling study. *TESL-EJ, 13*(4), 1–20.

Ashby, F. G., & Maddox, W. T. (2005). Human category learning. *Annual Review of Psychology, 56*, 149–178. doi:10.1146/annurev.psych.56.091103.070217

Ashby, F. G., & Valentin, V. V. (2017). Multiple systems of perceptual category learning: Theory and cognitive tests. In H. Cohen, & C. Lefebvre (Eds.), *Handbook of categorization in cognitive science* (2nd ed., pp. 157–188). Amsterdam, The Netherlands: Elsevier. doi:10.1016/B978-0-08-101107-2.00007-5

Association of Test Publishers. (2017). *Welcome to ATP: The intelligent voice for testing*. Retrieved from https://www.testpublishers.org/

Attali, Y. (2011). Sequential effects in essay ratings. *Educational and Psychological Measurement, 71*, 68–79. doi:10.1177/0013164410387344

Attali, Y. (2013). Validity and reliability of automated essay scoring. In M. D. Shermis, & J. Burstein (Eds.), *Handbook of automated essay evaluation: Current applications and new directions* (pp. 181–198). New York, NY: Routledge.

Attali, Y. (2016). A comparison of newly-trained and experienced raters on a standardized writing assessment. *Language Testing, 33*, 99–115. doi:10.1177/026553221 5582283

Attali, Y., Bridgeman, B., & Trapani, C. (2010). Performance of a generic approach in automated essay scoring. *The Journal of Technology, Learning, and Assessment, 10*(3), 1–17. Retrieved from https://ejournals.bc.edu/index.php/jtla/article/view/1603

Attali, Y., & Burstein, J. (2006). *Automated essay scoring with e-rater v.2.0* (Research Report No. RR-04-45). Princeton, NJ: Educational Testing Service. doi:10.1002/j.2333-8504.2004.tb01972.x

Baaijen, V. M., Galbraith, D., & de Glopper, K. (2012). Keystroke analysis: Reflections on procedures and measures. *Written Communication, 29*, 246–277. doi:10.1177/0741088312451108

'Babel' essay machine can generate prize-winning, but meaningless exams. (2014, April 30). *Huffington Post UK*. Retrieved from http://www.huffingtonpost. co.uk/2014/04/30/babel-essay-machine_n_5240272.html

Bacha, N. (2001). Writing evaluation: What can analytic versus holistic essay scoring tell us? *System, 29*, 371–383. doi:10.1016/S0346-251X(01)00025-2

Baird, J. A. (1998). What's in a name? Experiments with blind marking in A-level examinations. *Educational Research, 40*, 191–202. doi:10.1080/0013188980400207

Baker, R., Corbett, A., Gowda, S., Wagner, A., MacLaren, B., Kauffman, L., … Giguere, S. (2010b). Contextual slip and prediction of student performance after use of an intelligent tutor. In P. De Bra, P. Kobsa, & D. Chin (Eds.), *Proceedings of the 18th annual conference on user modeling, adaptation, and personalization, UMAP 2010*. LNCS 6075 (pp. 52–63). Big Island of Hawaii, HI: Springer-Verlag.

Baker, R. S. J. d., Corbett, A., & Wagner, A. Z. (2006, June). Human classification of low-fidelity replays of student actions. In *Proceedings of the educational data mining workshop at the 8th international conference on intelligent tutoring systems* (pp. 29–36).

Baker, R. S. J. d., D'Mello, S. K., Rodrigo, M. M. T., & Graesser, A. C. (2010). Better to be frustrated than bored: The incidence, persistence, and impact of learners' cognitive-affective states during interactions with three different computer-based learning environments. *International Journal of Human-Computer Studies, 68*, 223–241. doi:10.1016/j.ijhcs.2009.12.003

Baker, R., Mitrovic, A., & Mathews, M. (2010). Detecting gaming the system in constraint-based tutors. In P. De Bra, P. Kobsa, & D. Chin (Eds.), *Lecture notes in computer science: Vol. 6075. Proceedings of the 18th annual conference on user modeling, adaptation, and personalization* (pp. 267–278). New York, NY: Springer-Verlag.

Baker, R., & Yacef, K. (2009). The state of educational data mining in 2009: A review and future visions. *Journal of Educational Data Mining, 1*(1), 3–17.

Balf, T. (2014, March 6). The story behind the SAT overhaul. *The New York Times Magazine*. Retrieved from https://www.nytimes.com/2014/03/09/magazine/the-story-behind-the-sat-overhaul.html

Baltrušaitis, T., Ahuja, C., & Morency, L.-P. (2019). Multimodal machine learning: A survey and taxonomy. *IEEE Transactions on Pattern Analysis and Machine Intelligence, 41*, 423–443. doi:10.1109/TPAMI.2018.2798607

Bass, L., Clements, P., & Kazman, R. (2013). *Software architecture in practice* (3rd ed.). Westford, MA: Addison Wesley.

Bassett, L. (2015). *Introduction to JavaScript object notation: A to-the-point guide to JSON*. Sebastopol, CA: O'Reilly Media, Inc.

Bauer, M., Wylie, C., Jackson, T., Mislevy, B., Hoffman-John, E., John, M., & Corrigan, S. (2017). Why video games can be a good fit for formative assessment. *Journal of Applied Testing Technology, 18*(S1), 19–31.

Beauvais, C., Olive, T., & Passerault, J.-M. (2011). Why are some texts good and others not? Relationship between text quality and management of the writing processes. *Journal of Educational Psychology, 103*, 415–428. doi:10.1037/a0022545

Behrens, J. T., & Smith, M. L. (1997). Data and data analysis. In D. Berliner, & R. Calfee (Eds.), *The handbook of educational psychology* (pp. 945–989). New York, NY: MacMillan.

Beigman Klebanov, B., & Flor, M. (2013a). Argument-relevant metaphors in student essays. In *Meta4NLP 2013: Proceedings of the first workshop on metaphor in NLP* (pp. 11–20). Retrieved from https://www.aclweb.org/anthology/W13-09

Beigman Klebanov, B., & Flor, M. (2013b). Word association profiles and their use for automated scoring of essays. In *Proceedings of the 51st annual meeting of the Association for Computational Linguistics* (Vol. 1, pp. 1148–1158). Retrieved from https://www.aclweb.org/anthology/P13-1113

Beigman Klebanov, B., Gyawali, B., & Song, Y. (2017). Detecting good arguments in a non-topic-specific way: An oxymoron? In R. Barzilasy, & M.-Y. Kan (Eds.), *Proceedings of the 55th annual meeting of the Association for Computational Linguistics* (Vol. 2, pp. 244–249). doi:10.18653/v1/P17-2038

Bejar, I. I. (1984). Educational diagnostic assessment. *Journal of Educational Measurement, 21*, 175–189. doi:10.1111/j.1745-3984.1984.tb00228.x

Bejar, I. I. (2010, May). *Towards a quality control and assurance framework for automated scoring*. Paper presented at the annual meeting of the National Council of Measurement in Education, Denver, CO.

Bejar, I. I. (2011). A validity-based approach to quality control and assurance of automated scoring. *Assessment in Education: Principles, Policy & Practice, 18*, 319–341. doi:10.1080/0969594X.2011.555329

Bejar, I. I. (2012). Rater cognition: Implications for validity. *Educational Measurement: Issues and Practice, 31*(3), 2–9. doi:10.1111/j.1745-3992.2012.00238.x

Bejar, I. I. (2017). Threats to score meaning in automated scoring. In K. Ercikan, & J. W. Pellegrino (Eds.), *Validation of score meaning for the next generation of assessments* (pp. 75–84). New York, NY: Routledge. doi:10.4324/9781315708591-7

Bejar, I. I., Mislevy, R. J., & Zhang, M. (2017). Automated scoring with validity in mind. In A. A. Rupp, & J. P. Leighton (Eds.), *The Wiley handbook of cognition and assessment: Frameworks, methodologies, and applications* (pp. 226–246). West Sussex, UK: Wiley-Blackwell. doi:10.1002/9781118956588.ch10

Benetos, K., & Betrancourt, M. (2015). Visualization of computer-supported argument writing processes using C-SAW. *Revista Romana de Interactiune Om-Calculator, 8*, 281–302.

Bengio, Y., Ducharme, R., Vincent, P., & Jauvin, C. (2003). A neural probabilistic language model. *Journal of Machine Learning Research, 3*, 1137–1155.

Bennett, A., Bridglall, B. L., Cauce, A. M., Everson, H. T., Gordon, E. W., Lee, C. D., … Stewart, J. K. (2004). *All students reaching the top: Strategies for closing academic achievement gaps* (Report from the National Study Group for the Affirmative Development of Academic Ability). Retrieved from the ERIC datbase. (ED483170).

Bennett, R. E. (2004). *Moving the field forward: Some thoughts on validity and automated scoring* (Research Memorandum No. RM-04-01). Princeton, NJ: Educational Testing Service.

Bennett, R. E. (2011). *Automated scoring of constructed-response literacy and mathematics items*. Washington, DC: Arabella Advisors.

Bennett, R. E., & Bejar, I. I. (1998). Validity and automated scoring: It's not only the scoring. *Educational Measurement: Issues and Practice, 17*(4), 9–16. doi:10.1111/j.1745-3992.1998.tb00631.x

Bennett, R. E., Steffen, M., Singley, K., Morley, M., & Jacquemin. D. (1997). Evaluating an automatically scorable, open-ended response type for measuring mathematical reasoning in computer-adaptive tests. *Journal of Educational Measurement, 34*, 162–176. doi:10.1111/j.1745-3984.1997.tb00512.x

Bennett, R. E., & Zhang, M. (2016). Validity and automate scoring. In F. Drasgow (Ed.), *Technology and testing: Improving educational and psychological measurement* (pp. 142–173). New York, NY: Routledge.

Bereiter, C. B., & Scardamalia, M. (1987). *The psychology of written composition*. Hillsdale, NJ: Lawrence Erlbaum.

Berkeley Data Science Department. (2019). *What is a data science?* Retrieved from https://datascience.berkeley.edu/about/what-is-data-science/

Berninger, V. W., & Chanquoy, L. (2012). What writing is and how it changes across early and middle childhood development: a multidisciplinary perspective. In E. Grigorenko, E. Mambrino, & D. Preiss (Eds.), *Writing, a mosaic of new perspectives* (pp. 65–84). New York, NY: Psychology Press.

Berninger, V. W., Fuller, F., & Whitaker, D. (1996). A process model of writing development across the life-span. *Educational Psychology Review, 8*, 193–218. doi:10.1007/BF01464073

Bernstein, J., Cohen, M., Murveit, H., Rtischev, D., & Weintraub, M. (1990). Automatic evaluation and training in English pronunciation. In *First international conference on spoken language processing (ICSLP 90)* (pp. 1185–1188). Retrieved from https://www.isca-speech.org/archive/icslp_1990/i90_1185.html

Bernstein, J., Van Moere, A., & Cheng, J. (2010). Validating automated speaking tests. *Language Testing, 27*, 355–377. doi:10.1177/0265532210364404

Betts, B., & Smith, R. (2018). *The learning technology manager's guide to the xAPI*. Retrieved from https://www.ht2labs.com/xapi-guide-download/

Bhat, S., & Yoon, S.-Y. (2015). Automatic assessment of syntactic complexity for spontaneous speech scoring. *Speech Communication, 67*, 42–57. doi:10.1016/j.specom.2014.09.005

Biber, D., & Gray, B. (2014). *Discourse characteristics of writing and speaking task types on the TOEFL iBT test: A lexico-grammatical analysis* (Research Report No. RR-13-04). Princeton, NJ: Educational Testing Service. doi:10.1002/j.2333-8504.2013.tb02311.x

Bixler, R., & D'Mello, S. (2013). Detecting boredom and engagement during writing with keystroke analysis, task appraisals, and stable traits. In *Proceedings of the 2013 international conference on intelligent user interfaces* (pp. 225–234). doi:10.1145/2449396.2449426

Blalock, G. (2013, March 12). *Human readers: Professionals against machine scoring of student essays in high-stakes assessment.* Retrieved February 28, 2017 from http://humanreaders.org/petition/index.php

Blanz, V., & Vetter, T. (1999). A morphable model for the synthesis of 3D faces. In *Proceedings of the 26th annual conference on computer graphics and interactive techniques* (pp. 187–194). doi:10.1145/311535.311556

Blau, I., & Caspi, A. (2009). *What type of collaboration helps? Psychological ownership, perceived learning and outcome quality of collaboration using Google Docs.* In Y. Eshet-Alkalai, A. Caspi, S. Eden, N. Geri, & Y. Yair (Eds.), *Proceedings of the chais conference on instructional technologies research* (pp. 48–55). Raanana: The Open University of Israel.

Bojanowski, P., Grave, E., Joulin, A., & Mikolov, T. (2017). Enriching word vectors with subword information. *Transactions of the Association for Computational Linguistics, 5*, 135–146. doi:10.1162/tacl_a_00051

Bosch, N., Chen, H., Baker, R., Shute, V., & D'Mello, S. K. (2015). Accuracy vs. availability heuristic in multimodal affect detection in the wild. In *ICMI '15: Proceedings of the 2015 ACM international conference on multimodal interaction* (pp. 267–274). doi:10.1145/2818346.2820739

Bosch, N., D'Mello, S. K., Ocumpaugh, J., Baker, R., & Shute, V. (2016). Using video to automatically detect learner affect in computer-enabled classrooms. *ACM Transactions on Interactive Intelligent Systems, 6*(2), 17.1–17.26. doi:10.1145/2946837

Boyer, M. (2018). *Examining the effects of automated rater bias and variability on test equating solutions* (Unpublished doctoral dissertation). University of Massachusetts, Amherst, MA.

Boyer, M. (2019). Beyond faster and cheaper, could automated scoring produce better scores? Using advances in technology to improve the quality of educational assessment [Blog post]. Retrieved from the Center for Assessment Blog: https://www.nciea.org/blog/assessment/beyond-faster-and-cheaper-could-automated-scoring-produce-better-scores

Boyer, M., & Patz, R. J. (2019, April). *Understanding and mitigating rater drift: When does rater drift threaten score comparability?* Paper presented at the annual meeting of the National Council on Measurement in Education, Toronto, Canada.

Boyer, M., Patz, R., & Keller, L. (2018, April). *Examining the effects of automated rater bias and variability on test equating solutions.* Paper presented at the annual meeting of the National Council on Measurement in Education, New York, NY.

Braun, H. I., Bejar, I. I., & Williamson, D. M. (2006). Rule-based methods for automated scoring: Applications in a licensing context. In D. M. Williamson, R. J. Mislevy, & I. I. Bejar (Eds.), *Automated scoring of complex tasks in computer-based testing* (pp. 83–122). Mahwah, NJ: Lawrence Erlbaum.

Braun, H. I., Bennett, R. E., Frye, D., & Soloway, E. (1990). Scoring constructed responses using expert systems. *Journal of Educational Measurement, 27*, 93–108. doi:10.1111/j.1745-3984.1990.tb00736.x

Breetvelt, I., Van den Bergh, H., & Rijlaarsdam, G. (1994). Relations between writing processes and text quality: When and how? *Cognition and Instruction, 12*, 103–123. doi:10.1207/s1532690xci1202_2

Breiman, L. (2001). Random forests. *Machine Learning, 45*(1), 5–32. doi:10.1023/A:1010933404324

Breland, H. M., Jones, R. J., Jenkins, L., Paynter, M., Pollack, J., & Fong, Y. F. (1994). *The College Board vocabulary study* (Research Report No. RR-94-26). Princeton, NJ: Educational Testing Service. doi:10.1002/j.2333-8504.1994.tb01599.x

Brennan, R. L. (2001). *Generalizability theory*. New York, NY: Springer-Verlag.

Breyer, F. J., Rupp, A. A., & Bridgeman, B. (2017). *Implementing a contributory scoring approach for the GRE analytical writing section: A comprehensive empirical investigation* (Research Report No. RR-17-14). Princeton, NJ: Educational Testing Service. doi:10.1002/ets2.12142

Bridgeman, B. (2013). Human ratings and automated essay evaluation. In M. D. Shermis, & J. Burstein (Eds.), *Handbook of automated essay evaluation: Current applications and new directions* (pp. 222–232). New York, NY: Routledge.

Bridgeman, B., & Lewis, C. (1994). The relationship of essay and multiple-choice scores with grades in college courses. *Journal of Educational Measurement, 31*, 37–50. doi:10.1111/j.1745-3984.1994.tb00433.x

Bridgeman, B., Powers, D., Stone, E., & Mollaun, P. (2011). TOEFL iBT speaking test scores as indicators of oral communicative language proficiency. *Language Testing, 29*, 91–108. doi:10.1177/0265532211411078

Bridgeman, B., & Ramineni, C. (2017). Design and evaluation of automated writing evaluation models: Relationships with writing in naturalistic settings. *Assessing Writing, 34*, 62–71. doi:10.1016/j.asw.2017.10.001

Bridgeman, B., Ramineni, C., Deane, P., & Li, C. (2014, April). *Using an external validity criterion as an alternate basis for automated scoring*. Paper presented at the annual meeting of the National Council on Measurement in Education, Philadelphia, PA.

Bridgeman, B., & Trapani, C. (2011, April). *The question of validity of automated essay scores and differentially valued evidence*. Paper presented at the annual meeting of the National Council on Measurement in Education, New Orleans, LA.

Bridgeman, B., Trapani, C., & Attali, Y. (2012). Comparison of human and machine scoring of essays: Differences by gender, ethnicity, and country. *Applied Measurement in Education, 25*, 27–40. doi:10.1080/08957347.2012.635502

Britt, M. A., & Aglinskas, C. (2002). Improving students' ability to identify and use source information. *Cognition and Instruction, 20*, 485–522. doi:10.1207/s1532690xci2004_2

Brockwell, P. J., & Davis, R. A. (2002). *Introduction to time series and forecasting*. New York, NY: Springer.

Brooke, J. (1996). SUS—A quick and dirty usability scale. *Usability Evaluation in Industry, 189*(194), 4–7.

Bui, T. D., Heylen, D., Poel, M., & Nijholt, A. (2001). Generation of facial expressions from emotion using a fuzzy rule based system. In M. Stumptner, D. R. Corbett, & M. Brooks (Eds.), *Lecture notes in computer science: Vol. 2256. AI 2001: Advances in artificial intelligence* (pp. 83–94). New York, NY: Springer.

Bunch, M. B., Vaughn, D., & Miel, S. (2016). Automated scoring in assessment systems. In Y. Rosen, S. Ferrara, & M. Mosharrat (Eds.), *Handbook of research on technology tools for real-world skill development* (pp. 611–626). Hershey, PA: IGI Global.

Burrows, S., Gurevych, I., & Stein, B. (2015). The eras and trends of automatic short answer grading. *International Journal of Artificial Intelligence in Education, 25*, 60–117. doi:10.1007/s40593-014-0026-8

Burstein, J., Beigman Klebanov, B., Elliot, N., & Molloy, H. (2016). A left turn: Automated feedback & activity generation for student writers. In *Proceedings of the 2nd language teaching, language & technology workshop* (pp. 6–13). doi:10.21437/LTLT.2016-2

Burstein, J., & Chodorow, M. (1999). Automated essay scoring for nonnative English speakers. In *Proceedings of a symposium on computer mediated language assessment and evaluation in natural language processing* (pp. 68–75). Stroudsburg, PA: Association for Computational Linguistics.

Burstein, J., Chodorow, M., & Leacock, C. (2003). Criterion online essay evaluation: An application for automated evaluation of student essays. In J. Riedl, & R. Hill (Eds.), *Proceedings of the fifteenth conference on innovative applications of artificial intelligence* (pp. 3–10). doi:10.1007/978-3-030-19823-7

Burstein, J., Chodorow, M., & Leacock, C. (2004). Automated essay evaluation. *AI Magazine 25*(3), 27–36.

Burstein, J., Elliot, N., Beigman Klebanov, B., Madnani, N., Napolitano, D., Schwartz, M., … Molloy, H. (2018). Writing mentor: Writing progress using self-regulated writing support. *Journal of Writing Analytics, 2*, 285–313.

Burstein, J., Elliot, N., & Molloy, H. (2016). Informing automated writing evaluation using the lens of genre: Two studies. *CALICO Journal, 3*, 117–141. doi:10.1558/cj.v33i1.26374

Burstein, J., Kukich, K., Wolff, S., Lu, C., & Chodorow, M. (1998, April). *Computer analysis of essays*. Paper presented at the annual meeting of the National Council of Measurement in Education, San Diego, CA.

Burstein, J., Kukich, K., Wolff, S., Lu, C., Chodorow, M., Braden-Harder, L., & Harris, M. D. (1998, August). Automated scoring using a hybrid feature identification technique. In *Proceedings of the 36th annual meeting of the Association for Computational Linguistics and 17th international conference on computational linguistics* (Vol. 1, pp. 206–210). doi:10.3115/980845.980879

Burstein, J., Marcu, D., & Knight, K. (2003). Finding the WRITE stuff: Automatic identification of discourse structure in student essays. *IEEE Intelligent Systems, 18*(1), 32–39. doi:10.1109/MIS.2003.1179191

Burstein, J., McCaffrey, D., Beigman Klebanov, B., & Ling, G. (2017). Exploring relationships between writing and broader outcomes with automated writing evaluation. In *Proceedings of the 12th workshop on innovative use of NLP for building educational applications* (pp. 101–108). doi:10.18653/v1/W17-5011

Burstein, J., McCaffrey, D., Beigman Klebanov, B., Ling, G., & Holtzman, S. (2019). Exploring writing analytics and postsecondary success indicators. In *Companion proceedings of the 9th international conference on learning analytics & knowledge* (pp. 213–214). Retrieved from https://solaresearch.org/core/compantion_proceedings_lak19/

Burstein, J., Tetreault J., & Madnani, N. (2013). The e-rater automated essay scoring system. In M. D. Shermis, & J. Burstein (Eds.), *Handbook of automated essay evaluation: Current applications and new directions* (pp. 55–67). New York, NY: Routledge.

Butterfield, E. C., Hacker, D. J., & Albertson, L. R. (1996). Environmental, cognitive and metacognitive influences on text revision. *Educational Pschology Review, 8,* 239–297. doi:10.1007/BF01464075

Buzick, H., Oliveri, M. E., Attali, Y., & Flor, M. (2016). Comparing human and automated essay scoring for prospective graduate students with learning disabilities and/or ADHD. *Applied Measurement in Education, 29,* 161–172. doi:10.1080/089 57347.2016.1171765

Cahill, A. (2012). *Updated modules for detecting fragments and run-ons.* Unpublished manuscript.

Cai, Z., Gong, Y., Qiu, Q., Hu, X., & Graesser, A. (2016). Making *AutoTutor* agents smarter: *AutoTutor* answer clustering and iterative script authoring. In D. Traum, W. Swartout, P. Khooshabeh, S. Kopp, S. Scherer, & A. Leuski (Eds.), *Lecture notes in computer science: Vol. 10011. Intelligent virtual agents* (pp. 438–441). Berlin, Germany: Springer. doi:10.1007/978-3-319-47665-0_50

Cai, Z., Graesser, A. C., Forsyth, C., Burkett, C., Millis, K., Wallace, P., ... Butler, H. (2011). Trialog in ARIES: User input assessment in an intelligent tutoring system. In W. Chen, & S. Li (Eds.), *Proceedings of the 3rd IEEE international conference on intelligent computing and intelligent systems* (pp. 429–433). Piscataway, NJ: IEEE.

Cai, Z., Graesser, A. C., & Hu, X. (2015). ASAT: *AutoTutor* script authoring tool. In. R. Sottilare, A. C. Graesser, X. Hu, & K. Brawner (Eds.), *Design recommendations for intelligent tutoring systems: Vol. 3. Authoring tools and expert modeling techniques* (pp. 199–210). Orlando, FL: Army Research Laboratory.

Cai, Z., Rus, V., Kim, H. J., Susarla, S., Karnam, P., & Graesser, A. C. (2006). NLGML: A natural language generation markup language. In T. C. Reeves, & S. F. Yamashita (Eds.), *Proceedings of E-Learn 2006* (pp. 2747–2752). Waynesville, NC: AACE.

Carbonell, J. (1970). AI in CAI: An artificial intelligence approach to computer aided instruction. *IEEE Transactions on Man-Machine Systems, 11,* 190–202. doi:10.1109/TMMS.1970.299942

Carrell, P. L. (1995). The effect of writers' personalities and raters' personalities on the holistic evaluation of writing. *Assessing Writing, 2,* 153–190. doi:10.1016/1075-2935(95)90011-X

Carvell, T., Gurewitch, D., Oliver, J. (Writers), & Murphy, L. (Director). (2015). Standardised testing [Television series episode]. In T. Carvell, & J. Oliver (Producers), *Last week tonight with John Oliver.* New York, NY: Home Box Office.

Chafe, W., & Tannen, D. (1987). The relation between written and spoken language. *Annual Review of Anthropology, 16,* 383–407. doi:10.1146/annurev.an.16.100187.002123

Chang, T.-H., & Lee, C.-H. (2009). Automatic Chinese essay scoring using connections between concepts in paragraphs. In *Proceedings of the 2009 international conference on Asian language processing* (pp. 265–268). doi:10.1109/IALP.2009.63

Chanquoy, L. (2009). Revision processes. In R. Beard, D. Myhill, J. Riley, & M. Nystrand (Eds.), *The Sage handbook of writing development* (pp. 80–97). London, UK: Sage.

Chen, L., Evanini, K., & Sun, X. (2010). Assessment of non-native speech using vowel space characteristics. In *Proceedings of the 2010 IEEE Spoken Language Technology Workshop* (pp. 139–144). doi:10.1109/SLT.2010.5700836

Chen, J., Fife, J. H., Bejar, I. I., & Rupp, A. A. (2016). *Building e-rater scoring models using machine learning methods* (Research Report No. RR-16-04). Princeton, NJ: Educational Testing Service. doi:10.1002/ets2.12094

Chen, L., Feng, G., Leong, C. W., Joe, J., Kitchen, C., & Lee, C. M (2016). Designing an automated assessment of public speaking skills using multimodal cues. *Journal of Learning Analytics, 3,* 261–281. doi:10.18608/jla.2016.32.13

Chen, L., Tao, J., Ghaffarzadegan, S., & Qian, Y. (2018). End-to-end neural network based automated speech scoring. In *Proceedings of the international conference on acoustics, speech, and signal processing (ICASSP 2018)* (pp. 6234–6238). doi:10.1109/icassp.2018.8462562

Chen, L., & Yoon, S.-Y. (2011). Detecting structural events for assessing non-native speech. In *Proceedings of the 6th workshop on innovative use of NLP for buidling educational applications* (pp. 38–45). Stroudsburg, PA: Association for Computational Linguistics.

Chen, L., & Yoon, S.-Y. (2012). Application of structural events detected on ASR outputs for automated speaking assessment. In *Proceedings of 13th annual conference of the international speech communication association* (pp. 767–770). Retrieved from https://www.isca-speech.org/archive/archive_papers/interspeech_2012/i12_0767.pdf

Chen, L., & Zechner, K. (2011a). Applying rhythm features to automatically assess non-native speech. In *Proceedings of INTERSPEECH 2011, 12th annual conference of the international speech communication association* (pp. 1861–1864). Retrieved from https://www.isca-speech.org/archive/interspeech_2011/

Chen, M., & Zechner, K. (2011b). Computing and evaluating syntactic complexity features for automated scoring of spontaneous non-native speech. In *Proceedings of the 49th annual meeting of the Association for Computational Linguistics* (pp. 722–731). Retrieved from http://www.aclweb.org/anthology/P11-1073

Chen, L., Zechner, K., & Xi, X. (2009). Improved pronunciation features for construct-driven assessment of non-native spontaneous speech. In *NAACL '09 proceedings of human language technologies: the annual conference of the North American chapter of the ACL* (pp. 442–449). doi:10.3115/1620754.1620819

Chen, L., Zechner, K., Yoon, S.-Y., Evanini, K., Wang, X., Loukina, A., … Gyawali, B. (2018). *Automated scoring of nonnative speech using the SpeechRater v.5.0 engine* (Research Report No. RR-18-10). Princeton, NJ: Educational Testing Service. doi:10.1002/ets2.12198

Chen, M., & Zechner, K. (2012). Using an ontology for improved automated content scoring of spontaneous non-native speech. In *Proceedings of the 7th workshop on innovative use of NLP for building educational applications* (pp. 86–94). Retrieved from https://www.aclweb.org/anthology/W12-2010

Chen, T., & Guestrin, C. (2016). Xgboost: A scalable tree boosting system. In *Proceedings of the 22nd ACM SIGKDD international conference on knowledge discovery and data mining* (pp. 785–794). doi:10.1145/2939672.2939785

Chen, W., & Chang, W. (2004). Applying hidden Markov models to keystroke pattern analysis for password verification. In *Proceedings of the 2004 IEEE international conference on information reuse and integration* (pp. 467–474). doi:10.1109/IRI.2004.1431505

Chen, X. (2016). *Remedial coursetaking at U.S. public 2- and 4-year institutions: Scope, experience, and outcomes* (NCES 2016-405). Washington, DC: National Center for Education Statistics.

Cheng, J., Chen, X., & Metallinou, A. (2015). Deep neural network acoustic models for spoken assessment applications. *Speech Communication, 73,* 14–27. doi:10.1016/j.specom.2015.07.006

Chenu, F., Pellegrino, F., Jisa, H., & Fayol, M. (2014). Interword and intraword pause threshold in writing. *Frontiers in Psychology, 5,* 1–7. doi:10.3389/fpsyg.2014.00182

Cho, K., & MacArthur, C. (2010). Student revision with peer and expert reviewing. *Learning and Instruction, 20,* 328–338. doi:10.1016/j.learninstruc.2009.08.006

Cho, K., van Merrienboer, B., Gulcehre, C., Bahdanau, D., Bougares, F., Schwenk, H., & Bengio, Y. (2014). Learning phrase representations using RNN encoder-decoder for statistical machine translation. In *Proceedings of the 2014 conference on empirical methods in natural language processing (EMNLP)* (pp. 1724–1734). doi:10.3115/v1/D14-1

Cho, W., Oh, J., Lee, J., & Kim, Y.-S. (2005). An intelligent marking system based on semantic kernel and Korean WordNet. *The KIPS Transactions: Part A, 12A,* 539–546. doi:10.3745/KIPSTA.2005.12A.6.539

Chollet, F. (2016). Double sarsa and double expected sarsa with shallow and deep learning. *Journal of Data Analysis and Information Processing, 4,* 159–176.

Christensen, C. M., Raynor, M. E., & McDonald, R. (2015, December). What is disruptive innovation? *Harvard Business Review,* 44–53. Retrieved from https://hbr.org/2015/12/what-is-disruptive-innovation

Chu, S., Kennedy, D., & Mak, Y. (2009, December). *MediaWiki and Google Docs as online collaboration tools for group project co-construction.* Paper presented at ICKM 2009: the 6th international conference on knowledge management, Hong Kong.

Chukharev-Hudilainen, E. (2019). Empowering automated writing evaluation with keystroke logging. In E. Lindgren, & K. P. H. Sullivan (Eds.), *Observing writing: Insights from keystroke logging and handwriting* (pp. 125–142). doi:10.1163/9789004392526_007

Clarke-Midura, J., Code, J., Dede, C., Mayrath, M., & Zap, N. (2012). Thinking outside the bubble: Virtual performance assessments for measuring complex learning. In D. Robinson, J. Clarke-Midura, & M. Mayrath (Eds.), *Technology-based assessments for 21st century skills: Theoretical and practical implications from modern research* (pp. 125–148). Charlotte, NC: Information Age.

Clauser, B. E., Clyman, S. G., & Swanson, D. B. (1999). Components of rater error in a complex performance assessment. *Journal of Educational Measurement, 36,* 29–45. doi:10.1111/j.1745-3984.1999.tb00544.x

Clauser, B. E., Harik, P., & Clyman, S. G. (2000). The generalizability of scores for a performance assessment scored with a computer-automated scoring system. *Journal of Educational Measurement, 37,* 245–262. doi:10.1111/j.1745-3984.2000.tb01085.x

Clauser, B. E., Margolis, M. J., Clyman, S. G., & Ross, L. P. (1997). Development of automated scoring algorithms for complex performance assessments: A comparison of two approaches. *Journal of Educational Measurement, 34,* 141–161. doi:10.1111/j.1745-3984.1997.tb00511.x

Clauser, B. E., Ross, L. P., Clyman, S. G., Rose, K. M., Margolis, M. J., Nungester, R. J., … Malakoff, G. L. (1997). Development of a scoring algorithm to replace expert rating for scoring a complex performance based assessment. *Applied Measurement in Education, 10,* 345–358. doi:10.1207/s15324818ame1004_3

Clauser, B. E., Ross, L. P., Fan, V. Y., & Clyman, S. G. (1998). A comparison of two approaches for modeling expert judgment in scoring a performance assessment of physicians' patient-management skills. *Academic Medicine (RIME Supplement), 73*(10), S117–S119. doi:10.1097/00001888-199810000-00065

Clauser, B. E., Subhiyah, R., Nungester, R. J., Ripkey, D. R., Clyman, S. G., & McKinley, D. (1995). Scoring a performance-based assessment by modeling the judgments of experts. *Journal of Educational Measurement, 32*, 397–415. doi:10.1111/j.1745-3984.1995.tb00474.x

Clauser, B. E., Subhiyah, R., Piemme, T. E., Greenberg, L., Clyman, S. G., Ripkey, D., & Nungester, R. J. (1993). Using clinician ratings to model score weights for a computer simulation performance assessment. *Academic Medicine (RIME Supplement), 68*(10). doi:10.1097/00001888-199310000-00048

Clauser, B. E., Swanson, D. B., & Clyman, S. G. (1999). A comparison of the generalizability of scores produced by expert raters and automated scoring systems. *Applied Measurement in Education, 12*, 281–299. doi:10.1207/S15324818AME1203_4

Cohen, Y. (2015, April). *The third rater fallacy in essay rating: An empirical test.* Paper presented at the annual meeting of the National Council on Measurement in Education, Chicago, IL.

Cohen, Y., Levi, E., & Ben-Simon, A. (2018). Validating human and automated essay scoring against "true" scores. *Applied Measurement in Education, 31*, 241–250. doi:10.1080/08957347.2018.1464450

Coiro, J., Sparks, J. R., & Kulikowich, J. M. (2018a). Assessing online collaborative inquiry and social deliberation skills as learners navigate multiple sources and perspectives. In J. L. G. Braasch, I. Bråten, & M. T. McCrudden (Eds.), *Handbook of multiple source use* (pp. 485–501). New York, NY: Routledge.

Coiro, J., Sparks, J. R., & Kulikowich, J. M. (2018b). Assessing online reading comprehension, collaborative inquiry and social deliberation across multiple sources and perspectives. In J. L. G. Braasch, I. Braten, & M. T. McCrudden (Eds.), *Handbook of multiple source use* (pp. 485–517). New York: Routledge.

College Board. (n.d.). *Computer accommodation.* Retrieved from https://www.collegeboard.org/students-with-disabilities/typical-accommodations/computer

Complete College America. (2012). *Remediation: Higher education's bridge to nowhere.* Retrieved from https://completecollege.org/wp-content/uploads/2017/11/CCA-Remediation-final.pdf

Condon, W. (2013). Large-scale assessment, locally-developed measures, and automated scoring of essays: Fishing for red herrings? *Assessing Writing, 18*(1), 100–108. doi:10.1016/j.asw.2012.11.001

Conference on College Composition and Communication. (2004, February 25). A current challenge: Electronic rating. In *CCCC position statement on teaching, learning, and assessing writing in digital environments.* Retrieved March 7, 2019 from https://cccc.ncte.org/cccc/resources/positions/digitalenvironments

Coniam, D. (2009). A comparison of onscreen and paper-based marking in the Hong Kong public examination system. *Educational Research and Evaluation, 15*, 243–263. doi:10.1080/13803610902972940

Conley, D. T. (2014). *The common core state standards: Insight into their development and purpose.* Washington, DC: Council of Chief State School Officers.

Conway, M. E. (n.d.). Conway's law. Retrieved from http://www.melconway.com/Home/Conways_Law.html

Cook, C., Olney, A. M., Kelly, S., & D'Mello, S. K. (2018). An open vocabulary approach for estimating teacher use of authentic questions in classroom discourse. In K. E. Boyer, & M. Yudelson (Eds.), *Proceedings of the 11th international conference on educational data mining (EDM 2018)* (pp. 116–126). Retrieved from http://educationaldatamining.org/files/conferences/EDM2018/EDM2018_Preface_TOC_Proceedings.pdf

Cooper, W. E. (Ed.). (1993). *Cognitive aspects of skilled typewriting*. New York, NY: Springer.

Corbett, A., & Anderson, J. (1995). Knowledge-tracing: Modeling the acquisition of procedural knowledge. *User Modeling and User-Adapted Interaction, 4*, 253–278. doi:10.1007/BF01099821

Corter, J. E., & Gluck, M. A. (1992). Explaining basic categories: Feature predictability and information. *Psychological Bulletin, 111*, 291–303. doi:10.1037/0033-2909.111.2.291

Cortes, C., & Vapnik, V. (1995). Support-vector networks. *Machine Learning, 20*, 273–297. doi:10.1007/BF00994018

Council of Chief State School Officers. (2018, September 26). *States are leading with ESSA*. Retrieved from https://ccsso.org/blog/states-are-leading-essa

Council of Chief State School Officers & Association of Test Publishers. (2013). *Operational best practices for statewide large-scale assessment programs*. Washington, DC: Council of Chief State School Officers.

Council of Europe. (2001). *Common European framework of reference for languages*. Strasbourg, France: Author.

Cronbach, L. (1980). Validity on parole: How can we go straight? In *Proceedings of the 1979 ETS invitational conference: New directions for testing and measurement* (pp. 99–108). San Francisco, CA: Jossey-Bass.

Cronbach, L. J., & Meehl, P. E. (1955). Construct validity in psychological tests. *Psychological Bulletin, 52*, 281–302.

Cronbach, L. J., Gleser, G. C., Nanda, H., & Rajaratnam, N. (1972). *The dependability of behavioral measurements: Theory of generalizability for scores and profiles*. New York, NY: John Wiley.

Crossley, S. A., Allen, L. K., Snow, E. L., & McNamara, D. S. (2016). Incorporating learning characteristics into automatic essay scoring models: What individual differences and linguistic features tell us about writing quality. *Journal of Educational Data Mining 8*(2), 1–19.

Crossley, S. A., & McNamara, D. (2013). Applications of text analysis tools for spoken response grading. *Language Learning & Technology, 17*(2), 171–192.

Crossley, S. A., Roscoe, R., & McNamara, D. S. (2014). What is successful writing? An investigation into the multiple ways writers can write successful essays. *Written Communication, 31*, 184–214. doi:10.1177/0741088314526354

Cruz, D., Wieland, T., & Ziegler, A. (2006). Evaluation criteria for free/open source software products based upon project analysis. *Software Process Improvement and Practice, 11*, 107–122. doi:10.1002/spip.257

Cucchiarini, C., & Strik, H. (1999). Automatic assessment of second language learners' fluency. In *Proceedings of the 14th international congress of phonetic sciences* (pp. 759–762). Retrieved from https://www.internationalphoneticassociation.org/icphs-proceedings/ICPhS1999/papers/p14_0759.pdf

Cucchiarini, C., Strik, H., & Boves, L. (1997). Automatic evaluation of Dutch pronunciation by using speech recognition technology. In *1997 IEEE workshop on automatic speech recognition and understanding proceedings* (pp. 622–629). doi:10.1109/ASRU.1997.659144

D'Mello, S. K., Dowell, N., & Graesser, A. C. (2011). Does it really matter whether students' contributions are spoken versus typed in an intelligent tutoring system with natural language? *Journal of Experimental Psychology: Applied, 17*, 1–17. doi:10.1037/a0022674

D'Mello, S. K., & Graesser, A. C. (2012). *AutoTutor* and affective *AutoTutor*: Learning by alking with cognitively and emotionally intelligent computers that talk back. *ACM Transactions on Interactive Intelligent Systems, 2*(23), 1–38.

D'Mello, S. K., Kappas, A., & Gratch, J. (2018). The affective computing approach to affect measurement. *Emotion Review, 10*, 174–183. doi:10.1177/1754073917696583

D'Mello, S. K., & Kory, J. (2015). A review and meta-analysis of multimodal affect detection systems. *ACM Computing Surveys, 47*(3), 43:1–43:36. doi:10.1145/2682899

D'Mello, S. K., Lehman, S., Pekrun, R., & Graesser, A. (2014). Confusion can be beneficial for learning. *Learning and Instruction, 29*, 153–170. doi:10.1016/j.learninstruc.2012.05.003

Darling-Hammond, L. (2017). *Developing and measuring higher order skills: Models for sate performance assessment systems*. Retrieved from https://www.learningpolicyinstitute.org/sites/default/files/product-files/Models_State_Performance_Assessment_Systems_REPORT.pdf

Darling-Hammond, L., Herman, J., Pellegrino, J., Abedi, J., Aber, J. L., Baker, E., … Steele, C. M. (2013). *Criteria for high-quality assessment*. Stanford, CA: Stanford Center for Opportunity Policy in Education.

Daugherty, P. R., & Wilson, H. J. (2018). *Human + machine: Reimagining work in the age of AI*. Watertown, MA: Harvard Business Review Press.

Dawes, R. M., & Corrigan, B. (1974). Linear models in decision making. *Psychological Bulletin, 81*, 95–106. doi:10.1037/h0037613

de Ayala, R. J. (2009). *The theory and practice of item response theory*. New York, NY: The Guilford Press.

Deane, P. (2011). *Writing assessment and cognition* (Research Report No. RR-11-14). Princeton, NJ: Educational Testing Service. doi:10.1002/j.2333-8504.2011.tb02250.x

Deane, P. (2013a). Covering the construct: An approach to automated essay scoring motivated by a socio-cognitive framework for defining literacy skills. In M. D. Shermis, & J. Burstein (Eds.), *Handbook of automated essay evaluation: Current applications and new directions* (pp. 298–312). New York, NY: Routledge.

Deane, P. (2013b). On the relation between automated essay scoring and modern views of the writing construct. *Assessing Writing, 18*, 7–24. doi:10.1016/j.asw.2012.10.002

Deane, P. (2014). *Using writing process and product features to assess writing quality and explore how those features relate to other literacy tasks* (Research Report No. RR-14-03). Princeton, NJ: Educational Testing Service. doi:10.1002/ets2.12002

Deane, P. (2018). The challenges of writing in school: Conceptualizing writing development within a sociocognitive framework. *Educational Psychologist, 53*, 280–300. doi:10.1080/00461520.2018.1513844

Deane, P., Feng, G., Zhang, M., Hao, J., Bergner, Y., Flor, M., … Lederer, N. (2015). Generating scores and feedback for writing assessment and instruction using electronic process logs. *U.S. Patent Application No. 14/937,164*. Washington, DC: U.S. Patent and Trademark Office.

Deane, P., Roth, A., Litz, A., Goswami, V., Steck, F., Lewis, M., & Richter, T. (2018). *Behavioral differences between retyping, drafting, and editing: A writing process analysis* (Research Memorandum No. RM-18-06). Princeton, NJ: Educational Testing Service.

Deane, P., & Zhang, M. (2015). *Exploring the feasibility of using writing process features to assess text production skills* (Research Report No. RR-15-26). Princeton, NJ: Educational Testing Service. doi:10.1002/ets2.12071

Deerwester, S., Dumais, S. T., Furnas, G. W., Landauer, T. K., & Harshman, R. (1990). Indexing by latent semantic analysis. *Journal of the American Society for Information Science, 41*, 391–407. doi:10.1002/(SICI)1097-4571(199009)41:6<391::AID-ASI1>3.0.CO;2-9

Demsar, J. (2006). Statistical comparisons of classifiers over multiple data sets. *Journal of Machine Learning Research, 7*, 1–30.

DeVault, D., Rizzo, A., Benn, G., Dey, T., Fast, E., Gainer, A., ... Morency, L.-P. (2014). SimSensei Kiosk: A virtual human interviewer for healthcare decision support. In *Proceedings of the 13th international conference on autonomous agents and multiagent systems (AAMAS)* (pp. 1061–1068). New York, NY: ACM.

DiCerbo, K. E. (2015). Assessment of task persistence. In Y. Rosen, S. Ferrara, & M. Mosharraf (Eds.), *Handbook of research on computational tools for real-world skill development* (pp. 780–806). Hershey, PA: IGI Global.

DiCerbo, K. E. (2017). Building the evidentiary argument in game-based assessment. *Journal of Applied Testing Technology, 18*, 7–18.

DiCerbo, K., & Behrens, J. (2012). Implications of the digital ocean on current and future assessment. In R. Lissitz, & H. Jiao (Eds.), *Computers and their impact on state assessment: Recent history and predictions for the future* (pp. 273–306). Charlotte, NC: Information Age.

DiCerbo, K. E., Crowell, C., & John, M. (2015). Alice in Arealand. Worked example presented at Games+Learning+Society Conference, Madison, WI. In K. E. H. Caldwell, S. Seyler, A. Ochsner, & C. Steinkuehler (Eds.), *GLS11 conference proceedings* (pp. 382–385). Madison, WI: Games+Learning+Society.

DiCerbo, K. E., Xu, Y., Levy, R., Lai, E., & Holland, L. (2017). Modeling student cognition in digital and nondigital assessment environments. *Educational Assessment, 4*, 275–297. doi:10.1080/10627197.2017.1382343

Dickler, R., Li, H., & Gobert, J. (2019, April). *Examining the generalizability of an automated scoring method and identifying student difficulties with scientific explanations.* Paper presented at the annual meeting of the American Educational Research Association (AERA), Toronto, Canada.

Dietvorst, B. J. (2016). *Algorithm aversion* (Doctoral dissertation, University of Pennsylvania). Retrieved from https://repository.upenn.edu/edissertations/1686/

Dikli, S. (2006). An overview of automated essay scoring. *Journal of Technology, Learning, and Assessment, 5*(1), 1–36.

DLM Consortium. (2019). *About DLM tests.* Retrieved from https://dynamiclearningmaps.org/about/tests

Dong, F., & Zhang, Y. (2016). Automatic features for essay scoring—An empirical study. In J. Su, K. Duh, & X. Carreras (Ed.), *Proceedings of the 2016 conference on empirical methods in natural language processing* (pp. 1072–1077). doi:10.18653/v1/D16-1115

Dong, F., Zhang, Y., & Yang, J. (2017). Attention-based recurrent convolutional network for automatic essay scoring. In *Proceedings of the 21st conference on computational natural language learning, Vancouver, Canada* (pp. 153–163). Retrieved from https://www.aclweb.org/anthology/K17-1017

Donnelly, P., Blanchard, N., Samei, B., Olney, A. M., Sun, X., Ward, B., ... D'Mello, S. K. (2016). Multi-sensor modeling of teacher instructional segments in live classrooms. In *ICMI '16: Proceedings of the 18th ACM international conference on multimodal interaction* (pp. 177–184). doi:10.1145/2993148.2993158

Doran, D., Schulz, S., & Besold, T. R. (2017). *What does explainable AI really mean? A new conceptualization of perspectives.* Retrieved from https://arxiv.org/abs/1710.00794

Dorans, N. J. (2004). Using subpopulation invariance to assess test score equity. *Journal of Educational Measurement, 4,* 43–68. doi:10.1111/j.1745-3984.2004.tb01158.x

Downey, R., Rubin, D., Cheng, J., & Bernstein, J. (2011). Performance of automated scoring for children's oral reading. In J. Tetreault, J. Burstein, & C. Leacock, *Proceedings of the 6th workshop on innovative use of NLP for building educational applications* (pp. 46–55). Retrieved from https://www.aclweb.org/anthology/W11-1406

Driessen, V. (2010). *A successful git-branching method.* Retrieved from the nvie.com website: https://nvie.com/posts/a-successful-git-branching-model/

Dronen, N., Foltz, P. W., & Habermehl, K. (2014). Effective sampling for large-scale automated writing evaluation systems. In *Proceedings of the second ACM conference on learning @ scale* (pp. 3–10). doi:10.1145/2724660.2724661

Duong, M., Mostow, J., & Sitaram, S. (2011). Two methods for assessing oral reading prosody. *ACM Transactions on Speech and Language Processing, 7*(4), 1–22. doi:10.1145/1998384.1998388

Eadicicco, L. (2014, April 29). This software can write a grade-A college paper in less than a second. *Business Insider.* Retrieved from http://www.businessinsider.com/babel-generator-mit-2014-4

Economist. (2016, June 23). *Special report: The return of the machinery question.* Retrieved from https://www.economist.com/special-report/2016/06/23/the-return-of-the-machinery-question

EdTech. (2017, February 01). *More than 50 percent of teachers report 1:1 computing.* Retrieved from https://edtechmagazine.com/k12/article/2017/02/more-50-percent-teachers-report-11-computing

EdTech Strategies. (2015). *Pencils down: The shift to online & computer-based testing, U.S. K-8 market: 2015–2016 school year.* Retrieved from https://www.edtechstrategies.com/wp-content/uploads/2015/11/PencilsDownK-8_EdTech-StrategiesLLC.pdf

Educational Testing Service. (2018). *California assessment of student performance and progress smarter balanced technical report, 2016–17 administration.* Retrieved from the California Department of Education website: https://www.cde.ca.gov/ta/tg/ca/documents/sbac17techrpt.pdf

Egan, K. L., Schneider, M. C., & Ferrara, S. (2012). Performance level descriptors: History, practice and a proposed framework. In G. Cizek (Ed.), *Setting performance standards: Foundations, methods, and innovations* (2nd ed., pp. 79–106). New York, NY: Routledge.

Eide, B., & Eide, F. F. (2006). *The mislabeled child: Looking beyond behavior to find the true sources—and solutions—for children's learning challenges.* New York, NY: Hyperion.

Ekman, P., & Friesen, W. V. (1978). *Facial action coding system: A technique for the measurement of facial movement*. Palo Alto, CA: Consulting Psychologists Press.

Elliot, S. M. (1999). *Construct validity of IntelliMetric with international assessment*. Yardley, PA: Vantage Technologies.

Elliot, S. M. (2003). IntelliMetric: From here to validity. In M. D. Shermis, & J. Burstein, *Automated essay scoring: A cross-disciplinary perspective* (pp. 71–86). Mahwah, NJ: Lawrence Erlbaum.

Epstein, D. (2019). *Range: Why generalists triumph in a specialized world*. New York, NY: Riverhead Books.

Ericsson, A. K., Charness, N., Feltovich, P., & Hoffman, R. R. (Eds.). (2006). *Cambridge handbook on expertise and expert performance*. Cambridge, UK: Cambridge University Press. doi:10.1017/CBO9780511816796

Evanini, K., Hauck, M. C., & Hakuta, K. (2017). *Approaches to automated scoring of speaking for K–12 English language proficiency assessments* (Research Report No. RR-17-18). Princeton, NJ: Educational Testing Service. doi:10.1002/ets2.12147

Evanini, K., Heilman, M., Wang, X., & Blanchard, D. (2015). *Automated scoring for the TOEFL junior comprehensive writing and speaking test* (Research Report No. RR-15-09). Princeton, NJ: Educational Testing Service. doi:10.1002/ets2.12052

Evanini, K., Higgins, D., & Zechner, K. (2010). Using Amazon Mechanical Turk for transcription of non-native speech. In *CSLDAMT '10: Proceedings of the NAACL HLT 2010 workshop on creating speech and language data with Amazon's mechanical turk* (pp. 53–56). Stroudsburg, PA: Association for Computational Linguistics.

Evanini, K., & Wang, X. (2014). Automatic detection of plagiarized spoken responses. In *Proceedings of the ninth workshop on innovative use of NLP for building educational applications* (pp. 22–27). doi:10.3115/v1/W14-1803

Evanini, K., Xie, S., & Zechner, K. (2013). Prompt-based content scoring for automated spoken language assessment. In *Proceedings of the eighth workshop on innovative use of NLP for building educational applications* (pp. 157–162). Retrieved from https://www.aclweb.org/anthology/W13-1721

Evans, E. (2003). *Domain-driven design: Tacking complexity in the heart of software*. Westford, MA: Addison Wesley.

Evans, M. P., & Saultz, A. (2015, June 9). The opt-out movement is gaining momentum. *Education Week*. Retrieved from https://www.edweek.org/ew/articles/2015/06/10/the-opt-out-movement-is-gaining-momentum.html

Eveleth, R. (2013, April 8). Can a computer really grade an essay? *Smithsonian*. Retrieved from http://www.smithsonianmag.com/smart-news/can-a-computer-really-grade-an-essay-17394014/

Every Student Succeeds Act, 20 U.S.C. § 6301 *et seq*. (2015).

Farley, T. (2011, April 8). Standardized testing: A decade in review. *Huff Post*. Retrieved from http://www.huffingtonpost.com/todd-farley/standardized-testing-a-de_b_846044.html

Farrington, C. A., Roderick, M., Allensworth, E., Nagaoka, J., Keyes, T. S., Johnson, D. W., & Beechum, N. O. (2012). *Teaching adolescents to become learners: The role of noncognitive factors in shaping school performance: A critical literature review*. Chicago, IL: Consortium on Chicago School Research.

Fasold, R. W., & Connor-Linton, J. (2014). *An introduction to language and linguistics* (2nd ed.). Cambridge, UK: Cambridge University Press.

Feit, A. M., Weir, D., & Oulasvirta, A. (2016). How we type: Movement strategies and performance in everyday typing. In *Proceedings of the conference on human factors in computing systems (CHI 2016)* (pp. 4262–4273). doi:10.1145/2858036.2858233

Feng, G. (2018, April). *Analyzing writing process in NAEP Writing Assessment: Implementation and evidence extraction*. Paper presented at the National Council on Measurement in Education, New York, NY.

Ferrara, S., Lai, E., Reilly, A., & Nichols, P. D. (2017). Principled approaches to assessment design, development, and implementation. In A. A. Rupp, & J. P. Leighton (Eds.), *The Wiley handbook of cognition and assessment: Frameworks, methodologies, and applications* (pp. 41–74). West Sussex, UK: Wiley Blackwell. doi:10.1002/9781118956588.ch3

Ferrara, S., & Steedle, J. (2015, April). *Predicting item parameters using regression trees: Analyzing existing item data to understand and improve item writing*. Paper presented at the annual meeting of the National Council of Measurement in Education, Chicago, IL.

Fife, J. H. (2011, April). *Integrating item generation and automated scoring*. Paper presented at the annual meeting of the National Council of Measurement in Education, New Orleans, LA.

Finegan, E. (2014). *Language: Its structure and use* (7th ed.). Stamford, CT: Cengage Learning.

Finn, B. (2015). *Measuring motivation in low-stakes assessments* (Research Report No. RR-15-19). Princeton, NJ: Educational Testing Service. doi:10.1002/ets2.12067

Finn, B., & Metcalfe, J. (2010). Scaffolding feedback to maximize long-term error correction. *Memory & Cognition, 38*, 951–961. doi:10.3758/MC.38.7.951

Fiore, S. M., Graesser, A. C., & Greiff, S. (2018). Collaborative problem solving education for the 21st century workforce. *Nature Human Behavior, June*, 1–3.

Flower, L. (1994). *The construction of negotiated meaning: A social cognitive theory of writing*. Carbondale and Edwardsville: Southern Illinois University Press. doi:10.2307/358881

Flower, L., & Hayes, J. R. (1980). The dynamics of composing: Making plans and juggling constraints. In L. W. Gregg, & E. R. Steinberg (Eds.), *Cognitive processes in writing* (pp. 31–50). Hillsdale, NJ: Lawrence Erlbaum.

Flower, L., & Hayes, J. R. (1981). The pregnant pause: An inquiry into the nature of planning. *Research in the Teaching of English, 15*, 229–243.

Foltz, P. W. (2007). Discourse coherence and LSA. In T. K. Landauer, D. S. McNamura, S. Dennis, & W. Kintsch (Eds.), *Handbook of latent semantic analysis* (pp. 167–184).

Foltz, P. W. (in press). Practical considerations for implementing automated scoring. In H. Jiao, & R. Lissitz (Eds.), *Applications of artificial intelligence to assessment*. Charlotte, NC: Information Age Publisher.

Foltz, P. W., Kintsch, W., & Landauer, T. K. (1998). The measurement of textual coherence with latent semantic analysis. *Discourse Processes, 25*(2&3), 285–307. doi:10.1080/01638539809545029

Foltz, P. W., & Rosenstein, M. (2015). Analysis of a large-scale formative writing assessment system with automated feedback. In *Proceedings of the second ACM conference on learning @ scale* (pp. 339–342). doi:10.1145/2724660.2728688

Foltz, P. W., & Rosenstein, M. (2017). Data mining large-scale formative writing. In C. Lang, G. Siemens, A. Wise, & D. Gasevic (Eds.), *Handbook of learning analytics & educational data mining* (pp. 199–210). Sydney, Australia: Society for Learning Analytics Research. doi:10.18608/hla17.017

Foltz, P. W., Streeter, L. A., Lochbaum, K. E., & Landauer, T. K. (2013). Implementation and applications of the Intelligent Essay Assessor. In M. D. Shermis, & J. Burstein (Eds.), *Handbook of automated essay evaluation: Current applications and new directions* (pp. 68–88). New York, NY: Routledge.

Forbus, K., Chang, M., McLure, M., & Usher, M. (2017). The cognitive science of sketch worksheets. *Topics in Cognitive Science, 9*, 921–942. doi:10.1111/tops.12262

Ford, M. (1979). *Sentence planning units: Implications for the speaker's representation on meaningful relations underlying sentences.* Cambridge, MA: Massachusetts Institute of Technology, Center for Cognitive Science.

Ford, M., & Holmes, V. M. (1978). Planning units and syntax in sentence production. *Cognition, 6*, 35–53. doi:10.1016/0010-0277(78)90008-2

Franco, H., Ferrer, L., & Bratt, H. (2014). Adaptive and discriminative modeling for improved mispronunciation detection. In *Proceedings of the 2014 IEEE international conference on acoustics, speech and signal processing* (pp. 7709–7713). doi:10.1109/ICASSP.2014.6855100

Friedman, J. H. (2001). Greedy function approximation: A gradient boosting machine. *Annals of Statistics, 29*, 1189–1232. doi:10.1214/aos/1013203451

Galbraith, D., & Baaijen, V. M. (2019). Aligning keystrokes with cognitive processes in writing. In E. Lindgren, & K. Sullivan (Eds.), *Observing writing: Insights from keystroke logging and handwriting* (pp. 306–325). doi:10.1163/9789004392526_015

Gamon, M., Chodorow, M., Leacock, C., & Tetreault, J. (2013). Grammatical error detection in automated essay scoring and feedback. In M. D. Shermis, & J. Burstein, (Eds.), *Handbook of automated essay evaluation: Current applications and new directions* (pp. 251–267). New York, NY: Routledge.

Garcia, S., & Herrera, F. (2008). An extension on "statistical comparisons of classifiers over multiple data sets" for all pairwise comparisons. *Journal of Machine Learning Research, 9*, 2677–2694.

Gee, J. P. (1992). *The social mind: Language, ideology, and social practice.* South Hadley, MA: Bergin & Garvey.

Gerard, L. F., & Linn, M. C. (2017). Using automated scores of student essays to support instructor guidance in classroom inquiry. *Journal of Science Instructor Education, 27*(1), 111–129. doi:10.1007/s10972-016-9455-6

Gerard, L. F., Matuk, C., McElhaney, K., & Linn, M. C. (2015). Automated, adaptive guidance for K–12 education. *Educational Research Review, 15*, 41–58. doi:10.1016/j.edurev.2015.04.001

Gerard, L. F., Ryoo, K., McElhaney, K. W., Liu, O. L., Rafferty, A. N., & Linn, M. C. (2016). Automated guidance for student inquiry. *Journal of Educational Psychology, 108*(1), 60–81. doi:10.1037/edu0000052

Géron, A. (2017). *Hands-on machine learning with scikit-learn & TensorFlow.* Sebastol, CA: O'Reilly Media.

Gitomer, D. H., Steinberg, L. S., & Mislevy, R. J. (1995). *Diagnostic assessment of trouble-shooting skill in an intelligent tutoring system* (Report No. AD-A280 554). Fort Belvoir, VA: Defense Technical Information Center. doi:10.21236/ADA280554

Gobert, J. D. (2005). Leveraging technology and cognitive theory on visualization to promote students' science. In J. K. Gilbert (Ed.), *Visualization in science education* (pp. 73–90). Dordrecht, The Netherlands: Springer. doi:10.1007/1-4020-3613-2_6

Gobert, J., Buckley, B. C., Levy, S. T., & Wilensky, U. (2006, April). *Teasing apart domain-specific and domain general inquiry skills: Co-evolution, bootstrapping, or separate paths?* Paper presented at the annual meeting of the American Educational Research Association, Chicago, IL.

Gobert, J. D., Kim, Y. J., Sao Pedro, M. A., Kennedy, M., & Betts, C. G. (2015). Using educational data mining to assess students' skills at designing and conducting experiments within a complex systems microworld. *Thinking Skills and Creativity, 18*, 81–90. doi:10.1016/j.tsc.2015.04.008

Gobert, J. D., Moussavi, R., Li, H., Sao Pedro, M., & Dickler, R. (2018). Real-time scaffolding of students' online data interpretation during inquiry with Inq-ITS using educational data mining. In A. K. M. Azad, M. Auer, A. Edwards, & T. de Jong (Eds), *Cyber-physical laboratories in engineering and science education* (pp. 191–217) New York, NY: Springer.

Gobert, J. D., & Sao Pedro, M. (2017). Digital assessment environments for scientific inquiry practices. In A. A. Rupp, & J. P. Leighton (Eds.), *The Wiley handbook of cognition and assessment: Frameworks, methodologies, and applications* (pp. 508–534). doi:10.1002/9781118956588.ch21

Gobert, J. D., Sao Pedro, M., Baker, R. S., Toto, E., & Montalvo, O. (2012). Leveraging educational data mining for real time performance assessment of scientific inquiry skills within microworlds. *Journal of Educational Data Mining, 15*, 153–185.

Gobert, J. D., Sao Pedro, M., Raziuddin, J., & Baker, R. (2013a, April). *Developing and validating EDM (educational data mining)-based assessment measures for measuring science inquiry skill acquisition and transfer across science topics.* Paper presented at the annual meeting of the American Educational Research Association. San Francisco, CA.

Gobert, J. D., Sao Pedro, M., Raziuddin, J., & Baker, R. (2013b). From log files to assessment metrics for science inquiry using educational data mining. *Journal of the Learning Sciences, 22*, 521–563. doi:10.1080/10508406.2013.837391

Goldman, S. R., & Scardamalia, M. (2013). Managing, understanding, applying, and creating knowledge in the information age: Next-generation challenges and opportunities. *Cognition and Instruction, 31*, 255–269. doi:10.1080/10824669.2013.773217

Goldstein, A., Kapelner, A., Bleich, J., & Pitkin, E. (2015). Peeking inside the black box: Visualizing statistical learning with plots of individual conditional expectation. *Journal of Computational and Graphical Statistics, 24*, 44–65. doi:10.1080/10618600.2014.907095

Gong, T., Feng, G., Zhang, M., Gao, J., Persky, H., & Donahue, P. (2019, April). *Does keyboarding fluency limit writing performance in digital writing assessment?* Paper presented at the National Council on Measurement in Education, Toronto, Canada.

Goodfellow, I., Bengio, J., & Courville, A. (2016). *Deep learning.* Cambridge, MA: MIT Press.

Goodfellow, I., Pouget-Abadie, J., Mirza, M., Xu, B., Warde-Farley, D., Ozair, S., … Bengio, Y. (2014, December). *Generative adversarial nets.* Paper presented at the Neural Information Processing Systems Conference, Montreal, Canada.

Grabowski, J. (2008). The internal structure of university students' keyboard skills. *Journal of Writing Research, 1*, 27–52. doi:10.17239/jowr-2008.01.01.2

Graesser, A. C. (2016). Conversations with *AutoTutor* help students learn. *International Journal of Artificial Intelligence in Education, 26,* 124–132. doi:10.1007/s40593-015-0086-4

Graesser, A. C., Fiore, S. M., Greiff, S., Andrews-Todd, J., Foltz, P. W., & Hesse, F. W. (2018). Advancing the science of collaborative problem solving. *Psychological Science in the Public Interest, 19,* 59–92. doi:10.1177/1529100618808244

Graesser, A. C., Foltz, P. W., Rosen, Y., Shaffer, D. W., Forsyth, C., & Germany, M.-L. (2018). Challenges of assessing collaborative problem solving. In E. Care, P. Griffin, & M. Wilson (Eds.), *Assessment and teaching of 21st century skills: Research and applications* (pp. 75–91). Berlin, Germany: Springer. doi:10.1007/978-3-319-65368-6_5

Graesser, A. C., Forsyth, C. M., & Foltz, P. (2017). Assessing conversation quality, reasoning, and problem solving performance with computer agents. In B. Csapo, J. Funke, & A. Schleicher (Eds.), *On the nature of problem solving: A look behind PISA 2012 problem solving assessment* (pp. 275–297). Heidelberg, Germany: OECD Series.

Graesser, A. C., Forsyth, C., & Lehman, B. (2017). Two heads are better than one: Learning from agents in conversational trialogues. *Teachers College Record, 119,* 1–20.

Graesser, A. C., Hu, X., Nye, B. D., VanLehn, K., Kumar, R., Heffernan,. … Baer, W. (2018). ElectronixTutor: An intelligent tutoring system with multiple learning resources. *International Journal of STEM Education, 5,* 1–21. doi:10.1186/s40594-018-0110-y

Graesser, A. C., Li, H., & Forsyth, C. (2014). Learning by communicating in natural language with conversational agents. *Current Directions in Psychological Science, 23,* 374–380.

Graesser, A. C., McNamara, D. S., & VanLehn, K. (2005). Scaffolding deep comprehension strategies through Point&Query, AutoTutor, and iSTART. *Educational Psychologist, 40,* 225–234. doi:10.1207/s15326985ep4004_4

Graesser, A. C., & Olde, B. A. (2003). How does one know whether a person understands a device? The quality of the questions the person asks when the device breaks down. *Journal of Educational Psychology, 95,* 524–536. doi:10.1037/0022-0663.95.3.524

Graesser, A. C., Penumatsa, P., Ventura, M., Cai, Z., & Hu, X. (2007). Using LSA in AutoTutor: Learning through mixed initiative dialogue in natural language. In T. K. Landauer, D. S. McNamara, S. Dennis, & W. Kintsch (Eds.), *Handbook of latent semantic analysis* (pp. 243–262). Mahwah, NJ: Erlbaum.

Grafsgaard, J. F., Duran, N. D., Randall, A. K., Tao, C., & D'Mello, S. K. (2018). Generative multimodal models of nonverbal synchrony in close relationships In S. K. D'Mello, L. Yin, L. P. Morency, & M. Valstar (Eds.), *Proceedings of the 13th IEEE conference on automatic face and gesture recognition (FG 2018)* (pp. 195–202). doi:10.1109/FG.2018.00037

Graham, S., Bollinger, A., Olson, C. B., D'Aoust, C., MacArthur, C., McCutchen, D., & Olinghouse, N. (2012). *Teaching elementary school students to be effective writers: A practice guide* (No. NCEE 2012-4058). Washington, DC: U.S. Department of Education.

Graham, S., Bruch, J., Fitzgerald, J., Friedrich, L., Furgeson, J., Greene, K., … Wulsin, C. S. (2016). *Teaching secondary students to write effectively* (No. NCEE 2017-4002). Washington, DC: U.S. Department of Education.

Graham, S., & Harris, K. R. (2000). The role of self-regulation and transcription skills in writing and writing development. *Educational Psychologist, 35,* 3–12. doi:10.1207/S15326985EP3501_2

Graham, S., & Harris, K. R. (2012). The role of strategies, knowledge, will, and skills in a 30-year program of writing research (with homage to Hayes, Fayol, and Boscodo). In V. W. Berninger (Ed.), *Past, present and future contributions of cognitive writing research to congitive psychology* (pp. 177–196). New York, NY: Psychology Press.

Graham, S., Hebert, M., & Harris, K. R. (2015). Formative assessment and writing: A meta-analysis. *The Elementary School Journal, 115,* 523–547. doi:10.1086/681947

Graham, S., & Leone, P. (1987). Effects of behavioral disability labels, writing performance, and examiner's expertise on the evaluation of written products. *Journal of Experimental Education, 55,* 89–94. doi:10.1080/00220973.1987.10806439

Graham, S., & Perin, D. (2007). A meta-analysis of writing instruction for adolescent students. *Journal of Educational Psychology, 99,* 445–476. doi:10.1037/0022-0663.99.3.445

Greeno, J. G., & van de Sande, C. (2007). Perspectival understanding of conceptions and conceptual growth in interaction. *Educational Psychologist, 42,* 9–23. doi:10.1080/00461520709336915

Gregg, N., Coleman, C., Davis, M., & Chalk, J. C. (2007). Timed essay writing implications for high-stakes tests. *Journal of Learning Disabilities, 40,* 306–318. doi:10.1177/00222194070400040201

Griffen, P., McGaw, B., & Care, E. (2012). *Assessment and teaching of 21st century skills.* Dordecht: Springer.

Grossman, R., Bailey, S., Ramu, A., Malhi, B., Hallstrom, P., Pulleyn, I., & Qin, X. (1999). The management and mining of multiple predictive models using the modeling markup language. *Information and Software Technology, 41,* 589–595. doi:10.1016/S0950-5849(99)00022-1

Guo, H., Deane, P. D., van Rijn, P. W., Zhang, M., & Bennett, R. E. (2018). Modeling basic writing processes from keystroke logs. *Journal of Educational Measurement, 55,* 194–216. doi:10.1111/jedm.12172

Guo, H., Zhang, M., Deane, P., & Bennett, R. E. (2019). Writing process differences in subgroups reflected in keystroke logs. *Journal of Educational and Behavioral Statistics, 44,* 571–596. doi:10.3102/1076998619856590

Haberman, S. J., Yao, L., & Sinharay, S. (2015). Prediction of true test scores from observed item scores and ancillary data. *British Journal of Mathematical and Statistical Psychology, 68,* 363–385. doi:10.1111/bmsp.12052

Habib, M. (2000). The neurological basis of developmental dyslexia. *Brain, 123,* 2373–2399. doi:10.1093/brain/123.12.2373

Hagge, S. L., & Kolen, M. J. (2012). Effects of group differences on equating using operational and pseudo-tests. In M. J. Kolen, & W. Lee (Eds.), *Mixed-format tests: Psychometric properties with a primary focus on equating (Volume 2)* (CASMA Monograph Number 2.2; pp. 45–86). Iowa City, IA: Center for Advanced Studies in Measurement and Assessment.

Haimson, L. (2016, May 7). *The fraud of computer scoring on the Common Core exams.* Retrieved from the Network for Public Education Action website: https://npe-action.org/the-fraud-of-computer-scoring-on-the-common-core-exams-2/

Haimson, L., Stickland, R., Woestehoff, J., Burris, C., & Neill, M. (2016). *Updated: Which commissioners have responded to our questions on machine scoring of PARCC and SBAC exams?* Retrieved from https://www.studentprivacymatters.org/our-letter-to-the-education-commissioners-in-the-parcc-and-sbac-states/

Haisfield, L., & Yao, E. (2018, April). *Industry standards for an emerging technology: Automated scoring.* Paper presented at the National Council on Measurement in Education, New York, NY.

Haisfield, L., Yao, E., & Wood, S. (2017, October). *Building trust: Industry standards for automated scoring.* Paper presented at the Northern Rocky Mountain Educational Research Association, Boulder, CO.

Hall, D., Jurafsky, D., & Manning, C. D. (2008). Studying the history of ideas using topic models. In *Proceedings of the conference on empirical methods in natural language processing* (pp. 363–371). Retrieved from https://web.stanford.edu/~jurafsky/hallemnlp08.pdf

Hanley, J. A., & McNeil, B. J. (1982). The meaning and use of the area under a receiver operating characteristic (ROC) curve. *Radiology, 143,* 29–36. doi:10.1148/radiology.143.1.7063747

Hanson, S. J., & Bauer, M. (1989). Conceptual clustering, categorization, and polymorphy. *Machine Learning, 3,* 343–372. doi:10.1007/BF00116838

Hao, J., Shu, Z., & von Davier, A. (2015). Analyzing process data from game/scenario-based tasks: An edit distance approach. *Journal of Educational Data Mining, 7*(1), 33–50. Retrieved from https://jedm.educationaldatamining.org/index.php/JEDM/article/view/JEDM072

Harik, P., Clauser, B. E., & Baldwin, P. A. (2010, April). *Comparison of alternative scoring methods for a computerized performance assessment of clinical judgment.* Paper presented at the annual meeting of the National Council on Measurement in Education (NCME).

Harik, P., Clauser, B. E., & Baldwin, P. (2013). Comparison of alternative scoring methods for a computerized performance assessment of clinical judgment. *Applied Psychological Measurement, 37,* 587–597. doi:10.1177%2F0146621613493829

Hastie, T., Tibshirani, R., & Friedman, J. (2001). *The elements of statistical learning: Data mining, inference, and prediction.* New York, NY: Springer. doi:10.1007/978-0-387-21606-5

Hastie, T., Tibshirani, R., & Friedman, J. (2009). *The elements of statistical learning: Data mining, inference, and prediction* (2nd ed.). New York, NY: Springer-Verlag.

Haswell, R. H. (1998). Rubrics, prototypes, and exemplars: Categorization theory and systems of writing placement. *Assessing Writing, 5,* 231–268. doi:10.1016/S1075-2935(99)80014-2

Hayes, J. R. (2004). What triggers revision? In L. Allal, L. Chanquoy, & P. Largy (Eds.), *Revision: Cognitive and instructional processes* (pp. 9–20). Dordrecht, The Netherlands: Springer.

Hayes, J. R. (2012a). Evidence from language bursts, revision, and transcription for translation and its relation to other writing processes. In M. Fayol, D. Alamargot, & V. W. Berninger (Eds.), *Translation of thought to written text while composing.* New York, NY: Psychology Press.

Hayes, J. R. (2012b). Modeling and remodeling writing. *Written Communication, 29,* 369–388. doi:10.1177/0741088312451260

Hayes, J. R., & Flower, L. (1981). *Uncovering cognitive processes in writing. An Introduction to protocol analysis.* Retrieved from ERIC database. (ED202035)

He, K., Zhang, X., Ren, S., & Sun, J. (2016a). Deep residual learning for image recognition. In *Proceedings of the 2016 IEEE conference on computer vision and pattern recognition* (pp. 770–778). doi:10.1109/CVPR.2016.90

He, K., Zhang, X., Ren, S., & Sun, J. (2016b). Deep residual learning for image recognition. *Computer Vision and Pattern Recognition.* Retrieved from https://arxiv.org/pdf/1512.03385.pdf

Heilman, M., & Madnani, N. (2015). The impact of training data on automated short answer scoring performance. In *Proceedings of the 10th workshop on innovative use of NLP for building educational applications* (pp. 81–85). Retrieved from doi:10.3115/v1/W15-0610

Henderson, G., & Andrade, A. (2019, April). *Methods for detecting when examinees game an automated scoring engine.* Paper presented at the National Council on Measurement in Education, Toronto, Canada.

Herron, D., Menzel, W., Atwell, E., Bisiani, R., Daneluzzi, F., Morton, R., & Schmidt, J. A. (1999). Automatic localization and diagnosis of pronunciation errors for second-language learners of English. In *EUROSPEECH '99* (pp. 855–858). Retrieved from https://www.isca-speech.org/archive/eurospeech_1999/index.html

Higgins, D., & Heilman, M. (2014). Managing what we can measure: Quantifying the susceptibility of automated scoring systems to gaming behavior. *Educational Measurement: Issues and Practice, 33*(3), 36–46. doi:10.1111/emip.12036

Higgins, D., Xi, X., Zechner, K., & Williamson, D. (2011). A three-stage approach to the automated scoring of spontaneous spoken responses. *Computer Speech & Language, 25,* 282–306. doi:10.1016/j.csl.2010.06.001

Hinton, G. E. (2007). Learning multiple layers of representation. *Trends in Cognitive Sciences, 11,* 428–434. doi:10.1016/j.tics.2007.09.004

Hinton, G. E., Srivastava, N, Krizhevsky, A, Sutskever, I, & Salakhutdinov, R. R. (2012). *Improving neural networks by preventing co-adaptation of feature detectors.* Retrieved from https://arxiv.org/pdf/1207.0580.pdf

Hochreiter, S., & Schmidhuber, J. (1997). Long short-term memory. *Neural Computation, 9,* 1735–1780.

Hoff, T. (2019, March 4). How is software developed at Amazon? [Blog post]. Retrieved from http://highscalability.com/blog/2019/3/4/how-is-software-developed-at-amazon.html

Holmes, V. M. (1988). Hesitations and sentence planning. *Language and Cognitive Processes, 3,* 323–361. doi:10.1080/01690968808402093

Holstein, K., Wortman Vaughan, J., Daumé III, H., Dudik, M., & Wallach, H. (2019). Improving fairness in machine learning systems: What do industry practitioners need? In *Proceedings of 2019 CHI conference on human factors in computing systems* (pp. 1–16). doi:10.1145/3290605.3300830

Hönig, F., Bocklet, T., Riedhammer, K., Batliner, A., & Nöth, E. (2012). The automatic assessment of non-native prosody: combining classical prosodic analysis with acoustic modelling. In *Proceedings of 13th annual conference of the international speech communication association* (pp. 823–826). Retrieved from https://www.isca-speech.org/archive/interspeech_2012/i12_0823.html

Horbach, A., & Palmer, A. (2016). Investigating active learning for short-answer scoring. In *Proceedings of the 11th workshop on innovative use of NLP for building educational applications* (pp. 301–311). doi:10.18653/v1/W16-0535

Hu, W., Qian, Y., Soong, F. K., & Wang, Y. (2015). Improved mispronunciation detection with deep neural network trained acoustic models and transfer learning based logistic regression classifiers. *Speech Communication, 67,* 154–166. doi:10.1016/j.specom.2014.12.008

Hu, X., Cai, Z., Wiemer-Hastings, P., Graesser, A. C., & McNamara, D. S. (2007). Strengths, limitations, and extensions of LSA. In T. K. Landauer, D. S. McNamara, S. Dennis, & W. Kintsch (Eds.), *The handbook of latent semantic analysis* (pp. 401–426). Mahwah, NJ: Lawrence Erlbaum.

Hu, X., Nye, B. D., Gao, C., Huang, X., Xie, J., & Shubeck, K. (2014). Semantic representation analysis: A general framework for individualized, domain-specific and context-sensitive semantic processing. In D. D. Schmorrow, & C. M. Fidopiastis (Eds.), *Foundations of augmented cognition: Advancing human performance and decision-making through adaptive systems* (pp. 35–46). New York, NY: Springer International. doi:10.1007/978-3-319-07527-3_4

Huang, Y., & Khan, S. (2018a). Generating photorealistic facial expressions in dyadic interactions. In *Proceedings of the 2018 British machine vision conference* (pp. 1–12). Retrieved from http://bmvc2018.org/contents/papers/0590.pdf

Huang, Y., & Khan, S. (2018b). A generative approach for dynamically varying photorealistic facial expressions in human-agent interactions. In *Proceedings of the 20th ACM international conference on multimodal interaction* (pp. 437–445). doi:10.1145/3242969.3243031

Hubbard, J. P., & Levit, E. J. (1985). *The National Board of Medical Examiners: The first seventy years.* Philadelphia, PA: National Board of Medical Examiners.

Hunt, B. (2012, April 17). *What automated essay grading says to children* [Web log post]. Retrieved from https://budtheteacher.com/blog/2012/04/17/what-automated-essay-grading-says-to-children/

Hutt, S., Krasich, K., Mills, C., Bosch, N., White, S., Brockmole, J., & D'Mello, S. K. (2019). Automated gaze-based mind wandering detection during computerized learning in classrooms. *User Modeling & User-Adapted Interaction.* Advance online publication. doi:10.1007/s11257-019-09228-5

Hutt, S., Mills, C., Bosch, N., Krasich, K., Brockmole, J., & D'Mello, S. (2017). Out of the fr-eye-ing pan: Towards gaze-based models of attention during learning with technology in the classroom. In M. Bielikova, E. Herder, F. Cena, & M. Desmarais (Eds.), *UMAP '17: Proceedings of the 25th conference on user modeling, adaptation, and personalization* (pp. 94–103). doi:10.1145/3079628.3079669

Institute of Electrical and Electronics Engineers. (1998). *IEEE Std 830-1998—IEEE recommended practice for software requirements specifications.* New York, NY: Author.

Institute of Electrical and Electronics Engineers. (2009). *IEEE Std 1016-2009—IEEE standard for information technology—systems design—software design descriptions.* New York, NY: Author.

Ishioka, T., & Kameda, M. (2004). Automated Japanese Essay Scoring System: jess. In *Proceedings of the 15th international workshop on database and expert systems* (pp. 1–4). doi:10.1109/dexa.2004.1333440

Isola, P., Zhu, J.-Y., Zhou, T., & Efros, A. A. (2016). Image-to-image translation with conditional adversarial networks. In *Proceedings of the 29th IEEE conference on computer vision and pattern recognition (CVPR)* (pp. 5967–5976). doi:10.1109/cvpr.2017.632

Jackson, G. T., & Graesser, A. C. (2006). Applications of human tutorial dialog in *AutoTutor*: An intelligent tutoring system. *Revista Signos, 39,* 31–48.

James, G., Witten, D., Hastie, T., & Tibshirani, R. (2013), *An introduction to statistical learning with applications in R*. New York, NY: Springer.

Jang, E.-S., Kang, S.-S., Noh, E.-H., Kim, M.-H., Sung, K.-H., & Seong, T.-J. (2014). KASS: Korean Automatic Scoring System for short-answer questions. In *Proceedings of the 6th international conference on computer supported education, Barcelona, Spain* (Vol. 2, pp. 226–230). doi:10.5220/0004864302260230

Johansson, R., Johansson, V., & Wengelin, Å. (2012). Text production in handwriting versus computer typing. In M. Torrance, D. Alamargot, M. Castelló, F. Ganier, O. Kruse, A. Mangen, … L. can Waes (Eds.), *Learning to write effectively: Current trends in European research* (pp. 375–377). doi:10.1163/9781780529295_087

Johansson, R., Wengelin, Å., Johansson, V., & Holmqvist, K. (2010). Looking at the keyboard or the monitor: Relationships with text production processes. *Reading and Writing: An Interdisciplinary Journal, 23*, 835–851. doi:10.1007/s11145-009-9189-3

Johnson, J. W. (2000). A heuristic method for estimating the relative weight of predictor variables in multile regression. *Multivariate Behavioral Research, 35*, 1–19. doi:10.1207/S15327906MBR3501_1

Johnson, M. C. (1971). *Educational uses of the computer: An introduction*. Chicago: Rand McNally & Company.

Johnson, M., Hopkin, R., Shiell, H., & Bell, J. F. (2012). Extended essay marking on screen: Is examiner marking accuracy influenced by marking mode? *Educational Research and Evaluation, 18*, 107–124. doi:10.1080/13803611.2012.659932

Johnson, R. L., Penny, J., Gordon, B., Shumate, S. R., & Fisher, S. P. (2009). Resolving scoring differences in the rating of writing samples: Does discussion improve the accuracy of scores? *Language Assessment Quarterly: An International Journal, 2*(2), 117–146. doi:10.1207/x15434311laq0202_2

Johnson, R. P. (2015, June 9). The power of parents is on display in opt-outs. *Education Week*. Retrieved from https://www.edweek.org/ew/articles/2015/06/10/the-power-of-parents-is-on-display.html

Johnson, W. L., & Lester, J. C. (2016). Face-to-face interaction with pedagogical agents, twenty years later. *International Journal of Artificial Intelligence in Education, 26*, 25–36. doi:10.1007/s40593-015-0065-9

Jonassen, D. H., & Hung, W. (2006). Learning to troubleshoot: A new theory-based design architecture. *Educational Psychology Review, 18*, 77–114. doi:10.1007/s10648-006-9001-8

Jones, J. (2008). Patterns of revision in online writing: A study of Wikipedia's featured articles. *Written Communication, 25*, 262–289. doi:10.1177/0741088307312940

Jung, E.-M., Choi, M.-S., & Shim, J.-C. (2009). Design and implementation of automatic marking system for a subjectivity problem of the program. *Journal of Korea Multimedia Society, 12*, 767–776.

Jurafsky, D., & Martin, J. H. (2008). *Speech and language processing: An introduction to natural language processing, computational linguistics, and speech recognition*. Upper Saddle River, NJ: Prentice-Hall.

Jurafsky, D., & Martin, J. H. (2019). Language modeling with n-grams. In D. Jurafsky, & J. H. Martin (Eds.), *Speech and language processing* (3rd ed.). Retrieved from https://web.stanford.edu/~jurafsky/slp3/3.pdf.

Kaggle. (2012). *The Hewlett Foundation: Automated essay scoring*. Retrieved from https://www.kaggle.com/c/asap-aes

Kaggle. (2017). *The state of data science and machine learning*. Retrieved from: https://www.kaggle.com/surveys/2017

Kakkonen, T., & Sutinen, E. (2004). Automatic assessment of the content of essays based on course materials. In *ITRE 2004: 2nd international conference information technology: Research and education* (pp. 126–130). doi:10.1109/ITRE.2004.1393660

Kane, M. T. (1992). An argument-based approach to validity. *Psychological Bulletin, 112*, 527–535. doi:10.1037/0033-2909.112.3.527

Kane, M. T. (2006). Validation. In R. L. Brennan (Ed.), *Educational measurement* (4th ed.; pp. 17–64). Westport, CT: Praeger.

Kang, W.-S. (2011). Automatic grading system for subjective questions through analyzing question type. *The Journal of the Korea Contents Association, 11*(2), 13–21. doi:10.5392/JKCA.2011.11.2.013

Katz, I. R., & C. M. James. (1998). *Toward assessment of design skill in engineering* (Research Report No. RR-97-16). Princeton, NJ: Educational Testing Service. doi:10.1002/j.2333-8504.1997.tb01737.x |

Kellogg, R. T. (1987). Writing performance: Effects of cognitive strategies. *Written Communication, 4*, 269–298. doi:10.1177/0741088387004003003

Kellogg, R. T., Olive, T., & Piolat, A. (2007). Verbal, visual, and spatial working memory in written language production. *Acta Psychologica, 124*, 382–397. doi:10.1016/j.actpsy.2006.02.005

Kellogg, R. T., Whiteford, A. P., & Quinlan, T. (2010). Does automated feedback help students learn to write? *Journal of Educational Computing Research, 42*, 173–196. doi:10.2190/EC.42.2.c

Kelly, S., Olney, A. M., Donnelly, P., Nystrand, M., & D'Mello, S. K. (2018). Automatically measuring question authenticity in real-world classrooms. *Educational Researcher, 47*, 451–464. doi:10.3102/0013189X18785613

Keras [Computer software]. San Francisco, CA: GitHub.

Kieftenbeld, V., & Boyer, M. (2017). Statistically comparing the performance of multiple automated raters across multiple items. *Applied Measurement in Education, 30*, 117–128. doi:10.1080/08957347.2017.1283316

Kim, S., & Walker, M. E. (2009). *Evaluating subpopulation invariance of linking functions to determine the anchor composition of a mixed-format test* (Research Report No. RR-09-36). Princeton, NJ: Educational Testing Service. doi:10.1002/j.2333-8504.2009.tb02193.x

Kim, Y. J., Almond, R. G., & Shute, V. J. (2016). Applying evidence-centered design for development of game-based assessments in Physics Playground. *International Journal of Testing, 16*, 142–163.

Klebanov, B., Madnani, N., Burstein, J., & Somasundaran, S. (2014). Content importance models for scoring writing from sources. *Proceedings of the 52nd annual meeting of the association for computational linguistics (ACL 2014)* (Vol. 2, pp. 247–252).

Klein, J., & Pat El, L. (2003). Impairment of teacher efficiency during extended sessions of test correction. *European Journal of Teacher Education, 26*, 379–392. doi:10.1080/0261976032000128201

Knospe, Y., Sullivan, K. P., Malmqvist, A., & Valfridsson, I. (2019). Observing writing and website browsing: Swedish students write L3 German. In *Observing writing: Insights from keystroke logging and handwriting* (pp. 258–284). doi:10.1163/9789004392526_013

Koedinger, K. R., & Corbett, A. T. (2006). Cognitive tutors: Technology bringing learning science to the classroom. In K. Sawyer (Ed.), *The Cambridge handbook of the learning sciences* (pp. 61–78). Cambridge, UK: Cambridge University Press. doi:10.1017/CBO9780511816833.006

Koedinger, K. R., Corbett, A. C., & Perfetti, C. (2012). The knowledge-learn ing-instruction (KLI) framework: Bridging the science-practice chasm to enhance robust student learning. *Cognitive Science, 36*(5), 757–798. doi:10.1111/j.1551-6709.2012.01245.x

Kolakowska, A. (2013). A review of emotion recognition methods based on keystroke dynamics and mouse movements. In W. A. Paja, & B. M. Wilamowski (Eds.), *2013 6th international conference on human system interactions (HSI)* (pp. 548–555). doi:10.1109/HSI.2013.6577879

Kolen, M. J. (2011, April). *Comparability issues associated with assessments for the Common Core State Standards*. Paper presented at the annual meeting of the National Council on Measurement in Education, New Orleans, LA.

Koller, D., & Friedman, N. (2009). *Probabilistic graphical models: Principles and tech- niques*. Cambridge, MA: MIT Press.

Kolowich, S. (2012, April 13). A win for the robo-readers. *Inside Higher Ed.* Retrieved from https://www.insidehighered.com/news/2012/04/13/ large-study-shows-little-difference-between-human-and-robot-essay-graders

Kolowich, S. (2014, April 28). Writing instructor, skeptical of automated grading, pits machine vs. machine. *The Chronical of Higher Education.* Retrieved from http:// www.chronicle.com/article/Writing-Instructor-Skeptical/146211

Koziol, M., Singhal, P., & Cook, H. (2018, January 29). Computer says no: Governments scrap plan for 'robot marking' of NAPLAN essays. *The Sydney Morning Herald.* Retrieved from https://www.smh.com.au/politics/federal/ computer-says-no-governments-scrap-plan-for-robot-marking-of-naplan- essays-20180129-h0py6v.html

Krizhevsky, A., Sutskever, I., & Hinton, G. E. (2012). ImageNet classification with deep convolutional neural networks. *Communications of the ACM, 60*(6), 84–90. doi:10.1145/3065386

Krogh, A., & Hertz, J. A. (1992). A simple weight decay can improve generalization. *Conference on Neural Information Processing Systems.* Retrieved from https://papers. nips.cc/paper/563-a-simple-weight-decay-can-improve-generalization.pdf

Kusner, M. J., Loftus, J., Russell, C., & Silva, R. (2017). Counterfactual fairness. In I. Guyon, U. V. Luxburg, S. Bengio, H. Wallach, R. Fergus, S. Vishwanathan, & R. Garnett (Eds.), *Advances in neural information processing systems 30* (pp. 4066–4076). Retrieved from https://papers.nips.cc/paper/6995-counterfactual-fairness

Lacour, M., & Tissington, L. D. (2011). The effects of poverty on academic achieve- ment. *Educational Research and Reviews, 6*, 522–527.

Lai, C., Evanini, K., & Zechner, K. (2013). Applying rhythm metrics to non-native spontaneous speech. In P. Badin, T. Hueber, G. Bailly, & D. Demolin (Eds.), *Proceedings of the workshop on speech and language technology in education (SLaTE 2013)* (pp. 159–163). Retrieved from https://www.isca-speech.org/archive/ slate_2013/sl13_intro.pdf

Lai, E. R., Wolfe, E. W., & Vickers, D. (2015). Differentiation of illusory and true halo in writing scores. *Educational and Psychological Measurement, 75*, 102–125. doi:10.1177/0013164414530990

Lakoff, G. (1987). *Women, fire and dangerous things*. Chicago, IL: The University of Chicago Press.

Lambert, E., Alamargot, D., Larocque, D., & Caparossi, G. (2011). Dynamics of the spelling process during a copy task: Effect of regularity and frequency. *Canadian Journal of Experimental Psychology, 65*, 141–150. doi:10.1037/a0022538

Landauer, T. K., & Dumais, S. T. (1997). A solution to Plato's problem: The latent semantic analysis theory of acquisition, induction, and representation of knowledge. *Psychological Review, 104*(2), 211–240.

Landauer, T. K., Foltz, P. W., & Laham, D. (1998). Introduction to latent semantic analysis. *Discourse Processes, 25*, 259–284. doi:10.1080/01638539809545028

Landauer, T. K., Kireyev, K., & Panaccione, C. (2011). Word maturity: A new metric for word knowledge. *Scientific Studies of Reading, 15*, 92–108. doi:10.1080/10888 438.2011.536130

Landauer, T. K., Laham, D., & Foltz, P. W. (2001, February). *The intelligent essay assessor: Putting knowledge to the test.* Paper presented at the Association of Test Publishers Conference, Tucson, AZ.

Landauer, T. K., Laham, D., & Foltz, P. W. (2003). Automated scoring and annotation of essays with the Intelligent Essay Assessor. In M. D. Shermis, & J. Burstein (Eds.), *Automated essay scoring: A cross-disciplinary perspective* (pp. 87–112). Mahwah, NJ: Lawrence Erlbaum.

Landauer, T. K., McNamara, D. S., Dennis, S., & Kintsch, W. (Eds.). (2007). *Handbook of latent semantic analysis.* Mahwah, NJ: Erlbaum.

Lauttamus, T., Nerbonne, J., & Wiersma, W. (2010). Filled pauses as evidence of L2 proficiency: Finnish Australians speaking English. In B. Heselwood, & C. Upton (Eds.), *Proceedings of Methods XIII: Papers from the thirteenth international conference on methods in dialectology* (Vol. 54, pp. 240–251). Frankfurt, Germany: Peter Lang.

Lawson, C. L., & Hanson, R. J. (1995). *Solving least squares problems.* Philadelphia, PA: Society for International and Applied Mathematics. doi:10.1137/1.9781611971217

Leacock, C., & Chodorow, M. (2003). C-rater: Automated scoring of short-answer questions. *Computers and the Humanities, 37*, 389–405. doi:10.1023/A:1025779619903

Leacock, C., Gonzalez, E., & Conarroe, M. (2014). *Developing effective scoring rubrics for automated short-response scoring.* Monterey, CA: CTB/McGraw-Hill.

Leacock, C., Messineo, D., & Zhang, X. (2013, April). *Issues in prompt selection for automate scoring of short answer questions.* Paper presented at the annual meeting of the National Council on Measurement in Education, San Francisco, CA.

Lee, Y. W., Gentile, C., & Kantor, R. (2008). *Analytic scoring of TOEFL CBT essays: Scores from humans and e-rater* (TOEFL Research Report No. 81). Princeton, NJ: Educational Testing Service. doi:10.1002/j.2333-8504.2008.tb02087.x

Lehman, B., & Graesser, A. C. (2016). Arguing your way out of confusion. In F. Paglieri (Ed.), *The Psychology of argument: Cognitive approaches to argumentation and persuasion.* London, UK: College Publications.

Leijten, M., & Van Waes, L. (2013). Keystroke logging in writing research: Using Inputlog to analyze and visualize writing processes. *Written Communication, 30*, 358–392. doi:10.1177/0741088313491692

Leonard, D., & Coltea, C. (2013, May). Most change initiatives fail -- but they don't have to. *Gallup Business Journal.* Retrieved from http://news.gallup.com/businessjournal/162707/change-initiatives-fail-don.aspx

Levy, J. (2015). *UX strategy.* Sebastopol, CA: O'Reilly.

Levy, F., & Murnane, R. J. (2005). *The new division of labor: How computers are creating the next job market.* Princeton, NJ: Princeton University Press. doi:10.1515/9781400845927

Li, C., & Wand, M. (2016). Precomputed real-time texture synthesis with Markovian generative adversarial networks. In B. Leibe, J. Matas, N. Sebe, & M. Welling (Eds.), *Computer Vision—ECCV 2016* (pp. 702–716). doi:10.1007/978-3-319-46487-9_43

Li, H., Gobert, J., & Dickler, R. (2018). The relationship between scientific explana tions and the proficiencies of content, inquiry, and writing. In *Proceedings of the fifth annual ACM conference on learning at scale* (pp. 12–22). doi:10.1145/ 3231644.3231660

Limpo, T., & Alves, R. A. (2014a). Children's high-level writing skills: Development of planning and revising and their contribution to writing quality. *British Journal of Educational Psychology, 84,* 177–193. doi:10.1111/bjep.12020

Limpo, T., & Alves, R. A. (2014b). The role of transcription in establishing bursts of written language. *Handwriting Today, 2014,* 10–14.

Limpo, T., Alves, R. A., & Connelly, V. (2017). Examining the transcription-writing link: Effects of handwriting fluency and spelling accuracy on writing performance via planning and translating in middle grades. *Learning and Individual Differences, 53,* 26–36. doi:10.1016/j.lindif.2016.11.004

Linacre, J. M. (1989). *Many-faceted Rasch measurement.* Chicago, IL: MESA Press.

Lindgren, E., & Sullivan, K. P. (2002). The LS graph: A methodology for visualizing writing revision. *Language Learning, 52,* 565–595. doi:10.1111/1467-9922.00195

Linn, M. C. (2000). Designing the knowledge integration environment. *International Journal of Science Education, 22,* 781–796. doi:10.1080/095006900412275

Linn, M. C., Clark, D., & Slotta, J. D. (2003). WISE design for knowledge integration. *Science Education, 87,* 517–538. doi:10.1002/sce.10086

Linn, M. C., & Eylon, B.-S. (2011). *Science learning and instruction: Taking advantage of technology to promote knowledge integration.* New York, NY: Routledge. doi:10.4324/9780203806524

Linn, M. C., Gerard, L., Ryoo, K., McElhaney, K., Liu, O. L., & Rafferty, A. N. (2014). Computer-guided inquiry to improve science learning. *Science, 344*(6180), 155–156. doi:10.1126/science.1245980

Lipton, Z. C., Berkowitz, J., & Elkan, C. (2015). A critical review of recurrent neural networks for sequence learning. *Machine Learning.* Retrieved from https:// arxiv.org/pdf/1506.00019.pdf

Liu, L., Steinberg, J., Qureshi, F., Bejar, I., & Yan, F. (2016). Conversation-based assessments: An innovative approach to measure scientific reasoning. *Bulletin of the IEEE Technical Committee on Learning Technology, 18,* 10–13.

Liu, L., von Davier, A., Hao, J., Kyllonen, P., & Zapata-Rivera, D. (2015). A tough nut to crack: Measuring collaborative problem solving. In R. Yigal, S. Ferrara, & M. Mosharraf (Eds.), *Handbook of research on technology tools for real-world skill development* (pp. 344–359). Hershey, PA: IGI Global. doi:10.4018/978-1-4666-9441-5.ch013

Liu, O. L., Rios, J. A., Heilman, M., Gerard, L., & Linn, M. C. (2016). Validation of automated scoring of science assessments. *Journal of Research in Science Teaching, 53,* 215–233. doi:10.1002/tea.21299

Liu, S., & Zhang, M. (2017, July). *Examining the effects of a teacher professional development program on student writing processes.* Paper presented at the 2017 Society for Text & Discourse annual conference, Philadelphia, PA.

Lochbaum, K. E., Rosenstein, M., Foltz, P., & Derr, M. A. (2013, April). *Detection of gaming in automated scoring of essays with the IEA.* Paper presented at the annual meeting of the National Council on Measurement in Education, San Francisco, CA.

Loevinger, J. (1957). Objective tests as instruments of psychological theory. *Psychological Reports, 3,* 635–694. doi:10.2466/pr0.1957.3.3.635

Loh, B., Radinsky, J., Reiser, B. J., Gomez, L. M., Edelson, D. C., & Russell, E. (1997). The Progress Portfolio: Promoting reflective inquiry in complex investigation environments. In R. Hall, N. Miyake, & N. Enyedy (Eds.), *Proceedings of the 2nd international conference on computer support for collaborative learning* (pp. 176–185). New York, NY: ACM.

Loraksa, C., & Peachavanish, R. (2007). Automatic Thai-language essay scoring using neural network and latent semantic analysis. In *Proceedings of the first Asia international conference on modelling & simulation* (pp. 400–402). doi:10.1109/AMS.2007.19

Lottridge, S. (2018, April). *Methods for feature evaluation and selection.* Paper presented at the annual meeting of the National Council on Measurement in Education, New York, NY.

Lottridge, S., Winter, P., & Mugan, L. (2013, June). *The AS decision matrix: Using program stakes and item type to make informed decisions about automated scoring implementations.* Paper presented at the national conference on student assessment, National Harbor, MD.

Lottridge, S., Wood, S., & Shaw, D. (2018). The effectiveness of machine score-ability ratings in predicting automated scoring performance. *Applied Measurement in Education, 31,* 215–232. doi:10.1080/08957347.2018.1464452

Loukina, A., & Buzick, H. (2017). *Use of automated scoring in spoken language assessments for test takers with speech impairments* (Research Report Series No. RR-17-42). Princeton, NJ: Educational Testing Service. doi:10.1002/ets2.12170

Loukina, A., & Cahill, A. (2016). Automated scoring across different modalities. In *Proceedings of the 11th workshop on innovative use of NLP for building educational applications* (pp. 130–135). doi:10.18653/v1/W16-0514

Loukina, A., Madnani, N., & Zechner, K. (2019). The many dimensions of algorithmic fairness in educational applications. In H. Yannakoudakis, E. Kochmar, C. Leacock, N. Madnani, I. Pilán, & T. Zesch (Eds.), *Proceedings of the 14th workshop on innovative use of NLP for building educational applications* (pp. 1–10). doi:10.18653/v1/W19-4401

Loukina, A., Zechner, K., & Chen, L. (2014). Automatic evaluation of spoken summaries: the case of language assessment. In J. Tetreault, J. Burstein, & C. Leacock (Eds.), *Proceedings of the ninth workshop on innovative use of NLP for building educational applications* (pp. 68–78). doi:10.3115/v1/W14-1809

Loukina, A., Zechner, K., Chen, L., & Heilman, M. (2015). Feature selection for automated speech scoring. In *Proceedings of the tenth workshop on innovative use of NLP for building educational applications* (pp. 12–19). doi:10.3115/v1/W15-0602

Lu, X. (2010). The relationship of lexical richness to the quality of ESL learners' oral narratives. *The Modern Language Journal, 92,* 190–208. doi:10.1111/j.1540-4781.2011.01232_1.x

Luecht, R. M. (2012). An introduction to assessment engineering for automated item generation. In M. Gierl, & T. Haladyna (Eds.), *Automatic item generation* (pp. 59–76). New York, NY: Taylor-Francis/Routledge.

MacArthur, C. A., Philippakos, Z. A., & Graham, S. (2016). A multicomponent measure of writing motivation with basic college writers. *Learning Disability Quarterly, 39,* 31–43. doi:10.1177/0731948715583115

MacDonald, N. H., Frase, L. T., Gingrich, P., & Keenan, S. (1982). The Writer's Workbench: Computer aids for text analysis. *IEEE Transactions on Communications, 30*(1), 105–110. doi:10.1109/TCOM.1982.1095380

Madnani, N., Burstein, J., Sabatini, J., Biggers, K., & Andreyev, S. (2016). Language muse: Automated linguistic activity generation for English language learners. In *Proceedings of ACL-2016 system demonstrations* (pp. 79–84). Retrieved from doi:10.18653/v1/P16-4014

Madnani, N., Cahill, A., Blanchard, D., Andreyev, S., Napolitano, D., Gyawali, B., & Riordan, B. (2018). *A robust microservice architecture for scaling automated scoring applications* (Research Report No. 18-14). Princeton, NJ: Educational Testing Service. doi:10.1002/ets2.12202

Madnani, N., Cahill, A., & Riordan, B. (2016). Automatically scoring tests of proficiency in music instruction. In J. Tetreault, J. Burstein, C. Leacock, & H. Yannakoudakis (Eds.), *Proceedings of the 11th workshop on innovative use of NLP for building educational applications* (pp. 217–222). doi:10.18653/v1/W16-0524

Madnani, N., Loukina, A., von Davier, A., Burstein, J., & Cahill, A. (2017). Building better open-source tools to support fairness in automated scoring. In D. Hovy, S. Spruit, M. Mitchell, E. M. Bender, M. Strube, & H. Wallach (Eds.), *Proceedings of the first workshop on ethics in natural language processing* (pp. 41–52). doi:10.18653/v1/W17-1605

Manning, C., & Schütze, H. (1999). *Foundations of statistical natural language processing*. Cambridge, MA: MIT Press.

Mao, L., Liu, O. L., Roohr, K., Belur, V., Mulholland, M., Lee, H.-S., & Pallant, A. (2018). Validation of automated scoring for a formative assessment that employs scientific argumentation. *Educational Assessment, 23*(2), 121–138. doi:10.1080/106271 97.2018.1427570

Mathieu, M., Couprie, C., & LeCun, Y. (2015). *Deep multi-scale video prediction beyond mean square error*. Retrieved from https://arxiv.org/abs/1511.05440

Matuk, C., McElhaney, K. W., Chen, J. K., Lim-Breitbart, J., Kirkpatrick, D., & Linn, M. C. (2016). Iteratively refining a science explanation tool through classroom implementation and stakeholder partnerships. *International Journal of Designs for Learning, 7*(2), 93–110. doi:10.14434/ijdl.v7i2.20203

Mayer, R. (Ed.). (2005). *The Cambridge handbook of multimedia learning*. doi:10.1017/CBO9780511816819

Mayfield, E., Adamson, D., Woods, B., Miel, S., Butler, S., & Crivelli, J. (2018). Beyond automated essay scoring: Forecasting and improving outcomes in middle and high school writing. In Pardo, A., Bartimote, K., Lynch, G., Buckingham Shum, S., Ferguson, R., Merceron, A., & Ochoa, X. (Eds.), *Companion proceedings of the 8th international conference on learning analytics and knowledge* (pp. 1–8). Sydney, Australia: Society for Learning Analytics Research.

Mayfield, E., & Penstein Rosé, C. (2013). LightSIDE: Open source machine learning for text. In M. D. Shermis, & J. Burstein (Eds.), *Handbook of automated essay evaluation: Current applications and new directions* (pp. 124–135). doi:10.4324/9780203122761.ch8

McCaffrey, D. F., Ling, G., Burstein, J., & Beigman Klebanov, B. (2018, April). *Linking writing analytics and broader cognitive and intrapersonal outcomes*. Paper presented at the annual meeting of the National Council on Measurement in Education, New York, NY.

McClellan, C. (2010, May). *Quality assurance and control of human scoring*. Paper presented at the annual meeting of the National Council on Measurement in Education, Denver, CO.

McCutchen, D. (1996). A capacity theory of writing: Working memory in composition. *Educational Pschology Review, 8*, 299–325. doi:10.1007/BF01464076

McGraw-Hill Education CTB. (2014, December 24). *Smarter Balanced Assessment Consortium field test: Automated scoring research studies*. Retrieved from http://www. smarterapp.org/documents/FieldTest_AutomatedScoringResearchStudies.pdf

McNeill, K. L., & Krajcik, J. (2011). *Suporting grade 5 – 8 students in constructing explanation in science: The claim, evidence, reasoning and rebuttal framework for talk and writing*. New York, NY: Pearson.

Meadows, M., & Billington, L. (2007). *NAA enhancing the quality of marking project: Final report for research on marker selection*. Manchester, UK: AQA Centre for Education Research and Policy.

Meadows, M., & Billington, L. (2013). *The effect of marker background and training on the quality of marking in GCSE English*. Manchester, UK: AQA Centre for Education Research and Policy.

Medimorec, S., & Risko, E. F. (2017). Pauses in written composition: on the importance of where writers pause. *Reading and Writing: An Interdisciplinary Journal, 30*, 1–19. doi:10.1007/s11145-017-9723-7

Melzer, D. (2014). *Assignments across the curriculum: A national study of college writing*. Logan: Utah State University Press. doi:10.7330/9780874219401

Meredith, M. A., & Stein, B. E. (1983). Interactions among converging sensory inputs in the superior colliculus. *Science, 221*, 389–391. doi:10.1126/science.6867718

Messick, S. J. (1989). Validity. In R. L. Linn (Ed.), *Educational measurement* (3rd ed., pp. 13–103). New York, NY: Macmillan.

Messick, S. (1994). The interplay of evidence and consequences in the validation of performance assessments. *Educational Researcher, 23*(2), 13–23. doi:10.3102/0013189X023002013

Meyer, G. (Writer), & Lynch, J. (Director). (1992). Separate vocations [Television series episode]. In G. Meyer, R. Sakai, & D. Silverman (Producers), *The Simpsons*. Los Angeles, CA: Fox Broadcasting Company.

Midgette, E., Haria, P., & MacArthur, C. A. (2008). The effects of content and audience awareness goals for revision on the persuasive essays of fifth and eighth-grade students. *Reading and Writing, 21*, 131–151. doi:10.1007/s11145-007-9067-9

Mikolov, T., Chen, K., Corrado, G., & Dean, J. (2013). Efficient estimation of word representations in vector space. *Computation and Language*. Retrieved from https://arxiv.org/pdf/1301.3781.pdf

Millis, K., Forsyth, C., Wallace, P., Graesser, A. C., & Timmins, G. (2017). The impact of game-like features on learning from an intelligent tutoring system. *Technology, Knowledge, and Learning, 22*, 1–22. doi:10.1007/s10758-016-9289-5

Mills, C., Fridman, I., Soussou, W., Waghray, D., Olney, A., & D'Mello, S. K. (2017). Put your thinking cap on: Detecting cognitive load using EEG during learning. In I. Molenaar, O. X., & S. Dawson (Eds.), *Proceedings of the 7th international learning analytics and knowledge conference (LAK'17)* (pp. 80–89). doi:10.1145/3027385.3027431

Mirza, M., & Osindero, S. (2014). *Conditional generative adversarial nets*. Retrieved from https://arxiv.org/pdf/1411.1784.pdf

Mishkin, S. (2012, April 19). Can essay-marking software pass the test? *Financial Times*. Retrieved from https://www.ft.com/content/f8924a5a-88aa-11e1-a526-00144feab49a

Mislevy, R. J. (1995). *Probability-based inference in cognitive diagnosis* (Report No. AD-A280 553). Fort Belvoir, VA: Defense Technical Information Center. doi:10.21236/ADA280553

Mislevy, R. J. (2018). *Sociocognitive foundations of educational measurement.* New York, NY/London, UK: Routledge.

Mislevy, R. J., Almond, R. G., & Lukas, J. F. (2003). *A brief introduction to evidence-centered design* (Research Report 03-16). Princeton, NJ: Educational Testing Service.

Mislevy, R. J., Behrens, J. T., DiCerbo, K. E., & Levy, R. (2012). Design and discovery in educational assessment: Evidence-centered design, psychometrics, and data mining. *Journal of Educational Data Mining, 4,* 11–48.

Mislevy, R. J., Corrigan, S., Oranje, A., DiCerbo, K., John, M., Bauer, M. I., … Hao, J. (2014). *Psychometric considerations in game-based assessment.* New York, NY: Institute of Play.

Mislevy, R. J., & Gitomer, D. H. (1996). The role of probability-based inference in an intelligent tutoring system. *User-Mediated and User-Adapted Interaction, 5,* 253–282. doi:10.1007/BF01126112

Mislevy, R. J., & Haertel, G. D. (2006). Implications of evidence-centered design for educational testing. *Educational Measurement: Issues and Practice, 25*(4), 6–20. doi:10.1111/j.1745-3992.2006.00075.x

Mislevy, R. J., Hambel, L, Fried, R., Gaffney, T., Haertel, G., Hafter, A., … Quellmalz, E. (2003). *Design patterns for assessing science inquiry* (PADI Technical Report 1). Menlo Park, CA: SRI International.

Mislevy, R. J., Oranje, A., Bauer, M. I., von Davier, A., Hao, J., Corrigan, S., … John, M. (2014). *Psychometric considerations in game-based assessment* [White paper]. Retrieved from http://www.instituteofplay.org/work/projects/glasslab-research/

Mislevy, R. J., Oranje, A., Bauer, M. I., von Davier, A., Hao, J., Corrigan, S., … Joh, M. (2015). *Psychometric considerations in game-based assessment.* GlassLab: Institute of Play. Retrieved from http://www.instituteofplay.org/work/projects/glasslab-research/

Mislevy, R. J., & Riconscente, M. M. (2011). Evidence-centered assessment design. In *Handbook of test development* (pp. 75–104). New York, NY: Routledge.

Mislevy, R. J., Riconscente, M. M., & Rutstein, D. W. (2009). *Design patterns for assessing model-based reasoning* (Large-Scale Assessment Technical Report No. 6). Menlo Park, CA: SRI International.

Mislevy, R. J., Steinberg, L. S., & Almond, R. (2003). On the structure of educational assessments. *Measurement: Interdisciplinary Research and Perspectives, 1,* 3–67. doi:10.1207/S15366359MEA0101_02

Mizera, G. J. (2006). *Working memory and L2 oral fluency* (Unpublished doctoral dissertation). University of Pittsburgh, Pittsburgh, PA. Retrieved from http://d-scholarship.pitt.edu/7655/1/Mizera.Dissertation.pdf

Mohr, J. C. (2013). *Licensed to practice: The Supreme Court defines the American medical profession.* Baltimore, MD: Johns Hopkins University Press.

MongoDB, Inc. (2018). *The MongoDB 4.0 manual.* Retrieved from https://docs.mongodb.com/manual/

Morgan, B., Keshtkar, F., Graesser, A., & Shaffer, D. W. (2013). Automating the mentor in a serious game: A discourse analysis using finite state machines. In C. Stephanidis (Ed.), *Communications in Computer and Information Science: Vol. 374. HCI International 2013—Posters' extended abstracts* (pp. 591–595). Berlin, Germany: Springer. doi:10.1007/978-3-642-39476-8_119

Morris, J. S., Öhman, A., & Dolan, R. J. (1996). A subcortical pathway to the right amygdala mediating "unseen" fear. *PNAS: Proceedings of the National Academy of Sciences of the United States of America, 96*(4), 1680–1685. doi:10.1073/pnas.96.4.1680

Moxley, J. M., & Eubanks, D. (2015). On keeping score: Instructors' vs. students' rubric Ratings of 46,689 essays. *WPA: Writing Program Administration, 39*, 53–80.

Murphy, P. M. (2012). *Machine learning: A probabilistic perspective.* Cambridge, MA: MIT Press.

Musu-Gillette, L., de Brey, C., McFarland, J., Hussar, W., Sonnenberg, W., & Wilkinson-Flicker, S. (2017). *Status and trends in the education of racial and ethnic groups 2017* (NCES 2017-051). Retrieved from https://nces.ed.gov/pubs2017/2017051.pdf

Myford, C. M., & Wolfe, E. W. (2009). Monitoring rater performance over time: A framework for detecting differential accuracy and differential scale category use. *Journal of Educational Measurement, 46*, 371–389. doi:10.1111/j.1745-3984.2009.00088.x

National Center for Education Statistics. (2012). *Writing 2011: The National Assessment of Educational Progress at grades 8 and 11* (NCES-2012-470). Retrieved from https://nces.ed.gov/nationsreportcard/pubs/main2011/2012470.aspx

National Center for Education Statistics. (2016). *The condition of education 2016.* Retrieved from https://nces.ed.gov/pubs2016/2016144.pdf

National Council of Teachers of English. (2013, April). *NCTE position statement on machine scoring.* Retrieved from http://www.ncte.org/positions/statements/machine_scoring

Neumeyer, L., Franco, H., Digalakis, V., & Weintraub, M. (2000). Automatic scoring of pronunciation quality. *Speech Communication, 30*, 83–93. doi:10.1016/S0167-6393(99)00046-1

Newell, A., & Simon, H. A. (1972). *Human problem solving.* Englewood Cliffs, NJ: Prentice-Hall.

NGSS Lead States. (2013). *Next Generation Science Standards: For states, by states.* Washington, DC: The National Academic Press.

Nguyen, H., & Litman, D. (2016). Context-aware argumentative relation mining. In K. Erk, & N. A. Smith (Eds.), *Proceedings of the 54th annual meeting of the Association for Computational Linguistics* (Vol. 1, pp. 1127–1137). doi:10.18653/v1/P16-1

NiteRater [Computer software]. (2007). Jerusalem, Israel: National Institute for Testing & Evaluation.

No Child Left Behind Act of 2001, 20 U.S.C. § 6319 *et seq.* (2002).

Nye, B. D., Graesser, A. C., & Hu, X. (2014). AutoTutor and family: A review of 17 years of natural language tutoring. *International Journal of Artificial Intelligence in Education, 24*, 427–469. doi:10.1007/s40593-014-0029-5

Nye, B. D., Graesser, A. C., Hu, X., & Cai, Z. (2014). AutoTutor in the cloud: A service-oriented paradigm for an interoperable natural-language ITS. *Journal of Advanced Distributed Learning Technology, 2*(6), 35–48.

Nystrand, M. (1997). *Opening dialogue: Understanding the dynamics of language and learning in the English classroom.* New York, NY: Teachers College Press.

OECD. (2013). *PISA 2015 collaborative problem solving framework.* Paris: OECD. Retrieved from https://www.oecd.org/pisa/pisaproducts/Draft%20PISA%202015%20Collaborative%20Problem%20Solving%20Framework%20.pdf

OECD. (2015). *Collaborative problem solving framework.*

Oh, J. S., Cho, W. J., Kim, J., & Lee, J. Y. (2005). A descriptive question marking system based on semantic kernels. *The Journal of Korean Institute of Information Technology, 3*(4), 95–104.

Olney, A., Louwerse, M., Mathews, E., Marineau, J., Hite-Mitchell, H., & Graesser, A. (2003). Utterance classification in *AutoTutor.* In J. Burstein, & C. Leacock (Eds.), *Proceedings of the HLT-NAACL 03 workshop on building educational applications using natural language processing* (pp. 1–8). doi:10.3115/1118894.1118895

Ong, D., Razon, A. R., Perigrino, J. M., Guevara, R. C. L., & Naval Jr., P. C. (2011, March) *Automated Filipino Essay Grader with concept-indexing*. Paper presented at the 11th Philippine computing science congress, Naga City, Philippines.

Östling, R., Smolentzov, A., & Hinnerich, B. T. (2013). Automated essay scoring for Swedish. In *Proceedings of the eight workshop on innovative use of NLP for building educational applications* (pp. 42–47). Retrieved from http://www.aclweb.org/anthology/W13-1705

Ostry, D. J. (1983). Determinant of interkey times in typing. In W. E. Cooper (Ed.), *Cognitive aspects of skilled typewriting* (pp. 225–246). New York, NY: Springer.

Oviatt, S. (2017). Theoretical foundations of multimodal interfaces and systems. In S. Oviatt, B. Schuller, P. R. Cohen, D. Sonntag, G. Potamianos, & A. Krüger (Eds.), *The handbook of multimodal-multisensor interfaces: Volume 1. Foundations, user modeling, and common modality combinations* (pp. 19–50). San Rafael, CA: Morgan & Claypool. doi:10.1145/3015783.3015786

Oviatt, S., Schuller, B., Cohen, P. R., Sonntag, D., Potamianos, G., & Krüger, A. (Eds.). (2017). *The handbook of multimodal-multisensor interfaces: Volume 1. Foundations, user modeling, and common modality combinations*. San Rafael, CA: Morgan & Claypool. doi:10.1145/3015783

Paas, F., & Ayres, P. (2014). Cognitive load theory: A broader view on the role of memory in learning and education. *Educational Psychology Review, 26*, 191–195. doi:10.1007/s10648-014-9263-5

Page, E. B. (1966a). Grading essays by computer: Progress report. In *Proceedings of the 1966 invitational conference on testing, Princeton, NJ* (pp. 87–100). Princeton, NJ: Educational Testing Service.

Page, E. B. (1966b). The imminence of ... grading essays by computer. *Phi Delta Kappan, 47*(5), 238–243.

Page, E. B. (1968). The use of the computer in analyzing student essays. *International Review of Education, 14*, 210–225. doi:10.1007/BF01419938

Page, E. B. (2003). Project essay grade: PEG. In M. D. Shermis, & J. Burstein (Eds.), *Automated essay scoring: A cross-disciplinary perspective* (pp. 43–54). Mahwah, NJ: Lawrence Erlbaum.

Pardos, Z. A., Fan, Z., & Jiang, W. (2019). Connectionist recommendation in the wild: On the utility and scrutability of neural networks for personalized course guidance. *User Modeling and User-Adapted Interaction, 29*, 487–525. doi:10.1007/s11257-019-09218-7

Parrila, R., Georgiou, G., & Corkett, J. (2007). University students with a significant history of reading difficulties: What is and is not compensated? *Exceptionality Education Canada, 17*, 195–220.

Parslow, N. L. (2015). *Automated analysis of L2 French writing: A preliminary study* (Master's thesis). University of Paris Diderot. doi:10.13140/RG.2.1.2833.5204

Partan, S., & Marler, P. (1999). Communication goes multimodal. *Science, 283*, 1272–1273. doi:10.1126/science.283.5406.1272

Paszke, A., Gross, S., Chintala, S., Chanan, G., Tang, E., DeVito, Z., ... Lerer, A. (2017). Automatic differentiation in PyTorch. In *Proceedings of the 31st conference on neural information processing systems* (pp. 1–4). Retrieved from https://openreview.net/pdf?id=BJJsrmfCZ

Patchan, M. M., & Schunn, C. D. (2015). Understanding the benefits of providing peer feedback: How students respond to peers' texts of varying quality. *Instructional Science, 43*, 591–614. doi:10.1007/s11251-015-9353-x

Patchan, M., Schunn, C. D., & Correnti, R. (2016). The nature of feedback: How peer feedback features affect students' implementation rate and quality of revisions. *Journal of Educational Psychology, 108*(8), 1098–1120. doi:10.1037/edu0000103

Patz, R. J., Junker, B. W., Johnson, M. S., & Mariano, L. T. (2002). The hierarchical rater model for rated test items and its application to large-scale educational assessment data. *Journal of Educational and Behavioral Statistics, 27*, 341–384. doi:10.3102/10769986027004341

Pearlman, M., Berger, K., & Tyler, L. (1993). *An application of multimedia software to standardized testing in music* (Research Report No. RR-93-36). Princeton, NJ: Educational Testing Service. doi:10.1002/j.2333-8504.1993.tb01547.x

Pearson & Educational Testing Service. (2015, March 9). *Research results of PARCC automated scoring proof of concept study*. Retrieved from https://parcc-assessment.org/content/uploads/2015/06/PARCC_AI_Research_Report.pdf

Pearson, K. (1915). On the problem of sexing osteometric material. *Biometrika, 10*, 479–487. Retrieved from https://www.jstor.org/stable/2331836

Penfield, R. D. (2016). Fairness in test scoring. In N. J. Dorans, & L. L. Cook (Eds.), *Fairness in educational assessment and measurement*. New York, NY: Routledge.

Perelman, L. (2014). When "the state of the art" is counting words. *Assessing Writing, 21*, 104–111. doi:10.1016/j.asw.2014.05.001

Perin, D., Lauterbach, M., Raufman, J., & Kalamkarian, H. S. (2017). Text-based writing of low-skilled postsecondary students: Relation to comprehension, self-efficacy and instructor judgments. *Reading and Writing: An Interdisciplinary Journal, 30*(4), 887–915. doi:10.1007/s11145-016-9706-0

Perl, S. (1979). The composing processes of unskilled college writers. *Research in the Teaching of English, 13*, 317–336.

Perlstein, L. (1998, October 13). Software's essay test: Should it be grading? *The Washington Post*, p. A1.

Perrin, D. (2014, July 9). What would Mark Twain have thought of Common Core testing? *The Atlantic*. Retrieved from https://www.theatlantic.com/education/archive/2014/07/what-would-mark-twain-have-thought-of-the-common-core/374114/?utm_source=SFFB

Pinet, S., Ziegler, J. C., & Alario, F.-X. (2016). Typing is writing: Linguistic properties modulate typing execution. *Psychonomic Bulletin, 23*, 1898–1906. doi:10.3758/s13423-016-1044-3

Pothos, E. M., Edwards, D. J., & Perlman, A. (2011). Supervised versus unsupervised categorization: Two sides of the same coin? *The Quarterly Journal of Experimental Psychology, 64*, 1692–1713. doi:10.1080/17470218.2011.554990

Powers, D. E., Burstein, J. C., Chodorow, M., Fowles, M. E., & Kukich, K. (2001). *Stumping e-rater: Challenging the validity of automated essay scoring* (GRE Board Research Report No. 98-08bP). Princeton, NJ: Educational Testing Service. doi:10.1002/j.2333-8504.2001.tb01845.x

Powers, D. E., Escoffery, D. S., & Duchnowski, M. P. (2015). Validating automated essay scoring: A (modest) refinement of the "gold standard". *Applied Measurement in Education, 28*, 130–142. doi:10.1080/08957347.2014.1002920

Powers, D. E., Fowles, M. E., Farnum, M., & Ramsey, P. (1994). Will they think less of my handwritten essay if others word process theirs? Effects on essay scores of intermingling handwritten and word-processed essays. *Journal of Educational Measurement, 31*, 220–233. doi:10.1111/j.1745-3984.1994.tb00444.x

Prior, P., & Bilbro, R. (2012). Academic enculturation: Developing literate practices and disciplinary identities. In M. Castelló, & C. Donahue (Eds.), *University writing. Selves and texts in academic societies* (pp. 19–32). Bingley, UK: Emerald.

Purdy, J. P. (2009). When the tenets of composition go public: A study of writing in Wikipedia. *College Composition and Communication, 61*, W351–W373.

Quillen, I. (2012, April 24). Study supports essay-grading technology. *Education Week.* Retrieved from http://www.edweek.org/ew/articles/2012/04/25/29essays.h31.html

Quinlan, T., Loncke, M., & Leijten, M. (2012). Coordinating the cognitive processes of writing: The role of the monitor. *Written Communication, 29*, 345–368. doi:10.1177/0741088312451112

Quintana, C., Reiser, B. J., Davis, E. A., Krajcik, J., Fretz, E., Duncan, R. G., … Soloway, E. (2004). A scaffolding design framework for software to support science inquiry. *The Journal of the Learning Sciences, 13*, 337–386. doi:10.1207/s15327809jls1303_4

Quintana, C., Zhang, M., & Krajcik, J. (2005). A framework for supporting metacognitive aspects of online inquiry through software-based scaffolding. *Educational Psychologist, 40*, 235–244. doi:10.1207/s15326985ep4004_5

Raczynski, K. R., Cohen, A. S., Engelhard Jr., G., & Lu, Z. (2015). Comparing the effectiveness of self-paced and collaborative frame-of-reference training on rater accuracy in a large-scale writing assessment. *Journal of Educational Measurement, 52*, 301–318. doi:10.1111/jedm.12079

Radford, A., Metz, L., & Chintala, S. (2015). Unsupervised representation learning with deep convolutional generative adversarial networks. *Machine Learning.* Retrieved from https://arxiv.org/pdf/1511.06434.pdf

Ramanarayanan, V., Lange, P. L., Evanini, K., Molloy, H. R., Suendermann-Oeft, D. (2017). Human and automated scoring of fluency, pronunciation and intonation during human-machine dialog interactions. In *Proceedings of the 8th annual conference of the international speech communication association* (pp. 1711–1715). doi:10.21437/Interspeech.2017-1213

Ramineni, C., Trapani, C. S., & Williamson, D. M. (2015). *Evaluation of e-rater for the PRAXIS I Writing test* (Research Report No. RR-15-03). Princeton, NJ: Educational Testing Service. doi:10.1002/ets2.12047

Ramineni, C., Trapani, C. S., Williamson, D. M., Davey, T., & Bridgeman, B. (2012a). *Evaluation of the e-rater scoring engine for the GRE issue and argument prompts* (Research Report No. RR-12-02). Princeton, NJ: Educational Testing Service. doi:10.1002/j.2333-8504.2012.tb02284.x

Ramineni, C., Trapani, C. S., Williamson, D. M., Davey, T., & Bridgeman, B. (2012b). *Evaluation of the e-rater scoring engine for the TOEFL independent and integrated prompts* (Research Report No. RR-12-06). Princeton, NJ: Educational Testing Service. doi:10.1002/j.2333-8504.2012.tb02288.x

Ramineni, C., & Williamson, D. M. (2013). Automated essay scoring: Psychometric guidelines and practices. *Assessing Writing, 18*, 25–39. doi:10.1016/j.asw.2012.10.004

Ransdell, S. (1995). Generating thinking aloud protocols: Impact on the narrative writing of college students. *American Journal of Psychology, 108*, 89–98. doi:10.2307/1423102

Ravitch, D. (2014, September 3). Why computers should not grade student essays [Weblog post]. Retrieved from https://dianeravitch.net/2014/09/03/why-computers-should-not-grade-student-essays-2/

Read, J. (2000). *Assessing vocabulary*. Cambridge, UK: Cambridge University Press.

Ready, D. A. (2016, January). 4 things successful change leaders do well. *Harvard Business Review*. Retrieved from https://hbr.org/2016/01/4-things-successful-change-leaders-do-well

Reed, S., Akata, Z., Yan, X., Logeswaran, L., Schiele, B., & Lee, H. (2016). Generative adversarial text to image synthesis. In M. F. Balcan, & K. Q. Weinberger (Eds.), *Proceedings of the 33rd international conference on machine learning* (Vol. 48, pp. 1060–1069). Retrieved from http://proceedings.mlr.press/v48/

Reimann, P., & Yacef, K. (2013). Using process mining for understanding learning. In R. Luckin, S. Puntambekar, P. Goodyear, B. Grabowski, J. Underwood, & N. Winters (Eds.), *Handbook of design in educational technology* (pp. 472–481). New York, NY: Routledge.

Rezaee, A. A., & Kermani, E. (2011). Essay raters' personality types and rater reliability. *International Journal of Language Studies, 5*(4), 109–122.

Rich, C., Schneider, M. C., & D'Brot, J. (2013). Applications of automated essay evaluation in West Virginia. In M. D. Shermis, & J. Burstein (Eds.), *Handbook of automated essay evaluation: Current application and new directions* (pp. 99–123). New York, NY: Routledge.

Ricker-Pedley, K. L. (2019). *Determining the optimal validity insertion rate to minimize cost and scoring time in human constructed response scoring*. Manuscript in preparation.

Riconscente, M. M., Mislevy, R. J., & Corrigan, S. (2016). Evidence-centered design. In S. Lane, M. R. Raymond, & T. M. Haladyna (Eds.), *Handbook of test development* (2nd ed., pp. 40–63). New York, NY: Routledge.

Robinson, D. (2017, September 6). The incredible growth of Python [Blog post]. Retrieved from https://stackoverflow.blog/2017/09/06/incredible-growth-python/

Robinson, N. (2018, January 28). *NAPLAN: Robot marking of school tests scrapped by education ministers*. Retrieved from https://www.abc.net.au/news/2018-01-29/push-to-have-robots-mark-naplan-tests-scrapped/9370318

Romero, C., & Ventura, S. (2007). Educational data mining: A survey from 1995 to 2005. *Expert Systems with Applications, 33*, 135–146. doi:10.1016/j.eswa.2006.04.005

Rosch, E. (1999). Principles of categorization. In E. Margolis, & S. Laurence (Eds.), *Concepts: Core readings* (pp. 189–206). Cambridge, MA: MIT Press.

Rose, M. (1980). Rigid rules, inflexible plans, and the stifling of language: A cognitivist analysis of writer's block. *College Composition and Communication, 31*, 389–401. doi:10.2307/356589

Rosenberg, D., & Scott, K. (1991). *Use case driven object modeling with UML: A practical approach*. New York, NY: Addison-Wesley.

Rosenblatt, F. (1961). *Principles of neurodynamics: Perceptrons and the theory of brain mechanisms* (Report No. 1196-G-8). Buffalo, NY: Cornell Aeronautical Laboratory.

Rosenqvist, S. (2015). *Developing pause thresholds for keystroke logging analysis* (Unpublished bachelor's thesis). Umea University, Umea, Sweden.

Ross, L. P., Clauser, B. E., Fan, V. Y., & Clyman, S. G. (1998). *An examination of alternative models for automated scoring of a computer-delivered performance assessment*. Paper presented at the annual meeting of the American Educational Research Association, San Diego, CA.

Royal-Dawson, L., & Baird, J. (2009). Is teaching experience necessary for reliable scoring of extended English questions? *Educational Measurement: Issues and Practice, 28*(2), 2–8. doi:10.1111/j.1745-3992.2009.00142.x

Rubin, D. L., & Williams-James, M. (1997). The impact of writer nationality on mainstream teachers' judgments of composition quality. *Journal of Second Language Writing, 6*(2), 139–153.

Rumelhart, D. E., Hinton, G. E., & Williams, R. J. (1986). Learning representations by back-propagating errors. *Nature, 323*, 533–536. doi:10.1038/323533a0

Rupp, A. A. (2017, April). *Evaluating automated scoring systems with validity in mind: Methodological design decisions.* Paper presented at the annual meeting of the National Council on Measurement in Education, San Antonio, TX.

Rupp, A. A. (2018). Designing, evaluating, and deploying automated scoring systems with validity in mind: Methodological design decisions. *Applied Measurement in Education, 31*, 191–214. doi:10.1080/08957347.2018.1464448

Rupp, A. A., Templin, J., & Henson, R. J. (2010). *Diagnostic measurement: Theory, methods, and applications.* New York, NY: Guilford Press.

Rus, V., D'Mello, S., Hu, X., & Graesser, A. C. (2013). Recent advances in intelligent systems with conversational dialogue. *AI Magazine, 34*, 42–54. doi:10.1609/aimag.v34i3.2485

Rus, V., Lintean, M. C., Banjade, R., Niraula, N. B., & Stefanescu, D. (2013). SEMILAR: The Semantic Similarity Toolkit. In *Proceedings of the 51st meeting of the Association for Computational Linguistics* (pp. 163–168). https://pdfs.semanticscholar.org/b148/1e7bf167ccbfd5edf23e250006bd506076df.pdf

Rus, V., Moldovan, Graesser, A. C., & Niraula, N. (2012). Automatic discovery of speech act categories in educational games. In K. Yacef, O. Zaïane, H. Hershkovitz, M. Yudelson, & J. Stamper (Eds.). In *Proceedings of the 5th international conference on educational data mining* (pp. 25–32). Chania, Greece: International Educational Data Mining Society.

Rus, V., Niraula, N., Maharjan, N., & Banjade, R. (2015). Automated labelling of dialogue modes in tutorial dialogues. In I. Russel, & B. Eberle (Eds.), *Proceedings of the 28th international Florida artificial intelligence research society conference* (pp. 205–210). Palo Alto, CA: AAAI Press.

Sakaguchi, K., Heilman, M., & Madnani, N. (2015). Effective feature integration for automated short answer scoring. In *Proceedings of the 2015 Annual conference of the North American chapter of the Association for Computational Linguistics: Human language technologies* (pp. 1049–1054). doi:10.3115/v1/N15-1111

Samei, B., Li, H., Keshtkar, F., Rus, V., & Graesser, A. C. (2014). Context-based speech act classification in intelligent tutoring systems. In S. Trausan-Matu, K. Boyer, M. Crosby, & K. Panou (Eds.), *Lecture notes in computer science: Vol. 8474. Intelligent tutoring systems* (pp. 236–241). Berlin, Germany: Springer. doi:10.1007/978-3-319-07221-0_28

Sao Pedro, M. A., Baker, R. S. J. d., & Gobert, J. D. (2012). Improving construct validity yields better models of systematic inquiry, even with less information. In J. Masthoff, B. Mobasher, M. C. Desmarais, & R. Nkambou (Eds.), *Lecture Notes in Computer Science: Vol. 7379. User modeling, adaptation, and personalization* (pp. 249–260). New York, NY: Springer.

Sao Pedro, M., Baker, R., & Gobert, J. (2013a). Incorporating scaffolding and tutor context into Bayesian knowledge tracing to predict inquiry skill acquisition. In S. K. D'Mello, R. A. Calvo, & A. Olney (Eds.), *Proceedings of the 6th international conference on educational data mining* (pp. 185–192). Retrieved from http://www.educationaldatamining.org/EDM2013/proceedings/EDM2013Proceedings.pdf

Sao Pedro, M., Baker, R., & Gobert, J. (2013b). What different kinds of stratification can reveal about the generalizability of data-mined skill assessment models. In *Proceedings of the 3rd conference on learning analytics and knowledge* (pp. 190–194). doi:10.1145/2460296.2460334

Sao Pedro, M., Baker, R., Gobert, J., Montalvo, O., & Nakama, A. (2013). Leveraging machine-learned detectors of systematic inquiry behavior to estimate and predict transfer of inquiry skill. *User Modeling and User-Adapted Interaction, 23,* 1–39. doi:10.1007/s11257-011-9101-0

Sao Pedro, M., Gobert, J., & Betts, C. (2018). *Inq-Blotter: A dashboard that alerts teachers when students struggle during investigations with Inq-ITS online labs.* Poster presented at the U.S. Department of Education IES PI Meeting, Washington, DC.

Sao Pedro, M., Jiang, Y., Paquette, L., Baker, R. S., & Gobert, J. (2014). Identifying transfer of inquiry skills across physical science simulations using educational data mining. In *Proceedings of the 11th international conference of the learning sciences* (pp. 222–229). Boulder, CO: International Society of the Learning Sciences.

Scaled Agile. (2018, September 14). *Architectural runway.* Retrieved from https://www.scaledagileframework.com/architectural-runway/

Schneider, M. C., & Osleson, L. (2013, April). *Evaluating the comparability of engine and human scores over time.* Paper presented at the annual meeting of the National Council on Measurement in Education, San Francisco, CA.

Schneider, M. C., Egan, K., & Gong, B. (2017). Defining and challenging fairness in tests involving students with dyslexia: Key opportunities in test design and score interpretations. In H. Jiao, & R. W. Lissitz (Eds.), *Test fairness in a new generation of large-scale assessment* (pp. 209–232). Charlotte, NC: Information Age Publishing.

Schneider, M. C., Waters, B., & Wright, W. (2012, April). *Stability of automated essay scoring engines across time and subgroups when student ability changes.* Paper presented at the annual meeting of the National Council on Measurement in Education, Vancouver, Canada.

Schneider, U. (2014). *Frequency, chunks and hesitations: A usage-based analysis of chunking in English* (Unpublished doctoral thesis). Universirt Library Freiburg, Breisgau, Germany.

Schober, M. F., Rapp, D. N., & Britt, M. A. (Eds.). (2018). *Routledge handbook of discourse processes.* New York, NY: Routledge. doi:10.4324/9781315687384

Schriver, K. A. (2012). What we know about expertise in professional communication. In V. W. Berninger (Ed.), *Past, present, and future contributions of cognitive writing research to cognitive psychology* (pp. 275–312). New York, NY: Psychology Press.

Schultz, M. T. (2013). The IntelliMetric automated essay scoring engine—A review and an application to Chinese essay scoring. In M. D. Shermis, & J. Burstein (Eds.), *Handbook of automated essay evaluation: Current applications and new directions* (pp. 89–98). New York, NY: Routledge.

Schum, D. A. (1994). *The evidential foundations of probabilistic reasoning.* New York, NY: Wiley.

Schwanenflugel, P. J., Hamilton, A. M., Wisenbaker, J. M., Kuhn, M. R., & Stahl, S. A. (2004). Becoming a fluent reader: Reading skill and prosodic features in the oral reading of young readers. *Journal of Educational Psychology, 96,* 119–129. doi:10.1037/0022-0663.96.1.119

Schwartz, A. E. (1998, April 26). Graded by machine. *The Washington Post,* p. 47.

Shadish, W., Cook, T. D., & Campbell, D. T. (2002). *Experimental and quasi-experimental designs for generalized causal inference*. Boston, MA: Houghton Mifflin.

Shanahan, T. (2015). Common core state standards: A new role for writing. *The Elementary School Journal, 115*, 464–479. doi:10.1086/681130

Shepard, L. A. (1993). Evaluating test validity. *Review of Research in Education, 19*, 405–450. doi:10.2307/1167347

Shermis, M. D. (2014). State-of-the-art automated essay scoring: Competition, results, and future directions from a United States demonstration. *Assessing Writing, 20*, 53–76. doi:10.1016/j.asw.2013.04.001

Shermis, M. D. (2015). Contrasting state-of-the-art in the machine scoring of short-form constructed responses. *Educational Assessment, 20*, 46–65. doi:10.1080/106 27197.2015.997617

Shermis, M. D. (2018). Establishing a crosswalk between the Common European Framework for Languages (CEFR) and writing domains scored by automated essay scoring. *Applied Measurement in Education, 31*, 177–190. doi:10.1080/0895 7347.2018.1464451

Shermis, M. D., & Burstein, J. (Eds.). (2003). *Automated essay scoring: A cross-disciplinary perspective*. Mahwah, NJ: Lawrence Erlbaum.

Shermis, M. D., & Burstein, J. (Eds.). (2013). *Handbook of automated essay evaluation: Current applications and new directions*. New York, NY: Routledge. doi:10.4324/9780203122761

Shermis, M. D., Burstein, J., Brew, C., Higgins, D., & Zechner, K. (2015). Recent innovations in machine scoring. In S. Lane, T. Haladyna, & M. Raymond (Eds.), *Handbook of test development* (2nd ed., pp. 335–354). New York, NY: Routledge.

Shermis, M., Burstein, J., Elliot, N., Miel, S., & Foltz, P. (2015). Automated writing evaluation: An expanding body of knowledge. In C. A. McArthur, S. Graham, & J. Fitzgerald (Eds.), *Handbook of writing research* (2nd ed., pp. 395–409). New York, NY: Guilford.

Shermis, M., Burstein, J., Elliot, N., Miel, S., & Foltz, W. (2016). Automated writing evaluation. In C. Macarthur, S. Graham, & J. Fitzgerald (Eds.), *Handbook of writing research* (2nd ed., pp. 395–409). New York, NY: Guilford.

Shermis, M. D., & Hamner, B. (2013). Contrasting state-of-the-art automated scoring of essays. In M. D. Shermis, & J. Burstein (Eds.), *Handbook of automated essay evaluation: Current applications and new directions* (pp. 313–346). New York, NY: Routledge.

Shute, V. J. (2008). Focus on formative feedback. *Review of Educational Research, 78*, 153–189. doi:10.3102/0034654307313795

Shute, V. J. (2011). Stealth assessment in computer-based games to support learning. In S. Tobias, & J. D. Fletcher (Eds.), *Computer games and instruction* (pp. 503–524). Charlotte, NC: Information Age Publishers.

Shute, V. J., Ventura, M., & Kim, Y. J. (2013). Assessment and learning of informal physics in Newton's Playground. *Journal of Educational Research, 106*, 423–430.

Silver, N. (2012). *The signal and the noise: Why so many predictions fail and some don't*. New York, NY: Penguin Books.

Simonite, T. (2016, May 13). Moore's law is dead. Now what? *MIT Technology Review*. Retrieved from https://www.technologyreview.com/s/601441/moores-law-is-dead-now-what/

Sinclair, J., & Coulthard, M. (1975). *Towards an analysis of discourse*. Oxford, UK: Oxford University Press.

Sinharay, S., Zhang, M., & Deane, P. (2019). Prediction of essay scores from writing process and product features using data mining methods. *Applied Measurement in Education, 32,* 116–137. doi:10.1080/08957347.2019.1577245

Smith, E. E., & Medin, D. L. (1981). *Categories and concepts* (Vol. 9). Cambridge, MA: Harvard University Press.

Smolentzov, A. (2013). *Automated essay scoring: Scoring essays in Swedish* (Master's thesis). University of Stockholm. Retrieved from http://www.diva-portal.org/smash/record.jsf?pid=diva2:602025

Somasundaran, S., Flor, M., Chodorow, M., Molloy, H., Gyawali, B., & McCulla, L. (2018). Towards evaluating narrative quality in student writing. *Transactions of the Association for Computational Linguistics, 6,* 91–106. doi:10.1162/tacl_a_00007

Sommers, N. (1980). Revision strategies of student writers and experienced adult writers. *College Composition and Communication, 31,* 378–388. doi:10.2307/356588

Sottilare, R., Graesser, A. C., Hu, X., & Brawner, K. (Eds.). (2015). *Design recommendations for intelligent tutoring systems: Vol. 3. Authoring tools and expert modeling techniques.* Orlando, FL: Army Research Laboratory.

Spelman Miller, K. (2002). Units of production in writing: Evidence of topic 'framing' in on-line writing research. *Reading Working Papers in Linguistics, 6,* 244–272.

Sperber, D. (1996). *Explaining culture: A naturalistic approach.* Oxford, UK: Blackwell.

Sperber, D., & Wilson, D. (1995). *Relevance: Communication and cognition* (2nd ed.). Cambridge, MA: Blackwell Publishers.

Stab, C., & Gurevych, I. (2014). Identifying argumentative discourse structures in persuasive essays. In *Proceedings of the 2014 conference on empirical methods in natural language processing* (pp. 46–56). Retrieved from https://www.aclweb.org/anthology/D14-1006

Stapleton, P. (2010). Writing in an electronic age: A case study of L2 composing processes. *Journal of English for Academic Purposes, 9,* 295–307. doi:10.1016/j.jeap.2010.10.002

Steinberg, L. S., & Gitomer, D. G. (1996). Intelligent tutoring and assessment built on an understanding of a technical problem-solving task. *Instructional Science, 24,* 223–258. doi:10.1007/BF00119978

Stellmack, M. A., Keenan, N. K., Sandidge, R. R., Sippl, A. L., & Konheim-Kalkstein, Y. L. (2012). Review, revise, and resubmit: The effects of self-critique, peer review, and instructor feedback on student writing. *Teaching of Psychology, 39,* 235–244. doi:10.1177%2F0098628312456589

Steve. (2013, April 6). The pros(e) and cons of computers grading essays [Web log post]. Retrieved from http://whatdoino-steve.blogspot.com/2013/04/the-prose-and-cons-of-computers-grading.html

Stosich, E. L., Snyder, J., & Wilczak, K. (2018). How do states integrate performance assessment in their systems of assessment? *Education Policy Analysis Archives 26*(13), 1–31. doi:10.14507/epaa.26.2906

Strauss, V. (2016, May 5). Should you trust a computer to grade your child's writing on Common Core tests? *Washington Post.* Retrieved from https://www.washingtonpost.com/news/answer-sheet/wp/2016/05/05/should-you-trust-a-computer-to-grade-your-childs-writing-on-common-core-tests/?utm_term=.c33825bfd1aa

Strik, H., Truong, K., de Wet, F., & Cucchiarini, C. (2007). Comparing classifiers for pronunciation error detection. In *Proceedings of the 8th annual conference of the International Speech Communication Association* (pp. 1837–1840). Retrieved from https://www.isca-speech.org/archive/archive_papers/interspeech_2007/i07_1837.pdf

Sukkarieh, J. Z., &Blackmore, J. (2009, May). *c-rater: Automatic content scoring of short constructed responses*. Paper presented at the 22nd international conference for the Florida artificial intelligence research society, Sanibel, Florida.

Sun, X., & Evanini, K. (2011). Gaussian mixture modeling of vowel durations for automated assessment of non-native speech. In *Proceedings of the 2011 IEEE International conference on acoustics, speech and signal processing (ICASSP)* (pp. 5716–5719). doi:10.1109/ICASSP.2011.5947658

Suto, I. (2012). A critical review of some qualitative research methods used to explore rater cognition. *Educational Measurement: Issues and Practice 31*, 21–30. doi:10.1111/j.1745-3992.2012.00240.x

Szegedy, C., Liu, W., Jia, Y., Sermanet, P., Reed, S., Anguelov, D., … Rabinovich, A. (2014). Going deeper with convolutions. In *Proceedings of the 2015 IEEE conference on computer vision and pattern recognition (CVPR)* (pp. 1–9). doi:10.1109/CVPR.2015.7298594

Taghipour, K., & Ng, H. T. (2016). A neural approach to automated essay scoring. In J. Su, K. Duh, & X. Carreras (Eds.), *Proceedings of the 2016 conference on empirical methods in natural language processing* (pp. 1882–1891). Retrieved from https://www.aclweb.org/anthology/D16-1193

Tansomboon, C., Gerard, L. F., Vitale, J. M., & Linn, M. C. (2017). Designing automated guidance to promote productive revision of science explanations. *International Journal of Artificial Intelligence in Education, 27*, 729–757. doi:10.1007/s40593-017-0145-0

Tao, J., Chen, L., & Lee, C. M. (2016). DNN online with iVectors acoustic modeling and Doc2Vec distributed representations for improving automated speech scoring. In *Proceedings of the 17th annual conference of the International Speech Communication Association* (pp. 3117–3121). doi:10.21437/interspeech.2016-1457

Tao, J., Ghaffarzadegan, S., Chen, L., & Zechner, K. (2016). Exploring deep learning architectures for automatically grading non-native spontaneous speech. In *2016 IEEE International conference on acoustics, speech and signal processing (ICASSP)* (pp. 6140–6144). doi:10.1109/ICASSP.2016.7472857

Tawfik, A. A., Sanchez, L., & Saparova, D. (2014). The effects of case libraries in supporting collaborative problem-solving in an online learning environment. *Technology, Knowledge and Learning, 19*(3), 337–358. doi:10.1007/S10758-014-9230-8

Thakker, D., Schireson, M., & Nguyen-Huu, D. (2017, April 7). *Tracking the explosive growth of open source software*. Retrieved from the TechCrunch website: https://techcrunch.com/2017/04/07/tracking-the-explosive-growth-of-open-source-software/

Thies, J., Zollhofer, M., Stamminger, M., Theobalt, C., & Nießner, M. (2016). Face2face: Real-time face capture and reenactment of RGB videos. In *Proceedings of the 2016 IEEE conference on computer vision and pattern recognition (CVPR)* (pp. 2387–2395). doi:10.1109/CVPR.2016.262

Thorndike, E. L. (1918). Fundamental theorems in judging men. *Journal of Applied Psychology, 2*, 67–76. doi:10.1037/h0074876

Tibshirani, R. (1996). Regression shrinkage and selection via the lasso. *Journal of the Royal Statistical Society, Series B, 58*, 267–288. doi:10.1111/j.2517-6161.1996.tb02080.x

TIOBE. (2019, July). *TIOBE index for July 2019: Perl is one of the victims of Python's hype*. Retrieved from https://www.tiobe.com/tiobe-index/?source=post_elevate_sequence_page

TNTP. (2015). *The mirage: Confronting the hard truth about our quest for teacher development*. New York, NY: Author.

Toppo, G. (2012, April 23). Computer scoring of essays shows promise, analysis shows. *USA Today*. Retrieved from https://www.eschoolnews.com/2012/04/25/computer-scoring-of-essays-shows-promise-analysis-shows/

Toulmin, S. E. (1958). *The uses of argument*. Cambridge, UK: Cambridge University Press.

Trapani, C., Bridgeman, B., & Breyer, J. (2011, April). *Using automated scoring as a trend score: The implications of score separation over time*. Paper presented at the annual meeting of the National Council on Measurement in Education, New Orleans, LA.

Trentin, G. (2009). Using a wiki to evaluate individual contribution to a collaborative learning project. *Journal of Computer Assisted Learning, 25*, 43–55. doi:10.1111/j.1365-2729.2008.00276.x

Ujifusa, A. (2014, April 21). Resistance to the Common Core mounts. *Education Week*. Retrieved from https://www.edweek.org/ew/articles/2014/04/23/29cc-backlash.h33.html

Ujifusa, A. (2015, August 4). Tough choices for PARCC as states drop out. *Education Week*. Retrieved from https://www.edweek.org/ew/articles/2015/08/05/tough-choices-for-parcc-as-states-drop.html

United States Department of Education. (n.d.). *Our nation's English learners: What are their characteristics?* Retrieved from https://www2.ed.gov/datastory/el-characteristics/index.html

Van den Bergh, H., & Rijlaarsdam, G. (2007). The dynamics of idea generation during writing: An on-line study. In M. Torrance, L. Van Waes, & D. Galbraith (Eds.), *Writing and cognition: Research and applications* (Vol. 20, pp. 125–150).

Van Waes, L., & Leijten, M. (2015). Fluency in writing: A multidimensional perspective on writing fluency applied to L1 and L2. *Computers and Composition, 38*, 79–95. doi:10.1016/j.compcom.2015.09.012

Vander Hart, S. (2014, May 22). *Beware the robo-grader*. Retrieved from https://truthinamericaneducation.com/common-core-assessments/beware-the-robo-grader/

VanLehn, K., Graesser, A. C., Jackson, G. T., Jordan, P., Olney, A., & Rose, C. P. (2007). When are tutorial dialogues more effective than reading? *Cognitive Science, 31*, 3–62. doi:10.1080/03640210709336984

Vantage Learning. (2001). *A preliminary study of the efficacy of IntelliMetric for use in scoring Hebrew assessments*. Yardley, PA: Author.

Vantage Learning. (2002). *A study of IntelliMetric for responses in scoring Bahasa Malay*. Yardley, PA: Author.

von Davier, A. A. (2018). Computational psychometrics in support of collaborative educational assessments. *Journal of Educational Measurement, 54*, 3–11. doi:10.1111/jedm.12129

von Davier, A. A., Deonovic, B. E., Yudelson, M., Polyak, S., & Woo, A. (2019). Computational psychometrics approach to holistic learning and assessment systems. *Frontiers in Education, 4*. doi:10.3389/feduc.2019.00069

Vondrick, C., Pirsiavash, H., & Torralba, A. (2016). Generating videos with scene dynamics. In D. D. Lee, M. Sugiyama, U. V. Luxburg, I. Guyon, & R. Garnett (Eds.), *Advances in neural information processing systems 29* (pp. 1–9). Retrieved from https://papers.nips.cc/paper/6194-generating-videos-with-scene-dynamics.pdf

Wagner, V., Jescheniak, J. D., & Schriefers, H. (2010). On the flexibility of grammatical advance planning during sentence production: Effects of cognitive load on multiple lexical access. *Journal of Experimental Psychology: Learning, Memory, and Cognition, 36,* 423–440. doi:10.1037/a0018619

Wallot, S., & Grabowski, J. (2013). Typewriting dynamics: What distinguishes simple from complex writing tasks? *Ecological Psychology, 25,* 1–14. doi:10.1080/10407413.2013.810512

Wallot, S., O'Brien, B. A., & Van Orden, G. (2012). Fractal and recurrence analysis of psycholinguistic data. In G. Liben, G. Jarema, & C. Westbury (Eds.), *Methodological and analytical frontiers in lexical research* (pp. 395–430). Amsterdam, The Netherlands: Bejamins.

Walsh, M. M., Arslan, B., & Finn, B. (2019, April). *Computational cognitive modeling of human calibration and validity response scoring for GRE.* Paper presented at the annual meeting of the National Council for Measurement in Education, Toronto, Canada.

Wang, C., Song, T., Wang, Z., & Wolfe, E. (2016). Essay selection methods for adaptive rater monitoring. *Applied Psychological Measurement, 41,* 1–20. doi:10.1177/0146621616672855

Wang, E., Matsumura, L. C., & Correnti, R. (2017). Written feedback to support students' higher level thinking about texts in writing. *The Reading Instructor, 71,* 101–107. doi:10.1002/trtr.1584

Wang, X., Bruno, J., Molloy, H., Evanini, K., & Zechner, K. (2017). Discourse annotation of non-native spontaneous spoken responses using the rhetorical structure theory framework. *Proceedings of the 55th annual meeting of the Association for Computational Linguistics* (Vol. 2, pp. 263–268).

Wang, X., Evanini, K., & Yoon, S.-Y. (2015). Word-level F0 modeling for the automated assessment of read aloud speech. In S. Steidl, A. Batliner, & O. Jokisch (Eds.), *Proceedings of the sixth workshop on speech and language technology for education* (pp. 23–27). Retrieved from https://www.slate2015.org/files/SLaTE2015-Proceedings.pdf

Wang, X., Evanini, K., & Zechner, K. (2013). Coherence modeling for the automated assessment of spontaneous spoken responses. In *Proceedings of the 2013 conference of the North American Chapter of the Association for Computational Linguistics: Human language technologies* (pp. 814–819). Retrieved from https://www.aclweb.org/anthology/N13-1101

Wang, X., Evanini, K., Zechner, K., & Mulholland, M. (2017). Modeling discourse coherence for the automated scoring of spontaneous spoken responses. In O. Engwall, J. Lopes, & I. Leite (Eds.), *Proceedings of the seventh ISCA workshop on speech and language technology in education* (pp. 132–137). doi:10.21437/SLaTE.2017-23

Wang, X., Zechner, K., Evanini, K., & Mulholland, M. (2017). Modeling discourse coherence for the automated scoring of spontaneous spoken responses. *Proceedings from SLaTE* (pp. 132–137).

Webb, N. L., Alt, M., Ely, R., & Vesperman, B. (2005). *Web alignment tool (WAT): Training manual.* Madison, WI: University of Wisconsin–Madison, Wisconsin Center for Education Research.

Webster, G. D., Shea, J. A., Norcini, J. J., Grosso, L. J., & Swanson, D. B. (1988). Strategies in comparison of methods for scoring patient management problems. *Evaluation in the Health Professions, 11,* 231–248. doi:10.1177%2F016327878801100206

Weigel, S. C. (2011). *Validation of automated scores of TOEFL iBT tasks against non-test indicators of writing ability* (Research Report No. RR-11-24). Princeton, NJ: Educational Testing Service. doi:10.1002/j.2333-8504.2011.tb02260.x

Weigle, S. C. (1999). Investigating rater/prompt interactions in writing assessment: Quantitative and qualitative approaches. *Assessing Writing, 6*(2), 145–178.

Weiss, J. (2014, March 14). The man who killed the SAT essay. *The Boston Globe.* Retrieved from https://www2.bostonglobe.com/opinion/2014/03/13/the-man-who-killed-sat-essay/L9v3dbPXewKq8oAvOUqONM/story.html

Wengelin, Å., Torrance, M., Holmqvist, K., Simpson, S., Galbraith, D., Johansson, V., & Johansson, R. (2009). Combined eyetracking and keystroke-logging methods for studying cognitive processes in text production. *Behavior Research Methods, Instruments, & Computers, 41,* 337–351. doi:10.3758/BRM.41.2.337

White, S., Kim, Y. Y., Chen, J., & Liu, F. (2015). *Performance of fourth-grade students in the 2012 NAEP computer-based writing piilot assessment: Scores, text length, and use of editing tools* (NCES 2015-119). Washington, DC: National Center for Education Statistics.

Whitlock, J. W. (1964). *Automatic data processing in education.* New York: The Macmillan Company.

Whittaker, J., Arbon, J., & Carollo, J. (2012). *How Google tests software.* Upper Saddle River, NJ: Addison-Wesley.

Wild, F., Stahl, C., Stermsek, G., Penya, Y. K., & Neumann, G. (2005). Factors influencing effectiveness in automated essay scoring with LSA. In C.-K. Looi, G. McCalla, B. Bredeweg, & J. Breuker (Eds.), *Proceedings of the 2005 conference on artificial intelligence in education: Supporting learning through intelligent and socially informed technology* (pp. 947–949). Amsterdam, The Netherlands: IOS Press.

Will, U., Nottbusch, G., & Weingarten, R. (2006). Linguistic units in word typing: effects of word presentation modes and typing delay. *Written Language and Literacy, 9,* 156–173. doi:10.1075/wll.9.1.10wil

Williamson, D. M., Bejar, I. I., & Hone, A. S. (1999). 'Mental model' comparison of automated and human scoring. *Journal of Educational Measurement, 36,* 158–184. doi:10.1111/j.1745-3984.1999.tb00552.x

Williamson, D. M., Mislevy, R. J., & Bejar, I. I. (Eds.). (2006). *Automated scoring of complex tasks in computer-based testing.* Hillsdale, NJ: Lawrence Erlbaum.

Williamson, D. M., Xi, X., & Breyer, F. J. (2012). A framework for evaluation and use of automated scoring. *Educational Measurement: Issues and Practice, 31*(1), 2–13. doi:10.1111/j.1745-3992.2011.00223.x

Williamson, M. M. (2003). Validity of automated scoring: Prologue for a continuing discussion of machine scoring student writing. *Journal of Writing Assessment, 1*(2), 85–104.

Wilson, J. (2017). Associated effects of automated essay evaluation software on growth in writing quality for students with and without disabilities. *Reading and Writing, 30,* 691–718. doi:10.1007/s11145-016-9695-z

Wilson, J., & Andrada, G. N. (2016). Using automated feedback to improve writing quality: Opportunities and challenges. In Y. Rosen, S. Ferrara, & M. Mosharraf (Eds.), *Handbook of research on technology tools for real-world skill development* (pp. 678–703). Hershey, PA: IGI Global. doi:10.4018/978-1-4666-9441-5.ch026

Wilson, J., & Czik, A. (2016). Automated essay evaluation software in English Language Arts classrooms: Effects on teacher feedback, student motivation, and writing quality. *Computers & Education, 100,* 94–109. doi:10.1016/j.compedu.2016.05.004

Wilson, J., Olinghouse, N. G., & Andrada, G. N. (2014). Does automated feedback improve writing quality? *Learning Disabilities: A Contemporary Journal, 12,* 93–118.

Wilson, M. (2005). *Constructing measures.* Mahwah, NJ: Erlbaum.

Wilson, M., & Hoskens, M. (2001). The rater bundle model. *Journal of Educational and Behavioral Statistics, 26,* 283–306. doi:10.3102/10769986026003283

Winterhalter, B. (2013, September 30). *Computer grading will destroy our schools.* Retrieved from https://www.salon.com/2013/09/30/computer_grading_will_destroy_our_schools/

Witt, S., & Young, S. (2000). Phone-level pronunciation scoring and assessment for interactive language learning. *Speech Communication, 30*(2–3), 95–108. doi:10.1016/S0167-6393(99)00044-8

Wolfe, E. W. (2004). Identifying rater effects using latent trait models. *Psychology Science, 46,* 35–51.

Wolfe, E. W. (2014). *Methods for monitoring rating quality: Current practices and suggested changes.* Iowa City, IA: Pearson.

Wolfe, E. W., & Baird, J.-A. (2015, April). *A causal model of human scoring behavior in educational assessments.* Paper presented at the National Council on Measurement in Education, Chicago, IL.

Wolfe, E. W., & McVay, A. (2012). Applications of latent trait models to identifying substantively interesting raters. *Educational Measurement: Issues and Practice, 31*(3), 31–37. doi:10.1111/j.1745-3992.2012.00241.x

Wolfe, E. W., Kao, C.-W., & Ranney, M. (1998). Cognitive differences in proficient and nonproficient essay scorers. *Written Communication, 4,* 465–492. doi:10.1177/0741088398015004002

Wolfe, E. W., Matthews, S., & Vickers, D. (2010). The effectiveness and efficiency of distributed online, regional online, and regional face-to-face training for writing assessment raters. *Journal of Technology, Learning, and Assessment, 10*(1), 1–21.

Wolfe, E. W., Ng, D., & Baird, J.-A. (2018, April). *A conceptual framework for examining the human essay rating process.* Paper presented at the annual meeting of the National Council on Measurement in Education, New York, NY.

Woods, B., Adamson, D., Miel, S., & Mayfield, E. (2017, August). Formative essay feedback using predictive scoring models. In *Proceedings of the 23rd ACM SIGKDD international conference on knowledge discovery and data mining* (pp. 2071–2080). New York, NY: Association for Computing Machinery. Retrieved from doi:10.1145/3097983.3098160

Word, C. O., Zanna, M. P., & Cooper, J. (1974). The nonverbal mediation of self-fulfilling prophecies in interracial interaction. *Journal of Experimental Social Psychology, 10*(2), 109–120. doi:10.1016/0022-1031(74)90059-6

Worsley, A., Blikstein, G., Schneider, B., & Tissenbaum, M. (2016). Situating multimodal learning analytics. In C.-K. Looi, J. Polman, U. Cress, & P. Reimann (Eds.), *Transforming learning, empowering learners: The International Conference of the Learning Sciences (ICLS) 2016* (Vol. 2, pp. 1346–1349). Retrieved from https://www.isls.org/icls/2016/docs/ICLS2016_Volume_2.pdf

Wresch, W. (1993). The imminence of grading essays by computer—25 years later. *Computers and Composition, 10*(2), 45–58. doi:10.1016/S8755-4615(05)80058-1

Wu, J., Zhang, C., Xue, T., Freeman, W. T., & Tenenbaum, J. T. (2017). Learning a probabilistic latent space of object shapes via 3d generative-adversarial modeling. *Computer Vision and Pattern Recognition.* Retrieved from https://arxiv.org/pdf/1610.07584.pdf

Xi, X., Higgins, D., Zechner, K., & Williamson, D. M. (2008). *Automated scoring of spontaneous speech using SpeechRater v1.0* (Research Report No. RR-08-62). Princeton, NJ: Educational Testing Service. doi:10.1002/j.2333-8504.2008.tb02148.x

Xie, S., Evanini, K., & Zechner, K. (2012). Exploring content features for automated speech scoring. In *Proceedings of the 2012 conference of the North American Chapter of the Association for Computational Linguistics: Human language technologies* (pp. 103–111). Retrieved from https://www.aclweb.org/anthology/N12-1011

Xiong, W., Evanini, K., Zechner, K., & Chen, L. (2013). Automated content scoring of spoken responses containing multiple parts with factual information. In *Proceedings of the Interspeech 2013 satellite workshop on speech and language technology in education* (pp. 137–142). Retrieved from https://www.isca-speech.org/archive/slate_2013/sl13_intro.pdf

Yao, L., Haberman, S., & Zhang, M. (2019a). Penalized best linear prediction of true test scores. *Psychometrika, 84,* 186–211. doi:10.1007/s11336-018-9636-7

Yao, L., Haberman, S. J., & Zhang, M. (2019b). *Prediction of writing true scores in automated scoring of essays by best linear predictors and penalized best linear predictors* (Research Report No. RR-19-13). Princeton, NJ: Educational Testing Service. doi:10.1002/ets2.12248

Yoon, S.-Y., & Bhat, S. (2018). A comparison of grammatical proficiency measures in the automated assessment of spontaneous speech. *Speech Communication, 99,* 221–230. doi:10.1016/j.specom.2018.04.003

Yoon, S.-Y., & Higgins, D. (2011). Non-English response detection method for automated proficiency scoring system. *Proceedings of the 6th workshop on innovative use of nlp for building educational applications* (pp. 161–169). Association for Computational Linguistics. Retrieved from https://www.aclweb.org/anthology/W11-1420.pdf

Yoon, S.-Y., & Xie, S. (2014). Similarity-based non-scorable response detection for auto-mated speech scoring. In *Proceedings of the workshop on innovative use of NLP for building educational application* (pp. 116–123). doi:10.3115/v1/W14-1814

Yoon, S.-Y., & Zechner, K. (2017). Combining human and automated scores for the improved assessment of non-native speech. *Speech Communication, 93,* 43–52. doi:10.1016/j.specom.2017.08.001

Yoon, S.-Y., Bhat, S., & Zechner, K. (2012). Vocabulary profile as a measure of vocabulary sophistication. In J. Tetreault, J. Burstein, & C. Leacock (Eds.), *Proceedings of the 7th workshop on the innovative use of NLP for building educational applications* (pp. 180–189). Retrieved from https://www.aclweb.org/anthology/W12-2021

Yoon, S.-Y., Cahill, A., Loukina, A., Zechner, K. Riordan, B., & Madnani, N. (2018). Atypical inputs in educational applications. In S. Bangalore, J. Chu-Carroll, & Y. Li (Eds.), *Proceedings of the 2018 conference of the North American Chapter of the Association for Computational Linguistics: Human language technologies* (Vol. 3, pp. 60–67). doi:10.18653/v1/N18-3008

Yoon, S.-Y., Hasegawa-Johnson, M., & Sproat, R. (2010). Landmark-based automated pronunciation error detection. In *Proceedings of the 11th annual conference of the international speech communication association* (pp. 614–617). Retrieved from http://isle.illinois.edu/sst/pubs/2010/yoon10interspeech.pdf

Yoon, S.-Y., Lee, C. M., Choi, I., Wang, X., Mulholland, M., & Evanini, K. (2017). Off-topic spoken response detection with word embeddings. In *Proceedings of Interspeech 2017,* (pp. 2754–2758). doi:10.21437/Interspeech.2017-388

Zapata-Rivera, D., Jackson, G. T., & Katz, I. (2015). Authoring conversation-based assessment scenarios. In R. Sotttilare, X. Hu, A. Graesser, & K. Brawner (Eds.), *Design recommendations for adaptive intelligent tutoring systems: Vol. 3. Authoring tools and expert modeling techniques* (pp. 169–178). Orlando, FL: Army Research Laboratory.

Zapata-Rivera, D., Jackson, G. T., Liu, L., Bertling, M., Vezzu, M., & Katz, I. R., (2014). Science inquiry skills using trialogues. In S. Trausan-Matu, K. E. Boyer, M. Crosby, & K. Panourgia (Eds.), *ITS 2014: Intelligence tutoring systems* (pp. 625–626). doi:10.1007/978-3-319-07221-0_84

Zechner, K. (2009). What did they actually say? Agreement and disagreement among transcribers of non-native spontaneous speech responses in an English proficiency test. In *Proceedings of speech and language technology in education (SLaTE2009)* (pp. 3–6). Retrieved from https://www.isca-speech.org/archive/slate_2009/index.html

Zechner, K., & Bejar, I. I. (2006). Towards automatic scoring of non-native spontaneous speech. *Proceedings of the main conference on Human Language Technology conference of the North American chapter of the Association of Computational Linguistics (HLT-NAACL '06)* (pp. 216–223). Association for Computational Linguistics. doi:10.3115/1220835.1220863

Zechner, K., Chen, L., Davis, L., Evanini, K., Lee, C. M., Leong, C. W., … Yoon, S.-Y. (2015). *Automated scoring of speaking tasks in the Test of English-for-Teaching (TEFTTM)* (Research Report No. RR-15-31). Princeton, NJ: Educational Testing Service. doi:10.1002/ets2.12080

Zechner, K., Evanini, K., Yoon, S.-Y., Davis, L., Wang, X., Chen, L., … Leong, C. W. (2014). Automated scoring of speaking items in an assessment for teachers of English as a foreign language. In *Proceedings of the ninth workshop on innovative use of NLP for building educational applications* (pp. 134–142). doi:10.3115/v1/W14-1816

Zechner, K., Higgins, D., Xi, X., & Williamson, D. M. (2009). Automatic scoring of non-native spontaneous speech in tests of spoken English. *Speech Communication, 51*, 883–895. doi:10.1016/j.specom.2009.04.009

Zechner, K., Xi, X., & Chen, L. (2011). Evaluating prosodic features for automated scoring of non-native read speech. In *Proceedings of the 2011 IEEE workshop on automatic speech recognition & understanding* (pp. 461–466). doi:10.1109/ASRU.2011.6163975

Zechner, K., Yoon, S.-Y., Bhat, S., & Leong, C. W. (2017). Comparative evaluation of automated scoring of syntactic competence of non-native speakers. *Computers in Human Behavior, 76*, 672–682. doi:10.1016/j.chb.2017.01.060

Zesiger, P., Orliaguet, J.-P., Boe, L.-J., & Maounoud, P. (1994). The influence of syllabic structue in handwriting and typing production. In C. Faure, P. Keuss, G. Lorette, & A. Vinter (Eds.), *Advances in handwriting and drawing: A multidisciplinary approach*. Paris, France: Europia.

Zhang, B., Xiao, Y., & Luo, J. (2015). Rater reliability and score discrepancy under holistic and analytic scoring of second language writing. *Language Testing in Asia, 5*(5), 1–9. doi:10.1186/s40468-015-0014-4

Zhang, M. (2013). Contrasting automated and human scoring of essays. *R&D Connections, 21*, 1–11. Retrieved from https://www.ets.org/Media/Research/pdf/RD_Connections_21.pdf

Zhang, M., Bennett, R. E., Deane, P., & van Rijn, P. W. (2019). Are there gender differences in how students write their essays? An analysis of writing processes. *Educational Measurement: Issues and Practice, 38*(2), 14–26. doi:10.1111/emip.12249

Zhang, M., Bridgeman, B., & Davis, L. (in press). Validity considerations for using automated scoring in speaking assessment. In K. Zechner, & K. Evanini (Eds.), *Automated speech technologies for speaking assessment.* New York, NY: Routledge.

Zhang, M., & Deane, P. (2015). *Process features in writing: Internal structure and incremental value over product features* (Research Report No. RR-15-27). Princeton, NJ: Educational Testing Service. doi:10.1002/ets2.12075

Zhang, M., Deane, P., Feng, G., & Guo, H. (2019, July). *Investigating an approach to evaluating keyboarding fluency.* Paper presented at the society for text and discourse, New York, NY.

Zhang, M., Dorans, N. J., Li, C., & Rupp, A. (2017). Differential feature functioning in automated essay scoring. In H. Jiao, & R. Lissitz (Eds.), *Test fairness in the new generation of large-scale assessment* (pp. 185–208). Charlotte, NC: Information Age.

Zhang, M., Hao, J., Li, C., & Deane, P. (2016). Classification of writing patterns using keystroke logs. In L. A. van der Ark, D. M. Bolt, W.-C. Wang, J. A. Douglas, & M. Wiberg (Eds.), *Quantitative psychology research* (pp. 299–314). New York, NY: Springer.

Zhang, M., Hao, J., Li, C., & Deane, P. D. (2018), Defining personalized writing burst measures of translation using keystroke logs. In K. E. Boyer & M. Yudelson (Eds.), *Proceedings of the 11th international conference on educational data mining* (pp. 549–552). Buffalo, NY.

Zhang, M., Zhu, M., Deane, P., & Guo, H. (2019). Analyzing editing behaviors in writing using keystroke logs. In M. Wiberg, S. Culpepper, R. Janssen, J. González, & D. Molenaar (Eds.), *Quantitative psychology: Vol. 265. The 83rd annual meeting of the psychometric society* (pp. 367–381) New York, NY: Springer.

Zhang, M., Zou, D., Wu, A. D., & Deane, P. (2017). An investigation of writing processes employed in scenario-based assessment. In B. D. Zumbo, & A. M. Hubley (Eds.), *Understanding and investigating response processes in validation research* (pp. 321–339). New York, NY: Springer.

Zhu, J.-Y., Krähenbühl, P., Shechtman, E., & Efros, A. A. (2016). Generative visual manipulation on the natural image manifold. In *Proceedings of the 14th European conference on computer vision, Amsterdam, The Netherlands.* Retrieved from https://www.philkr.net/papers/2016-10-01-eccv/2016-10-01-eccv.pdf

Zhu, M., Lee, H.-S., Wang, T., Liu, O. L., Belur, V., & Pallant, A. (2017). Investigating the impact of automated feedback on students' scientific argumentation. *International Journal of Science Education, 39*(12), 1648–1668. doi:10.1080/09500693.2017.1347303

Zhu, M., Zhang, M., & Deane, P. (2019a). *Analysis of keystroke sequences in writing logs* (Research Report No. RR-19-11). doi:10.1002/ets2.12247

Zhu, M., Zhang, M., & Deane, P. (2019b, April). *Sequence analysis of keystroke sequences in writing logs: An investigation of composition strategies in timed-writing assessments.* Paper presented at the annual meeting of the National Council on Measurement in Education, Toronto, Canada.

Zinkevich, M. (2019, June 12). *Rules of machine learning: Best practices for ML engineering.* Retrieved from https://developers.google.com/machine-learning/guides/rules-of-ml/

Zupanc, K., & Bosnic, Z. (2015). Advances in the field of automated essay evaluation. *Informatica, 39,* 383–395.

Index